Digital Signal Processing

Digital Signal Processing

Fundamentals and Applications

Li Tan
DeVry University
Decatur, Georgia

AMSTERDAM • BOSTON • HEIDELBERG • LONDON
NEW YORK • OXFORD • PARIS • SAN DIEGO
SAN FRANCISCO • SINGAPORE • SYDNEY • TOKYO

Academic Press is an imprint of Elsevier

Academic Press is an imprint of Elsevier

30 Corporate Drive, Suite 400, Burlington, MA 01803, USA
525 B Street, Suite 1900, San Diego, California 92101-4495, USA
84 Theobald's Road, London WC1X 8RR, UK

This book is printed on acid-free paper. ∞

Copyright © 2008, Elsevier Inc. All rights reserved.

No part of this publication may be reproduced or transmitted in any form or by any means, electronic or mechanical, including photocopy, recording, or any information storage and retrieval system, without permission in writing from the publisher.

Permissions may be sought directly from Elsevier's Science & Technology Rights Department in Oxford, UK: phone: (+44) 1865 843830, fax: (+44) 1865 853333, E-mail: permissions@elsevier.com. You may also complete your request on-line via the Elsevier homepage (http://elsevier.com), by selecting "Support & Contact" then "Copyright and Permission" and then "Obtaining Permissions."

Library of Congress Cataloging-in-Publication Data
Tan, Li., 1963-
　Digital signal processing: fundamentals and applications / Li Tan.
　　　p. cm.
　Includes bibliographical references and index.
　ISBN-13: 978-0-12-374090-8 (alk. paper)　1. Signal processing–Digital techniques.　I. Title.
　TK5102.9.T36 2008
　621.382′2–dc22　　　　　　　　　　　　　　　　　　　　　　　　　　　2007025363

British Library Cataloguing-in-Publication Data
A catalogue record for this book is available from the British Library.

ISBN: 978-0-12-374090-8

For information on all Academic Press publications
visit our Web site at www.books.elsevier.com

Printed in the United States of America
07　08　09　10　11　9　8　7　6　5　4　3　2　1

Working together to grow
libraries in developing countries

www.elsevier.com | www.bookaid.org | www.sabre.org

ELSEVIER　　BOOK AID International　　Sabre Foundation

Contents

Preface		xiii
About the Author		xvii

1 Introduction to Digital Signal Processing — 1
- 1.1 Basic Concepts of Digital Signal Processing — 1
- 1.2 Basic Digital Signal Processing Examples in Block Diagrams — 3
    - 1.2.1 Digital Filtering — 3
    - 1.2.2 Signal Frequency (Spectrum) Analysis — 4
- 1.3 Overview of Typical Digital Signal Processing in Real-World Applications — 6
    - 1.3.1 Digital Crossover Audio System — 6
    - 1.3.2 Interference Cancellation in Electrocardiography — 7
    - 1.3.3 Speech Coding and Compression — 7
    - 1.3.4 Compact-Disc Recording System — 9
    - 1.3.5 Digital Photo Image Enhancement — 10
- 1.4 Digital Signal Processing Applications — 11
- 1.5 Summary — 12

2 Signal Sampling and Quantization — 13
- 2.1 Sampling of Continuous Signal — 13
- 2.2 Signal Reconstruction — 20
    - 2.2.1 Practical Considerations for Signal Sampling: Anti-Aliasing Filtering — 25
    - 2.2.2 Practical Considerations for Signal Reconstruction: Anti-Image Filter and Equalizer — 29
- 2.3 Analog-to-Digital Conversion, Digital-to-Analog Conversion, and Quantization — 35
- 2.4 Summary — 49
- 2.5 MATLAB Programs — 50
- 2.6 Problems — 51

3 Digital Signals and Systems — 57
- 3.1 Digital Signals — 57
    - 3.1.1 Common Digital Sequences — 58
    - 3.1.2 Generation of Digital Signals — 62

	3.2	Linear Time-Invariant, Causal Systems	64
		3.2.1 Linearity	64
		3.2.2 Time Invariance	65
		3.2.3 Causality	67
	3.3	Difference Equations and Impulse Responses	68
		3.3.1 Format of Difference Equation	68
		3.3.2 System Representation Using Its Impulse Response	69
	3.4	Bounded-in-and-Bounded-out Stability	72
	3.5	Digital Convolution	74
	3.6	Summary	82
	3.7	Problems	83

4 Discrete Fourier Transform and Signal Spectrum — 87

	4.1	Discrete Fourier Transform	87
		4.1.1 Fourier Series Coefficients of Periodic Digital Signals	88
		4.1.2 Discrete Fourier Transform Formulas	92
	4.2	Amplitude Spectrum and Power Spectrum	98
	4.3	Spectral Estimation Using Window Functions	110
	4.4	Application to Speech Spectral Estimation	117
	4.5	Fast Fourier Transform	120
		4.5.1 Method of Decimation-in-Frequency	121
		4.5.2 Method of Decimation-in-Time	127
	4.6	Summary	131
	4.7	Problems	131

5 The z-Transform — 135

	5.1	Definition	135
	5.2	Properties of the z-Transform	139
	5.3	Inverse z-Transform	142
		5.3.1 Partial Fraction Expansion Using MATLAB	148
	5.4	Solution of Difference Equations Using the z-Transform	151
	5.5	Summary	155
	5.6	Problems	156

6 Digital Signal Processing Systems, Basic Filtering Types, and Digital Filter Realizations — 159

	6.1	The Difference Equation and Digital Filtering	159
	6.2	Difference Equation and Transfer Function	165
		6.2.1 Impulse Response, Step Response, and System Response	169
	6.3	The z-Plane Pole-Zero Plot and Stability	171
	6.4	Digital Filter Frequency Response	179
	6.5	Basic Types of Filtering	188

6.6	Realization of Digital Filters	195
	6.6.1 Direct-Form I Realization	195
	6.6.2 Direct-Form II Realization	196
	6.6.3 Cascade (Series) Realization	197
	6.6.4 Parallel Realization	198
6.7	Application: Speech Enhancement and Filtering	202
	6.7.1 Pre-Emphasis of Speech	202
	6.7.2 Bandpass Filtering of Speech	205
6.8	Summary	208
6.9	Problems	209

7 Finite Impulse Response Filter Design — 215

7.1	Finite Impulse Response Filter Format	215
7.2	Fourier Transform Design	217
7.3	Window Method	229
7.4	Applications: Noise Reduction and Two-Band Digital Crossover	253
	7.4.1 Noise Reduction	253
	7.4.2 Speech Noise Reduction	255
	7.4.3 Two-Band Digital Crossover	256
7.5	Frequency Sampling Design Method	260
7.6	Optimal Design Method	268
7.7	Realization Structures of Finite Impulse Response Filters	280
	7.7.1 Transversal Form	280
	7.7.2 Linear Phase Form	282
7.8	Coefficient Accuracy Effects on Finite Impulse Response Filters	283
7.9	Summary of Finite Impulse Response (FIR) Design Procedures and Selection of FIR Filter Design Methods in Practice	287
7.10	Summary	290
7.11	MATLAB Programs	291
7.12	Problems	294

8 Infinite Impulse Response Filter Design — 303

8.1	Infinite Impulse Response Filter Format	303
8.2	Bilinear Transformation Design Method	305
	8.2.1 Analog Filters Using Lowpass Prototype Transformation	306
	8.2.2 Bilinear Transformation and Frequency Warping	310
	8.2.3 Bilinear Transformation Design Procedure	317
8.3	Digital Butterworth and Chebyshev Filter Designs	322
	8.3.1 Lowpass Prototype Function and Its Order	322
	8.3.2 Lowpass and Highpass Filter Design Examples	326
	8.3.3 Bandpass and Bandstop Filter Design Examples	336

8.4		Higher-Order Infinite Impulse Response Filter Design Using the Cascade Method	343
8.5		Application: Digital Audio Equalizer	346
8.6		Impulse Invariant Design Method	350
8.7		Polo-Zero Placement Method for Simple Infinite Impulse Response Filters	358
	8.7.1	Second-Order Bandpass Filter Design	359
	8.7.2	Second-Order Bandstop (Notch) Filter Design	360
	8.7.3	First-Order Lowpass Filter Design	362
	8.7.4	First-Order Highpass Filter Design	364
8.8		Realization Structures of Infinite Impulse Response Filters	365
	8.8.1	Realization of Infinite Impulse Response Filters in Direct-Form I and Direct-Form II	366
	8.8.2	Realization of Higher-Order Infinite Impulse Response Filters via the Cascade Form	368
8.9		Application: 60-Hz Hum Eliminator and Heart Rate Detection Using Electrocardiography	370
8.10		Coefficient Accuracy Effects on Infinite Impulse Response Filters	377
8.11		Application: Generation and Detection of Dual-Tone Multifrequency Tones Using Goertzel Algorithm	381
	8.11.1	Single-Tone Generator	382
	8.11.2	Dual-Tone Multifrequency Tone Generator	384
	8.11.3	Goertzel Algorithm	386
	8.11.4	Dual-Tone Multifrequency Tone Detection Using the Modified Goertzel Algorithm	391
8.12		Summary of Infinite Impulse Response (IIR) Design Procedures and Selection of the IIR Filter Design Methods in Practice	396
8.13		Summary	401
8.14		Problems	402

9 Hardware and Software for Digital Signal Processors — 413

9.1		Digital Signal Processor Architecture	413
9.2		Digital Signal Processor Hardware Units	416
	9.2.1	Multiplier and Accumulator	416
	9.2.2	Shifters	417
	9.2.3	Address Generators	418
9.3		Digital Signal Processors and Manufactures	419
9.4		Fixed-Point and Floating-Point Formats	420
	9.4.1	Fixed-Point Format	420
	9.4.2	Floating-Point Format	429
	9.4.3	IEEE Floating-Point Formats	434

		9.4.5	Fixed-Point Digital Signal Processors	437
		9.4.6	Floating-Point Processors	439
	9.5	Finite Impulse Response and Infinite Impulse Response Filter Implementation in Fixed-Point Systems		441
	9.6	Digital Signal Processing Programming Examples		447
		9.6.1	Overview of TMS320C67x DSK	447
		9.6.2	Concept of Real-Time Processing	451
		9.6.3	Linear Buffering	452
		9.6.4	Sample C Programs	455
	9.7	Summary		460
	9.8	Problems		461

10 Adaptive Filters and Applications — 463

	10.1	Introduction to Least Mean Square Adaptive Finite Impulse Response Filters		463
	10.2	Basic Wiener Filter Theory and Least Mean Square Algorithm		467
	10.3	Applications: Noise Cancellation, System Modeling, and Line Enhancement		473
		10.3.1	Noise Cancellation	473
		10.3.2	System Modeling	479
		10.3.3	Line Enhancement Using Linear Prediction	484
	10.4	Other Application Examples		486
		10.4.1	Canceling Periodic Interferences Using Linear Prediction	487
		10.4.2	Electrocardiography Interference Cancellation	488
		10.4.3	Echo Cancellation in Long-Distance Telephone Circuits	489
	10.5	Summary		491
	10.6	Problems		491

11 Waveform Quantization and Compression — 497

	11.1	Linear Midtread Quantization		497
	11.2	μ-law Companding		501
		11.2.1	Analog μ-Law Companding	501
		11.2.2	Digital μ-Law Companding	506
	11.3	Examples of Differential Pulse Code Modulation (DPCM), Delta Modulation, and Adaptive DPCM G.721		510
		11.3.1	Examples of Differential Pulse Code Modulation and Delta Modulation	510
		11.3.2	Adaptive Differential Pulse Code Modulation G.721	515
	11.4	Discrete Cosine Transform, Modified Discrete Cosine Transform, and Transform Coding in MPEG Audio		522
		11.4.1	Discrete Cosine Transform	522

		11.4.2	Modified Discrete Cosine Transform	525
		11.4.3	Transform Coding in MPEG Audio	530
	11.5	Summary		533
	11.6	MATLAB Programs		534
	11.7	Problems		550

12 Multirate Digital Signal Processing, Oversampling of Analog-to-Digital Conversion, and Undersampling of Bandpass Signals — 557

- 12.1 Multirate Digital Signal Processing Basics — 557
  - 12.1.1 Sampling Rate Reduction by an Integer Factor — 558
  - 12.1.2 Sampling Rate Increase by an Integer Factor — 564
  - 12.1.3 Changing Sampling Rate by a Non-Integer Factor L/M — 570
  - 12.1.4 Application: CD Audio Player — 575
  - 12.1.5 Multistage Decimation — 578
- 12.2 Polyphase Filter Structure and Implementation — 583
- 12.3 Oversampling of Analog-to-Digital Conversion — 589
  - 12.3.1 Oversampling and Analog-to-Digital Conversion Resolution — 590
  - 12.3.2 Sigma-Delta Modulation Analog-to-Digital Conversion — 593
- 12.4 Application Example: CD Player — 599
- 12.5 Undersampling of Bandpass Signals — 601
- 12.6 Summary — 609
- 12.7 Problems — 610

13 Image Processing Basics — 617

- 13.1 Image Processing Notation and Data Formats — 617
  - 13.1.1 8-Bit Gray Level Images — 618
  - 13.1.2 24-Bit Color Images — 619
  - 13.1.3 8-Bit Color Images — 620
  - 13.1.4 Intensity Images — 621
  - 13.1.5 Red, Green, Blue Components and Grayscale Conversion — 622
  - 13.1.6 MATLAB Functions for Format Conversion — 624
- 13.2 Image Histogram and Equalization — 625
  - 13.2.1 Grayscale Histogram and Equalization — 625
  - 13.2.2 24-Bit Color Image Equalization — 632
  - 13.2.3 8-Bit Indexed Color Image Equalization — 633
  - 13.2.4 MATLAB Functions for Equalization — 636
- 13.3 Image Level Adjustment and Contrast — 637
  - 13.3.1 Linear Level Adjustment — 638
  - 13.3.2 Adjusting the Level for Display — 641

		13.3.3	Matlab Functions for Image Level Adjustment	642
	13.4	Image Filtering Enhancement		642
		13.4.1	Lowpass Noise Filtering	643
		13.4.2	Median Filtering	646
		13.4.3	Edge Detection	651
		13.4.4	MATLAB Functions for Image Filtering	655
	13.5	Image Pseudo-Color Generation and Detection		657
	13.6	Image Spectra		661
	13.7	Image Compression by Discrete Cosine Transform		664
		13.7.1	Two-Dimensional Discrete Cosine Transform	666
		13.7.2	Two-Dimensional JPEG Grayscale Image Compression Example	669
		13.7.3	JPEG Color Image Compression	671
	13.8	Creating a Video Sequence by Mixing Two Images		677
	13.9	Video Signal Basics		677
		13.9.1	Analog Video	678
		13.9.2	Digital Video	685
	13.10	Motion Estimation in Video		687
	13.11	Summary		690
	13.12	Problems		692

Appendix A Introduction to the MATLAB Environment — **699**

A.1	Basic Commands and Syntax	699
A.2	MATLAB Array and Indexing	703
A.3	Plot Utilities: Subplot, Plot, Stem, and Stair	704
A.4	MATLAB Script Files	704
A.5	MATLAB Functions	705

Appendix B Review of Analog Signal Processing Basics — **709**

B.1	Fourier Series and Fourier Transform		709
	B.1.1	Sine-Cosine Form	709
	B.1.2	Amplitude-Phase Form	710
	B.1.3	Complex Exponential Form	711
	B.1.4	Spectral Plots	714
	B.1.5	Fourier Transform	721
B.2	Laplace Transform		726
	B.2.1	Laplace Transform and Its Table	726
	B.2.2	Solving Differential Equations Using Laplace Transform	727
	B.2.3	Transfer Function	730
B.3	Poles, Zeros, Stability, Convolution, and Sinusoidal Steady-State Response		731

	B.3.1	Poles, Zeros, and Stability	731
	B.3.2	Convolution	733
	B.3.3	Sinusoidal Steady-State Response	735
B.4	Problems		736

Appendix C Normalized Butterworth and Chebyshev Fucntions   **741**
    C.1   Normalized Butterworth Function   741
    C.2   Normalized Chebyshev Function   744

Appendix D Sinusoidal Steady-State Response of Digital Filters   **749**
    D.1   Sinusoidal Steady-State Response   749
    D.2   Properties of the Sinusoidal Steady-State Response   751

Appendix E Finite Impulse Response Filter Design Equations by the Frequency Sampling Design Method   **753**

Appendix F Some Useful Mathematical Formulas   **757**

Bibliography   **761**

Answers to Selected Problems   **765**

Index   **791**

Preface

Technologies such as microprocessors, microcontrollers, and digital signal processors have become so advanced that they have had a dramatic impact on the disciplines of electronics engineering, computer engineering, and biomedical engineering. Technologists need to become familiar with digital signals and systems and basic digital signal processing (DSP) techniques. The objective of this book is to introduce students to the fundamental principles of these subjects and to provide a working knowledge such that they can apply DSP in their engineering careers.

The book is suitable for a sequence of two-semester courses at the senior level in undergraduate electronics, computer, and biomedical engineering technology programs. Chapters 1 to 8 provide the topics for a one semester course, and a second course can complete the rest of the chapters. This textbook can also be used in an introductory DSP course at the junior level in undergraduate electrical engineering programs at traditional colleges. Additionally, the book should be useful as a reference for undergraduate engineering students, science students, and practicing engineers.

The material has been tested in two consecutive courses in signal processing sequence at DeVry University on the Decatur campus in Georgia. With the background established from this book, students can be well prepared to move forward to take other senior-level courses that deal with digital signals and systems for communications and controls.

The textbook consists of 13 chapters, organized as follows:

- Chapter 1 introduces concepts of DSP and presents a general DSP block diagram. Application examples are included.

- Chapter 2 covers the sampling theorem described in time domain and frequency domain and also covers signal reconstruction. Some practical considerations for designing analog anti-aliasing lowpass filters and anti-image lowpass filters are included. The chapter ends with a section dealing with analog-to-digital conversion (ADC) and digital-to-analog conversion (DAC), as well as signal quantization and encoding.

- Chapter 3 introduces digital signals, linear time-invariant system concepts, difference equations, and digital convolutions.

- Chapter 4 introduces the discrete Fourier transform (DFT) and digital signal spectral calculations using the DFT. Applying the DFT to estimate the speech spectrum is demonstrated. The chapter ends with a section dedicated to illustrating fast Fourier transform (FFT) algorithms.

- Chapter 5 is devoted to the z-transform and difference equations.

- Chapter 6 covers digital filtering using difference equations, transfer functions, system stability, digital filter frequency responses, and implementation methods such as the direct form I and direct form II.

- Chapter 7 deals with various methods of finite impulse response (FIR) filter design, including the Fourier transform method for calculating FIR filter coefficients, window method, frequency sampling design, and optimal design. Chapter 7 also includes applications using FIR filters for noise reduction and digital crossover system design.

- Chapter 8 covers various methods of infinite impulse response (IIR) filter design, including the bilinear transformation (BLT) design, impulse invariant design, and pole-zero placement design. Applications using IIR filters include audio equalizer design, biomedical signal enhancement, dual-tone multifrequency (DTMF) tone generation and detection with the Goertzel algorithm.

- Chapter 9 introduces DSP architectures, software and hardware, and fixed-point and floating-point implementations of digital filters.

- Chapter 10 covers adaptive filters with applications such as noise cancellation, system modeling, line enhancement, cancellation of periodic interferences, echo cancellation, and 60-Hz interference cancellation in biomedical signals.

- Chapter 11 is devoted to speech quantization and compression, including pulse code modulation (PCM) coding, mu-law compression, adaptive differential pulse code modulation (ADPCM) coding, windowed modified discrete cosine transform (W-MDCT) coding, and MPEG audio format, specifically MP3 (MPEG-1, layer 3).

- Chapter 12 covers topics pertaining to multirate DSP and applications, as well as principles of oversampling ADC, such as sigma-delta modulation. Undersampling for bandpass signals is also examined.

- Finally, Chapter 13 covers image enhancement using histogram equalization and filtering methods, including edge detection. The chapter also explores pseudo-color image generation and detection, two-dimensional spectra, JPEG compression using DCT, and the mixing of two images to create a video sequence. Finally, motion compensation of the video sequence is explored, which is a key element of video compression used in MPEG.

MATLAB programs are listed wherever they are possible. Therefore, a MATLAB tutorial should be given to students who are new to the MATLAB environment.

- Appendix A serves as a MATLAB tutorial.

- Appendix B reviews key fundamentals of analog signal processing. Topics include Fourier series, Fourier transform, Laplace transform, and analog system basics.

- Appendixes C, D, and E overview Butterworth and Chebyshev filters, sinusoidal steady-state responses in digital filters, and derivation of the FIR filter design equation via the frequency sampling method, respectively.

- Appendix F offers general useful mathematical formulas.

Instructor support, including solutions, can be found at http://textbooks.elsevier.com. MATLAB programs and exercises for students, plus Real-time C programs, can be found at http://books.elsevier.com/companions/9780123740908.

The author wishes to thank Dr. Samuel D. Stearns (professor at the University of New Mexico; Sandia National Laboratories, Albuquerque, NM), Dr. Delores M. Etler (professor at the United States Naval Academy at Annapolis, MD) and Dr. Neeraj Magotra (Texas Instruments, former professor at the University of New Mexico) for inspiration, guidance, and sharing of their insight into DSP over the years. A special thanks goes to Dr. Jean Jiang (professor at DeVry University in Decatur) for her encouragement, support, insightful suggestions, and testing of the manuscript in her DSP course.

Special thanks go to Tim Pitts (senior commissioning editor), Rick Adams (senior acquistions editor), and Rachel Roumeliotis (acquisitions editor) and to the team members at Elsevier Science publishing for their encouragement and guidance in developing the complete manuscript.

I also wish to thank Jamey Stegmaier (publishing project manager) at SPi for coordinating the copyediting of the manuscript.

Thanks to all the faculty and staff at DeVry University, Decatur, for their encouragement and support.

The book has benefited from many constructive comments and suggestions from the following reviewers and anonymous reviewers. The author takes this opportunity to thank them for their significant contributions:

Professor Mateo Aboy, Oregon Institute of Technology, Klamath Falls, OR
Professor Jean Andrian, Florida International University, Miami, FL
Professor Rabah Aoufi, DeVry University, Irving, TX
Professor Larry Bland, John Brown University, Siloam Springs, AR

Professor Phillip L. De Leon, New Mexico State University, Las Cruces, NM
Professor Mohammed Feknous, New Jersey Institute of Technology, Newark, NJ
Professor Richard L. Henderson, DeVry University, Kansas City, MO
Professor Ling Hou, St. Cloud State University, St. Cloud, MN
Professor Robert C. (Rob) Maher, Montana State University, Bozeman, MT
Professor Abdulmagid Omar, DeVry University, Tinley Park, IL
Professor Ravi P. Ramachandran, Rowan University, Glassboro, NJ
Professor William (Bill) Routt, Wake Technical Community College, Raleigh, NC
Professor Samuel D. Stearns, University of New Mexico; Sandia National Laboratories, Albuquerque, NM
Professor Les Thede, Ohio Northern University, Ada, OH
Professor Igor Tsukerman, University of Akron, Akron, OH
Professor Vijay Vaidyanathan, University of North Texas, Denton, TX
Professor David Waldo, Oklahoma Christian University, Oklahoma City, OK

Finally, I am immensely grateful to my wife, Jean, and my children, Ava, Alex, and Amber, for their extraordinary patience and understanding during the entire preparation of this book.

Li Tan
DeVry University
Decatur, Georgia
May 2007

About the Author

Dr. Li Tan is a Professor of Electronics Engineering Technology at DeVry University, Decatur, Georgia. He received his M.S. and Ph.D. degrees in Electrical Engineering from the University of New Mexico. He has extensively taught analog and digital signal processing and analog and digital communications for many years. Before teaching at DeVry University, Dr. Tan worked in the DSP and communications industry.

Dr. Tan is a senior member of the Institute of Electronic and Electronic Engineers (IEEE). His principal technical areas include digital signal processing, adaptive signal processing, and digital communications. He has published a number of papers in these areas.

1

Introduction to Digital Signal Processing

Objectives:

This chapter introduces concepts of digital signal processing (DSP) and reviews an overall picture of its applications. Illustrative application examples include digital noise filtering, signal frequency analysis, speech and audio compression, biomedical signal processing such as interference cancellation in electrocardiography, compact-disc recording, and image enhancement.

1.1 Basic Concepts of Digital Signal Processing

Digital signal processing (DSP) technology and its advancements have dramatically impacted our modern society everywhere. Without DSP, we would not have digital/Internet audio or video; digital recording; CD, DVD, and MP3 players; digital cameras; digital and cellular telephones; digital satellite and TV; or wire and wireless networks. Medical instruments would be less efficient or unable to provide useful information for precise diagnoses if there were no digital electrocardiography (ECG) analyzers or digital x-rays and medical image systems. We would also live in many less efficient ways, since we would not be equipped with voice recognition systems, speech synthesis systems, and image and video editing systems. Without DSP, scientists, engineers, and technologists would have no powerful tools to analyze and visualize data and perform their design, and so on.

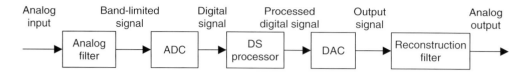

FIGURE 1.1   A digital signal processing scheme.

The concept of DSP is illustrated by the simplified block diagram in Figure 1.1, which consists of an analog filter, an analog-to-digital conversion (ADC) unit, a digital signal (DS) processor, a digital-to-analog conversion (DAC) unit, and a reconstruction (anti-image) filter.

As shown in the diagram, the analog input signal, which is continuous in time and amplitude, is generally encountered in our real life. Examples of such analog signals include current, voltage, temperature, pressure, and light intensity. Usually a transducer (sensor) is used to convert the nonelectrical signal to the analog electrical signal (voltage). This analog signal is fed to an analog filter, which is applied to limit the frequency range of analog signals prior to the sampling process. The purpose of filtering is to significantly attenuate *aliasing distortion,* which will be explained in the next chapter. The band-limited signal at the output of the analog filter is then sampled and converted via the ADC unit into the digital signal, which is discrete both in time and in amplitude. The DS processor then accepts the digital signal and processes the digital data according to DSP rules such as lowpass, highpass, and bandpass digital filtering, or other algorithms for different applications. Notice that the DS processor unit is a special type of digital computer and can be a general-purpose digital computer, a microprocessor, or an advanced microcontroller; furthermore, DSP rules can be implemented using software in general.

With the DS processor and corresponding software, a processed digital output signal is generated. This signal behaves in a manner according to the specific algorithm used. The next block in Figure 1.1, the DAC unit, converts the processed digital signal to an analog output signal. As shown, the signal is continuous in time and discrete in amplitude (usually a sample-and-hold signal, to be discussed in Chapter 2). The final block in Figure 1.1 is designated as a function to smooth the DAC output voltage levels back to the analog signal via a reconstruction (anti-image) filter for real-world applications.

In general, the analog signal process does not require software, an algorithm, ADC, and DAC. The processing relies wholly on electrical and electronic devices such as resistors, capacitors, transistors, operational amplifiers, and integrated circuits (ICs).

DSP systems, on the other hand, use software, digital processing, and algorithms; thus they have a great deal of flexibility, less noise interference, and no signal distortion in various applications. However, as shown in Figure 1.1, DSP systems still require minimum analog processing such as the anti-aliasing and reconstruction filters, which are musts for converting real-world information into digital form and digital form back into real-world information.

Note that there are many real-world DSP applications that do not require DAC, such as data acquisition and digital information display, speech recognition, data encoding, and so on. Similarly, DSP applications that need no ADC include CD players, text-to-speech synthesis, and digital tone generators, among others. We will review some of them in the following sections.

1.2 Basic Digital Signal Processing Examples in Block Diagrams

We first look at digital noise filtering and signal frequency analysis, using block diagrams.

1.2.1 Digital Filtering

Let us consider the situation shown in Figure 1.2, depicting a digitized noisy signal obtained from digitizing analog voltages (sensor output) containing a useful low-frequency signal and noise that occupies all of the frequency range. After ADC, the digitized noisy signal $x(n)$, where n is the sample number, can be enhanced using digital filtering.

Since our useful signal contains the low-frequency component, the high-frequency components above that of our useful signal are considered as noise, which can be removed by using a digital lowpass filter. We set up the DSP block in Figure 1.2 to operate as a simple digital lowpass filter. After processing the digitized noisy signal $x(n)$, the digital lowpass filter produces a clean digital signal $y(n)$. We can apply the cleaned signal $y(n)$ to another DSP algorithm for a different application or convert it to the analog signal via DAC and the reconstruction filter.

The digitized noisy signal and clean digital signal, respectively, are plotted in Figure 1.3, where the top plot shows the digitized noisy signal, while the bottom plot demonstrates the clean digital signal obtained by applying the digital lowpass filter. Typical applications of noise filtering include acquisition of clean

FIGURE 1.2 **The simple digital filtering block.**

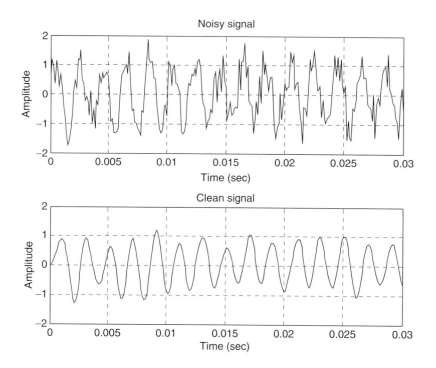

FIGURE 1.3 (*Top*) Digitized noisy signal. (*Bottom*) Clean digital signal using the digital lowpass filter.

digital audio and biomedical signals and enhancement of speech recording, among others (Embree, 1995; Rabiner and Schafer, 1978; Webster, 1998).

1.2.2 Signal Frequency (Spectrum) Analysis

As shown in Figure 1.4, certain DSP applications often require that time domain information and the frequency content of the signal be analyzed. Figure 1.5 shows a digitized audio signal and its calculated signal spectrum (frequency content), defined as the signal amplitude versus its corresponding frequency for the time being via a DSP algorithm, called *fast Fourier transform* (FFT), which will be studied in Chapter 4. The plot in Figure 1.5 (a) is a time domain display of the recorded audio signal with a frequency of 1,000 Hz sampled at 16,000 samples per second, while the frequency content display of plot (b) displays the calculated signal spectrum versus frequencies, in which the peak amplitude is clearly located at 1,000 Hz. Plot (c) shows a time domain display of an audio signal consisting of one signal of 1,000 Hz and another of 3,000 Hz sampled at 16,000 samples per second. The frequency content display shown in Plot (d)

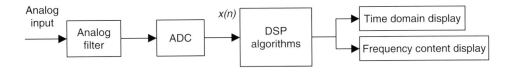

FIGURE 1.4   Signal spectral analysis.

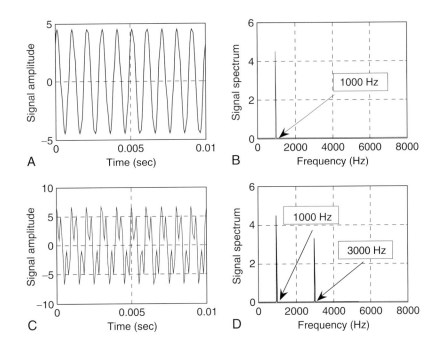

FIGURE 1.5   Audio signals and their spectrums.

gives two locations (1,000 Hz and 3,000 Hz) where the peak amplitudes reside, hence the frequency content display presents clear frequency information of the recorded audio signal.

As another practical example, we often perform spectral estimation of a digitally recorded speech or audio (music) waveform using the FFT algorithm in order to investigate spectral frequency details of speech information. Figure 1.6 shows a speech signal produced by a human in the time domain and frequency content displays. The top plot shows the digital speech waveform versus its digitized sample number, while the bottom plot shows the frequency content information of speech for a range from 0 to 4,000 Hz. We can observe that there are about ten spectral peaks, called *speech formants,* in the range between 0 and 1,500 Hz. Those identified speech formants can be used for

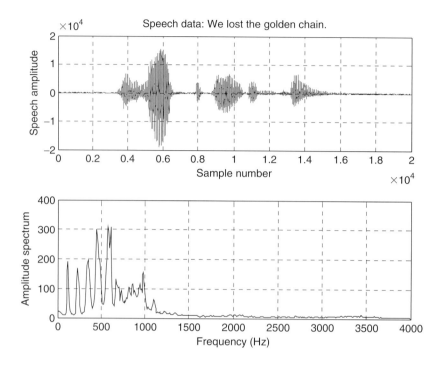

FIGURE 1.6 **Speech sample and speech spectrum.**

applications such as speech modeling, speech coding, speech feature extraction for speech synthesis and recognition, and so on (Deller et al., 1993).

1.3 Overview of Typical Digital Signal Processing in Real-World Applications

1.3.1 Digital Crossover Audio System

An audio system is required to operate in an entire audible range of frequencies, which may be beyond the capability of any single speaker driver. Several drivers, such as the speaker cones and horns, each covering a different frequency range, are used to cover the full audio frequency range.

Figure 1.7 shows a typical two-band digital crossover system consisting of two speaker drivers: a woofer and a tweeter. The woofer responds to low frequencies, while the tweeter responds to high frequencies. The incoming digital audio signal is split into two bands by using a digital lowpass filter and a digital highpass filter in parallel. Then the separated audio signals are amplified. Finally, they are sent to their corresponding speaker drivers. Although the

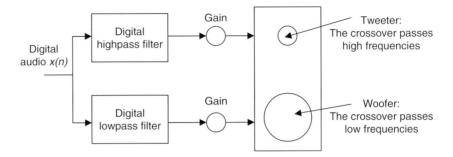

FIGURE 1.7 Two-band digital crossover.

traditional crossover systems are designed using the analog circuits, the digital crossover system offers a cost-effective solution with programmable ability, flexibility, and high quality. This topic is taken up in Chapter 7.

1.3.2 Interference Cancellation in Electrocardiography

In ECG recording, there often is unwanted 60-Hz interference in the recorded data (Webster, 1998). The analysis shows that the interference comes from the power line and includes magnetic induction, displacement currents in leads or in the body of the patient, effects from equipment interconnections, and other imperfections. Although using proper grounding or twisted pairs minimizes such 60-Hz effects, another effective choice can be use of a digital notch filter, which eliminates the 60-Hz interference while keeping all the other useful information. Figure 1.8 illustrates a 60-Hz interference eliminator using a digital notch filter. With such enhanced ECG recording, doctors in clinics can give accurate diagnoses for patients. This technique can also be applied to remove 60-Hz interferences in audio systems. This topic is explored in depth in Chapter 8.

1.3.3 Speech Coding and Compression

One of the speech coding methods, called *waveform coding*, is depicted in Figure 1.9(a), describing the encoding process, while Figure 1.9(b) shows the decoding process. As shown in Figure 1.9(a), the analog signal is first filtered by analog lowpass to remove high-frequency noise components and is then passed through the ADC unit, where the digital values at sampling instants are captured by the DS processor. Next, the captured data are compressed using data compression rules to save the storage requirement. Finally, the compressed digital information is sent to storage media. The compressed digital information

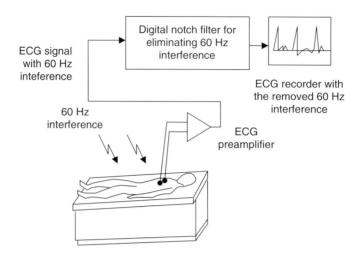

FIGURE 1.8 Elimination of 60-Hz interference in electrocardiography (ECG).

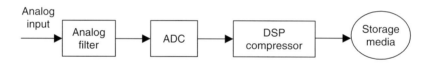

FIGURE 1.9A Simplified data compressor.

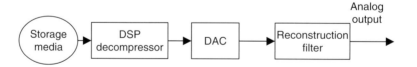

FIGURE 1.9B Simplified data expander (decompressor).

can also be transmitted efficiently, since compression reduces the original data rate. Digital voice recorders, digital audio recorders, and MP3 players are products that use compression techniques (Deller et al., 1993; Li and Drew, 2004; Pan, 1985).

To retrieve the information, the reverse process is applied. As shown in Figure 1.9b, the DS processor decompresses the data from the storage media and sends the recovered digital data to DAC. The analog output is acquired by filtering the DAC output via the reconstruction filter.

1.3.4 Compact-Disc Recording System

A compact-disc (CD) recording system is described in Figure 1.10a. The analog audio signal is sensed from each microphone and then fed to the anti-aliasing lowpass filter. Each filtered audio signal is sampled at the industry standard rate of 44.1 kilo-samples per second, quantized, and coded to 16 bits for each digital sample in each channel. The two channels are further multiplexed and encoded, and extra bits are added to provide information such as playing time and track number for the listener. The encoded data bits are modulated for storage, and more synchronized bits are added for subsequent recovery of sampling frequency. The modulated signal is then applied to control a laser beam that illuminates the photosensitive layer of a rotating glass disc. When the laser turns on and off, the digital information is etched onto the photosensitive layer as a pattern of pits and lands in a spiral track. This master disc forms the basis for mass production of the commercial CD from the thermoplastic material.

During playback, as illustrated in Figure 1.10b, a laser optically scans the tracks on a CD to produce a digital signal. The digital signal is then

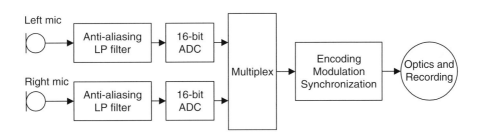

FIGURE 1.10A Simplified encoder of the CD recording system.

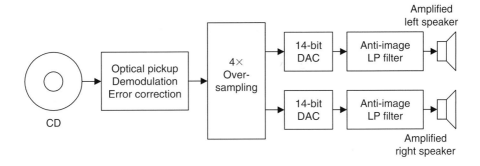

FIGURE 1.10B Simplified decoder of the CD recording system.

demodulated. The demodulated signal is further oversampled by a factor of 4 to acquire a sampling rate of 176.4 kHz for each channel and is then passed to the 14-bit DAC unit. For the time being, we can consider the oversampling process as interpolation, that is, adding three samples between every two original samples in this case, as we shall see in Chapter 12. After DAC, the analog signal is sent to the anti-image analog filter, which is a lowpass filter to smooth the voltage steps from the DAC unit. The output from each anti-image filter is fed to its amplifier and loudspeaker. The purpose of the oversampling is to relieve the higher-filter-order requirement for the anti-image lowpass filter, making the circuit design much easier and economical (Ambardar, 1999).

Software audio players that play music from CDs, such as Windows Media Player and RealPlayer, installed on computer systems, are examples of DSP applications. The audio player has many advanced features, such as a graphical equalizer, which allows users to change audio with sound effects such as boosting low-frequency content or emphasizing high-frequency content to make music sound more entertaining (Ambardar, 1999; Embree, 1995; Ifeachor and Jervis, 2002).

1.3.5 Digital Photo Image Enhancement

We can look at another example of signal processing in two dimensions. Figure 1.11(a) shows a picture of an outdoor scene taken by a digital camera on a cloudy day. Due to this weather condition, the image was improperly exposed in natural light and came out dark. The image processing technique called *histogram equalization* (Gonzalez and Wintz, 1987) can stretch the light intensity of an

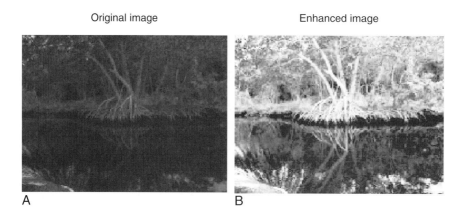

FIGURE 1.11 **Image enhancement.**

image using the digital information (pixels) to increase image contrast so that detailed information in the image can clearly be seen, as we can see in Figure 1.11(b). We will study this technique in Chapter 13.

1.4 Digital Signal Processing Applications

Applications of DSP are increasing in many areas where analog electronics are being replaced by DSP chips, and new applications are depending on DSP techniques. With the cost of DS processors decreasing and their performance increasing, DSP will continue to affect engineering design in our modern daily life. Some application examples using DSP are listed in Table 1.1.

However, the list in the table by no means covers all DSP applications. Many more areas are increasingly being explored by engineers and scientists. Applications of DSP techniques will continue to have profound impacts and improve our lives.

TABLE 1.1   Applications of digital signal processing.

Digital audio and speech	Digital audio coding such as CD players, digital crossover, digital audio equalizers, digital stereo and surround sound, noise reduction systems, speech coding, data compression and encryption, speech synthesis and speech recognition
Digital telephone	Speech recognition, high-speed modems, echo cancellation, speech synthesizers, DTMF (dual-tone multifrequency) generation and detection, answering machines
Automobile industry	Active noise control systems, active suspension systems, digital audio and radio, digital controls
Electronic communications	Cellular phones, digital telecommunications, wireless LAN (local area networking), satellite communications
Medical imaging equipment	ECG analyzers, cardiac monitoring, medical imaging and image recognition, digital x-rays and image processing
Multimedia	Internet phones, audio, and video; hard disk drive electronics; digital pictures; digital cameras; text-to-voice and voice-to-text technologies

1.5 Summary

1. An analog signal is continuous in both time and amplitude. Analog signals in the real world include current, voltage, temperature, pressure, light intensity, and so on. The digital signal is the digital values converted from the analog signal at the specified time instants.

2. Analog-to-digital signal conversion requires an ADC unit (hardware) and a lowpass filter attached ahead of the ADC unit to block the high-frequency components that ADC cannot handle.

3. The digital signal can be manipulated using arithmetic. The manipulations may include digital filtering, calculation of signal frequency content, and so on.

4. The digital signal can be converted back to an analog signal by sending the digital values to DAC to produce the corresponding voltage levels and applying a smooth filter (reconstruction filter) to the DAC voltage steps.

5. Digital signal processing finds many applications in the areas of digital speech and audio, digital and cellular telephones, automobile controls, communications, biomedical imaging, image/video processing, and multimedia.

References

Ambardar, A. (1999). *Analog and Digital Signal Processing,* 2nd ed. Pacific Grove, CA: Brooks/Cole Publishing Company.

Deller, J. R., Proakis, J. G., and Hansen, J. H. L. (1993). *Discrete-Time Processing of Speech Signals.* New York: Macmillian Publishing Company.

Embree, P. M. (1995). *C Algorithms for Real-Time DSP.* Upper Saddle River, NJ: Prentice Hall.

Gonzalez, R. C., and Wintz, P. (1987). *Digital Image Processing*, 2nd ed. Reading, MA: Addison-Wesley Publishing Company.

Ifeachor, E. C., and Jervis, B. W. (2002). *Digital Signal Processing: A Practical Approach,* 2nd ed. Upper Saddle River, NJ: Prentice Hall.

Li, Z.-N., and Drew, M. S. (2004). *Fundamentals of Multimedia.* Upper Saddle River, NJ: Pearson Prentice Hall.

Pan, D. (1995). A tutorial on MPEG/audio compression. *IEEE Multimedia*, 2, 60–74.

Rabiner, L. R., and Schafer, R. W. (1978). *Digital Processing of Speech Signals.* Englewood Cliffs, NJ: Prentice Hall.

Webster, J. G. (1998). *Medical Instrumentation: Application and Design*, 3rd ed. New York: John Wiley & Sons, Inc.

2

Signal Sampling and Quantization

Objectives:

This chapter investigates the sampling process, sampling theory, and the signal reconstruction process. It also includes practical considerations for anti-aliasing and anti-image filters and signal quantization.

2.1 Sampling of Continuous Signal

As discussed in Chapter 1, Figure 2.1 describes a simplified block diagram of a digital signal processing (DSP) system. The analog filter processes the analog input to obtain the band-limited signal, which is sent to the analog-to-digital conversion (ADC) unit. The ADC unit samples the analog signal, quantizes the sampled signal, and encodes the quantized signal levels to the digital signal.

Here we first develop concepts of sampling processing in time domain. Figure 2.2 shows an analog (continuous-time) signal (solid line) defined at every point over the time axis (horizontal line) and amplitude axis (vertical line). Hence, the analog signal contains an infinite number of points.

It is impossible to digitize an infinite number of points. Furthermore, the infinite points are not appropriate to be processed by the digital signal (DS) processor or computer, since they require infinite amount of memory and infinite amount of processing power for computations. Sampling can solve such a problem by taking samples at the fixed time interval, as shown in Figure 2.2 and Figure 2.3, where the time T represents the sampling interval or sampling period in seconds.

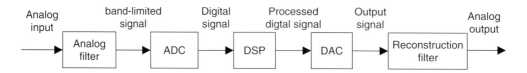

FIGURE 2.1 **A digital signal processing scheme.**

As shown in Figure 2.3, each sample maintains its voltage level during the sampling interval T to give the ADC enough time to convert it. This process is called *sample and hold*. Since there exists one amplitude level for each sampling interval, we can sketch each sample amplitude level at its corresponding sampling time instant shown in Figure 2.2, where 14 samples at their sampling time instants are plotted, each using a vertical bar with a solid circle at its top.

For a given sampling interval T, which is defined as the time span between two sample points, the sampling rate is therefore given by

$$f_s = \frac{1}{T} \text{ samples per second (Hz)}.$$

For example, if a sampling period is $T = 125$ microseconds, the sampling rate is determined as $f_s = 1/125 \mu s = 8{,}000$ samples per second (Hz).

After the analog signal is sampled, we obtain the sampled signal whose amplitude values are taken at the sampling instants, thus the processor is able to handle the sample points. Next, we have to ensure that samples are collected at a rate high enough that the original analog signal can be reconstructed or recovered later. In other words, we are looking for a minimum sampling rate to acquire a complete reconstruction of the analog signal from its sampled version.

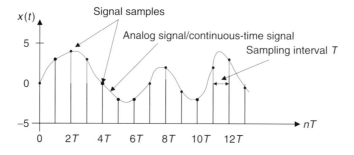

FIGURE 2.2 **Display of the analog (continuous) signal and display of digital samples versus the sampling time instants.**

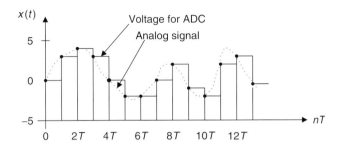

FIGURE 2.3   Sample-and-hold analog voltage for ADC.

If an analog signal is not appropriately sampled, *aliasing* will occur, which causes unwanted signals in the desired frequency band.

The sampling theorem guarantees that an analog signal can be in theory perfectly recovered as long as the sampling rate is at least twice as large as the highest-frequency component of the analog signal to be sampled. The condition is described as

$$f_s \geq 2f_{max},$$

where f_{max} is the maximum-frequency component of the analog signal to be sampled. For example, to sample a speech signal containing frequencies up to 4 kHz, the minimum sampling rate is chosen to be at least 8 kHz, or 8,000 samples per second; to sample an audio signal possessing frequencies up to 20 kHz, at least 40,000 samples per second, or 40 kHz, of the audio signal are required.

Figure 2.4 illustrates sampling of two sinusoids, where the sampling interval between sample points is $T = 0.01$ second, thus the sampling rate is $f_s = 100$ Hz. The first plot in the figure displays a sine wave with a frequency of 40 Hz and its sampled amplitudes. The sampling theorem condition is satisfied, since $2f_{max} = 80$ Hz $< f_s$. The sampled amplitudes are labeled using the circles shown in the first plot. We notice that the 40-Hz signal is adequately sampled, since the sampled values clearly come from the analog version of the 40-Hz sine wave. However, as shown in the second plot, the sine wave with a frequency of 90 Hz is sampled at 100 Hz. Since the sampling rate of 100 Hz is relatively low compared with the 90-Hz sine wave, the signal is undersampled due to $2f_{max} = 180$ Hz $> f_s$. Hence, the condition of the sampling theorem is not satisfied. Based on the sample amplitudes labeled with the circles in the second plot, we cannot tell whether the sampled signal comes from sampling a 90-Hz sine wave (plotted using the solid line) or from sampling a 10-Hz sine wave (plotted using the dot-dash line). They are not distinguishable. Thus they

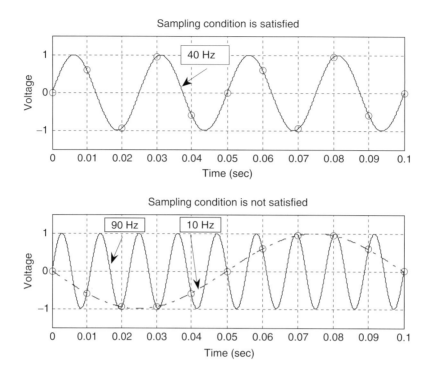

FIGURE 2.4 Plots of the appropriately sampled signals and nonappropriately sampled (aliased) signals.

are *aliases* of each other. We call the 10-Hz sine wave the aliasing noise in this case, since the sampled amplitudes actually come from sampling the 90-Hz sine wave.

Now let us develop the sampling theorem in frequency domain, that is, the minimum sampling rate requirement for an analog signal. As we shall see, in practice this can help us design the anti-aliasing filter (a lowpass filter that will reject high frequencies that cause aliasing) to be applied before sampling, and the anti-image filter (a reconstruction lowpass filter that will smooth the recovered sample-and-hold voltage levels to an analog signal) to be applied after the digital-to-analog conversion (DAC).

Figure 2.5 depicts the sampled signal $x_s(t)$ obtained by sampling the continuous signal $x(t)$ at a sampling rate of f_s samples per second.

Mathematically, this process can be written as the product of the continuous signal and the sampling pulses (pulse train):

$$x_s(t) = x(t)p(t), \tag{2.1}$$

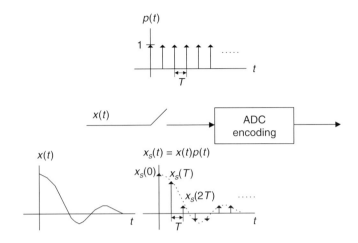

FIGURE 2.5 The simplified sampling process.

where $p(t)$ is the pulse train with a period $T = 1/f_s$. From spectral analysis, the original spectrum (frequency components) $X(f)$ and the sampled signal spectrum $X_s(f)$ in terms of Hz are related as

$$X_s(f) = \frac{1}{T} \sum_{n=-\infty}^{\infty} X(f - nf_s), \qquad (2.2)$$

where $X(f)$ is assumed to be the original baseband spectrum, while $X_s(f)$ is its sampled signal spectrum, consisting of the original baseband spectrum $X(f)$ and its replicas $X(f \pm nf_s)$. Since Equation (2.2) is a well-known formula, the derivation is omitted here and can be found in well-known texts (Ahmed and Natarajan, 1983; Alkin, 1993; Ambardar, 1999; Oppenheim and Schafer, 1975; Proakis and Manolakis, 1996).

Expanding Equation (2.2) leads to the sampled signal spectrum in Equation (2.3):

$$X_s(f) = \cdots + \frac{1}{T}X(f + f_s) + \frac{1}{T}X(f) + \frac{1}{T}X(f - f_s) + \cdots \qquad (2.3)$$

Equation (2.3) indicates that the sampled signal spectrum is the sum of the scaled original spectrum and copies of its shifted versions, called *replicas*. The sketch of Equation (2.3) is given in Figure 2.6, where three possible sketches are classified. Given the original signal spectrum $X(f)$ plotted in Figure 2.6(a), the sampled signal spectrum according to Equation (2.3) is plotted in Figure 2.6(b), where the replicas, $\frac{1}{T}X(f), \frac{1}{T}X(f - f_s), \frac{1}{T}X(f + f_s), \ldots$, have separations between them. Figure 2.6(c) shows that the baseband spectrum and its replicas, $\frac{1}{T}X(f), \frac{1}{T}X(f - f_s), \frac{1}{T}X(f + f_s), \ldots$, are just connected, and finally, in Figure

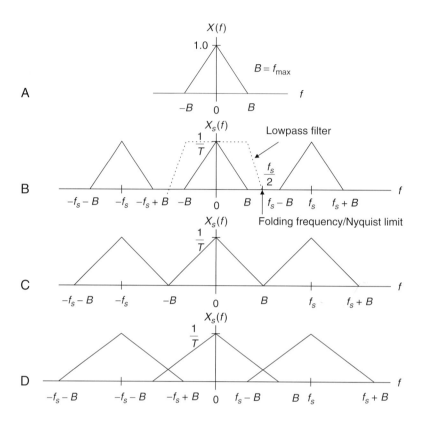

FIGURE 2.6 Plots of the sampled signal spectrum.

2.6(d), the original spectrum $\frac{1}{T}X(f)$ and its replicas $\frac{1}{T}X(f-f_s), \frac{1}{T}X(f+f_s), \ldots,$ are overlapped; that is, there are many overlapping portions in the sampled signal spectrum.

From Figure 2.6, it is clear that the sampled signal spectrum consists of the scaled baseband spectrum centered at the origin and its replicas centered at the frequencies of $\pm n f_s$ (multiples of the sampling rate) for each of $n = 1, 2, 3, \ldots$.

If applying a lowpass reconstruction filter to obtain exact reconstruction of the original signal spectrum, the following condition must be satisfied:

$$f_s - f_{\max} \geq f_{\max}. \tag{2.4}$$

Solving Equation (2.4) gives

$$f_s \geq 2 f_{\max}. \tag{2.5}$$

In terms of frequency in radians per second, Equation (2.5) is equivalent to

$$\omega_s \geq 2\omega_{\max}. \tag{2.6}$$

2.1 Sampling of Continuous Signal

This fundamental conclusion is well known as the **Shannon sampling theorem,** which is formally described below:

> For a uniformly sampled DSP system, an analog signal can be perfectly recovered as long as the sampling rate is at least twice as large as the highest-frequency component of the analog signal to be sampled.

We summarize two key points here.

1. Sampling theorem establishes a minimum sampling rate for a given band-limited analog signal with the highest-frequency component f_{max}. If the sampling rate satisfies Equation (2.5), then the analog signal can be recovered via its sampled values using the lowpass filter, as described in Figure 2.6(b).

2. Half of the sampling frequency $f_s/2$ is usually called the *Nyquist frequency* (Nyquist limit), or *folding frequency*. The sampling theorem indicates that a DSP system with a sampling rate of f_s can ideally sample an analog signal with its highest frequency up to half of the sampling rate without introducing spectral overlap (aliasing). Hence, the analog signal can be perfectly recovered from its sampled version.

Let us study the following example.

Example 2.1.

Suppose that an analog signal is given as

$$x(t) = 5\cos(2\pi \cdot 1000t), \text{ for } t \geq 0$$

and is sampled at the rate of 8,000 Hz.
    a. Sketch the spectrum for the original signal.
    b. Sketch the spectrum for the sampled signal from 0 to 20 kHz.

Solution:

    a. Since the analog signal is sinusoid with a peak value of 5 and frequency of 1,000 Hz, we can write the sine wave using Euler's identity:

$$5\cos(2\pi \times 1000t) = 5 \cdot \left(\frac{e^{j2\pi \times 1000t} + e^{-j2\pi \times 1000t}}{2} \right) = 2.5e^{j2\pi \times 1000t} + 2.5e^{-j2\pi \times 1000t},$$

which is a Fourier series expansion for a continuous periodic signal in terms of the exponential form (see Appendix B). We can identify the Fourier series coefficients as

$$c_1 = 2.5, \text{ and } c_{-1} = 2.5.$$

Using the magnitudes of the coefficients, we then plot the two-sided spectrum as

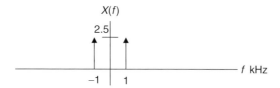

FIGURE 2.7A Spectrum of the analog signal in Example 2.1.

b. After the analog signal is sampled at the rate of 8,000 Hz, the sampled signal spectrum and its replicas centered at the frequencies $\pm nf_s$, each with the scaled amplitude being $2.5/T$, are as shown in Figure 2.7b:

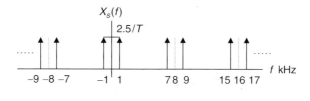

FIGURE 2.7B Spectrum of the sampled signal in Example 2.1

Notice that the spectrum of the sampled signal shown in Figure 2.7b contains the images of the original spectrum shown in Figure 2.7a; that the images repeat at multiples of the sampling frequency f_s (for our example, 8 kHz, 16 kHz, 24 kHz, ...); and that all images must be removed, since they convey no additional information.

2.2 Signal Reconstruction

In this section, we investigate the recovery of analog signal from its sampled signal version. Two simplified steps are involved, as described in Figure 2.8. First, the digitally processed data $y(n)$ are converted to the ideal impulse train $y_s(t)$, in which each impulse has its amplitude proportional to digital output $y(n)$, and two consecutive impulses are separated by a sampling period of T; second, the analog reconstruction filter is applied to the ideally recovered sampled signal $y_s(t)$ to obtain the recovered analog signal.

To study the signal reconstruction, we let $y(n) = x(n)$ for the case of no DSP, so that the reconstructed sampled signal and the input sampled signal are ensured to be the same; that is, $y_s(t) = x_s(t)$. Hence, the spectrum of the sampled signal $y_s(t)$ contains the same spectral content as the original spectrum $X(f)$,

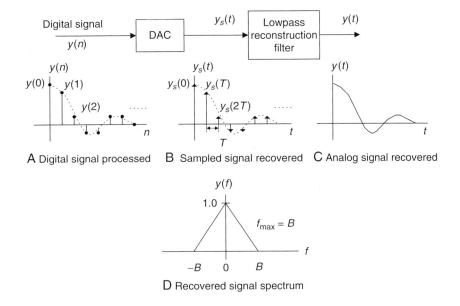

FIGURE 2.8 Signal notations at reconstruction stage.

that is, $Y(f) = X(f)$, with a bandwidth of $f_{max} = B$ Hz (described in Figure 2.8(d) and the images of the original spectrum (scaled and shifted versions). The following three cases are discussed for recovery of the original signal spectrum $X(f)$.

Case 1: $f_s = 2f_{max}$

As shown in Figure 2.9, where the Nyquist frequency is equal to the maximum frequency of the analog signal $x(t)$, an ideal lowpass reconstruction filter is required to recover the analog signal spectrum. This is an impractical case.

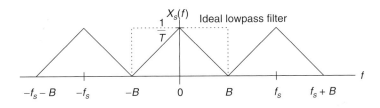

FIGURE 2.9 Spectrum of the sampled signal when $f_s = 2f_{max}$.

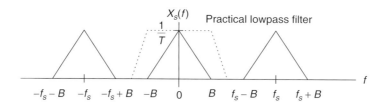

FIGURE 2.10 Spectrum of the sampled signal when $f_s > 2f_{max}$.

Case 2: $f_s > 2f_{max}$

In this case, as shown in Figure 2.10, there is a separation between the highest-frequency edge of the baseband spectrum and the lower edge of the first replica. Therefore, a practical lowpass reconstruction (anti-image) filter can be designed to reject all the images and achieve the original signal spectrum.

Case 3: $f_s < 2f_{max}$

Case 3 violates the condition of the Shannon sampling theorem. As we can see, Figure 2.11 depicts the spectral overlapping between the original baseband spectrum and the spectrum of the first replica and so on. Even when we apply an ideal lowpass filter to remove these images, in the baseband there is still some foldover frequency components from the adjacent replica. This is aliasing, where the recovered baseband spectrum suffers spectral distortion, that is, contains an aliasing noise spectrum; in time domain, the recovered analog signal may consist of the aliasing noise frequency or frequencies. Hence, the recovered analog signal is incurably distorted.

Note that if an analog signal with a frequency f is undersampled, the aliasing frequency component f_{alias} in the baseband is simply given by the following expression:

$$f_{alias} = f_s - f.$$

The following examples give a spectrum analysis of the signal recovery.

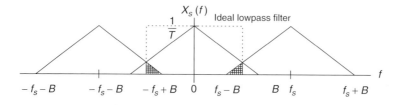

FIGURE 2.11 Spectrum of the sampled signal when $f_s < 2f_{max}$.

Example 2.2.

Assuming that an analog signal is given by

$$x(t) = 5\cos(2\pi \cdot 2000t) + 3\cos(2\pi \cdot 3000t), \text{ for } t \geq 0$$

and it is sampled at the rate of 8,000 Hz,
  a. Sketch the spectrum of the sampled signal up to 20 kHz.
  b. Sketch the recovered analog signal spectrum if an ideal lowpass filter with a cutoff frequency of 4 kHz is used to filter the sampled signal ($y(n) = x(n)$ in this case) to recover the original signal.

Solution: Using Euler's identity, we get

$$x(t) = \frac{3}{2}e^{-j2\pi \cdot 3000t} + \frac{5}{2}e^{-j2\pi \cdot 2000t} + \frac{5}{2}e^{j2\pi \cdot 2000t} + \frac{3}{2}e^{j2\pi \cdot 3000t}.$$

The two-sided amplitude spectrum for the sinusoids is displayed in Figure 2.12:
  a.

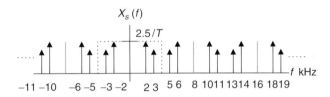

FIGURE 2.12 Spectrum of the sampled signal in Example 2.2.

  b. Based on the spectrum in (a), the sampling theorem condition is satisfied; hence, we can recover the original spectrum using a reconstruction lowpass filter. The recovered spectrum is shown in Figure 2.13:

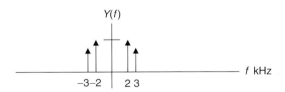

FIGURE 2.13 Spectrum of the recovered signal in Example 2.2.

Example 2.3.

Given an analog signal

$$x(t) = 5\cos(2\pi \times 2000t) + 1\cos(2\pi \times 5000t), \text{ for } t \geq 0,$$

which is sampled at a rate of 8,000 Hz,

a. Sketch the spectrum of the sampled signal up to 20 kHz.

b. Sketch the recovered analog signal spectrum if an ideal lowpass filter with a cutoff frequency of 4 kHz is used to recover the original signal ($y(n) = x(n)$ in this case).

Solution:

a. The spectrum for the sampled signal is sketched in Figure 2.14:

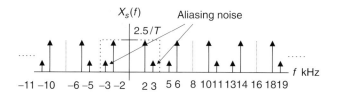

FIGURE 2.14 Spectrum of the sampled signal in Example 2.3.

b. Since the maximum frequency of the analog signal is larger than that of the Nyquist frequency—that is, twice the maximum frequency of the analog signal is larger than the sampling rate—the sampling theorem condition is violated. The recovered spectrum is shown in Figure 2.15, where we see that aliasing noise occurs at 3 kHz.

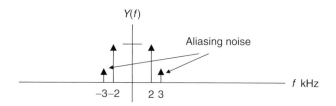

FIGURE 2.15 Spectrum of the recovered signal in Example 2.3.

2.2.1 Practical Considerations for Signal Sampling: Anti-Aliasing Filtering

In practice, the analog signal to be digitized may contain other frequency components in addition to the folding frequency, such as high-frequency noise. To satisfy the sampling theorem condition, we apply an anti-aliasing filter to limit the input analog signal, so that all the frequency components are less than the folding frequency (half of the sampling rate). Considering the worst case, where the analog signal to be sampled has a flat frequency spectrum, the band-limited spectrum $X(f)$ and sampled spectrum $X_s(f)$ are depicted in Figure 2.16, where the shape of each replica in the sampled signal spectrum is the same as that of the anti-aliasing filter magnitude frequency response.

Due to nonzero attenuation of the magnitude frequency response of the anti-aliasing lowpass filter, the aliasing noise from the adjacent replica still appears in the baseband. However, the level of the aliasing noise is greatly reduced. We can also control the aliasing noise level by either using a higher-order lowpass filter or increasing the sampling rate. For illustrative purposes, we use a Butterworth filter. The method can also be extended to other filter types such as the Chebyshev filter. The Butterworth magnitude frequency response with an order of n is given by

$$|H(f)| = \frac{1}{\sqrt{1 + \left(\frac{f}{f_c}\right)^{2n}}}. \tag{2.7}$$

For a second-order Butterworth lowpass filter with the unit gain, the transfer function (which will be discussed in Chapter 8) and its magnitude frequency response are given by

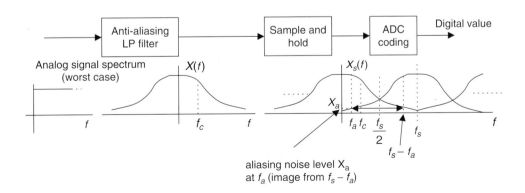

FIGURE 2.16 Spectrum of the sampled analog signal with a practical anti-aliasing filter.

2 SIGNAL SAMPLING AND QUANTIZATION

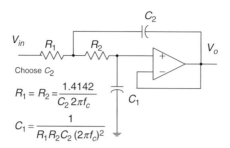

FIGURE 2.17 Second-order unit gain Sallen-Key lowpass filter.

$$H(s) = \frac{(2\pi f_c)^2}{s^2 + 1.4142 \times (2\pi f_c)s + (2\pi f_c)^2} \quad (2.8)$$

$$\text{and} \quad |H(f)| = \frac{1}{\sqrt{1 + \left(\frac{f}{f_c}\right)^4}}. \quad (2.9)$$

A unit gain second-order lowpass filter using a Sallen-Key topology is shown in Figure 2.17. Matching the coefficients of the circuit transfer function to that of the second-order Butterworth lowpass transfer function in Equation (2.10) gives the design formulas shown in Figure 2.17, where for a given cutoff frequency of f_c in Hz, and a capacitor value of C_2, we can determine the other elements using the formulas listed in the figure.

$$\frac{\frac{1}{R_1 R_2 C_1 C_2}}{s^2 + \left(\frac{1}{R_1 C_2} + \frac{1}{R_2 C_2}\right)s + \frac{1}{R_1 R_2 C_1 C_2}} = \frac{(2\pi f_c)^2}{s^2 + 1.4142 \times (2\pi f_c)s + (2\pi f_c)^2} \quad (2.10)$$

As an example, for a cutoff frequency of 3,400 Hz, and by selecting $C_2 = 0.01$ micro-farad (uF), we can get

$$R_1 = R_2 = 6620 \, \Omega, \text{ and } C_1 = 0.005 \, uF.$$

Figure 2.18 shows the magnitude frequency response, where the absolute gain of the filter is plotted. As we can see, the absolute attenuation begins at the level of 0.7 at 3,400 Hz and reduces to 0.3 at 6,000 Hz. Ideally, we want the gain attenuation to be zero after 4,000 Hz if our sampling rate is 8,000 Hz. Practically speaking, aliasing will occur anyway to some degree. We will study achieving the higher-order analog filter via Butterworth and Chebyshev prototype function tables in Chapter 8. More details of the circuit realization for the analog filter can be found in Chen (1986).

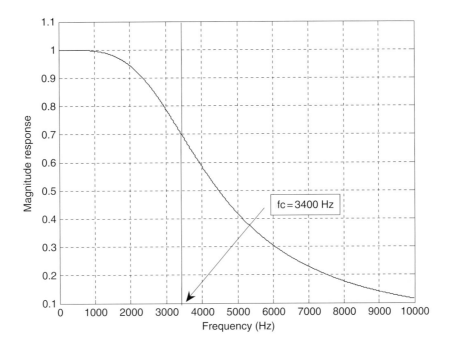

FIGURE 2.18 Magnitude frequency response of the second-order Butterworth lowpass filter.

According to Figure 2.16, we can derive the percentage of the aliasing noise level using the symmetry of the Butterworth magnitude function and its first replica. It follows that

$$\text{aliasing noise level \%} = \frac{X_a}{X(f)|_{f=f_a}} = \frac{|H(f)|_{f=f_s-f_a}}{|H(f)|_{f=f_a}}$$

$$= \frac{\sqrt{1 + \left(\frac{f_a}{f_c}\right)^{2n}}}{\sqrt{1 + \left(\frac{f_s-f_a}{f_c}\right)^{2n}}} \quad \text{for } 0 \leq f \leq f_c. \quad (2.11)$$

With Equation (2.11), we can estimate the aliasing noise level, or choose a higher-order anti-aliasing filter to satisfy the requirement for the percentage of aliasing noise level.

Example 2.4.

Given the DSP system shown in Figures 2.16 to 2.18, where a sampling rate of 8,000 Hz is used and the anti-aliasing filter is a second-order Butterworth lowpass filter with a cutoff frequency of 3.4 kHz,

a. Determine the percentage of aliasing level at the cutoff frequency.

b. Determine the percentage of aliasing level at the frequency of 1,000 Hz.

Solution:

$$f_s = 8000, f_c = 3400, \text{ and } n = 2.$$

a. Since $f_a = f_c = 3400$ Hz, we compute

$$\text{aliasing noise level } \% = \frac{\sqrt{1 + \left(\frac{3.4}{3.4}\right)^{2\times 2}}}{\sqrt{1 + \left(\frac{8-3.4}{3.4}\right)^{2\times 2}}} = \frac{1.4142}{2.0858} = 67.8\%.$$

b. With $f_a = 1000$ Hz, we have

$$\text{aliasing noise level } \% = \frac{\sqrt{1 + \left(\frac{1}{3.4}\right)^{2\times 2}}}{\sqrt{1 + \left(\frac{8-1}{3.4}\right)^{2\times 2}}} = \frac{1.03007}{4.3551} = 23.05\%.$$

Let us examine another example with an increased sampling rate.

Example 2.5.

a. Given the DSP system shown in Figures 2.16 to 2.18, where a sampling rate of 16,000 Hz is used and the anti-aliasing filter is a second-order Butterworth lowpass filter with a cutoff frequency of 3.4 kHz, determine the percentage of aliasing level at the cutoff frequency.

Solution:

$$f_s = 16000, f_c = 3400, \text{ and } n = 2.$$

a. Since $f_a = f_c = 3400$ Hz, we have

$$\text{aliasing noise level } \% = \frac{\sqrt{1 + \left(\frac{3.4}{3.4}\right)^{2\times 2}}}{\sqrt{1 + \left(\frac{16-3.4}{3.4}\right)^{2\times 2}}} = \frac{1.4142}{13.7699} = 10.26\%.$$

As a comparison with the result in Example 2.4, increasing the sampling rate can reduce the aliasing noise level.

The following example shows how to choose the order of the anti-aliasing filter.

Example 2.6.

a. Given the DSP system shown in Figure 2.16, where a sampling rate of 40,000 Hz is used, the anti-aliasing filter is a Butterworth lowpass filter with a cutoff frequency of 8 kHz, and the percentage of aliasing level at the cutoff frequency is required to be less than 1%, determine the order of the anti-aliasing lowpass filter.

Solution:

a. Using $f_s = 40,000$, $f_c = 8000$, and $f_a = 8000 \text{ Hz}$, we try each of the following filters with the increasing number of the filter order.

$$n = 1, \text{ aliasing noise level } \% = \frac{\sqrt{1 + \left(\frac{8}{8}\right)^{2 \times 1}}}{\sqrt{1 + \left(\frac{40-8}{8}\right)^{2 \times 1}}} = \frac{1.4142}{\sqrt{1 + (4)^2}} = 34.30\%$$

$$n = 2, \text{ aliasing noise level } \% = \frac{1.4142}{\sqrt{1 + (4)^4}} = 8.82\%$$

$$n = 3, \text{ aliasing noise level } \% = \frac{1.4142}{\sqrt{1 + (4)^6}} = 2.21\%$$

$$n = 4, \text{ aliasing noise level } \% = \frac{1.4142}{\sqrt{1 + (4)^8}} = 0.55\% < 1\%$$

To satisfy 1% aliasing noise level, we choose $n = 4$.

2.2.2 Practical Considerations for Signal Reconstruction: Anti-Image Filter and Equalizer

The analog signal recovery for a practical DSP system is illustrated in Figure 2.19.

As shown in Figure 2.19, the DAC unit converts the processed digital signal $y(n)$ to a sampled signal $y_s(t)$, and then the hold circuit produces the sample-and-hold voltage $y_H(t)$. The transfer function of the hold circuit can be derived to be

$$H_h(s) = \frac{1 - e^{-sT}}{s}. \tag{2.12}$$

We can obtain the frequency response of the DAC with the hold circuit by substituting $s = j\omega$ into Equation (2.12). It follows that

$$H_h(\omega) = e^{-j\omega T/2} \frac{\sin(\omega T/2)}{\omega T/2}. \tag{2.13}$$

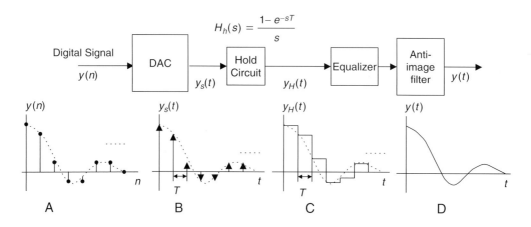

FIGURE 2.19 Signal notations at the practical reconstruction stage. (a) Processed digital signal. (b) Recovered ideal sampled signal. (c) Recovered sample-and-hold voltage. (d) Recovered analog signal.

The magnitude and phase responses are given by

$$|H_h(\omega)| = \left|\frac{\sin(\omega T/2)}{\omega T/2}\right| = \left|\frac{\sin(x)}{x}\right| \qquad (2.14)$$

$$\angle H_h(\omega) = -\omega T/2, \qquad (2.15)$$

where $x = \omega T/2$. In terms of Hz, we have

$$|H_h(f)| = \left|\frac{\sin(\pi f T)}{\pi f T}\right| \qquad (2.16)$$

$$\angle H_h(f) = -\pi f T. \qquad (2.17)$$

The plot of the magnitude effect is shown in Figure 2.20.

The magnitude frequency response acts like lowpass filtering and shapes the sampled signal spectrum of $Y_s(f)$. This shaping effect distorts the sampled signal spectrum $Y_s(f)$ in the desired frequency band, as illustrated in Figure 2.21. On the other hand, the spectral images are attenuated due to the lowpass effect of $\sin(x)/x$. This sample-and-hold effect can help us design the anti-image filter.

As shown in Figure 2.21, the percentage of distortion in the desired frequency band is given by

$$\text{distortion \%} = (1 - H_h(f)) \times 100\%$$

$$= \left(1 - \frac{\sin(\pi f T)}{\pi f T}\right) \times 100\% \qquad (2.18)$$

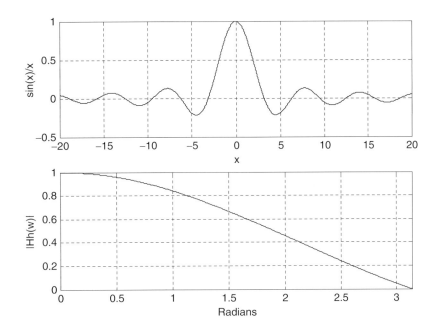

FIGURE 2.20 Sample-and-hold lowpass filtering effect.

Let us look at Example 2.7.

Example 2.7.

Given a DSP system with a sampling rate of 8,000 Hz and a hold circuit used after DAC,

a. Determine the percentage of distortion at the frequency of 3,400 Hz.

b. Determine the percentage of distortion at the frequency of 1,000 Hz.

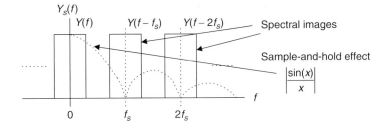

FIGURE 2.21 Sample-and-hold effect and distortion.

Solution:

a. Since $fT = 3400 \times 1/8000 = 0.425$,

$$\text{distortion \%} = \left(1 - \frac{\sin(0.425\pi)}{0.425\pi}\right) \times 100\% = 27.17\%.$$

b. Since $fT = 1000 \times 1/8000 = 0.125$,

$$\text{distortion \%} = \left(1 - \frac{\sin(0.125\pi)}{0.125\pi}\right) \times 100\% = 2.55\%.$$

To overcome the sample-and-hold effect, the following methods can be applied.

1. We can compensate the sample-and-hold shaping effect using an equalizer whose magnitude response is opposite to the shape of the hold circuit magnitude frequency response, which is shown as the solid line in Figure 2.22.

2. We can increase the sampling rate using oversampling and interpolation methods when a higher sampling rate is available at the DAC. Using the interpolation will increase the sampling rate without affecting the signal bandwidth, so that the baseband spectrum and its images are separated farther apart and a lower-order anti-image filter can be used. This subject will be discussed in Chapter 12.

3. We can change the DAC configuration and perform digital pre-equalization using the flexible digital filter whose magnitude frequency response is against the spectral shape effect due to the hold circuit. Figure 2.23 shows a possible implementation. In this way, the spectral shape effect can be balanced before the sampled signal passes through the hold circuit. Finally, the anti-image filter will remove the rest of the images and recover the desired analog signal.

The following practical example will illustrate the design of an anti-image filter using a higher sampling rate while making use of the sample-and-hold effect.

Example 2.8.

a. Determine the cutoff frequency and the order for the anti-image filter given a DSP system with a sampling rate of 16,000 Hz and specifications for the anti-image filter as shown in Figure 2.24.

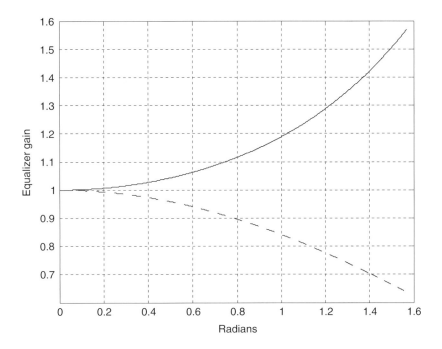

FIGURE 2.22 Ideal equalizer magnitude frequency response to overcome the distortion introduced by the sample-and-hold process.

Design requirements:

- Maximum allowable gain variation from 0 to 3,000 Hz = 2 dB
- 33 dB rejection at the frequency of 13,000 Hz
- Butterworth filter assumed for the anti-image filter

Solution:

a. We first determine the spectral shaping effects at $f = 3000$ Hz and $f = 13,000$ Hz; that is,

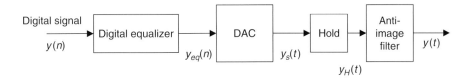

FIGURE 2.23 Possible implementation using the digital equalizer.

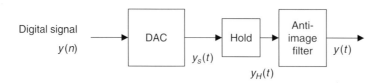

FIGURE 2.24 DSP recovery system for Example 2.8.

$$f = 3000\,\text{Hz},\ fT = 3000 \times 1/16000 = 0.1785$$
$$\text{gain} = \frac{\sin(0.1785\pi)}{0.1785\pi} = 0.9484 = -0.46\,\text{dB}$$

and

$$f = 13000\,\text{Hz},\ fT = 13000 \times 1/16000 = 0.8125$$
$$\text{gain} = \frac{\sin(0.8125\pi)}{0.8125\pi} = 0.2177 \approx -13\,\text{dB}.$$

This gain would help the attenuation requirement.

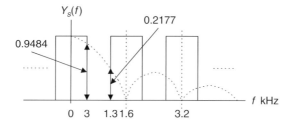

FIGURE 2.25 Spectral shaping by the sample-and-hold effect in Example 2.8.

Hence, the design requirements for the anti-image filter are:

- Butterworth lowpass filter
- Maximum allowable gain variation from 0 to 3,000 Hz = $(2 - 0.46)$ = 1.54 dB
- $(33 - 13) = 20$ dB rejection at frequency 13,000 Hz.

We set up equations using log operations of the Butterworth magnitude function as

$$20\log\left(1 + (3000/f_c)^{2n}\right)^{1/2} \le 1.54$$
$$20\log\left(1 + (13000/f_c)^{2n}\right)^{1/2} \ge 20.$$

From these two equations, we have to satisfy

$$(3000/f_c)^{2n} = 10^{0.154} - 1$$
$$(13000/f_c)^{2n} = 10^2 - 1.$$

Taking the ratio of these two equations yields

$$\left(\frac{13000}{3000}\right)^{2n} = \frac{10^2 - 1}{10^{0.154} - 1}.$$

Then $n = \frac{1}{2} \log((10^2 - 1)/(10^{0.154} - 1))/\log(13000/3000) = 1.86 \approx 2$.

Finally, the cutoff frequency can be computed as

$$f_c = \frac{13000}{(10^2 - 1)^{1/(2n)}} = \frac{13000}{(10^2 - 1)^{1/4}} = 4121.30 \text{ Hz}$$

$$f_c = \frac{3000}{(10^{0.154} - 1)^{1/(2n)}} = \frac{3000}{(10^{0.154} - 1)^{1/4}} = 3714.23 \text{ Hz}.$$

We choose the smaller one, that is,

$$f_c = 3714.23 \text{ Hz}.$$

With the filter order and cutoff frequency, we can realize the anti-image (reconstruction) filter using the second-order unit gain Sallen-Key lowpass filter described in Figure 2.17.

Note that the specifications for anti-aliasing filter designs are similar to those for anti-image (reconstruction) filters, except for their stopband edges. The anti-aliasing filter is designed to block the frequency components beyond the folding frequency before the ADC operation, while the reconstruction filter is to block the frequency components beginning at the lower edge of the first image after the DAC.

2.3 Analog-to-Digital Conversion, Digital-to-Analog Conversion, and Quantization

During the ADC process, amplitudes of the analog signal to be converted have infinite precision. The continuous amplitude must be converted into digital data with finite precision, which is called the *quantization*. Figure 2.26 shows that quantization is a part of ADC.

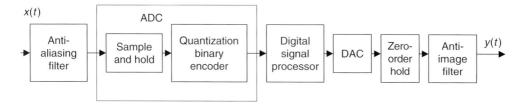

FIGURE 2.26   A block diagram for a DSP system.

There are several ways to implement ADC. The most common ones are

- flash ADC,
- successive approximation ADC, and
- sigma-delta ADC.

In this chapter, we will focus on a simple 2-bit flash ADC unit, described in Figure 2.27, for illustrative purposes. Sigma-delta ADC will be studied in Chapter 12.

As shown in Figure 2.27, the 2-bit flash ADC unit consists of a serial reference voltage created by the equal value resistors, a set of comparators, and logic units. As an example, the reference voltages in the figure are 1.25 volts, 2.5 volts, 3.75 volts, and 5 volts. If an analog sample-and-hold voltage is $V_{in} = 3$ volts, then the lower two comparators will each output logic 1. Through the logic units, only the line labeled 10 is actively high, and the rest of the lines are actively low. Hence, the encoding logic circuit outputs a 2-bit binary code of 10.

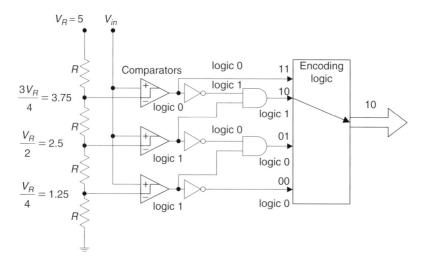

FIGURE 2.27   An example of a 2-bit flash ADC.

Flash ADC offers the advantage of high conversion speed, since all bits are acquired at the same time. Figure 2.28 illustrates a simple 2-bit DAC unit using an R-2R ladder. The DAC contains the R-2R ladder circuit, a set of single-throw switches, a summer, and a phase shifter. If a bit is logic 0, the switch connects a 2R resistor to ground. If a bit is logic 1, the corresponding 2R resistor is connected to the branch to the input of the operational amplifier (summer). When the operational amplifier operates in a linear range, the negative input is virtually equal to the positive input. The summer adds all the currents from all branches. The feedback resistor R in the summer provides overall amplification. The ladder network is equivalent to two 2R resistors in parallel. The entire network has a total current of $I = \frac{V_R}{R}$ using Ohm's law, where V_R is the reference voltage, chosen to be 5 volts for our example. Hence, half of the total current flows into the b_1 branch, while the other half flows into the rest of the network. The halving process repeats for each branch successively to the lower bit branches to get lower bit weights. The second operational amplifier acts like a phase shifter to cancel the negative sign of the summer output. Using the basic electric circuit principle, we can determine the DAC output voltage as

$$V_0 = V_R \left(\frac{1}{2^1} b_1 + \frac{1}{2^2} b_0 \right),$$

where b_1 and b_0 are bits in the 2-bit binary code, with b_0 as the least significant bit (LSB).

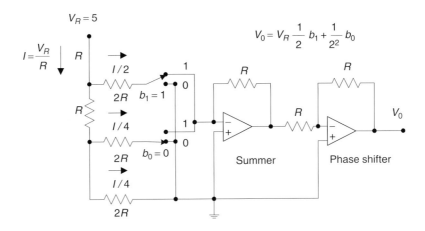

FIGURE 2.28 R-2R ladder DAC.

As an example shown in Figure 2.28, where we set $V_R = 5$ and $b_1 b_0 = 10$, the ADC output is expected to be

$$V_0 = 5 \times \left(\frac{1}{2^1} \times 1 + \frac{1}{2^2} \times 0 \right) = 2.5 \text{ volts}.$$

As we can see, the recovered voltage of $V_0 = 2.5$ volts introduces voltage error as compared with $V_{in} = 3$, discussed in the ADC stage. This is due to the fact that in the flash ADC unit, we use only four (i.e., finite) voltage levels to represent continuous (infinitely possible) analog voltage values. The introduction is called *quantization error*, obtained by subtracting the original analog voltage from the recovered analog voltage. For our example, we have the quantization error as

$$V_0 - V_{in} = 2.5 - 3 = -0.5 \text{ volts}.$$

Next, we focus on quantization development. The process of converting analog voltage with infinite precision to finite precision is called the *quantization process*. For example, if the digital processor has only a 3-bit word, the amplitudes can be converted into eight different levels.

A *unipolar quantizer* deals with analog signals ranging from 0 volt to a positive reference voltage, and a *bipolar quantizer* has an analog signal range from a negative reference to a positive reference. The notations and general rules for quantization are:

$$\Delta = \frac{(x_{max} - x_{min})}{L} \tag{2.19}$$

$$L = 2^m \tag{2.20}$$

$$i = round\left(\frac{x - x_{min}}{\Delta} \right) \tag{2.21}$$

$$x_q = x_{min} + i\Delta, \text{ for } i = 0, 1, \ldots, L - 1, \tag{2.22}$$

where x_{max} and x_{min} are the maximum and minimum values, respectively, of the analog input signal x. The symbol L denotes the number of quantization levels, which is determined by Equation (2.20), where m is the number of bits used in ADC. The symbol Δ is the step size of the quantizer or the ADC resolution. Finally, x_q indicates the quantization level, and i is an index corresponding to the binary code.

Figure 2.29 depicts a 3-bit unipolar quantizer and corresponding binary codes. From Figure 2.29, we see that $x_{min} = 0$, $x_{max} = 8\Delta$, and $m = 3$. Applying Equation (2.22) gives each quantization level as follows: $x_q = 0 + i\Delta, i = 0, 1, \ldots, L - 1$, where $L = 2^3 = 8$ and i is the integer corresponding to the 3-bit binary code. Table 2.1 details quantization for each input signal subrange.

2.3 Analog-to-Digital Conversion, Digital-to-Analog Conversion, and Quantization

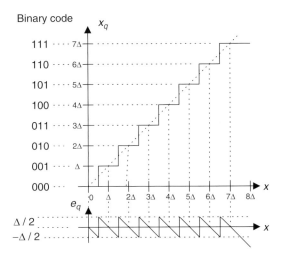

FIGURE 2.29 Characteristics of the unipolar quantizer.

Similarly, a 3-bit bipolar quantizer and binary codes are shown in Figure 2.30, where we have $x_{\min} = -4\Delta$, $x_{\max} = 4\Delta$, and $m = 3$. The corresponding quantization table is given in Table 2.2.

Example 2.9.

Assuming that a 3-bit ADC channel accepts analog input ranging from 0 to 5 volts, determine the following:

a. number of quantization levels

b. step size of the quantizer or resolution

TABLE 2.1 Quantization table for the 3-bit unipolar quantizer (step size = $\Delta = (x_{\max} - x_{\min})/2^3$, $x_{\max}$ = maximum voltage, and $x_{\min} = 0$).

Binary Code	Quantization Level x_q (V)	Input Signal Subrange (V)
0 0 0	0	$0 \leq x < 0.5\Delta$
0 0 1	Δ	$0.5\Delta \leq x < 1.5\Delta$
0 1 0	2Δ	$1.5\Delta \leq x < 2.5\Delta$
0 1 1	3Δ	$2.5\Delta \leq x < 3.5\Delta$
1 0 0	4Δ	$3.5\Delta \leq x < 4.5\Delta$
1 0 1	5Δ	$4.5\Delta \leq x < 5.5\Delta$
1 1 0	6Δ	$5.5\Delta \leq x < 6.5\Delta$
1 1 1	7Δ	$6.5\Delta \leq x < 7.5\Delta$

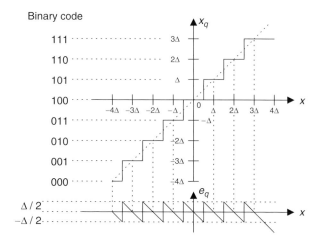

FIGURE 2.30 Characteristics of the bipolar quantizer.

c. quantization level when the analog voltage is 3.2 volts

d. binary code produced by the ADC

Solution:

Since the range is from 0 to 5 volts and the 3-bit ADC is used, we have

$$x_{\min} = 0 \text{ volt}, \ x_{\max} = 5 \text{ volts, and } m = 3 \text{ bits.}$$

a. Using Equation (2.20), we get the number of quantization levels as

TABLE 2.2 Quantization table for the 3-bit bipolar quantizer (step size $= \Delta = (x_{\max} - x_{\min})/2^3$, $x_{\max}$ = maximum voltage, and $x_{\min} = -x_{\max}$).

Binary Code	Quantization Level x_q (V)	Input Signal Subrange (V)
000	-4Δ	$-4\Delta \leq x < -3.5\Delta$
001	-3Δ	$-3.5\Delta \leq x < -2.5\Delta$
010	-2Δ	$-2.5\Delta \leq x < -1.5\Delta$
011	$-\Delta$	$-1.5\Delta \leq x < -0.5\Delta$
100	0	$-0.5\Delta \leq x < 0.5\Delta$
101	Δ	$0.5\Delta \leq x < 1.5\Delta$
110	2Δ	$1.5\Delta \leq x < 2.5\Delta$
111	3Δ	$2.5\Delta \leq x < 3.5\Delta$

$$L = 2^m = 2^3 = 8.$$

b. Applying Equation (2.19) yields

$$\Delta = \frac{5-0}{8} = 0.625 \text{ volt}.$$

c. When $x = 3.2 \frac{\Delta}{0.625} = 5.12\Delta$, from Equation (2.21) we get

$$i = round\left(\frac{x - x_{\min}}{\Delta}\right) = round(5.12) = 5.$$

From Equation (2.22), we determine the quantization level as

$$x_q = 0 + 5\Delta = 5 \times 0.625 = 3.125 \text{ volts}.$$

d. The binary code is determined as 101, from either Figure 2.29 or Table 2.1.

After quantizing the input signal x, the ADC produces binary codes, as illustrated in Figure 2.31.

The DAC process is shown in Figure 2.32. As shown in the figure, the DAC unit takes the binary codes from the DS processor. Then it converts the binary code using the zero-order hold circuit to reproduce the sample-and-hold signal. Assuming that the spectrum distortion due to sample-and-hold effect can be ignored for our illustration, the recovered sample-and-hold signal is further processed using the anti-image filter. Finally, the analog signal is yielded.

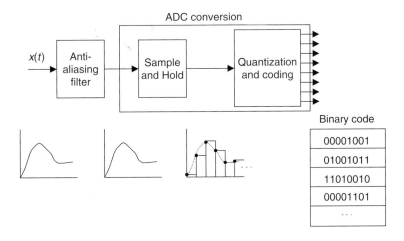

FIGURE 2.31   **Typical ADC process.**

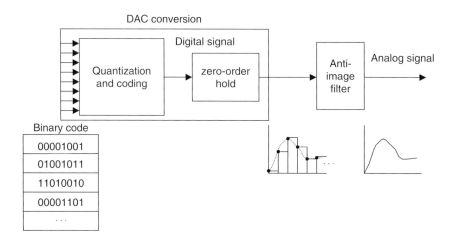

FIGURE 2.32 Typical DAC process.

When the DAC outputs the analog amplitude x_q with finite precision, it introduces the quantization error, defined as

$$e_q = x_q - x. \tag{2.23}$$

The quantization error as shown in Figure 2.29 is bounded by half of the step size, that is,

$$-\frac{\Delta}{2} \leq e_q \leq \frac{\Delta}{2}, \tag{2.24}$$

where Δ is the quantization step size, or the ADC resolution. We also refer to Δ as $V_{\min}$ (minimum detectable voltage) or the LSB value of the ADC.

Example 2.10.

a. Using Example 2.9, determine the quantization error when the analog input is 3.2 volts.

Solution:

a. Using Equation (2.23), we obtain

$$e_q = x_q - x = 3.125 - 3.2 = -0.075 \, \text{volt}.$$

Note that the quantization error is less than half of the step size, that is,

$$|e_q| = 0.075 < \Delta/2 = 0.3125 \, \text{volt}.$$

In practice, we can empirically confirm that the quantization error appears in uniform distribution when the step size is much smaller than the dynamic range of the signal samples and we have a sufficiently large number of samples. Based on theory of probability and random variables, the power of quantization noise is related to the quantization step and given by

$$E\left(e_q^2\right) = \frac{\Delta^2}{12}, \tag{2.25}$$

where $E()$ is the expectation operator, which actually averages the squared values of the quantization error (the reader can get more information from the texts by Roddy and Coolen (1997); Tomasi (2004); and Stearns and Hush (1990)). The ratio of signal power to quantization noise power (SNR) due to quantization can be expressed as

$$SNR = \frac{E(x^2)}{E\left(e_q^2\right)}. \tag{2.26}$$

If we express the SNR in terms of decibels (dB), we have

$$SNR_{dB} = 10 \cdot \log_{10}(SNR) \text{ dB}. \tag{2.27}$$

Substituting Equation (2.25) and $E(x^2) = x_{rms}$ into Equation (2.27), we achieve

$$SNR_{dB} = 10.79 + 20 \cdot \log_{10}\left(\frac{x_{rms}}{\Delta}\right), \tag{2.28}$$

where x_{rms} is the RMS (root mean squared) value of the signal to be quantized x.

Practically, the SNR can be calculated using the following formula:

$$SNR = \frac{\frac{1}{N}\sum_{n=0}^{N-1} x^2(n)}{\frac{1}{N}\sum_{n=0}^{N-1} e_q^2(n)} = \frac{\sum_{n=0}^{N-1} x^2(n)}{\sum_{n=0}^{N-1} e_q^2(n)}, \tag{2.29}$$

where $x(n)$ is the nth sample amplitude and $e_q(n)$ is the quantization error from quantizing $x(n)$.

Example 2.11.

a. If the analog signal to be quantized is a sinusoidal waveform, that is,

$$x(t) = A \sin(2\pi \times 1000t),$$

and if the bipolar quantizer uses m bits, determine the SNR in terms of m bits.

Solution:

a. Since $x_{rms} = 0.707A$ and $\Delta = 2A/2^m$, substituting x_{rms} and Δ into Equation (2.28) leads to

$$SNR_{dB} = 10.79 + 20 \cdot \log_{10}\left(\frac{0.707A}{2A/2^m}\right)$$

$$= 10.79 + 20 \cdot \log_{10}(0.707/2) + 20m \cdot \log_{10} 2.$$

After simplifying the numerical values, we get

$$SNR_{dB} = 1.76 + 6.02m \text{ dB}. \tag{2.30}$$

Example 2.12.

For a speech signal, if a ratio of the RMS value over the absolute maximum value of the analog signal (Roddy and Coolen, 1997) is given, that is, $\left(\frac{x_{rms}}{|x|_{max}}\right)$, and the ADC quantizer uses m bits, determine the SNR in terms of m bits.

Solution:

Since

$$\Delta = \frac{x_{max} - x_{min}}{L} = \frac{2|x|_{max}}{2^m},$$

substituting Δ in Equation (2.28) achieves

$$SNR_{dB} = 10.79 + 20 \cdot \log_{10}\left(\frac{x_{rms}}{2|x|_{max}/2^m}\right).$$

$$= 10.79 + 20 \cdot \log_{10}\left(\frac{x_{rms}}{|x|_{max}}\right) + 20m \log_{10} 2 - 20 \log_{10} 2.$$

Thus, after numerical simplification, we have

$$SNR_{dB} = 4.77 + 20 \cdot \log_{10}\left(\frac{x_{rms}}{|x|_{max}}\right) + 6.02m. \tag{2.31}$$

From Examples 2.11 and 2.12, we observed that increasing 1 bit of the ADC quantizer can improve SNR due to quantization by 6 dB.

Example 2.13.

Given a sinusoidal waveform with a frequency of 100 Hz,

$$x(t) = 4.5 \cdot \sin(2\pi \times 100t),$$

sampled at 8,000 Hz,

a. Write a MATLAB program to quantize the $x(t)$ using 4 bits to obtain and plot the quantized signal x_q, assuming that the signal range is between -5 and 5 volts.

b. Calculate the SNR due to quantization.

Solution:

a. Program 2.1. MATLAB program for Example 2.13.

```
%Example 2.13
clear all;close all
disp('Generate 0.02-second sine wave of 100 Hz and Vp=5');
fs = 8000;              % Sampling rate
T = 1/fs;               % Sampling interval
t = 0:T:0.02;           % Duration of 0.02 second
sig = 4.5*sin(2*pi*100*t);       % Generate the sinusoid
bits = input('input number of bits =>');
lg = length(sig);       % Length of the signal vector sig
for x = 1:lg
[Index(x) pq] = biquant(bits,-5,5,sig(x));    % Output the quantized index
end
% transmitted
% received
for x = 1:lg
qsig(x) = biqtdec(bits,-5,5,Index(x));    %Recover the quantized value
end
qerr = qsig-sig;        %Calculate the quantized errors
stairs(t,qsig);hold % Plot the signal in a staircase style
plot(t,sig); grid;      % Plot the signal
xlabel('Time (sec.)');ylabel('Quantized x(n)')
disp('Signal to noise power ratio due to quantization')
snr(sig,qsig);
```

b. Theoretically, applying Equation (2.30) leads to
$$SNR_{dB} = 1.76 + 6.02 \cdot 4 = 25.84 \, dB.$$

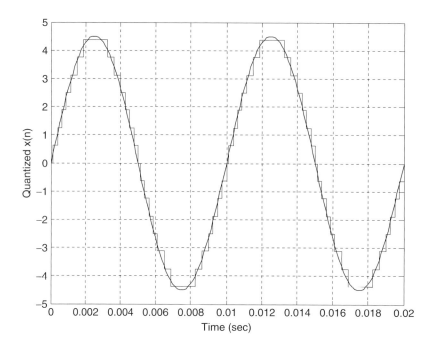

FIGURE 2.33 Comparison of the quantized signal and the original signal.

Practically, using Equation (2.29), the simulated result is obtained as

$$SNR_{dB} = 25.78 \, \text{dB}.$$

It is clear from this example that the ratios of signal power to noise power due to quantization achieved from theory and from simulation are very close. Next, we look at an example for quantizing a speech signal.

Example 2.14.

Given a speech signal sampled at 8,000 Hz in the file we.dat,

a. Write a MATLAB program to quantize the $x(t)$ using 4-bit quantizers to obtain the quantized signal x_q, assuming that the signal range is from -5 to 5 volts.

b. Plot the original speech, quantized speech, and quantization error, respectively.

c. Calculate the SNR due to quantization using the MATLAB program.

2.3 Analog-to-Digital Conversion, Digital-to-Analog Conversion, and Quantization

Solution:

a. Program 2.2. MATLAB program for Example 2.14.

```
%Example 2.14
clear all; close all
disp('load speech: We');
load we.dat% Load speech data at the current folder
sig = we;            % Provided by the instructor
fs=8000;             % Sampling rate
lg=length(sig);      % Length of the signal vector
T=1/fs;              % Sampling period
t = [0:1:lg − 1]*T;  % Time instants in second
sig= 4.5*sig/max(abs(sig)); %Normalizes speech in the range from −4.5 to 4.5
Xmax = max(abs(sig));        % Maximum amplitude
Xrms = sqrt( sum(sig .* sig) / length(sig)) % RMS value
disp('Xrms/Xmax')
k=Xrms/Xmax
disp('20*log10(k) =>');
k = 20*log10(k)
bits = input('input number of bits =>');
lg = length(sig);
for x = 1:lg
[Index(x) pq] =biquant(bits, −5,5, sig(x));    %Output the quantized index.
end
% Transmitted
% Received
for x = 1:lg
qsig(x) = biqtdec(bits, −5,5, Index(x));    %Recover the quantized value
end
qerr = sig-qsig;            %Calculate the quantized errors
subplot(3,1,1);plot(t,sig);
ylabel('Original speech');Title('we.dat: we');
subplot(3,1,2);stairs(t, qsig);grid
ylabel('Quantized speech')
subplot(3,1,3);stairs(t, qerr);grid
ylabel('Quantized error')
xlabel('Time (sec.)');axis([0 0.25 −1 1]);
disp('signal to noise ratio due to quantization noise')
snr(sig,qsig); % Signal to noise power ratio in dB: sig = signal vector,
% qsig =quantized signal vector
```

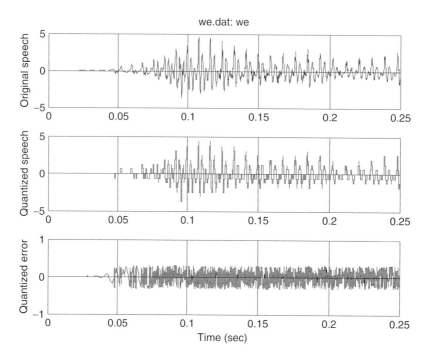

FIGURE 2.34 Original speech, quantized speech using the 4-bit bipolar quantizer, and quantization error.

b. In Figure 2.34, the top plot shows the speech wave to be quantized, while the middle plot displays the quantized speech signal using 4 bits. The bottom plot shows the quantization error. It also shows that the absolute value of the quantization error is uniformly distributed in a range between -0.3125 and 0.3125.

c. From the MATLAB program, we have $\frac{x_{rms}}{|x|_{max}} = 0.203$. Theoretically, from Equation (2.31), it follows that

$$SNR_{dB} = 4.77 + 20\log_{10}\left(\frac{x_{rms}}{|x|_{max}}\right) + 6.02 \cdot 4$$
$$= 4.77 + 20\log_{10}(0.203) + 6.02 \cdot 4 = 15\,dB.$$

On the other hand, the simulated result using Equation (2.29) gives

$$SNR_{dB} = 15.01\,\text{dB}.$$

Results for SNRs from Equations (2.31) and (2.29) are very close in this example.

2.4 Summary

1. Analog signal is sampled at a fixed time interval so the ADC will convert the sampled voltage level to a digital value; this is called the sampling process.

2. The fixed time interval between two samples is the sampling period, and the reciprocal of the sampling period is the sampling rate. Half of the sampling rate is the folding frequency (Nyquist limit).

3. The sampling theorem condition that the sampling rate be larger than twice the highest frequency of the analog signal to be sampled must be met in order to have the analog signal be recovered.

4. The sampled spectrum is explained using the well-known formula

$$X_s(f) = \cdots + \frac{1}{T}X(f+f_s) + \frac{1}{T}X(f) + \frac{1}{T}X(f-f_s) + \cdots,$$

that is, the sampled signal spectrum is a scaled and shifted version of its analog signal spectrum and its replicas centered at the frequencies that are multiples of the sampling rate.

5. The analog anti-aliasing lowpass filter is used before ADC to remove frequency components having high frequencies larger than the folding frequency to avoid aliasing.

6. The reconstruction (analog lowpass) filter is adopted after DAC to remove the spectral images that exist in the sample-and-hold signal and obtain the smoothed analog signal. The sample-and-hold DAC effect may distort the baseband spectrum, but it also reduces image spectrum.

7. Quantization means that the ADC unit converts the analog signal amplitude with infinite precision to digital data with finite precision (a finite number of codes).

8. When the DAC unit converts a digital code to a voltage level, quantization error occurs. The quantization error is bounded by half of the quantization step size (ADC resolution), which is a ratio of the full range of the signal over the number of the quantization levels (number of the codes).

9. The performance of the quantizer in terms of the signal to quantization noise ratio (SNR), in dB, is related to the number of bits in ADC. Increasing 1 bit used in each ADC code will improve 6 dB SNR due to quantization.

2.5 MATLAB Programs

Program 2.3. MATLAB function for uniform quantization encoding.

```
function [ I, pq] = biquant(NoBits, Xmin, Xmax, value)
% function pq = biquant(NoBits, Xmin, Xmax, value)
% This routine is created for simulation of the uniform quantizer.
%
% NoBits: number of bits used in quantization.
% Xmax: overload value.
% Xmin: minimum value
% value: input to be quantized.
% pq: output of the quantized value
% I: coded integer index
L = 2^NoBits;
delta=(Xmax-Xmin)/L;
I=round((value-Xmin)/delta);
if ( I==L)
I=I-1;
end
if I <0
I = 0;
end
pq=Xmin+I*delta;
```

Program 2.4. MATLAB function for uniform quantization decoding.

```
function pq = biqtdec(NoBits, Xmin, Xmax, I)
% function pq = biqtdec(NoBits, Xmin, Xmax, I)
% This routine recovers the quantized value.
%
% NoBits: number of bits used in quantization.
% Xmax: overload value
% Xmin: minimum value
% pq: output of the quantized value
% I: coded integer index
L= 2^NoBits;
delta=(Xmax-Xmin)/L;
pq=Xmin+I*delta;
```

Program 2.5. MATLAB function for calculation of signal to quantization ratio.

```
function snr = calcsnr(speech, qspeech)
% function snr = calcsnr(speech, qspeech)
% this routine is created for calculation of SNR
%
% speech: original speech waveform.
% qspeech: quantized speech.
% snr: output SNR in dB.
%
qerr = speech-qspeech;
snr= 10* log 10 (sum(speech.*speech)/sum(qerr.*qerr))
```

2.6 Problems

2.1. Given an analog signal

$$x(t) = 5\cos(2\pi \cdot 1500t), \text{ for } t \geq 0,$$

sampled at a rate of 8,000 Hz,

a. sketch the spectrum of the original signal;

b. sketch the spectrum of the sampled signal from 0 kHz to 20 kHz.

2.2. Given an analog signal

$$x(t) = 5\cos(2\pi \cdot 2500t) + 2\cos(2\pi \cdot 3200t), \text{ for } t \geq 0,$$

sampled at a rate of 8,000 Hz,

a. sketch the spectrum of the sampled signal up to 20 kHz;

b. sketch the recovered analog signal spectrum if an ideal lowpass filter with a cutoff frequency of 4 kHz is used to filter the sampled signal in order to recover the original signal.

2.3. Given an analog signal

$$x(t) = 5\cos(2\pi \cdot 2500t) + 2\cos(2\pi \cdot 4500t), \text{ for } t \geq 0,$$

sampled at a rate of 8,000 Hz,

a. sketch the spectrum of the sampled signal up to 20 kHz;

2 SIGNAL SAMPLING AND QUANTIZATION

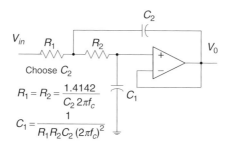

FIGURE 2.35 Filter circuit in Problem 2.5.

    b. sketch the recovered analog signal spectrum if an ideal lowpass filter with a cutoff frequency of 4 kHz is used to filter the sampled signal in order to recover the original signal;

    c. determine the frequency/frequencies of aliasing noise.

2.4. Assuming a continuous signal is given as

$$x(t) = 10\cos(2\pi \cdot 5500t) + 5\sin(2\pi \cdot 7500t), \text{ for } t \geq 0,$$

sampled at a sampling rate of 8,000 Hz,

    a. sketch the spectrum of the sampled signal up to 20 kHz;

    b. sketch the recovered analog signal spectrum if an ideal lowpass filter with a cutoff frequency of 4 kHz is used to filter the sampled signal in order to recover the original signal;

    c. determine the frequency/frequencies of aliasing noise.

2.5. Given the following second-order anti-aliasing lowpass filter, which is a Butterworth type, determine the values of circuit elements if we want the filter to have a cutoff frequency of 1,000 Hz.

2.6. From Problem 2.5, determine the percentage of aliasing level at the frequency of 500 Hz, assuming that the sampling rate is 4,000 Hz.

2.7. Given a DSP system in which a sampling rate of 8,000 Hz is used and the anti-aliasing filter is a second-order Butterworth lowpass filter with a cutoff frequency of 3.2 kHz, determine

    a. the percentage of aliasing level at the cutoff frequency;

    b. the percentage of aliasing level at the frequency of 1,000 Hz.

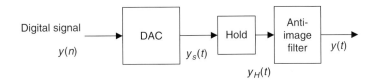

FIGURE 2.36 Analog signal reconstruction in Problem 2.10.

2.8. Given a DSP system in which a sampling rate of 8,000 Hz is used and the anti-aliasing filter is a Butterworth lowpass filter with a cutoff frequency of 3.2 kHz, determine the order of the Butterworth lowpass filter for the percentage of aliasing level at the cutoff frequency required to be less than 10%.

2.9. Given a DSP system with a sampling rate of 8,000 Hz and assuming that the hold circuit is used after DAC, determine

  a. the percentage of distortion at the frequency of 3,200 Hz;

  b. the percentage of distortion at the frequency of 1,500 Hz.

2.10. A DSP system is given with the following specifications:

Design requirements:

- Sampling rate 20,000 Hz
- Maximum allowable gain variation from 0 to 4,000 Hz = 2 dB

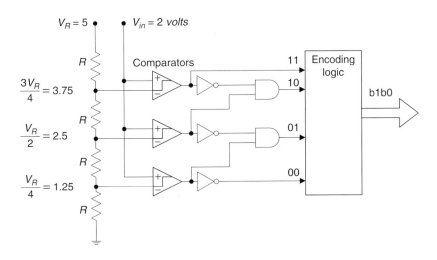

FIGURE 2.37 2-bit flash ADC in Problem 2.11.

SIGNAL SAMPLING AND QUANTIZATION

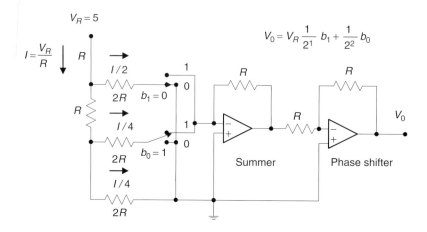

FIGURE 2.38 2-bit R-2R DAC in Problem 2.12.

- 40 dB rejection at the frequency of 16,000 Hz
- Butterworth filter assumed

Determine the cutoff frequency and order for the anti-image filter.

2.11. Given the 2-bit flash ADC unit with an analog sample-and-hold voltage of 2 volts shown in Figure 2.37, determine the output bits.

2.12. Given the R-2R DAC unit with a 2-bit value of $b_1 b_0 = 01$ shown in Figure 2.38, determine the converted voltage.

2.13. Assuming that a 4-bit ADC channel accepts analog input ranging from 0 to 5 volts, determine the following:

   a. number of quantization levels;

   b. step size of the quantizer or resolution;

   c. quantization level when the analog voltage is 3.2 volts;

   d. binary code produced by the ADC;

   e. quantization error.

2.14. Assuming that a 3-bit ADC channel accepts analog input ranging from −2.5 to 2.5 volts, determine the following:

   a. number of quantization levels;

   b. step size of the quantizer or resolution;

c. quantization level when the analog voltage is −1.

d. binary code produced by the ADC;

e. quantization error.

2.15. If the analog signal to be quantized is a sinusoidal

$$x(t) = 9.5\sin(2000 \times \pi t),$$

and if the bipolar quantizer uses 6 bits, determine

a. number of quantization levels;

b. quantization step size or resolution, Δ, assuming that the signal range is from −10 to 10 volts;

c. the signal power to quantization noise power ratio.

2.16. For a speech signal, if the ratio of the RMS value over the absolute maximum value of the signal is given, that is, $\left(\frac{x_{rms}}{|x|_{max}}\right) = 0.25$, and the ADC bipolar quantizer uses 6 bits, determine

a. number of quantization levels;

b. quantization step size or resolution, Δ, if the signal range is 5 volts;

c. the signal power to quantization noise power ratio.

Computer Problems with MATLAB: Use the MATLAB programs in Section 2.5 to solve the following problems.

2.17. Given a sinusoidal waveform of 100 Hz,

$$x(t) = 4.5\sin(2\pi \times 100t)$$

sample it at 8,000 samples per second and

a. write a MATLAB program to quantize $x(t)$ using a 6-bit bipolar quantizer to obtain the quantized signal x_q, assuming the signal range to be from −5 to 5 volts;

b. plot the original signal and the quantized signal;

c. calculate the SNR due to quantization using the MATLAB program.

2.18. Given a speech signal sampled at 8,000 Hz, as shown in Example 2.14,

a. write a MATLAB program to quantize $x(t)$ using a 6-bit bipolar quantizer to obtain the quantized signal x_q, assuming that the signal range is from -5 to 5 volts;

b. plot the original speech waveform, quantized speech, and quantization error;

c. calculate the SNR due to quantization using the MATLAB program.

References

Ahmed, N., and Natarajan, T. (1983). *Discrete-Time Signals and Systems.* Reston, VA: Reston Publishing Co.

Alkin, O. (1993). *Digital Signal Processing: A Laboratory Approach Using PC-DSP.* Englewood Cliffs, NJ: Prentice Hall.

Ambardar, A. (1999). *Analog and Digital Signal Processing,* 2nd ed. Pacific Grove, CA: Brooks/Cole Publishing Company.

Chen, W. (1986). *Passive and Active Filters: Theory and Implementations.* New York: John Wiley & Sons.

Oppenheim, A. V., and Schafer, R. W. (1975). *Discrete-Time Signal Processing.* Englewood Cliffs, NJ: Prentice Hall.

Proakis, J. G., and Manolakis, D. G. (1996). *Digital Signal Processing: Principles, Algorithms, and Applications,* 3rd ed. Upper Saddle River, NJ: Prentice Hall.

Roddy, D., and Coolen, J. (1997). *Electronic Communications,* 4th ed. Englewood Cliffs, NJ: Prentice Hall.

Stearns, S. D., and Hush, D. R. (1990). *Digital Signal Analysis,* 2nd ed. Englewood Cliffs, NJ: Prentice Hall.

Tomasi, W. (2004). *Electronic Communications Systems: Fundamentals Through Advanced,* 5th ed. Upper Saddle River, NJ: Pearson/Prentice Hall.

3

Digital Signals and Systems

Objectives:

This chapter introduces notations for digital signals and special digital sequences that are widely used in this book. The chapter continues to study some properties of linear systems such as time invariance, BIBO (bounded-in-and-bounded-out) stability, causality, impulse response, difference equation, and digital convolution.

3.1 Digital Signals

In our daily lives, analog signals appear as speech, audio, seismic, biomedical, and communications signals. To process an analog signal using a digital signal processor, the analog signal must be converted into a digital signal; that is, analog-to-digital conversion (ADC) must take place, as discussed in Chapter 2. Then the digital signal is processed via digital signal processing (DSP) algorithm(s).

A typical digital signal $x(n)$ is shown in Figure 3.1, where both the time and the amplitude of the digital signal are discrete. Notice that the amplitudes of digital signal samples are given and sketched only at their corresponding time indices, where $x(n)$ represents the amplitude of the nth sample and n is the time index or sample number. From Figure 3.1, we learn that

$x(0)$: zero-th sample amplitude at the sample number $n = 0$,
$x(1)$: first sample amplitude at the sample number $n = 1$,
$x(2)$: second sample amplitude at the sample number $n = 2$,
$x(3)$: third sample amplitude at the sample number $n = 3$, and so on.

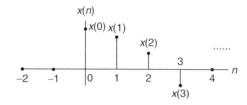

FIGURE 3.1   **Digital signal notation.**

Furthermore, Figure 3.2 illustrates the digital samples whose amplitudes are the discrete encoded values represented in the DS processor. Precision of the data is based on the number of bits used in the DSP system. The encoded data format can be either an integer if a fixed-point DS processor is used or a floating-point number if a floating-point DS processor is used. As shown in Figure 3.2 for the floating-point DS processor, we can identify the first five sample amplitudes at their time indices as follows:

$x(0) = 2.25$
$x(1) = 2.0$
$x(2) = 1.0$
$x(3) = -1.0$
$x(4) = 0.0$
......

Again, note that each sample amplitude is plotted using a vertical bar with a solid dot. This notation is well accepted in the DSP literature.

3.1.1 Common Digital Sequences

Let us study some special digital sequences that are widely used. We define and plot each of them as follows:

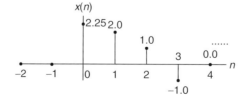

FIGURE 3.2   **Plot of the digital signal samples.**

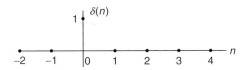

FIGURE 3.3   **Unit-impulse sequence.**

Unit-impulse sequence (digital unit-impulse function):

$$\delta(n) = \begin{cases} 1 & n = 0 \\ 0 & n \neq 0 \end{cases} \quad (3.1)$$

The plot of the unit-impulse function is given in Figure 3.3. The unit-impulse function has the unit amplitude at only $n = 0$ and zero amplitudes at other time indices.

Unit-step sequence (digital unit-step function):

$$u(n) = \begin{cases} 1 & n \geq 0 \\ 0 & n < 0 \end{cases} \quad (3.2)$$

The plot is given in Figure 3.4. The unit-step function has the unit amplitude at $n = 0$ and for all the positive time indices, and amplitudes of zero for all the negative time indices.

The shifted unit-impulse and unit-step sequences are displayed in Figure 3.5.

As shown in Figure 3.5, the shifted unit-impulse function $\delta(n - 2)$ is obtained by shifting the unit-impulse function $\delta(n)$ to the right by two samples, and the

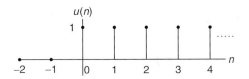

FIGURE 3.4   **Unit-step sequence.**

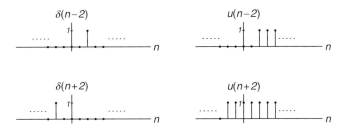

FIGURE 3.5   **Shifted unit-impulse and unit-step sequences.**

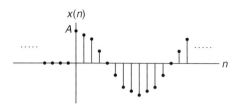

FIGURE 3.6 Plot of samples of the sinusoidal function.

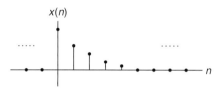

FIGURE 3.7 Plot of samples of the exponential function.

shifted unit-step function $u(n-2)$ is achieved by shifting the unit-step function $u(n)$ to the right by two samples; similarly, $\delta(n+2)$ and $u(n+2)$ are acquired by shifting $\delta(n)$ and $u(n)$ via two samples to the left, respectively.

Sinusoidal and exponential sequences are depicted in Figures 3.6 and 3.7, respectively.

For the sinusoidal sequence for the case of $x(n) = A\cos(0.125\pi n)u(n)$, and $A = 10$, we can calculate the digital values for the first eight samples and list their values in Table 3.1.

TABLE 3.1 Sample values calculated from the sinusoidal function.

n	$x(n) = 10\cos(0.125\pi n)u(n)$
0	10.0000
1	9.2388
2	7.0711
3	3.8628
4	0.0000
5	−3.8628
6	−7.0711
7	−9.2388

TABLE 3.2 Sample values calculated from the exponential function.

n	$10(0.75)^n u(n)$
0	10.0000
1	7.5000
2	5.6250
3	4.2188
4	3.1641
5	2.3730
6	1.7798
7	1.3348

For the exponential sequence for the case of $x(n) = A(0.75)^n u(n)$, the calculated digital values for the first eight samples with $A = 10$ are listed in Table 3.2.

Example 3.1.

Given the following,

$$x(n) = \delta(n+1) + 0.5\delta(n-1) + 2\delta(n-2),$$

a. Sketch this sequence.

Solution:

a. According to the shift operation, $\delta(n+1)$ is obtained by shifting $\delta(n)$ to the left by one sample, while $\delta(n-1)$ and $\delta(n-2)$ are yielded by shifting $\delta(n)$ to the right by one sample and two samples, respectively. Using the amplitude of each impulse function, we yield the following sketch.

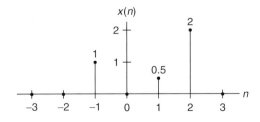

FIGURE 3.8 Plot of digital sequence in Example 3.1.

3.1.2 Generation of Digital Signals

Given the sampling rate of a DSP system to sample the analytical function of an analog signal, the corresponding digital function or digital sequence (assuming its sampled amplitudes are encoded to have finite precision) can be found. The digital sequence is often used to

1. calculate the encoded sample amplitude for a given sample number n;
2. generate the sampled sequence for simulation.

The procedure to develop the digital sequence from its analog signal function is as follows. Assuming that an analog signal $x(t)$ is uniformly sampled at the time interval of $\Delta t = T$, where T is the sampling period, the corresponding digital function (sequence) $x(n)$ gives the *instant encoded values* of the analog signal $x(t)$ at all the time instants $t = n\Delta t = nT$ and can be achieved by substituting time $t = nT$ into the analog signal $x(t)$, that is,

$$x(n) = x(t)|_{t=nT} = x(nT). \tag{3.3}$$

Also notice that for sampling the unit-step function $u(t)$, we have

$$u(t)|_{t=nT} = u(nT) = u(n). \tag{3.4}$$

The following example will demonstrate the use of Equations (3.3) and (3.4).

Example 3.2.

Assuming a DSP system with a sampling time interval of 125 microseconds,
  a. Convert each of the following analog signals $x(t)$ to the digital signal $x(n)$.
      1. $x(t) = 10e^{-5000t}u(t)$
      2. $x(t) = 10\sin(2000\pi t)u(t)$
  b. Determine and plot the sample values from each obtained digital function.

Solution:

  a. Since $T = 0.000125$ seconds in Equation (3.3), substituting $t = nT = n \times 0.000125 = 0.000125n$ into the analog signal $x(t)$ expressed in (a) leads to the digital sequence

   1. $x(n) = x(nT) = 10e^{-5000 \times 0.000125n}u(nT) = 10e^{-0.625n}u(n)$.

   Similarly, the digital sequence for (b) is achieved as follows:

   2. $x(n) = x(nT) = 10\sin(2000\pi \times 0.000125n)u(nT) = 10\sin(0.25\pi n)u(n)$.

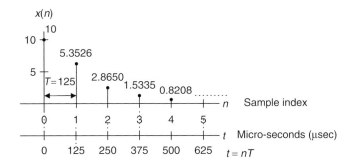

FIGURE 3.9 Plot of the digital sequence for (1) in Example 3.2.

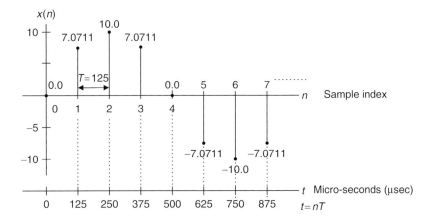

FIGURE 3.10 Plot of the digital sequence for (2) in Example 3.2.

b.1. The first five sample values are calculated below and plotted in Figure 3.9.

$$x(0) = 10e^{-0.625 \times 0}u(0) = 10.0$$
$$x(1) = 10e^{-0.625 \times 1}u(1) = 5.3526$$
$$x(2) = 10e^{-0.625 \times 2}u(2) = 2.8650$$
$$x(3) = 10e^{-0.625 \times 3}u(3) = 1.5335$$
$$x(4) = 10e^{-0.625 \times 4}u(4) = 0.8208$$

2. The first eight amplitudes are computed below and sketched in Figure 3.10.

$$x(0) = 10\sin(0.25\pi \times 0)u(0) = 0$$
$$x(1) = 10\sin(0.25\pi \times 1)u(1) = 7.0711$$
$$x(2) = 10\sin(0.25\pi \times 2)u(2) = 10.0$$

$$x(3) = 10\sin(0.25\pi \times 3)u(3) = 7.0711$$
$$x(4) = 10\sin(0.25\pi \times 4)u(4) = 0.0$$
$$x(5) = 10\sin(0.25\pi \times 5)u(5) = -7.0711$$
$$x(6) = 10\sin(0.25\pi \times 6)u(6) = -10.0$$
$$x(7) = 10\sin(0.25\pi \times 7)u(7) = -7.0711$$

3.2 Linear Time-Invariant, Causal Systems

In this section, we study linear time-invariant causal systems and focus on properties such as linearity, time invariance, and causality.

3.2.1 Linearity

A linear system is illustrated in Figure 3.11, where $y_1(n)$ is the system output using an input $x_1(n)$, and $y_2(n)$ is the system output using an input $x_2(n)$.

Figure 3.11 illustrates that the system output due to the weighted sum inputs $\alpha x_1(n) + \beta x_2(n)$ is equal to the same weighted sum of the individual outputs obtained from their corresponding inputs, that is,

$$y(n) = \alpha y_1(n) + \beta y_2(n), \tag{3.5}$$

where α and β are constants.

For example, assuming a digital amplifier as $y(n) = 10x(n)$, the input is multiplied by 10 to generate the output. The inputs $x_1(n) = u(n)$ and $x_2(n) = \delta(n)$ generate the outputs

$$y_1(n) = 10u(n) \text{ and } y_2(n) = 10\delta(n), \text{ respectively.}$$

If, as described in Figure 3.11, we apply to the system using the combined input $x(n)$, where the first input is multiplied by a constant 2 while the second input is multiplied by a constant 4,

$$x(n) = 2x_1(n) + 4x_2(n) = 2u(n) + 4\delta(n),$$

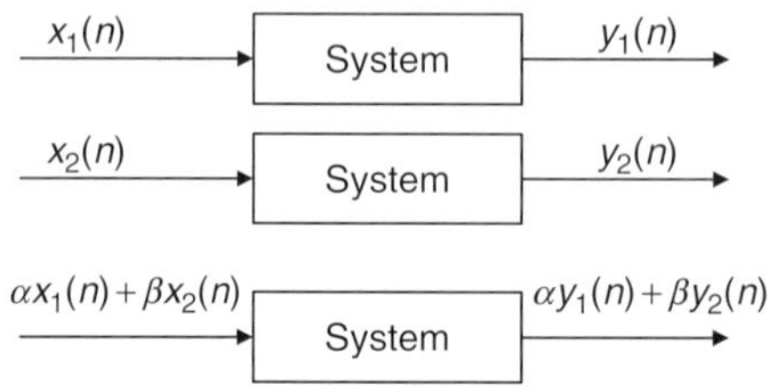

FIGURE 3.11 Digital linear system.

then the system output due to the combined input is obtained as

$$y(n) = 10x(n) = 10(2u(n) + 4\delta(n)) = 20u(n) + 40\delta(n). \qquad (3.6)$$

If we verify the weighted sum of the individual outputs, we see

$$2y_1(n) + 4y_2(n) = 20u(n) + 40\delta(n). \qquad (3.7)$$

Comparing Equations (3.6) and (3.7) verifies

$$y(n) = 2y_1(n) + 4y_2(n). \qquad (3.8)$$

Hence, the system $y(n) = 10x(n)$ is a linear system. Linearity means that the system obeys the superposition, as shown in Equation (3.8). Let us verify a system whose output is a square of its input:

$$y(n) = x^2(n).$$

Applying to the system with the inputs $x_1(n) = u(n)$ and $x_2(n) = \delta(n)$ leads to

$$y_1(n) = u^2(n) = u(n) \text{ and } y_2(n) = \delta^2(n) = \delta(n).$$

It is very easy to verify that $u^2(n) = u(n)$ and $\delta^2(n) = \delta(n)$.

We can determine the system output using a combined input, which is the weighed sum of the individual inputs with constants 2 and 4, respectively. Working on algebra, we see that

$$\begin{aligned} y(n) = x^2(n) &= (4x_1(n) + 2x_2(n))^2 \\ &= (4u(n) + 2\delta(n))^2 = 16u^2(n) + 16u(n)\delta(n) + 4\delta^2(n) \\ &= 16u(n) + 20\delta(n). \end{aligned} \qquad (3.9)$$

Note that we use a fact that $u(n)\delta(n) = \delta(n)$, which can be easily verified.

Again, we express the weighted sum of the two individual outputs with the same constants 2 and 4 as

$$4y_1(n) + 2y_2(n) = 4u(n) + 2\delta(n). \qquad (3.10)$$

It is obvious that

$$y(n) \neq 4y_1(n) + 2y_2(n). \qquad (3.11)$$

Hence, the system is a nonlinear system, since the linear property, superposition, does not hold, as shown in Equation (3.11).

3.2.2 Time Invariance

A time-invariant system is illustrated in Figure 3.12, where $y_1(n)$ is the system output for the input $x_1(n)$. Let $x_2(n) = x_1(n - n_0)$ be the shifted version of $x_1(n)$

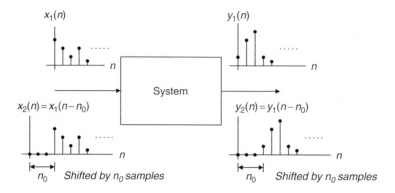

FIGURE 3.12 Illustration of the linear time-invariant digital system.

by n_0 samples. The output $y_2(n)$ obtained with the shifted input $x_2(n) = x_1(n - n_0)$ is equivalent to the output $y_2(n)$ acquired by shifting $y_1(n)$ by n_0 samples, $y_2(n) = y_1(n - n_0)$.

This can simply be viewed as the following:

> If the system is time invariant and $y_1(n)$ is the system output due to the input $x_1(n)$, then the shifted system input $x_1(n - n_0)$ will produce a shifted system output $y_1(n - n_0)$ by the same amount of time n_0.

Example 3.3.

Given the linear systems

a. $y(n) = 2x(n - 5)$

b. $y(n) = 2x(3n)$,

determine whether each of the following systems is time invariant.

Solution:

a. Let the input and output be $x_1(n)$ and $y_1(n)$, respectively; then the system output is $y_1(n) = 2x_1(n - 5)$. Again, let $x_2(n) = x_1(n - n_0)$ be the shifted input and $y_2(n)$ be the output due to the shifted input. We determine the system output using the shifted input as

$$y_2(n) = 2x_2(n - 5) = 2x_1(n - n_0 - 5).$$

Meanwhile, shifting $y_1(n) = 2x_1(n - 5)$ by n_0 samples leads to

$$y_1(n - n_0) = 2x_1(n - 5 - n_0).$$

We can verify that $y_2(n) = y_1(n - n_0)$. Thus the shifted input of n_0 samples causes the system output to be shifted by the same n_0 samples, thus the system is time invariant.

b. Let the input and output be $x_1(n)$ and $y_1(n)$, respectively; then the system output is $y_1(n) = 2x_1(3n)$. Again, let the input and output be $x_2(n)$ and $y_2(n)$, where $x_2(n) = x_1(n - n_0)$, a shifted version, and the corresponding output is $y_2(n)$. We get the output due to the shifted input $x_2(n) = x_1(n - n_0)$ and note that $x_2(3n) = x_1(3n - n_0)$:

$$y_2(n) = 2x_2(3n) = 2x_1(3n - n_0).$$

On the other hand, if we shift $y_1(n)$ by n_0 samples, which replaces n in $y_1(n) = 2x_1(3n)$ by $n - n_0$, we yield

$$y_1(n - n_0) = 2x_1(3(n - n_0)) = 2x_1(3n - 3n_0).$$

Clearly, we know that $y_2(n) \neq y_1(n - n_0)$. Since the system output $y_2(n)$ using the input shifted by n_0 samples is not equal to the system output $y_1(n)$ shifted by the same n_0 samples, the system is not time invariant.

3.2.3 Causality

A causal system is one in which the output $y(n)$ at time n depends only on the current input $x(n)$ at time n, its past input sample values such as $x(n - 1)$, $x(n - 2), \ldots$. Otherwise, if a system output depends on the future input values, such as $x(n + 1)$, $x(n + 2), \ldots$, the system is noncausal. The noncausal system cannot be realized in real time.

Example 3.4.

Given the following linear systems,

a. $y(n) = 0.5x(n) + 2.5x(n - 2)$, for $n \geq 0$

b. $y(n) = 0.25x(n - 1) + 0.5x(n + 1) - 0.4y(n - 1)$, for $n \geq 0$,

determine whether each is causal.

Solution:

a. Since for $n \geq 0$, the output $y(n)$ depends on the current input $x(n)$ and its past value $x(n - 2)$, the system is causal.

b. Since for $n \geq 0$, the output $y(n)$ depends on the current input $x(n)$ and its future value $x(n + 2)$, the system is noncausal.

3.3 Difference Equations and Impulse Responses

Now we study the difference equation and its impulse response.

3.3.1 Format of Difference Equation

A causal, linear, time-invariant system can be described by a difference equation having the following general form:

$$y(n) + a_1 y(n-1) + \ldots + a_N y(n-N)$$
$$= b_0 x(n) + b_1 x(n-1) + \ldots + b_M x(n-M), \quad (3.12)$$

where $a_1, \ldots, a_N$ and $b_0, b_1, \ldots, b_M$ are the coefficients of the difference equation. Equation (3.12) can further be written as

$$\begin{aligned} y(n) = & -a_1 y(n-1) - \ldots - a_N y(n-N) \\ & + b_0 x(n) + b_1 x(n-1) + \ldots + b_M x(n-M) \end{aligned} \quad (3.13)$$

or

$$y(n) = -\sum_{i=1}^{N} a_i y(n-i) + \sum_{j=0}^{M} b_j x(n-j). \quad (3.14)$$

Notice that $y(n)$ is the current output, which depends on the past output samples $y(n-1), \ldots, y(n-N)$, the current input sample $x(n)$, and the past input samples, $x(n-1), \ldots, x(n-N)$.

We will examine the specific difference equations in the following examples.

Example 3.5.

Given the following difference equation:

$$y(n) = 0.25 y(n-1) + x(n),$$

a. Identify the nonzero system coefficients.

Solution:

a. Comparison with Equation (3.13) leads to

$$b_0 = 1$$
$$-a_1 = 0.25,$$

that is, $a_1 = -0.25$.

FIGURE 3.13 Unit-impulse response of the linear time-invariant system.

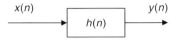

FIGURE 3.14 Representation of a linear time-invariant system using the impulse response.

Example 3.6.

Given a linear system described by the difference equation

$$y(n) = x(n) + 0.5x(n-1),$$

a. Determine the nonzero system coefficients.

Solution:

a. By comparing Equation (3.13), we have

$$b_0 = 1, \text{ and } b_1 = 0.5.$$

3.3.2 System Representation Using Its Impulse Response

A linear time-invariant system can be completely described by its unit-impulse response, which is defined as the system response due to the impulse input $\delta(n)$ with zero initial conditions, depicted in Figure 3.13.

With the obtained unit-impulse response $h(n)$, we can represent the linear time-invariant system in Figure 3.14.

Example 3.7.

Given the linear time-invariant system

$$y(n) = 0.5x(n) + 0.25x(n-1) \text{ with an initial condition } x(-1) = 0,$$

a. Determine the unit-impulse response h(n).

b. Draw the system block diagram.

c. Write the output using the obtained impulse response.

Solution:

a. According to Figure 3.13, let $x(n) = \delta(n)$, then

$$h(n) = y(n) = 0.5x(n) + 0.25x(n-1) = 0.5\delta(n) + 0.25\delta(n-1).$$

Thus, for this particular linear system, we have

$$h(n) = \begin{cases} 0.5 & n=0 \\ 0.25 & n=1 \\ 0 & \text{elsewhere} \end{cases}$$

b. The block diagram of the linear time-invariant system is shown as

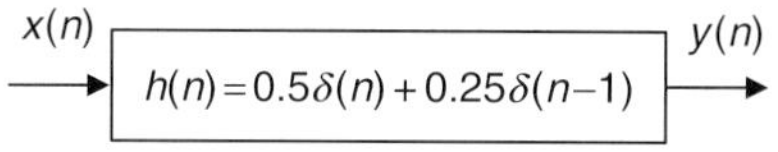

FIGURE 3.15 The system block diagram in Example 3.7.

c. The system output can be rewritten as

$$y(n) = h(0)x(n) + h(1)x(n-1).$$

From this result, it is noted that if the difference equation without the past output terms, $y(n-1), \ldots, y(n-N)$, that is, the corresponding coefficients $a_1, \ldots, a_N$, are zeros, the impulse response $h(n)$ has a finite number of terms. We call this a *finite impulse response* (FIR) system. In general, we can express the output sequence of a linear time-invariant system from its impulse response and inputs as

$$y(n) = \ldots + h(-1)x(n+1) + h(0)x(n) + h(1)x(n-1) + h(2)x(n-2) + \ldots. \quad (3.15)$$

Equation (3.15) is called the *digital convolution sum,* which will be explored in a later section. We can verify Equation (3.15) by substituting the impulse sequence $x(n) = \delta(n)$ to get the impulse response

$$h(n) = \ldots + h(-1)\delta(n+1) + h(0)\delta(n) + h(1)\delta(n-1) + h(2)\delta(n-2) + \ldots,$$

where ... $h(-1)$, $h(0)$, $h(1)$, $h(2)$... are the amplitudes of the impulse response at the corresponding time indices. Now let us look at another example.

Example 3.8.

Given the difference equation

$$y(n) = 0.25y(n-1) + x(n) \text{ for } n \geq 0 \text{ and } y(-1) = 0,$$

a. Determine the unit-impulse response $h(n)$.
b. Draw the system block diagram.
c. Write the output using the obtained impulse response.
d. For a step input $x(n) = u(n)$, verify and compare the output responses for the first three output samples using the difference equation and digitial convolution sum (Equation 3.15).

Solution:

a. Let $x(n) = \delta(n)$, then

$$h(n) = 0.25h(n-1) + \delta(n).$$

To solve for $h(n)$, we evaluate

$$h(0) = 0.25h(-1) + \delta(0) = 0.25 \times 0 + 1 = 1$$
$$h(1) = 0.25h(0) + \delta(1) = 0.25 \times 1 + 0 = 0.25$$
$$h(2) = 0.25h(1) + \delta(2) = 0.25 \times 0.5 + 0 = 0.0625$$
...

With the calculated results, we can predict the impulse response as

$$h(n) = (0.25)^n u(n) = \delta(n) + 0.25\delta(n-1) + 0.0625\delta(n-2) + \ldots.$$

b. The system block diagram is given in Figure 3.16.

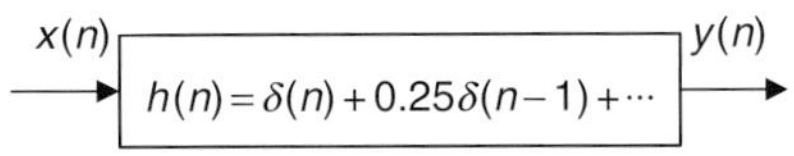

FIGURE 3.16   The system block diagram in Example 3.8.

c. The output sequence is a sum of infinite terms expressed as

$$y(n) = h(0)x(n) + h(1)x(n-1) + h(2)x(n-2) + \ldots$$
$$= x(n) + 0.25x(n-1) + 0.0625x(n-2) + \ldots$$

d. From the difference equation and using the zero-initial condition, we have

$$y(n) = 0.25y(n-1) + x(n) \text{ for } n \geq 0 \text{ and } y(-1) = 0$$
$$n = 0, y(0) = 0.25y(-1) + x(0) = u(0) = 1$$
$$n = 1, y(1) = 0.25y(0) + x(1) = 0.25 \times u(0) + u(1) = 1.25$$
$$n = 2, y(2) = 0.25y(1) + x(2) = 0.25 \times 1.25 + u(2) = 1.3125$$
$$\ldots$$

Applying the convolution sum in Equation (3.15) yields

$$y(n) = x(n) + 0.25x(n-1) + 0.0625x(n-2) + \ldots$$

$$n = 0, \ y(0) = x(0) + 0.25x(-1) + 0.0625x(-2) + \ldots$$
$$= u(0) + 0.25 \times u(-1) + 0.125 \times u(-2) + \ldots = 1$$

$$n = 1, \ y(1) = x(1) + 0.25x(0) + 0.0625x(-1) + \ldots$$
$$= u(1) + 0.25 \times u(0) + 0.125 \times u(-1) + \ldots = 1.25$$

$$n = 2, \ y(2) = x(2) + 0.25x(1) + 0.0625x(0) + \ldots$$
$$= u(2) + 0.25 \times u(1) + 0.0625 \times u(0) + \ldots = 1.3125$$
$$\ldots$$

Comparing the results, we verify that a linear time-invariant system can be represented by the convolution sum using its impulse response and input sequence. Note that we verify only the causal system for simplicity, and the principle works for both causal and noncausal systems.

Notice that this impulse response $h(n)$ contains an infinite number of terms in its duration due to the past output term $y(n-1)$. Such a system as described in the preceding example is called an *infinite impulse response* (IIR) system, which will be studied in later chapters.

3.4 Bounded-in-and-Bounded-out Stability

We are interested in designing and implementing stable linear systems. A stable system is one for which every bounded input produces a bounded output

(BIBO). There are many other stability definitions. To find the stability criterion, consider a linear time-invariant representation with all the inputs reaching the maximum value M for the worst case. Equation (3.15) becomes

$$y(n) = M(\ldots + h(-1) + h(0) + h(1) + h(2) + \ldots). \tag{3.16}$$

Using the absolute values of the impulse response leads to

$$y(n) < M(\ldots + |h(-1)| + |h(0)| + |h(1)| + |h(2)| + \ldots). \tag{3.17}$$

If the absolute sum in Equation (3.17) is a finite number, the product of the absolute sum and the maximum input value is therefore a finite number. Hence, we have a bounded input and a bounded output. In terms of the impulse response, a linear system is stable if the sum of its absolute impulse response coefficients is a finite number. We can apply Equation (3.18) to determine whether a linear time-invariant system is stable or not stable, that is,

$$S = \sum_{k=-\infty}^{\infty} |h(k)| = \ldots + |h(-1)| + |h(0)| + |h(1)| + \ldots < \infty. \tag{3.18}$$

Figure 3.17 describes a linear stable system, where the impulse response decreases to zero in finite amount of time so that the summation of its absolute impulse response coefficients is guaranteed to be finite.

Example 3.9.

Given the linear system in Example 3.8,

$$y(n) = 0.25y(n-1) + x(n) \text{ for } n \geq 0 \text{ and } y(-1) = 0,$$

which is described by the unit-impulse response

$$h(n) = (0.25)^n u(n),$$

a. Determine whether this system is stable or not.

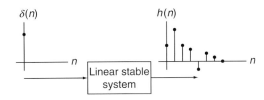

FIGURE 3.17 Illustration of stability of the digital linear system.

Solution:

a. Using Equation (3.18), we have

$$S = \sum_{k=-\infty}^{\infty} |h(k)| = \sum_{k=-\infty}^{\infty} |(0.25)^k u(k)|.$$

Applying the definition of the unit-step function $u(k) = 1$ for $k \geq 0$, we have

$$S = \sum_{k=0}^{\infty} (0.25)^k = 1 + 0.25 + 0.25^2 + \ldots.$$

Using the formula for a sum of the geometric series (see Appendix F),

$$\sum_{k=0}^{\infty} a^k = \frac{1}{1-a},$$

where $a = 0.25 < 1$, we conclude

$$S = 1 + 0.25 + 0.25^2 + \ldots = \frac{1}{1 - 0.25} = \frac{4}{3} < \infty.$$

Since the summation is a finite number, the linear system is stable.

3.5 Digital Convolution

Digital convolution plays an important role in digital filtering. As we verify in the last section, a linear time-invariant system can be represented by using a digital convolution sum. Given a linear time-invariant system, we can determine its unit-impulse response $h(n)$, which relates the system input and output. To find the output sequence $y(n)$ for any input sequence $x(n)$, we write the digital convolution as shown in Equation (3.15) as:

$$y(n) = \sum_{k=-\infty}^{\infty} h(k)x(n-k)$$
$$= \ldots + h(-1)x(n+1) + h(0)x(n) + h(1)x(n-1) + h(2)x(n-2) + \ldots$$

(3.19)

The sequences $h(k)$ and $x(k)$ in Equation (3.19) are interchangeable. Hence, we have an alternative form as

$$y(n) = \sum_{k=-\infty}^{\infty} x(k)h(n-k)$$
$$= \ldots + x(-1)h(n+1) + x(0)h(n) + x(1)h(n-1) + x(2)h(n-2) + \ldots$$

(3.20)

Using a conventional notation, we express the digital convolution as

$$y(n) = h(n)*x(n). \tag{3.21}$$

Note that for a causal system, which implies its impulse response

$$h(n) = 0 \text{ for } n < 0,$$

the lower limit of the convolution sum begins at 0 instead of ∞, that is

$$y(n) = \sum_{k=0}^{\infty} h(k)x(n-k) = \sum_{k=0}^{\infty} x(k)h(n-k). \tag{3.22}$$

We will focus on evaluating the convolution sum based on Equation (3.20). Let us examine first a few outputs from Equation (3.20):

$$y(0) = \sum_{k=-\infty}^{\infty} x(k)h(-k) = \ldots + x(-1)h(1) + x(0)h(0) + x(1)h(-1) + x(2)h(-2) + \ldots$$

$$y(1) = \sum_{k=-\infty}^{\infty} x(k)h(1-k) = \ldots + x(-1)h(2) + x(0)h(1) + x(1)h(0) + x(2)h(-1) + \ldots$$

$$y(2) = \sum_{k=-\infty}^{\infty} x(k)h(2-k) = \ldots + x(-1)h(3) + x(0)h(2) + x(1)h(1) + x(2)h(0) + \ldots$$

...

We see that the convolution sum requires the sequence $h(n)$ to be reversed and shifted. The graphical, formula, and table methods will be discussed for evaluating the digital convolution via the several examples. To begin with evaluating the convolution sum graphically, we need to apply the reversed sequence and shifted sequence. The reversed sequence is defined as follows: If $h(n)$ is the given sequence, $h(-n)$ is the reversed sequence. The reversed sequence is a mirror image of the original sequence, assuming the vertical axis as the mirror. Let us study the reversed sequence and shifted sequence via the following example.

Example 3.10.

Given a sequence,

$$h(k) = \begin{cases} 3, & k = 0,1 \\ 1, & k = 2,3 \\ 0 & elsewhere \end{cases}$$

where k is the time index or sample number,

a. Sketch the sequence $h(k)$ and reversed sequence $h(-k)$.

b. Sketch the shifted sequences $h(k+3)$ and $h(-k-2)$.

Solution:

a. Since $h(k)$ is defined, we plot it in Figure 3.18. Next, we need to find the reversed sequence $h(-k)$. We examine the following for

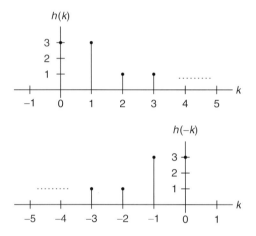

FIGURE 3.18 Plots of the digital sequence and its reversed sequence in Example 3.10.

$$k > 0, \; h(-k) = 0$$
$$k = 0, \; h(-0) = h(0) = 3$$
$$k = -1, \; h(-k) = h(-(-1)) = h(1) = 3$$
$$k = -2, \; h(-k) = h(-(-2)) = h(2) = 1$$
$$k = -3, \; h(-k) = h(-(-3)) = h(3) = 1$$

One can verify that $k \leq -4, h(-k) = 0$. Then the reversed sequence $h(-k)$ is shown as the second plot in Figure 3.18.

As shown in the sketches, $h(-k)$ is just a mirror image of the original sequence $h(k)$.

b. Based on the definition of the original sequence, we know that $h(0) = h(1) = 3, h(2) = h(3) = 1$, and the others are zeros. The time indices correspond to the following:

$$-k + 3 = 0, \; k = 3$$
$$-k + 3 = 1, \; k = 2$$
$$-k + 3 = 2, \; k = 1$$
$$-k + 3 = 3, \; k = 0.$$

Thus we can sketch $h(-k+3)$, as shown in Figure 3.19.

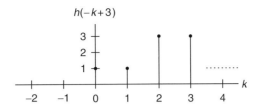

FIGURE 3.19 Plot of the sequence h(−k + 3) in Example 3.10.

Similarly, $h(-k-2)$ is yielded in Figure 3.20.

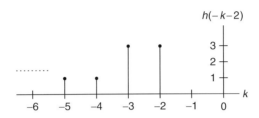

FIGURE 3.20 Plot of the sequence h(−k − 2) in Example 3.10.

We can get $h(-k+3)$ by shifting $h(-k)$ to the right by three samples, and we can obtain $h(-k-2)$ by shifting $h(-k)$ to the left by two samples.

In summary, given $h(-k)$, we can obtain $h(n-k)$ by shifting $h(-k)$ n samples to the right or the left, depending on whether n is positive or negative.

Once we understand the shifted sequence and reversed sequence, we can perform digital convolution of two sequences $h(k)$ and $x(k)$, defined in Equation (3.20) graphically. From that equation, we see that each convolution value $y(n)$ is the sum of the products of two sequences $x(k)$ and $h(n-k)$, the latter of which is the shifted version of the reversed sequence $h(-k)$ by $|n|$ samples. Hence, we can summarize the graphical convolution procedure in Table 3.3.

We illustrate digital convolution sum via the following example.

Example 3.11.

Using the following sequences defined in Figure 3.21, evaluate the digital convolution

$$y(n) = \sum_{k=-\infty}^{\infty} x(k)h(n-k)$$

TABLE 3.3 Digital convolution using the graphical method.

Step 1. Obtain the reversed sequence $h(-k)$.

Step 2. Shift $h(-k)$ by $|n|$ samples to get $h(n-k)$. If $n \geq 0$, $h(-k)$ will be shifted to the right by n samples; but if $n < 0$, $h(-k)$ will be shifted to the left by $|n|$ samples.

Step 3. Perform the convolution sum that is the sum of the products of two sequences $x(k)$ and $h(n-k)$ to get $y(n)$.

Step 4. Repeat steps 1 to 3 for the next convolution value $y(n)$.

a. By the graphical method.

b. By applying the formula directly.

Solution:

a. To obtain $y(0)$, we need the reversed sequence $h(-k)$; and to obtain $y(1)$, we need the reversed sequence $h(1-k)$, and so on. Using the technique we have discussed, sequences $h(-k)$, $h(-k+1)$, $h(-k+2)$, $h(-k+3)$, and $h(-k+4)$ are achieved and plotted in Figure 3.22, respectively. Again, using the information in Figures 3.21 and 3.22, we can compute the convolution sum as:

sum of product of $x(k)$ and $h(-k)$: $y(0) = 3 \times 3 = 9$
sum of product of $x(k)$ and $h(1-k)$: $y(1) = 1 \times 3 + 3 \times 2 = 9$
sum of product of $x(k)$ and $h(2-k)$: $y(2) = 2 \times 3 + 1 \times 2 + 3 \times 1 = 11$
sum of product of $x(k)$ and $h(3-k)$: $y(3) = 2 \times 2 + 1 \times 1 = 5$
sum of product of $x(k)$ and $h(4-k)$: $y(4) = 2 \times 1 = 2$
sum of product of $x(k)$ and $h(5-k)$: $y(n) = 0$ for $n > 4$, since sequences $x(k)$ and $h(n-k)$ do not overlap.

Finally, we sketch the output sequence $y(n)$ in Figure 3.23.

b. Applying Equation (3.20) with zero initial conditions leads to

$$y(n) = x(0)h(n) + x(1)h(n-1) + x(2)h(n-2)$$

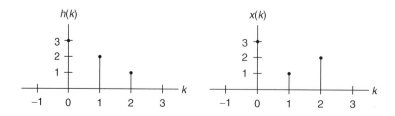

FIGURE 3.21 Plots of digital input sequence and impulse sequence in Example 3.11.

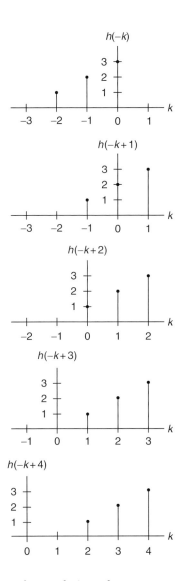

FIGURE 3.22 Illustration of convolution of two sequences x(k) and h(k) in Example 3.11.

$n = 0$, $y(0) = x(0)h(0) + x(1)h(-1) + x(2)h(-2) = 3 \times 3 + 1 \times 0 + 2 \times 0 = 9$,

$n = 1$, $y(1) = x(0)h(1) + x(1)h(0) + x(2)h(-1) = 3 \times 2 + 1 \times 3 + 2 \times 0 = 9$,

$n = 2$, $y(2) = x(0)h(2) + x(1)h(1) + x(2)h(0) = 3 \times 1 + 1 \times 2 + 2 \times 3 = 11$,

$n = 3$, $y(3) = x(0)h(3) + x(1)h(2) + x(2)h(1) = 3 \times 0 + 1 \times 1 + 2 \times 2 = 5$.

$n = 4$, $y(4) = x(0)h(4) + x(1)h(3) + x(2)h(2) = 3 \times 0 + 1 \times 0 + 2 \times 1 = 2$,

$n \geq 5$, $y(n) = x(0)h(n) + x(1)h(n-1) + x(2)h(n-2) = 3 \times 0 + 1 \times 0 + 2 \times 0 = 0$.

80 3 DIGITAL SIGNALS AND SYSTEMS

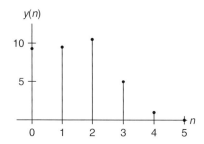

FIGURE 3.23 Plot of the convolution sum in Example 3.11.

In simple cases such as Example 3.11, it is not necessary to use the graphical or formula methods. We can compute the convolution by treating the input sequence and impulse response as number sequences and sliding the reversed impulse response past the input sequence, cross-multiplying, and summing the nonzero overlap terms at each step. The procedure and calculated results are listed in Table 3.4.

We can see that the calculated results using all the methods are consistent. The steps using the table method are concluded in Table 3.5.

Example 3.12.

Given the following two rectangular sequences,

$$x(n) = \begin{cases} 1 & n = 0,1,2 \\ 0 & otherwise \end{cases} \text{ and } h(n) = \begin{cases} 0 & n = 0 \\ 1 & n = 1,2 \\ 0 & otherwise \end{cases}$$

a. Convolve them using the table method.

Solution:

a. Using Table 3.5 as a guide, we list the operations and calculations in Table 3.6.

TABLE 3.4 Convolution sum using the table method.

k:	-2	-1	0	1	2	3	4	5	
$x(k)$:			3	1	2				
$h(-k)$:	1	2	3						$y(0) = 3 \times 3 = 9$
$h(1-k)$		1	2	3					$y(1) = 3 \times 2 + 1 \times 3 = 9$
$h(2-k)$			1	2	3				$y(2) = 3 \times 1 + 1 \times 2 + 2 \times 3 = 11$
$h(3-k)$				1	2	3			$y(3) = 1 \times 1 + 2 \times 2 = 5$
$h(4-k)$					1	2	3		$y(4) = 2 \times 1 = 2$
$h(5-k)$						1	2	3	$y(5) = 0$ (no overlap)

TABLE 3.5 Digital convolution steps via the table.

Step 1. List the index k covering a sufficient range.
Step 2. List the input $x(k)$.
Step 3. Obtain the reversed sequence $h(-k)$, and align the rightmost element of $h(n-k)$ to the leftmost element of $x(k)$.
Step 4. Cross-multiply and sum the nonzero overlap terms to produce $y(n)$.
Step 5. Slide $h(n-k)$ to the right by one position.
Step 6. Repeat step 4; stop if all the output values are zero or if required.

Note that the output should show the trapezoidal shape.
Let us examine convolving a finite long sequence with an infinite long sequence.

Example 3.13.

A system representation using the unit-impulse response for the linear system

$$y(n) = 0.25y(n-1) + x(n) \text{ for } n \geq 0 \text{ and } y(-1) = 0$$

is determined in Example 3.8 as

$$y(n) = \sum_{k=-\infty}^{\infty} x(k)h(n-k),$$

where $h(n) = (0.25)^n u(n)$. For a step input $x(n) = u(n)$,

a. Determine the output response for the first three output samples using the table method.

Solution:

a. Using Table 3.5 as a guide, we list the operations and calculations in Table 3.7.
   As expected, the output values are the same as those obtained in Example 3.8.

TABLE 3.6 Convolution sum in Example 3.12.

k:	-2	-1	0	1	2	3	4	5	...	
$x(k)$:			1	1	1				...	
$h(-k)$:	1	1	0							$y(0) = 0$ (no overlap)
$h(1-k)$		1	1	0						$y(1) = 1 \times 1 = 1$
$h(2-k)$			1	1	0					$y(2) = 1 \times 1 + 1 \times 1 = 2$
$h(3-k)$				1	1	0				$y(3) = 1 \times 1 + 1 \times 1 = 2$
$h(4-k)$					1	1	0			$y(4) = 1 \times 1 = 1$
$h(n-k)$						1	1	0		$y(n) = 0, n \geq 5$ (no overlap)
										Stop

TABLE 3.7   **Convolution sum in Example 3.13.**

k:	−2	−1	0	1	2	3 ...	
x(k):			1	1	1	1 ...	
h(−k):	0.0625	0.25	1				$y(0) = 1 \times 1 = 1$
h(1−k):		0.0625	0.25	1			$y(1) = 1 \times 0.25 + 1 \times 1 = 1.25$
h(2−k):			0.0625	0.25	1		$y(2) = 1 \times 0.0625 + 1 \times 0.25 + 1 \times 1$
							$= 1.3125$
							Stop as required

3.6 Summary

1. Digital signal samples are sketched using their encoded amplitudes versus sample numbers with vertical bars topped by solid circles located at their sampling instants, respectively. Impulse sequence, unit-step sequence, and their shifted versions are sketched in this notation.

2. The analog signal function can be sampled to its digital (discrete-time) version by substituting time $t = nT$ into the analog function, that is,

$$x(n) = x(t)|_{t=nT} = x(nT).$$

   The digital function values can be calculated for the given time index (sample number).

3. The DSP system we wish to design must be a linear, time-invariant, causal system. Linearity means that the superposition principle exists. Time invariance requires that the shifted input generates the corresponding shifted output with the same amount of time. Causality indicates that the system output depends on only its current input sample and past input sample(s).

4. The difference equation describing a linear, time-invariant system has a format such that the current output depends on the current input, past input(s), and past output(s) in general.

5. The unit-impulse response can be used to fully describe a linear, time-invariant system. Given the impulse response, the system output is the sum of the products of the impulse response coefficients and corresponding input samples, called the digital convolution sum.

6. BIBO is a type of stability in which a bounded input will produce a bounded output. The condition for a BIBO system requires that the sum of the absolute impulse response coefficients be a finite number.

7. Digital convolution sum, which represents a DSP system, is evaluated in three ways: the graphical method, evaluation of the formula, and the table method. The table method is found to be most effective.

3.7 Problems

3.1. Sketch each of the following special digital sequences:

   a. $5\delta(n)$

   b. $-2\delta(n-5)$

   c. $-5u(n)$

   d. $5u(n-2)$

3.2. Calculate the first eight sample values and sketch each of the following sequences:

   a. $x(n) = 0.5^n u(n)$

   b. $x(n) = 5\sin(0.2\pi n)u(n)$

   c. $x(n) = 5\cos(0.1\pi n + 30^0)u(n)$

   d. $x(n) = 5(0.75)^n \sin(0.1\pi n)u(n)$

3.3. Sketch the following sequences:

   a. $x(n) = 3\delta(n+2) - 0.5\delta(n) + 5\delta(n-1) - 4\delta(n-5)$

   b. $x(n) = \delta(n+1) - 2\delta(n-1) + 5u(n-4)$

3.4. Given the digital signals $x(n)$ in Figures 3.24 and 3.25, write an expression for each digital signal using the unit-impulse sequence and its shifted sequences.

3.5. Assuming that a DS processor with a sampling time interval of 0.01 second converts each of the following analog signals $x(t)$ to the digital signal $x(n)$, determine the digital sequences for each of the following analog signals.

   a. $x(t) = e^{-50t}u(t)$

   b. $x(t) = 5\sin(20\pi t)u(t)$

c. $x(t) = 10\cos(40\pi t + 30^0)u(t)$

d. $x(t) = 10e^{-100t}\sin(15\pi t)u(t)$

3.6. Determine which of the following is a linear system.

  a. $y(n) = 5x(n) + 2x^2(n)$

  b. $y(n) = x(n-1) + 4x(n)$

  c. $y(n) = 4x^3(n-1) - 2x(n)$

3.7. Given the following linear systems, find which one is time invariant.

  a. $y(n) = -5x(n-10)$

  b. $y(n) = 4x(n^2)$

3.8. Determine which of the following linear systems is causal.

  a. $y(n) = 0.5x(n) + 100x(n-2) - 20x(n-10)$

  b. $y(n) = x(n+4) + 0.5x(n) - 2x(n-2)$

3.9. Determine the causality for each of the following linear systems.

  a. $y(n) = 0.5x(n) + 20x(n-2) - 0.1y(n-1)$

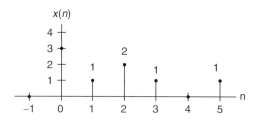

FIGURE 3.24   The first digitial signal in Problem 3.4.

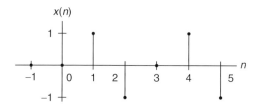

FIGURE 3.25   The second digitial signal in Problem 3.4.

b. $y(n) = x(n+2) - 0.4y(n-1)$

c. $y(n) = x(n-1) + 0.5y(n+2)$

3.10. Find the unit-impulse response for each of the following linear systems.

a. $y(n) = 0.5x(n) - 0.5x(n-2)$; for $n \geq 0$, $x(-2) = 0$, $x(-1) = 0$

b. $y(n) = 0.75y(n-1) + x(n)$; for $n \geq 0$, $y(-1) = 0$

c. $y(n) = -0.8y(n-1) + x(n-1)$; for $n \geq 0$, $x(-1) = 0$, $y(-1) = 0$

3.11. For each of the following linear systems, find the unit-impulse response, and draw the block diagram.

a. $y(n) = 5x(n-10)$

b. $y(n) = x(n) + 0.5x(n-1)$

3.12. Determine the stability for the following linear system.

$$y(n) = 0.5x(n) + 100x(n-2) - 20x(n-10)$$

3.13. Determine the stability for each of the following linear systems.

a. $y(n) = \sum_{k=0}^{\infty} 0.75^k x(n-k)$

b. $y(n) = \sum_{k=0}^{\infty} 2^k x(n-k)$

3.14. Given the sequence

$$h(k) = \begin{cases} 2, & k = 0,1,2 \\ 1, & k = 3,4 \\ 0 & \text{elsewhere,} \end{cases}$$

where k is the time index or sample number,

a. sketch the sequence $h(k)$ and the reverse sequence $h(-k)$;

b. sketch the shifted sequences $h(-k+2)$ and $h(-k-3)$.

3.15. Using the following sequence definitions,

$$h(k) = \begin{cases} 2, & k = 0,1,2 \\ 1, & k = 3,4 \\ 0 & \text{elsewhere} \end{cases} \quad \text{and} \quad x(k) = \begin{cases} 2, & k = 0 \\ 1, & k = 1,2 \\ 0 & \text{elsewhere,} \end{cases}$$

evaluate the digital convolution

$$y(n) = \sum_{k=-\infty}^{\infty} x(k)h(n-k)$$

a. using the graphical method;

b. using the table method;

c. applying the convolution formula directly.

3.16. Using the sequence definitions

$$x(k) = \begin{cases} -2, & k = 0,1,2 \\ 1, & k = 3,4 \\ 0 & elsewhere \end{cases} \quad \text{and} \quad h(k) = \begin{cases} 2, & k = 0 \\ -1, & k = 1,2 \\ 0 & elsewhere, \end{cases}$$

evaluate the digital convolution

$$y(n) = \sum_{k=-\infty}^{\infty} h(k)x(n-k)$$

a. using the graphical method;

b. using the table method;

c. applying the convolution formula directly.

3.17. Convolve the following two rectangular sequences:

$$x(n) = \begin{cases} 1 & n = 0,1 \\ 0 & otherwise \end{cases} \quad \text{and} \quad h(n) = \begin{cases} 0 & n = 0 \\ 1 & n = 1,2 \\ 0 & otherwise \end{cases}$$

using the table method.

4

Discrete Fourier Transform and Signal Spectrum

Objectives:

This chapter investigates discrete Fourier transform (DFT) and fast Fourier transform (FFT) and their properties; introduces the DFT/FFT algorithms to compute signal amplitude spectrum and power spectrum; and uses the window function to reduce spectral leakage. Finally, the chapter describes the FFT algorithm and shows how to apply it to estimate a speech spectrum.

4.1 Discrete Fourier Transform

In time domain, representation of digital signals describes the signal amplitude versus the sampling time instant or the sample number. However, in some applications, signal frequency content is very useful otherwise than as digital signal samples. The representation of the digital signal in terms of its frequency component in a frequency domain, that is, the signal spectrum, needs to be developed. As an example, Figure 4.1 illustrates the time domain representation of a 1,000-Hz sinusoid with 32 samples at a sampling rate of 8,000 Hz; the bottom plot shows the signal spectrum (frequency domain representation), where we can clearly observe that the amplitude peak is located at the frequency of 1,000 Hz in the calculated spectrum. Hence, the spectral plot better displays frequency information of a digital signal.

The algorithm transforming the time domain signal samples to the frequency domain components is known as the *discrete Fourier transform*, or DFT. The DFT also establishes a relationship between the time domain representation and

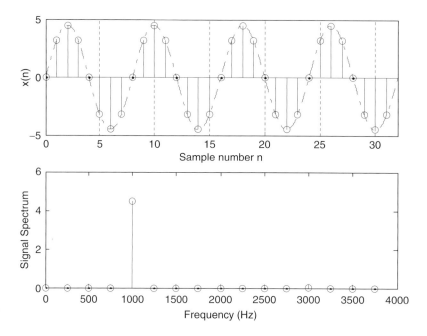

FIGURE 4.1 Example of the digital signal and its amplitude spectrum.

the frequency domain representation. Therefore, we can apply the DFT to perform frequency analysis of a time domain sequence. In addition, the DFT is widely used in many other areas, including spectral analysis, acoustics, imaging/video, audio, instrumentation, and communications systems.

To be able to develop the DFT and understand how to use it, we first study the spectrum of periodic digital signals using the Fourier series. (Detailed discussion of Fourier series is in Appendix B.)

4.1.1 Fourier Series Coefficients of Periodic Digital Signals

Let us look at a process in which we want to estimate the spectrum of a periodic digital signal $x(n)$ sampled at a rate of f_s Hz with the fundamental period $T_0 = NT$, as shown in Figure 4.2, where there are N samples within the duration of the fundamental period and $T = 1/f_s$ is the sampling period. For the time being, we assume that the periodic digital signal is band limited to have all harmonic frequencies less than the folding frequency $f_s/2$ so that aliasing does not occur.

According to Fourier series analysis (Appendix B), the coefficients of the Fourier series expansion of a periodic signal $x(t)$ in a complex form is

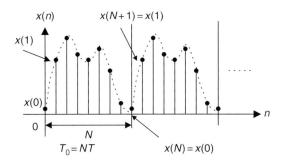

FIGURE 4.2   Periodic digital signal.

$$c_k = \frac{1}{T_0} \int_{T_0} x(t) e^{-jk\omega_0 t} dt \qquad -\infty < k < \infty, \qquad (4.1)$$

where k is the number of harmonics corresponding to the harmonic frequency of kf_0 and $\omega_0 = 2\pi/T_0$ and $f_0 = 1/T_0$ are the fundamental frequency in radians per second and the fundamental frequency in Hz, respectively. To apply Equation (4.1), we substitute $T_0 = NT$, $\omega_0 = 2\pi/T_0$ and approximate the integration over one period using a summation by substituting $dt = T$ and $t = nT$. We obtain

$$c_k = \frac{1}{N} \sum_{n=0}^{N-1} x(n) e^{-j\frac{2\pi kn}{N}}, \qquad -\infty < k < \infty. \qquad (4.2)$$

Since the coefficients c_k are obtained from the Fourier series expansion in the complex form, the resultant spectrum c_k will have two sides. There is an important feature of Equation (4.2) in which the Fourier series coefficient c_k is periodic of N. We can verify this as follows

$$c_{k+N} = \frac{1}{N} \sum_{n=0}^{N-1} x(n) e^{-j\frac{2\pi(k+N)n}{N}} = \frac{1}{N} \sum_{n=0}^{N-1} x(n) e^{-j\frac{2\pi kn}{N}} e^{-j2\pi n}. \qquad (4.3)$$

Since $e^{-j2\pi n} = \cos(2\pi n) - j\sin(2\pi n) = 1$, it follows that

$$c_{k+N} = c_k. \qquad (4.4)$$

Therefore, the two-sided line amplitude spectrum $|c_k|$ is periodic, as shown in Figure 4.3.

We note the following points:

a. As displayed in Figure 4.3, only the line spectral portion between the frequency $-f_s/2$ and frequency $f_s/2$ (folding frequency) represents the frequency information of the periodic signal.

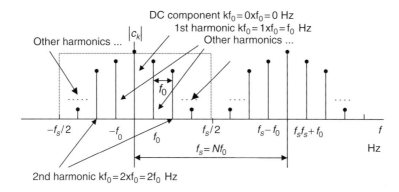

FIGURE 4.3 Amplitude spectrum of the periodic digital signal.

b. Notice that the spectral portion from $f_s/2$ to f_s is a copy of the spectrum in the negative frequency range from $-f_s/2$ to 0 Hz due to the spectrum being periodic for every Nf_0 Hz. Again, the amplitude spectral components indexed from $f_s/2$ to f_s can be folded at the folding frequency $f_s/2$ to match the amplitude spectral components indexed from 0 to $f_s/2$ in terms of $f_s - f$ Hz, where f is in the range from $f_s/2$ to f_s. For convenience, we compute the spectrum over the range from 0 to f_s Hz with nonnegative indices, that is,

$$c_k = \frac{1}{N}\sum_{n=0}^{N-1} x(n)e^{-j\frac{2\pi kn}{N}}, \quad k = 0,1,\ldots,N-1. \tag{4.5}$$

We can apply Equation (4.4) to find the negative indexed spectral values if they are required.

c. For the kth harmonic, the frequency is

$$f = kf_0 \quad \text{Hz}. \tag{4.6}$$

The frequency spacing between the consecutive spectral lines, called the frequency resolution, is f_0 Hz.

Example 4.1.

The periodic signal

$$x(t) = \sin(2\pi t)$$

is sampled using the rate $f_s = 4$ Hz.

a. Compute the spectrum c_k using the samples in one period.

b. Plot the two-sided amplitude spectrum $|c_k|$ over the range from -2 to 2 Hz.

Solution:

a. From the analog signal, we can determine the fundamental frequency $\omega_0 = 2\pi$ radians per second and $f_0 = \frac{\omega_0}{2\pi} = \frac{2\pi}{2\pi} = 1$ Hz, and the fundamental period $T_0 = 1$ second. Since using the sampling interval $T = 1/f_s = 0.25$ second, we get the sampled signal as

$$x(n) = x(nT) = \sin(2\pi nT) = \sin(0.5\pi n)$$

and plot the first eight samples as shown in Figure 4.4.

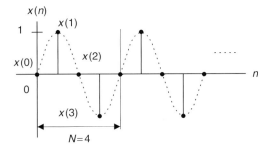

FIGURE 4.4 Periodic digital signal.

Choosing the duration of one period, $N = 4$, we have the sample values as follows

$$x(0) = 0; \; x(1) = 1; \; x(2) = 0; \text{ and } x(3) = -1.$$

Using Equation (4.5),

$$c_0 = \frac{1}{4}\sum_{n=0}^{3} x(n) = \frac{1}{4}(x(0) + x(1) + x(2) + x(3)) = \frac{1}{4}(0 + 1 + 0 - 1) = 0$$

$$c_1 = \frac{1}{4}\sum_{n=0}^{3} x(n)e^{-j2\pi \times 1n/4} = \frac{1}{4}\left(x(0) + x(1)e^{-j\pi/2} + x(2)e^{-j\pi} + x(3)e^{-j3\pi/2}\right)$$

$$= \frac{1}{4}(x(0) - jx(1) - x(2) + jx(3)) = 0 - j(1) - 0 + j(-1)) = -j0.5.$$

Similarly, we get

$$c_2 = \frac{1}{4}\sum_{k=0}^{3} x(n)e^{-j2\pi \times 2n/4} = 0, \text{ and } c_3 = \frac{1}{4}\sum_{n=0}^{3} x(k)e^{-j2\pi \times 3n/4} = j0.5.$$

Using periodicity, it follows that

$$c_{-1} = c_3 = j0.5, \text{ and } c_{-2} = c_2 = 0.$$

b. The amplitude spectrum for the digital signal is sketched in Figure 4.5.

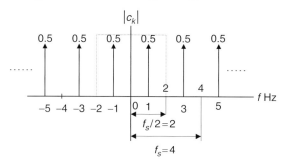

FIGURE 4.5 Two-sided spectrum for the periodic digital signal in Example 4.1.

As we know, the spectrum in the range of -2 to 2 Hz presents the information of the sinusoid with a frequency of 1 Hz and a peak value of $2|c_1| = 1$, which is converted from two sides to one side by doubling the spectral value. Note that we do not double the direct-current (DC) component, that is, c_0.

4.1.2 Discrete Fourier Transform Formulas

Now, let us concentrate on development of the DFT. Figure 4.6 shows one way to obtain the DFT formula.

First, we assume that the process acquires data samples from digitizing the interested continuous signal for a duration of T seconds. Next, we assume that a periodic signal $x(n)$ is obtained by copying the acquired N data samples with the duration of T to itself repetitively. Note that we assume continuity between the N data sample frames. This is not true in practice. We will tackle this problem in Section 4.3. We determine the Fourier series coefficients using one-period N data samples and Equation (4.5). Then we multiply the Fourier series coefficients by a factor of N to obtain

$$X(k) = Nc_k = \sum_{n=0}^{N-1} x(n) e^{-j\frac{2\pi k n}{N}}, \quad k = 0, 1, \ldots, N-1,$$

where $X(k)$ constitutes the DFT coefficients. Notice that the factor of N is a constant and does not affect the relative magnitudes of the DFT coefficients $X(k)$. As shown in the last plot, applying DFT with N data samples of $x(n)$ sampled at a rate of f_s (sampling period is $T = 1/f_s$) produces N complex DFT

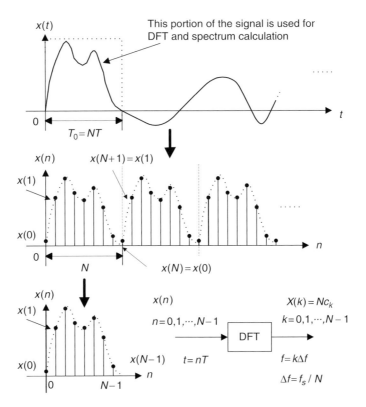

FIGURE 4.6 Development of DFT formula.

coefficients $X(k)$. The index n is the time index representing the sample number of the digital sequence, whereas k is the frequency index indicating each calculated DFT coefficient, and can be further mapped to the corresponding signal frequency in terms of Hz.

Now let us conclude the DFT definition. Given a sequence $x(n)$, $0 \leq n \leq N-1$, its DFT is defined as

$$X(k) = \sum_{n=0}^{N-1} x(n) e^{-j2\pi kn/N} = \sum_{n=0}^{N-1} x(n) W_N^{kn}, \text{ for } k = 0, 1, \ldots, N-1. \quad (4.7)$$

Equation (4.7) can be expanded as

$$X(k) = x(0) W_N^{k0} + x(1) W_N^{k1} + x(2) W_N^{k2} + \ldots + x(N-1) W_N^{k(N-1)}, \quad (4.8)$$
$$\text{for } k = 0, 1, \ldots, N-1,$$

where the factor W_N (called the twiddle factor in some textbooks) is defined as

$$W_N = e^{-j2\pi/N} = \cos\left(\frac{2\pi}{N}\right) - j\sin\left(\frac{2\pi}{N}\right). \qquad (4.9)$$

The inverse DFT is given by

$$x(n) = \frac{1}{N}\sum_{k=0}^{N-1} X(k)e^{j2\pi kn/N} = \frac{1}{N}\sum_{k=0}^{N-1} X(k)W_N^{-kn}, \text{ for } n = 0,1,\ldots, N-1. \qquad (4.10)$$

Proof can be found in Ahmed and Natarajan (1983); Proakis and Manolakis (1996); Oppenheim, Schafer, and Buck (1999); and Stearns and Hush (1990).

Similar to Equation (4.7), the expansion of Equation (4.10) leads to

$$x(n) = \frac{1}{N}\left(X(0)W_N^{-0n} + X(1)W_N^{-1n} + X(2)W_N^{-2n} + \ldots + X(N-1)W_N^{-(N-1)n}\right),$$
for $n = 0, 1, \ldots, N-1$.

$$(4.11)$$

As shown in Figure 4.6, in time domain we use the sample number or time index n for indexing the digital sample sequence $x(n)$. However, in frequency domain, we use index k for indexing N calculated DFT coefficients $X(k)$. We also refer to k as the frequency bin number in Equations (4.7) and (4.8).

We can use MATLAB functions **fft()** and **ifft()** to compute the DFT coefficients and the inverse DFT with the following syntax:

TABLE 4.1 MATLAB FFT functions.

X = fft(x)	% Calculate DFT coefficients
x = ifft(X)	% Inverse DFT
x = input vector	
X = DFT coefficient vector	

The following examples serve to illustrate the application of DFT and the inverse of DFT.

Example 4.2.

Given a sequence $x(n)$ for $0 \le n \le 3$, where $x(0) = 1$, $x(1) = 2$, $x(2) = 3$, and $x(3) = 4$,

a. Evaluate its DFT $X(k)$.

Solution:

a. Since $N = 4$ and $W_4 = e^{-j\frac{\pi}{2}}$, using Equation (4.7) we have a simplified formula,

$$X(k) = \sum_{n=0}^{3} x(n) W_4^{kn} = \sum_{n=0}^{3} x(n) e^{-j\frac{\pi k n}{2}}.$$

Thus, for $k = 0$

$$X(0) = \sum_{n=0}^{3} x(n) e^{-j0} = x(0) e^{-j0} + x(1) e^{-j0} + x(2) e^{-j0} + x(3) e^{-j0}$$
$$= x(0) + x(1) + x(2) + x(3)$$
$$= 1 + 2 + 3 + 4 = 10$$

for $k = 1$

$$X(1) = \sum_{n=0}^{3} x(n) e^{-j\frac{\pi n}{2}} = x(0) e^{-j0} + x(1) e^{-j\frac{\pi}{2}} + x(2) e^{-j\pi} + x(3) e^{-j\frac{3\pi}{2}}$$
$$= x(0) - jx(1) - x(2) + jx(3)$$
$$= 1 - j2 - 3 + j4 = -2 + j2$$

for $k = 2$

$$X(2) = \sum_{n=0}^{3} x(n) e^{-j\pi n} = x(0) e^{-j0} + x(1) e^{-j\pi} + x(2) e^{-j2\pi} + x(3) e^{-j3\pi}$$
$$= x(0) - x(1) + x(2) - x(3)$$
$$= 1 - 2 + 3 - 4 = -2$$

and for $k = 3$

$$X(3) = \sum_{n=0}^{3} x(n) e^{-j\frac{3\pi n}{2}} = x(0) e^{-j0} + x(1) e^{-j\frac{3\pi}{2}} + x(2) e^{-j3\pi} + x(3) e^{-j\frac{9\pi}{2}}$$
$$= x(0) + jx(1) - x(2) - jx(3)$$
$$= 1 + j2 - 3 - j4 = -2 - j2$$

Let us verify the result using the MATLAB function **fft()**:

```
≫ X = fft([1 2 3 4])
X = 10.0000   -2.0000 + 2.0000i   -2.0000   -2.0000 - 2.0000i
```

Example 4.3.

Using the DFT coefficients $X(k)$ for $0 \leq k \leq 3$ computed in Example 4.2,

a. Evaluate its inverse DFT to determine the time domain sequence $x(n)$.

Solution:

a. Since $N = 4$ and $W_4^{-1} = e^{j\frac{\pi}{2}}$, using Equation (4.10) we achieve a simplified formula,

$$x(n) = \frac{1}{4}\sum_{k=0}^{3} X(k)W_4^{-nk} = \frac{1}{4}\sum_{k=0}^{3} X(k)e^{j\frac{\pi k n}{2}}.$$

Then for $n = 0$

$$x(0) = \frac{1}{4}\sum_{k=0}^{3} X(k)e^{j0} = \frac{1}{4}\left(X(0)e^{j0} + X(1)e^{j0} + X(2)e^{j0} + X(3)e^{j0}\right)$$

$$= \frac{1}{4}(10 + (-2 + j2) - 2 + (-2 - j2)) = 1$$

for $n = 1$

$$x(1) = \frac{1}{4}\sum_{k=0}^{3} X(k)e^{j\frac{k\pi}{2}} = \frac{1}{4}\left(X(0)e^{j0} + X(1)e^{j\frac{\pi}{2}} + X(2)e^{j\pi} + X(3)e^{j\frac{3\pi}{2}}\right)$$

$$= \frac{1}{4}(X(0) + jX(1) - X(2) - jX(3))$$

$$= \frac{1}{4}(10 + j(-2 + j2) - (-2) - j(-2 - j2)) = 2$$

for $n = 2$

$$x(2) = \frac{1}{4}\sum_{k=0}^{3} X(k)e^{jk\pi} = \frac{1}{4}\left(X(0)e^{j0} + X(1)e^{j\pi} + X(2)e^{j2\pi} + X(3)e^{j3\pi}\right)$$

$$= \frac{1}{4}(X(0) - X(1) + X(2) - X(3))$$

$$= \frac{1}{4}(10 - (-2 + j2) + (-2) - (-2 - j2)) = 3$$

and for $n = 3$

$$x(3) = \frac{1}{4}\sum_{k=0}^{3} X(k)e^{j\frac{k\pi 3}{2}} = \frac{1}{4}\left(X(0)e^{j0} + X(1)e^{j\frac{3\pi}{2}} + X(2)e^{j3\pi} + X(3)e^{j\frac{9\pi}{2}}\right)$$

$$= \frac{1}{4}(X(0) - jX(1) - X(2) + jX(3))$$

$$= \frac{1}{4}(10 - j(-2 + j2) - (-2) + j(-2 - j2)) = 4$$

This example actually verifies the inverse DFT. Applying the MATLAB function **ifft()** achieves:

$$\gg x = \text{ifft}([10 \ -2+2j \ -2 \ -2-2j])$$
$$x = 1 \quad 2 \quad 3 \quad 4.$$

Now we explore the relationship between the frequency bin k and its associated frequency. Omitting the proof, the calculated N DFT coefficients $X(k)$ represent the frequency components ranging from 0 Hz (or radians/second) to f_s Hz (or ω_s radians/second), hence we can map the frequency bin k to its corresponding frequency as follows:

$$\omega = \frac{k\omega_s}{N} \text{ (radians per second)}, \quad (4.12)$$

or in terms of Hz,

$$f = \frac{kf_s}{N} \text{ (Hz)}, \quad (4.13)$$

where $\omega_s = 2\pi f_s$.

We can define the frequency resolution as the frequency step between two consecutive DFT coefficients to measure how fine the frequency domain presentation is and achieve

$$\Delta\omega = \frac{\omega_s}{N} \text{ (radians per second)}, \quad (4.14)$$

or in terms of Hz, it follows that

$$\Delta f = \frac{f_s}{N} \text{ (Hz)}. \quad (4.15)$$

Let us study the following example.

Example 4.4.

In Example 4.2, given a sequence $x(n)$ for $0 \le n \le 3$, where $x(0) = 1$, $x(1) = 2$, $x(2) = 3$, and $x(3) = 4$, we have computed four DFT coefficients $X(k)$ for $0 \le k \le 3$ as $X(0) = 10$, $X(1) = -2 + j2$, $X(2) = -2$, and $X(3) = -2 - j2$. If the sampling rate is 10 Hz,

    a. Determine the sampling period, time index, and sampling time instant for a digital sample $x(3)$ in time domain.

    b. Determine the frequency resolution, frequency bin number, and mapped frequency for each of the DFT coefficients $X(1)$ and $X(3)$ in frequency domain.

Solution:

a. In time domain, we have the sampling period calculated as

$$T = 1/f_s = 1/10 = 0.1 \text{ second}.$$

For data $x(3)$, the time index is $n = 3$ and the sampling time instant is determined by

$$t = nT = 3 \cdot 0.1 = 0.3 \text{ second}.$$

b. In frequency domain, since the total number of DFT coefficients is four, the frequency resolution is determined by

$$\Delta f = \frac{f_s}{N} = \frac{10}{4} = 2.5 \text{ Hz}.$$

The frequency bin number for $X(1)$ should be $k = 1$ and its corresponding frequency is determined by

$$f = \frac{k f_s}{N} = \frac{1 \times 10}{4} = 2.5 \text{ Hz}.$$

Similarly, for $X(3)$ and $k = 3$,

$$f = \frac{k f_s}{N} = \frac{3 \times 10}{4} = 7.5 \text{ Hz}.$$

Note that from Equation (4.4), $k = 3$ is equivalent to $k - N = 3 - 4 = -1$, and $f = 7.5$ Hz is also equivalent to the frequency $f = (-1 \times 10)/4 = -2.5$ Hz, which corresponds to the negative side spectrum. The amplitude spectrum at 7.5 Hz after folding should match the one at $f_s - f = 10.0 - 7.5 = 2.5$ Hz. We will apply these developed notations in the next section for amplitude and power spectral estimation.

4.2 Amplitude Spectrum and Power Spectrum

One of the DFT applications is transformation of a finite-length digital signal $x(n)$ into the spectrum in frequency domain. Figure 4.7 demonstrates such an application, where A_k and P_k are the computed amplitude spectrum and the power spectrum, respectively, using the DFT coefficients $X(k)$.

First, we achieve the digital sequence $x(n)$ by sampling the analog signal $x(t)$ and truncating the sampled signal with a data window with a length $T_0 = NT$, where T is the sampling period and N the number of data points. The time for data window is

$$T_0 = NT. \tag{4.16}$$

4.2 Amplitude Spectrum and Power Spectrum

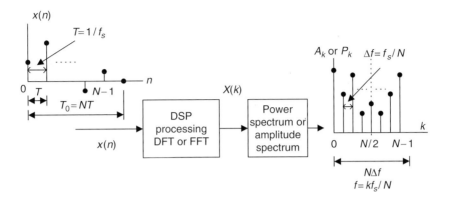

FIGURE 4.7 Applications of DFT/FFT.

For the truncated sequence $x(n)$ with a range of $n = 0, 1, 2, \ldots, N - 1$, we get

$$x(0), x(1), x(2), \ldots, x(N-1). \tag{4.17}$$

Next, we apply the DFT to the obtained sequence, $x(n)$, to get the N DFT coefficients

$$X(k) = \sum_{n=0}^{N-1} x(n) W_N^{nk}, \text{ for } k = 0, 1, 2, \ldots, N-1. \tag{4.18}$$

Since each calculated DFT coefficient is a complex number, it is not convenient to plot it versus its frequency index. Hence, after evaluating Equation (4.18), the magnitude and phase of each DFT coefficient (we refer to them as the amplitude spectrum and phase spectrum, respectively) can be determined and plotted versus its frequency index. We define the amplitude spectrum as

$$A_k = \frac{1}{N}|X(k)| = \frac{1}{N}\sqrt{(\text{Real}[X(k)])^2 + (\text{Imag}[X(k)])^2},$$
$$k = 0, 1, 2, \ldots, N-1. \tag{4.19}$$

We can modify the amplitude spectrum to a one-sided amplitude spectrum by doubling the amplitudes in Equation (4.19), keeping the original DC term at $k = 0$. Thus we have

$$\bar{A}_k = \begin{cases} \frac{1}{N}|X(0)|, & k = 0 \\ \frac{2}{N}|X(k)|, & k = 1, \ldots, N/2 \end{cases}. \tag{4.20}$$

We can also map the frequency bin k to its corresponding frequency as

$$f = \frac{kf_s}{N}. \tag{4.21}$$

Correspondingly, the phase spectrum is given by

$$\varphi_k = \tan^{-1}\left(\frac{\text{Imag}[X(k)]}{\text{Real}[X(k)]}\right), \ k = 0, 1, 2, \ldots, N - 1. \tag{4.22}$$

Besides the amplitude spectrum, the power spectrum is also used. The DFT power spectrum is defined as

$$P_k = \frac{1}{N^2}|X(k)|^2 = \frac{1}{N^2}\left\{(\text{Real}[X(k)])^2 + (\text{Imag}[X(k)])^2\right\},$$
$$k = 0, 1, 2, \ldots, N - 1. \tag{4.23}$$

Similarly, for a one-sided power spectrum, we get

$$\bar{P}_k = \begin{cases} \frac{1}{N^2}|X(0)|^2 & k = 0 \\ \frac{2}{N^2}|X(k)|^2 & k = 0, 1, \ldots, N/2 \end{cases} \tag{4.24}$$

and
$$f = \frac{kf_s}{N}. \tag{4.25}$$

Again, notice that the frequency resolution, which denotes the frequency spacing between DFT coefficients in frequency domain, is defined as

$$\Delta f = \frac{f_s}{N} \ (\text{Hz}). \tag{4.26}$$

It follows that better frequency resolution can be achieved by using a longer data sequence.

Example 4.5.

Consider the sequence

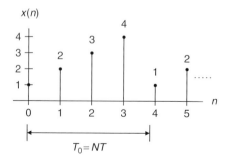

FIGURE 4.8 Sampled values in Example 4.5.

Assuming that $f_s = 100 \, \text{Hz}$,

a. Compute the amplitude spectrum, phase spectrum, and power spectrum.

Solution:

a. Since $N = 4$, and using the DFT shown in Example 4.1, we find the DFT coefficients to be

$$X(0) = 10$$
$$X(1) = -2 + j2$$
$$X(2) = -2$$
$$X(3) = -2 - j2.$$

The amplitude spectrum, phase spectrum, and power density spectrum are computed as follows.

For $k = 0$, $f = k \cdot f_s/N = 0 \times 100/4 = 0 \, \text{Hz}$,

$$A_0 = \frac{1}{4}|X(0)| = 2.5, \quad \varphi_0 = \tan^{-1}\left(\frac{\text{Imag}[X(0)]}{\text{Real}([X(0)]}\right) = 0^0,$$

$$P_0 = \frac{1}{4^2}|X(0)|^2 = 6.25.$$

For $k = 1$, $f = 1 \times 100/4 = 25 \, \text{Hz}$,

$$A_1 = \frac{1}{4}|X(1)| = 0.7071, \quad \varphi_1 = \tan^{-1}\left(\frac{\text{Imag}[X(1)]}{\text{Real}[X(1)]}\right) = 135^0,$$

$$P_1 = \frac{1}{4^2}|X(1)|^2 = 0.5000.$$

For $k = 2$, $f = 2 \times 100/4 = 50 \, \text{Hz}$,

$$A_2 = \frac{1}{4}|X(2)| = 0.5, \quad \varphi_2 = \tan^{-1}\left(\frac{\text{Imag}[X(2)]}{\text{Real}[X(2)]}\right) = 180^0,$$

$$P_2 = \frac{1}{4^2}|X(2)|^2 = 0.2500.$$

Similarly, for $k = 3$, $f = 3 \times 100/4 = 75 \, \text{Hz}$,

$$A_3 = \frac{1}{4}|X(3)| = 0.7071, \quad \varphi_3 = \tan^{-1}\left(\frac{\text{Imag}[X(3)]}{\text{Real}[X(3)]}\right) = -135^0,$$

$$P_3 = \frac{1}{4^2}|X(3)|^2 = 0.5000.$$

Thus, the sketches for the amplitude spectrum, phase spectrum, and power spectrum are given in Figure 4.9.

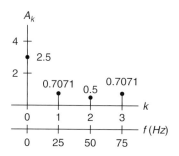

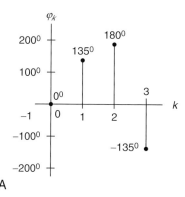

FIGURE 4.9A Amplitude spectrum and phase spectrum in Example 4.5.

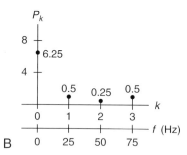

FIGURE 4.9B Power density spectrum in Example 4.5.

Note that the folding frequency in this example is 50 Hz and the amplitude and power spectrum values at 75 Hz are each image counterparts (corresponding negative-indexed frequency components). Thus values at 0, 25, and 50 Hz correspond to the positive-indexed frequency components.

We can easily find the one-sided amplitude spectrum and one-sided power spectrum as

$$\bar{A}_0 = 2.5, \ \bar{A}_1 = 1.4141, \ \bar{A}_2 = 1 \text{ and}$$
$$\bar{P}_0 = 6.25, \ \bar{P}_1 = 2, \ \bar{P}_2 = 1.$$

We plot the one-sided amplitude spectrum for comparison:

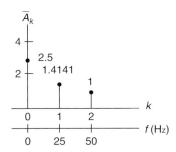

FIGURE 4.10 One-sided amplitude spectrum in Example 4.5.

Note that in the one-sided amplitude spectrum, the negative-indexed frequency components are added back to the corresponding positive-indexed frequency components; thus each amplitude value other than the DC term is doubled. It represents the frequency components up to the folding frequency.

Example 4.6.

Consider a digital sequence sampled at the rate of 10 kHz. If we use a size of 1,024 data points and apply the 1,024-point DFT to compute the spectrum,

a. Determine the frequency resolution.

b. Determine the highest frequency in the spectrum.

Solution:

a. $\Delta f = \frac{f_s}{N} = \frac{10000}{1024} = 9.776 \text{ Hz}.$

b. The highest frequency is the folding frequency, given by

$$f_{\max} = \frac{N}{2} \Delta f = \frac{f_s}{2}$$

$$= 512 \cdot 9.776 = 5000 \text{ Hz}$$

As shown in Figure 4.7, the DFT coefficients may be computed via a *fast Fourier transform* (FFT) algorithm. The FFT is a very efficient algorithm for computing DFT coefficients. The FFT algorithm requires the time domain sequence $x(n)$ to have a length of data points equal to a power of 2; that is, 2^m samples, where m is a positive integer. For example, the number of samples in $x(n)$ can be $N = 2, 4, 8, 16$, etc.

In the case of using the FFT algorithm to compute DFT coefficients, where the length of the available data is not equal to a power of 2 (required by the FFT), we can pad the data sequence with zeros to create a new sequence with a larger number of samples, $\overline{N} = 2^m > N$. The modified data sequence for applying FFT, therefore, is

$$\bar{x}(n) = \begin{cases} x(n) & 0 \leq n \leq N-1 \\ 0 & N \leq n \leq \overline{N}-1 \end{cases}. \qquad (4.27)$$

It is very important to note that the signal spectra obtained via zero-padding the data sequence in Equation (4.27) does not add any new information and does not contain more accurate signal spectral presentation. In this situation, the frequency spacing is reduced due to more DFT points, and the achieved spectrum is an interpolated version with "better display." We illustrate the zero padding effect via the following example instead of theoretical analysis. A theoretical discussion of zero padding in FFT can be found in Proakis and Manolakis (1996).

Figure 4.11a shows the 12 data samples from an analog signal containing frequencies of 10 Hz and 25 Hz at a sampling rate of 100 Hz, and the amplitude

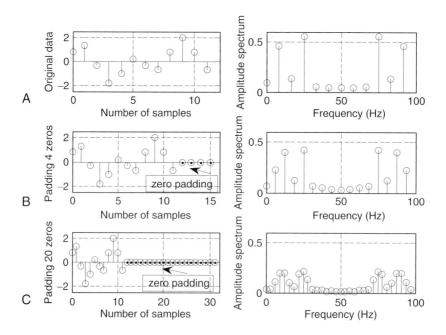

FIGURE 4.11 Zero padding effect by using FFT.

spectrum obtained by applying the DFT. Figure 4.11b displays the signal samples with padding of four zeros to the original data to make up a data sequence of 16 samples, along with the amplitude spectrum calculated by FFT. The data sequence padded with 20 zeros and its calculated amplitude spectrum using FFT are shown in Figure 4.11c. It is evident that increasing the data length via zero padding to compute the signal spectrum does not add basic information and does not change the spectral shape but gives the "interpolated spectrum" with the reduced frequency spacing. We can get a better view of the two spectral peaks described in this case.

The only way to obtain the detailed signal spectrum with a fine frequency resolution is to apply more available data samples, that is, a longer sequence of data. Here, we choose to pad the least number of zeros possible to satisfy the minimum FFT computational requirement. Let us look at another example.

Example 4.7.

We use the DFT to compute the amplitude spectrum of a sampled data sequence with a sampling rate $f_s = 10 \, \text{kHz}$. Given that it requires the frequency resolution to be less than 0.5 Hz,

a. Determine the number of data points by using the FFT algorithm, assuming that the data samples are available.

Solution:

$$\Delta f = 0.5 \, \text{Hz}$$

$$N = \frac{f_s}{\Delta f} = \frac{10000}{0.5} = 20000$$

a. Since we use the FFT to compute the spectrum, the number of the data points must be a power of 2, that is,

$$N = 2^{15} = 32768.$$

And the resulting frequency resolution can be recalculated as

$$\Delta f = \frac{f_s}{N} = \frac{10000}{32768} = 0.31 \, \text{Hz}.$$

Next, we study a MATLAB example.

Example 4.8.

Given the sinusoid

$$x(n) = 2 \cdot \sin\left(2000\pi \frac{n}{8000}\right)$$

obtained by sampling the analog signal

$$x(t) = 2 \cdot \sin(2000\pi t)$$

with a sampling rate of $f_s = 8,000$ Hz,

a. Use the MATLAB DFT to compute the signal spectrum with the frequency resolution to be equal to or less than 8 Hz.

b. Use the MATLAB FFT and zero padding to compute the signal spectrum, assuming that the data samples are available in (1).

Solution:

a. The number of data points is found to be $N = \frac{f_s}{\Delta f} = \frac{8000}{8} = 1000$. There is no zero padding needed if we use the DFT formula. Detailed implementation is given in Program 4.1. The first and second plots in Figure 4.12 show the two-sided amplitude and power spectra, respectively, using the DFT, where each frequency counterpart at 7,000 Hz appears. The third and fourth plots are the one-side amplitude and power spectra, where the true frequency contents are displayed from 0 Hz to the Nyquist frequency of 4 kHz (folding frequency).

b. If the FFT is used, the number of data points must be a power of 2.

Hence we choose

$$N = 2^{10} = 1024.$$

Assuming there are only 1,000 data samples available in (a), we need to pad 24 zeros to the original 1,000 data samples before applying the FFT algorithm, as required. Thus the calculated frequency resolution is $\Delta f = f_s/N = 8000/1024 = 7.8125$ Hz. Note that this is an interpolated

Program 4.1. MATLAB program for Example 4.8

```
% Example 4.8
close all;clear all
% Generate the sine wave sequence
fs = 8000;              %Sampling rate
N = 1000;               % Number of data points
x = 2*sin(2000*pi*[0:1:N – 1]/fs);
% Apply the DFT algorithm
figure(1)
xf = abs(fft(x))/N;          %Compute the amplitude spectrum
```

```
P = xf.*xf;            %Compute the power spectrum
f = [0:1:N − 1]*fs/N;      %Map the frequency bin to the frequency (Hz)
subplot(2,1,1); plot(f,xf);grid
xlabel('Frequency (Hz)'); ylabel('Amplitude spectrum (DFT)');
subplot(2,1,2);plot(f,P);grid
xlabel('Frequency (Hz)'); ylabel('Power spectrum (DFT)');
figure(2)
% Convert it to one-sided spectrum
xf(2:N) = 2*xf(2:N);       % Get the single-sided spectrum
P = xf.*xf; % Calculate the power spectrum
f = [0:1:N/2]*fs/N % Frequencies up to the folding frequency
subplot(2,1,1); plot(f,xf(1:N/2+1));grid
xlabel('Frequency (Hz)'); ylabel('Amplitude spectrum (DFT)');
subplot(2,1,2);plot(f,P(1:N/2+1));grid
xlabel('Frequency (Hz)'); ylabel('Power spectrum (DFT)');
figure(3)
% Zero padding to the length of 1024
x = [x, zeros(1,24)];
N = length(x);
xf = abs(fft(x))/N;        %Compute the amplitude spectrum with zero padding
P = xf.*xf;           %Compute the power spectrum
f = [0:1:N − 1]*fs/N;      %Map frequency bin to frequency (Hz)
subplot(2,1,1); plot(f,xf);grid
xlabel('Frequency (Hz)'); ylabel('Amplitude spectrum (FFT)');
subplot(2,1,2);plot(f,P);grid
xlabel('Frequency (Hz)'); ylabel('Power spectrum (FFT)');
figure(4)
% Convert it to one-sided spectrum
xf(2:N) = 2*xf(2:N);
P = xf.*xf;
f = [0:1:N/2]*fs/N;
subplot(2,1,1); plot(f,xf(1:N/2+1));grid
xlabel('Frequency (Hz)'); ylabel('Amplitude spectrum (FFT)');
subplot(2,1,2);plot(f,P(1:N/2+1));grid
xlabel('Frequency (Hz)'); ylabel('Power spectrum (FFT)');
```

frequency resolution by using zero padding. The zero padding actually interpolates a signal spectrum and carries no additional frequency information. Figure 4.13 shows the spectral plots using FFT. The detailed implementation is given in Program 4.1.

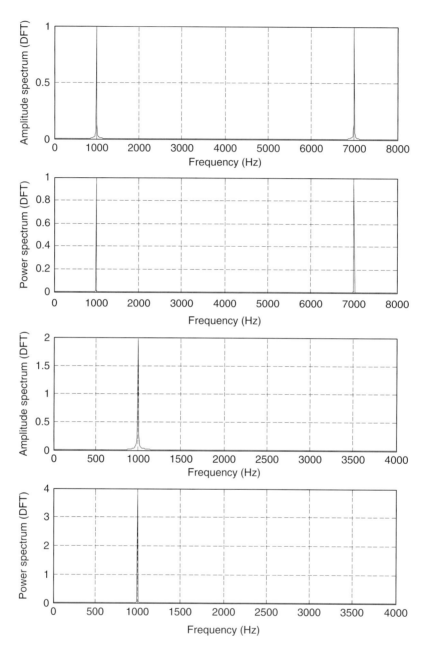

FIGURE 4.12 Amplitude spectrum and power spectrum using DFT for Example 4.8.

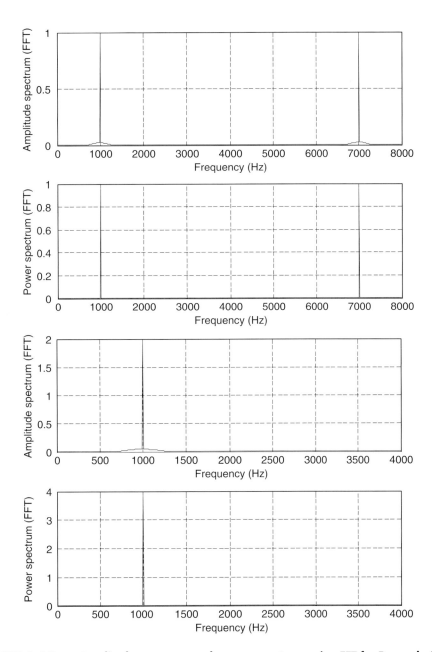

FIGURE 4.13 Amplitude spectrum and power spectrum using FFT for Example 4.8.

4.3 Spectral Estimation Using Window Functions

When we apply DFT to the sampled data in the previous section, we theoretically imply the following assumptions: first, that the sampled data are periodic to themselves (repeat themselves), and second, that the sampled data are continuous to themselves and band limited to the folding frequency. The second assumption is often violated, thus the discontinuity produces undesired harmonic frequencies. Consider the pure 1-Hz sine wave with 32 samples shown in Figure 4.14.

As shown in the figure, if we use a window size of $N = 16$ samples, which is a multiple of the two waveform cycles, the second window repeats with continuity. However, when the window size is chosen to be 18 samples, which is not a multiple of the waveform cycles (2.25 cycles), the second window repeats the first window with discontinuity. It is this discontinuity that produces harmonic frequencies that are not present in the original signal. Figure 4.15 shows the spectral plots for both cases using the DFT/FFT directly.

The first spectral plot contains a single frequency, as we expected, while the second spectrum has the expected frequency component plus many harmonics, which do not exist in the original signal. We call such an effect *spectral leakage*.

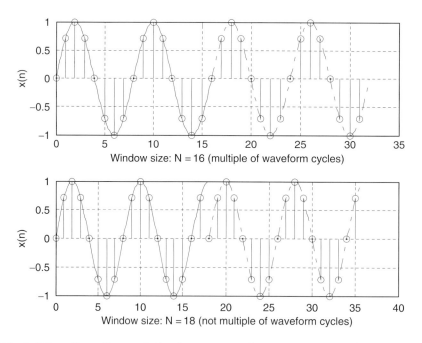

FIGURE 4.14 Sampling a 1-Hz sine wave using (*top*) 16 samples per cycle and (*bottom*) 18 samples per cycle.

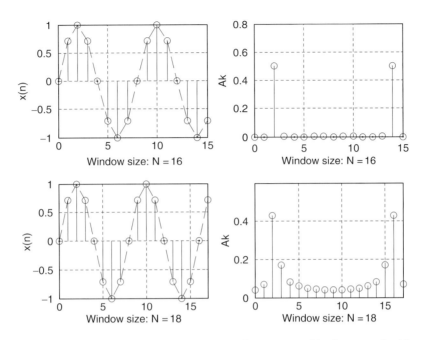

FIGURE 4.15 Signal samples and spectra without spectral leakage and with spectral leakage.

The amount of spectral leakage shown in the second plot is due to amplitude discontinuity in time domain. The bigger the discontinuity, the more the leakage. To reduce the effect of spectral leakage, a window function can be used whose amplitude tapers smoothly and gradually toward zero at both ends. Applying the window function $w(n)$ to a data sequence $x(n)$ to obtain a windowed sequence $x_w(n)$ is better illustrated in Figure 4.16 using Equation (4.28):

$$x_w(n) = x(n)w(n), \text{ for } n = 0, 1, \ldots, N - 1. \quad (4.28)$$

The top plot is the data sequence $x(n)$, and the middle plot is the window function $w(n)$. The bottom plot in Figure 4.16 shows that the windowed sequence $x_w(n)$ is tapped down by a window function to zero at both ends such that the discontinuity is dramatically reduced.

Example 4.9.

In Figure 4.16, given

- $x(2) = 1$ and $w(2) = 0.2265$;
- $x(5) = -0.7071$ and $w(5) = 0.7008$,

a. Calculate the windowed sequence data points $x_w(2)$ and $x_w(5)$.

Solution:

a. Applying the window function operation leads to

$$x_w(2) = x(2) \times w(2) = 1 \times 0.2265 = 0.2265 \text{ and}$$
$$x_w(5) = x(5) \times w(5) = -0.7071 \times 0.7008 = -0.4956,$$

which agree with the values shown in the bottom plot in the Figure 4.16. Using the window function shown in Example 4.9, the spectral plot is reproduced. As a result, spectral leakage is greatly reduced, as shown in Figure 4.17.

The common window functions are listed as follows.
The rectangular window (no window function):

$$w_R(n) = 1 \qquad 0 \leq n \leq N - 1 \qquad (4.29)$$

The triangular window:

$$w_{tri}(n) = 1 - \frac{|2n - N + 1|}{N - 1}, \; 0 \leq n \leq N - 1 \qquad (4.30)$$

The Hamming window:

$$w_{hm}(n) = 0.54 - 0.46 \cos\left(\frac{2\pi n}{N - 1}\right), \; 0 \leq n \leq N - 1 \qquad (4.31)$$

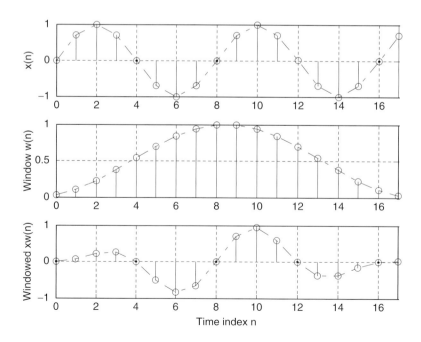

FIGURE 4.16 Illustration of the window operation.

4.3 Spectral Estimation Using Window Functions

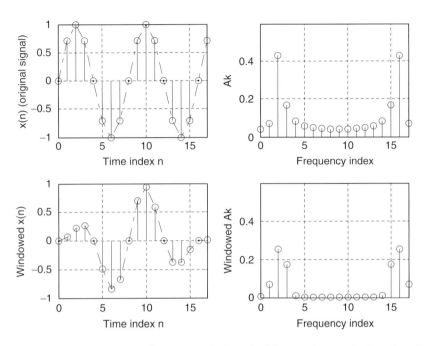

FIGURE 4.17 Comparison of spectra calculated without using a window function and using a window function to reduce spectral leakage.

The Hanning window:

$$w_{hn}(n) = 0.5 - 0.5\cos\left(\frac{2\pi n}{N-1}\right), \quad 0 \leq n \leq N-1 \quad (4.32)$$

Plots for each window function for a size of 20 samples are shown in Figure 4.18.

The following example details each step for computing the spectral information using the window functions.

Example 4.10.

Considering the sequence $x(0) = 1$, $x(1) = 2$, $x(2) = 3$, and $x(3) = 4$, and given $f_s = 100\,\text{Hz}$, $T = 0.01$ seconds, compute the amplitude spectrum, phase spectrum, and power spectrum

  a. Using the triangular window function.

  b. Using the Hamming window function.

Solution:

  a. Since $N = 4$, from the triangular window function, we have

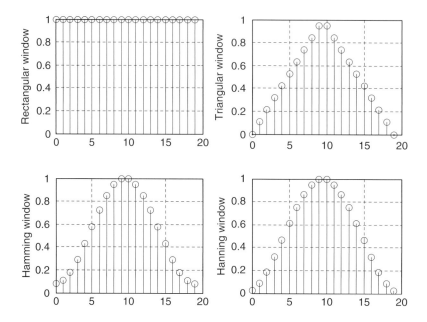

FIGURE 4.18   Plots of window sequences.

$$w_{tri}(0) = 1 - \frac{|2 \times 0 - 4 + 1|}{4 - 1} = 0$$

$$w_{tri}(1) = 1 - \frac{|2 \times 1 - 4 + 1|}{4 - 1} = 0.6667.$$

Similarly, $w_{tri}(2) = 0.6667$, $w_{tri}(3) = 0$. Next, the windowed sequence is computed as

$$x_w(0) = x(0) \times w_{tri}(0) = 1 \times 0 = 0$$
$$x_w(1) = x(1) \times w_{tri}(1) = 2 \times 0.6667 = 1.3334$$
$$x_w(2) = x(2) \times w_{tri}(2) = 3 \times 0.6667 = 2$$
$$x_w(3) = x(3) \times w_{tri}(3) = 4 \times 0 = 0.$$

Applying DFT Equation (4.8) to $x_w(n)$ for $k = 0, 1, 2, 3$, respectively,

$$X(k) = x_w(0)W_4^{k \times 0} + x(1)W_4^{k \times 1} + x(2)W_4^{k \times 2} + x(3)W_4^{k \times 3}.$$

We have the following results:

$$X(0) = 3.3334$$
$$X(1) = -2 - j1.3334$$
$$X(2) = 0.6666$$

$$X(3) = -2 + j1.3334$$

$$\Delta f = \frac{1}{NT} = \frac{1}{4 \cdot 0.01} = 25\,\text{Hz}$$

Applying Equations (4.19), (4.22), and (4.23) leads to

$$A_0 = \frac{1}{4}|X(0)| = 0.8334, \quad \varphi_0 = \tan^{-1}\left(\frac{0}{3.3334}\right) = 0^0,$$

$$P_0 = \frac{1}{4^2}|X(0)|^2 = 0.6954$$

$$A_1 = \frac{1}{4}|X(1)| = 0.6009, \quad \varphi_1 = \tan^{-1}\left(\frac{-1.3334}{-2}\right) = -146.31^0,$$

$$P_1 = \frac{1}{4^2}|X(1)|^2 = 0.3611$$

$$A_2 = \frac{1}{4}|X(2)| = 0.1667, \quad \varphi_2 = \tan^{-1}\left(\frac{0}{0.6666}\right) = 0^0,$$

$$P_2 = \frac{1}{4^2}|X(2)|^2 = 0.0278.$$

Similarly,

$$A_3 = \frac{1}{4}|X(3)| = 0.6009, \quad \varphi_3 = \tan^{-1}\left(\frac{1.3334}{-2}\right) = 146.31^0,$$

$$P_3 = \frac{1}{4^2}|X(3)|^2 = 0.3611.$$

b. Since $N = 4$, from the Hamming window function, we have

$$w_{hm}(0) = 0.54 - 0.46\cos\left(\frac{2\pi \times 0}{4-1}\right) = 0.08$$

$$w_{hm}(1) = 0.54 - 0.46\cos\left(\frac{2\pi \times 1}{4-1}\right) = 0.77.$$

Similarly, $w_{hm}(2) = 0.77$, $w_{hm}(3) = 0.08$. Next, the windowed sequence is computed as

$$x_w(0) = x(0) \times w_{hm}(0) = 1 \times 0.08 = 0.08$$
$$x_w(1) = x(1) \times w_{hm}(1) = 2 \times 0.77 = 1.54$$
$$x_w(2) = x(2) \times w_{hm}(2) = 3 \times 0.77 = 2.31$$
$$x_w(0) = x(3) \times w_{hm}(3) = 4 \times 0.08 = 0.32.$$

Applying DFT Equation (4.8) to $x_w(n)$ for $k = 0, 1, 2, 3$, respectively,

$$X(k) = x_w(0)W_4^{k\times 0} + x(1)W_4^{k\times 1} + x(2)W_4^{k\times 2} + x(3)W_4^{k\times 3}.$$

We yield the following:

$$X(0) = 4.25$$
$$X(1) = -2.23 - j1.22$$
$$X(2) = 0.53$$
$$X(3) = -2.23 + j1.22$$
$$\Delta f = \frac{1}{NT} = \frac{1}{4 \cdot 0.01} = 25 \text{ Hz}$$

Using Equations (4.19), (4.22), and (4.23), we achieve

$$A_0 = \frac{1}{4}|X(0)| = 1.0625, \quad \varphi_0 = \tan^{-1}\left(\frac{0}{4.25}\right) = 0^\circ,$$

$$P_0 = \frac{1}{4^2}|X(0)|^2 = 1.1289$$

$$A_1 = \frac{1}{4}|X(1)| = 0.6355, \quad \varphi_1 = \tan^{-1}\left(\frac{-1.22}{-2.23}\right) = -151.32^\circ,$$

$$P_1 = \frac{1}{4^2}|X(1)|^2 = 0.4308$$

$$A_2 = \frac{1}{4}|X(2)| = 0.1325, \quad \varphi_2 = \tan^{-1}\left(\frac{0}{0.53}\right) = 0^\circ,$$

$$P_2 = \frac{1}{4^2}|X(2)|^2 = 0.0176.$$

Similarly,

$$A_3 = \frac{1}{4}|X(3)| = 0.6355, \quad \varphi_3 = \tan^{-1}\left(\frac{1.22}{-2.23}\right) = 151.32^\circ,$$

$$P_3 = \frac{1}{4^2}|X(3)|^2 = 0.4308.$$

Example 4.11.

Given the sinusoid

$$x(n) = 2 \cdot \sin\left(2000\pi \frac{n}{8000}\right)$$

obtained by using a sampling rate of $f_s = 8{,}000$ Hz, use the DFT to compute the spectrum with the following specifications:

a. Compute the spectrum of a triangular window function with a window size = 50.

b. Compute the spectrum of a Hamming window function with a window size = 100.

c. Compute the spectrum of a Hanning window function with a window size = 150 and one-sided spectrum.

The MATLAB program is listed in Program 4.2, and the results are plotted in Figures 4.19 to 4.21. As compared with the no-windowed (rectangular window) case, all three windows are able to effectively reduce spectral leakage, as shown in the figures.

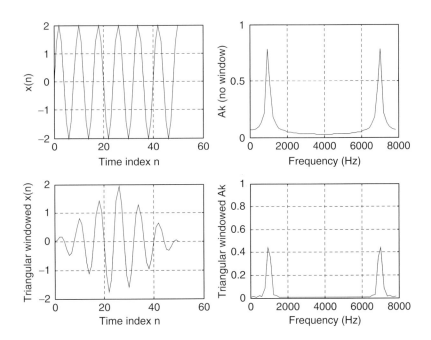

FIGURE 4.19 Comparison of a spectrum without using a window function and a spectrum using a triangular window of size of 50 samples in Example 4.11.

Program 4.2. MATLAB program for Example 4.11

```
%Example 4.11
close all;clear all
% Generate the sine wave sequence
fs = 8000; T = 1/fs;          % Sampling rate and sampling period
```

(*Continued*)

```matlab
x = 2*sin(2000*pi*[0:1:50]*T);          %Generate the 51 2000-Hz samples
% Apply the FFT algorithm
N = length(x);
index_t = [0:1:N-1];
f = [0:1-1]*8000/N;                     %Map the frequency bin to the frequency (Hz)
xf=abs(fft(x))/N;                       %Calculate the amplitude spectrum
figure(1)
%Using the Bartlett window
x_b = x.*bartlett(N)';                  %Apply the triangular window function
xf_b = abs(fft(x_b))/N;                 %Calculate the amplitude spectrum
subplot(2,2,1);plot(index_t,x);grid
xlabel('Time index n'); ylabel('x(n)');
subplot(2,2,3); plot(index_t,x_b);grid
xlabel('Time index n'); ylabel('Triangular windowed x(n)');
subplot(2,2,2);plot(f,xf);grid;axis([0 8000 0 1]);
xlabel('Frequency (Hz)'); ylabel('Ak (no window)');
subplot(2,2,4); plot(f,xf_b);grid; axis([0 8000 0 1]);
xlabel('Frequency (Hz)'); ylabel('Triangular windowed Ak');
figure(2)
% Generate the sine wave sequence
x = 2*sin(2000*pi*[0:1:100]*T);         %Generate the 101 2000-Hz samples.
% Apply the FFT algorithm
N=length(x);
index_t = [0:1:N-1];
f = [0:1:N-1]*fs/N;
xf = abs(fft(x))/N;
%Using the Hamming window
x_hm = x.*hamming(N)';                  %Apply the Hamming window function
xf_hm=abs(fft(x_hm))/N;                 %Calculate the amplitude spectrum
subplot(2,2,1);plot(index_t,x);grid
xlabel('Time index n'); ylabel('x(n)');
subplot(2,2,3); plot(index_t,x_hm);grid
xlabel('Time index n'); ylabel('Hamming windowed x(n)');
subplot(2,2,2);plot(f,xf);grid;axis([0 fs 0 1]);
xlabel('Frequency (Hz)'); ylabel('Ak (no window)');
subplot(2,2,4); plot(f,xf_hm);grid;axis([0 fs 0 1]);
xlabel('Frequency (Hz)'); ylabel('Hamming windowed Ak');
figure(3)
% Generate the sine wave sequence
x = 2*sin(2000*pi*[0:1:150]*T);         % Generate the 151 2-kHz samples
% Apply the FFT algorithm
N=length(x);
```

```
index_t = [0:1:N − 1];
f = [0:1:N − 1]*fs/N;
xf = 2*abs(fft(x))/N;xf(1) = xf(1)/2;          % Single-sided spectrum
%Using the Hanning window
x_hn = x.*hanning(N)';;
xf_hn = 2*abs(fft(x_hn))/N;xf_hn(1) = xf_hn(1)/2; %Single-sided spectrum
subplot(2,2,1);plot(index_t,x);grid
xlabel('Time index n'); ylabel('x(n)');
subplot(2,2,3); plot(index_t,x_hn);grid
xlabel('Time index n'); ylabel('Hanning windowed x(n)');
subplot(2,2,2);plot(f(1:(N-1)/2),xf(1:(N-1)/2));grid;axis([0 fs/2 0 1]);
xlabel('Frequency (Hz)'); ylabel('Ak (no window)');
subplot(2,2,4); plot(f(1:(N-1)/2),xf_hn(1:(N-1)/2));grid;axis([0 fs/2 0 1]);
xlabel('Frequency (Hz)'); ylabel('Hanning windowed Ak');
```

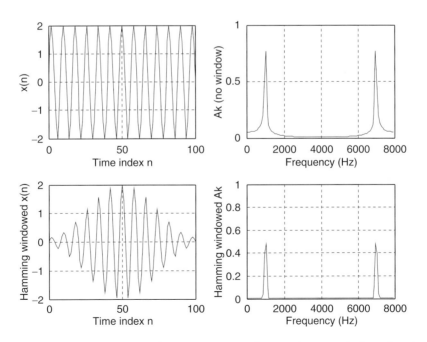

FIGURE 4.20 Comparison of a spectrum without using a window function and a spectrum using a Hamming window of size of 100 samples in Example 4.11.

120   **4   DISCRETE FOURIER TRANSFORM**

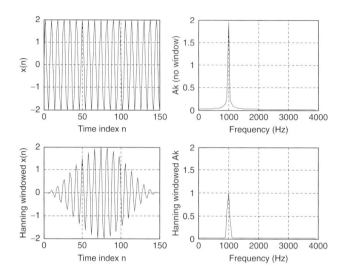

FIGURE 4.21   Comparison of a one-sided spectrum without using the window function and a one-sided spectrum using a Hanning window of size of 150 samples in Example 4.11.

4.4 Application to Speech Spectral Estimation

The following plots show the comparisons of amplitude spectral estimation for speech data (we.dat) with 2,001 samples and a sampling rate of 8,000 Hz using the rectangular window (no window) function and the Hamming window function. As demonstrated in Figure 4.22 (two-sided spectrum) and Figure 4.23 (one-sided spectrum), there is little difference between the amplitude spectrum using the Hamming window function and the spectrum without using the window function. This is due to the fact that when the data length of the sequence (e.g., 2,001 samples) increases, the frequency resolution will be improved and spectral leakage will become less significant. However, when data length is short, reduction of spectral leakage using a window function will come to be prominent.

4.5 Fast Fourier Transform

Now we study FFT in detail. FFT is a very efficient algorithm in computing DFT coefficients and can reduce a very large amount of computational complexity (multiplications). Without loss of generality, we consider the digital

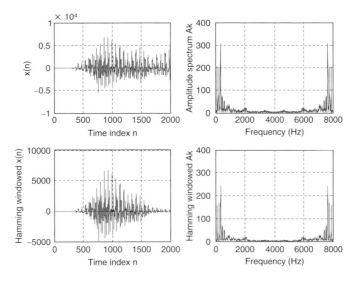

FIGURE 4.22 Comparison of a spectrum without using a window function and a spectrum using the Hamming window for speech data.

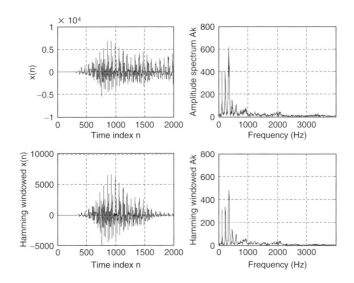

FIGURE 4.23 Comparison of a one-sided spectrum without using a window function and a one-sided spectrum using the Hamming window for speech data.

sequence $x(n)$ consisting of 2^m samples, where m is a positive integer—the number of samples of the digital sequence $x(n)$ is a power of 2, $N = 2, 4, 8, 16$, etc. If $x(n)$ does not contain 2^m samples, then we simply append it with zeros until the number of the appended sequence is equal to an integer of a power of 2 data points.

In this section, we focus on two formats. One is called the decimation-in-frequency algorithm, while the other is the decimation-in-time algorithm. They are referred to as the *radix*-2 FFT algorithms. Other types of FFT algorithms are the radix-4 and the split radix and their advantages can be exploited (see Proakis and Manolakis, 1996).

4.5.1 Method of Decimation-in-Frequency

We begin with the definition of DFT studied in the opening section of this chapter as follows:

$$X(k) = \sum_{n=0}^{N-1} x(n) W_N^{kn} \text{ for } k = 0, 1, \ldots, N - 1, \qquad (4.33)$$

where $W_N = e^{-j\frac{2\pi}{N}}$ is the twiddle factor, and $N = 2, 4, 8, 16, \ldots$ Equation (4.33) can be expanded as

$$X(k) = x(0) + x(1)W_N^k + \ldots + x(N-1)W_N^{k(N-1)}. \qquad (4.34)$$

Again, if we split Equation (4.34) into

$$X(k) = x(0) + x(1)W_N^k + \ldots + x\left(\frac{N}{2} - 1\right)W_N^{k(N/2-1)}$$

$$+ x\left(\frac{N}{2}\right)W^{kN/2} + \ldots + x(N-1)W_N^{k(N-1)} \qquad (4.35)$$

then we can rewrite as a sum of the following two parts

$$X(k) = \sum_{n=0}^{(N/2)-1} x(n) W_N^{kn} + \sum_{n=N/2}^{N-1} x(n) W_N^{kn}. \qquad (4.36)$$

Modifying the second term in Equation (4.36) yields

$$X(k) = \sum_{n=0}^{(N/2)-1} x(n) W_N^{kn} + W_N^{(N/2)k} \sum_{n=0}^{(N/2)-1} x\left(n + \frac{N}{2}\right) W_N^{kn}. \qquad (4.37)$$

Recall $W_N^{N/2} = e^{-j\frac{2\pi(N/2)}{N}} = e^{-j\pi} = -1$; then we have

$$X(k) = \sum_{n=0}^{(N/2)-1} \left(x(n) + (-1)^k x\left(n + \frac{N}{2}\right) \right) W_N^{kn}. \qquad (4.38)$$

Now letting $k = 2m$ as an even number achieves

$$X(2m) = \sum_{n=0}^{(N/2)-1} \left(x(n) + x\left(n + \frac{N}{2}\right) \right) W_N^{2mn}, \qquad (4.39)$$

while substituting $k = 2m + 1$ as an odd number yields

$$X(2m+1) = \sum_{n=0}^{(N/2)-1} \left(x(n) - x\left(n + \frac{N}{2}\right) \right) W_N^n W_N^{2mn}. \qquad (4.40)$$

Using the fact that $W_N^2 = e^{-j\frac{2\pi \times 2}{N}} = e^{-j\frac{2\pi}{(N/2)}} = W_{N/2}$, it follows that

$$X(2m) = \sum_{n=0}^{(N/2)-1} a(n) W_{N/2}^{mn} = DFT\{a(n) \text{ with } (N/2) \text{ points}\} \qquad (4.41)$$

$$X(2m+1) = \sum_{n=0}^{(N/2)-1} b(n) W_N^n W_{N/2}^{mn} = DFT\{b(n) W_N^n \text{ with } (N/2) \text{ points}\}, \qquad (4.42)$$

where $a(n)$ and $b(n)$ are introduced and expressed as

$$a(n) = x(n) + x\left(n + \frac{N}{2}\right), \text{ for } n = 0, 1 \ldots, \frac{N}{2} - 1 \qquad (4.43a)$$

$$b(n) = x(n) - x\left(n + \frac{N}{2}\right), \text{ for } n = 0, 1, \ldots, \frac{N}{2} - 1. \qquad (4.43b)$$

Equations (4.33), (4.41), and (4.42) can be summarized as

$$DFT\{x(n) \text{ with } N \text{ points}\} = \begin{cases} DFT\{a(n) \text{ with } (N/2) \text{ points}\} \\ DFT\{b(n) W_N^n \text{ with } (N/2) \text{ points}\} \end{cases} \qquad (4.44)$$

The computation process can be illustrated in Figure 4.24. As shown in this figure, there are three graphical operations, which are illustrated in Figure 4.25. If we continue the process described by Figure 4.24, we obtain the block diagrams shown in Figures 4.26 and 4.27.

Figure 4.27 illustrates the FFT computation for the eight-point DFT, where there are 12 complex multiplications. This is a big saving as compared with the eight-point DFT with 64 complex multiplications. For a data length of N, the number of complex multiplications for DFT and FFT, respectively, are determined by

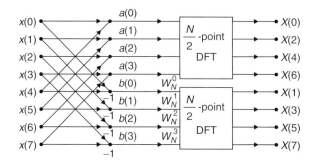

FIGURE 4.24 The first iteration of the eight-point FFT.

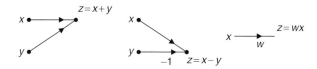

FIGURE 4.25 Definitions of the graphical operations.

$$\text{Complex multiplications of DFT} = N^2, \text{ and}$$
$$\text{Complex multiplications of FFT} = \frac{N}{2} \log_2(N).$$

To see the effectiveness of FFT, let us consider a sequence with 1,024 data points. Applying DFT will require $1{,}024 \times 1{,}024 = 1{,}048{,}576$ complex multiplications; however, applying FFT will need only $(1{,}024/2)\log_2(1{,}024) = 5{,}120$ complex multiplications. Next, the index (bin number) of the eight-point DFT coefficient $X(k)$ becomes 0, 4, 2, 6, 1, 5, 3, and 7, respectively, which are not in the natural order. This can be fixed by index matching. Index matching between the input sequence and the output frequency bin number by applying reversal bits is described in Table 4.2.

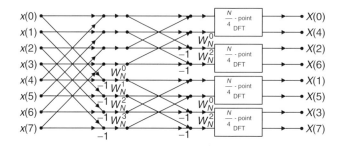

FIGURE 4.26 The second iteration of the eight-point FFT.

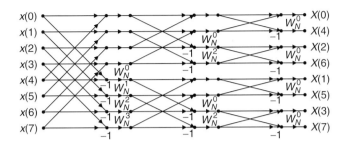

FIGURE 4.27 Block diagram for the eight-point FFT (total twelve multiplications).

TABLE 4.2 Index mapping for fast Fourier transform.

Input Data	Index Bits	Reversal Bits	Output Data
$x(0)$	000	000	$X(0)$
$x(1)$	001	100	$X(4)$
$x(2)$	010	010	$X(2)$
$x(3)$	011	110	$X(6)$
$x(4)$	100	001	$X(1)$
$x(5)$	101	101	$X(5)$
$x(6)$	110	011	$X(3)$
$x(7)$	111	111	$X(7)$

Figure 4.28 explains the bit reversal process. First, the input data with indices 0, 1, 2, 3, 4, 5, 6, 7 are split into two parts. The first half contains even indices—0, 2, 4, 6—while the second half contains odd indices. The first half with indices 0, 2, 4, 6 at the first iteration continues to be split into even indices 2, 4 and odd indices 4, 6, as shown in the second iteration. The second half with indices 1, 3, 5,

Binary	index	1st split	2nd split	3rd split	Bit reversal
000	0	0	0	0	000
001	1	2	4	4	100
010	2	4	2	2	010
011	3	6	6	6	011
100	4	1	1	1	001
101	5	3	5	5	101
110	6	5	3	3	011
111	7	7	7	7	111

FIGURE 4.28 Bit reversal process in FFT.

7 at the first iteration is split into even indices 1, 5 and odd indices 3, 7 in the second iteration. The splitting process continues to the end at the third iteration. The bit patterns of the output data indices are just the respective reversed bit patterns of the input data indices. Although Figure 4.28 illustrates the case of an eight-point FFT, this bit reversal process works as long as N is a power of 2.

The inverse FFT is defined as

$$x(n) = \frac{1}{N}\sum_{k=0}^{N-1} X(k)W_N^{-kn} = \frac{1}{N}\sum_{k=0}^{N-1} X(k)\tilde{W}_N^{kn}, \text{ for } k = 0, 1, \ldots, N-1. \quad (4.45)$$

Comparing Equation (4.45) with Equation (4.33), we notice the difference as follows: The twiddle factor W_N is changed to be $\tilde{W}_N = W_N^{-1}$, and the sum is multiplied by a factor of $1/N$. Hence, by modifying the FFT block diagram as shown in Figure 4.27, we achieve the inverse FFT block diagram shown in Figure 4.29.

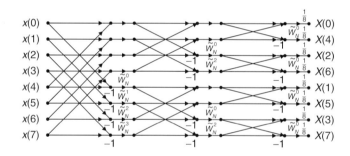

FIGURE 4.29 Block diagram for the inverse of eight-point FFT.

Example 4.12.

Given a sequence $x(n)$ for $0 \leq n \leq 3$, where $x(0) = 1$, $x(1) = 2$, $x(2) = 3$, and $x(3) = 4$,

a. Evaluate its DFT $X(k)$ using the decimation-in-frequency FFT method.

b. Determine the number of complex multiplications.

Solution:

a. Using the FFT block diagram in Figure 4.27, the result is shown in Figure 4.30.

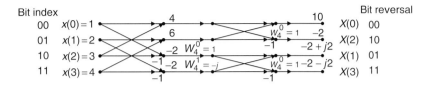

FIGURE 4.30 Four-point FFT block diagram in Example 4.12.

b. From Figure 4.30, the number of complex multiplications is four, which can also be determined by

$$\frac{N}{2} \log_2(N) = \frac{4}{2} \log_2(4) = 4.$$

Example 4.13.

Given the DFT sequence $X(k)$ for $0 \leq k \leq 3$ computed in Example 4.12,

a. Evaluate its inverse DFT $x(n)$ using the decimation-in-frequency FFT method.

Solution:

a. Using the inverse FFT block diagram in Figure 4.28, we have

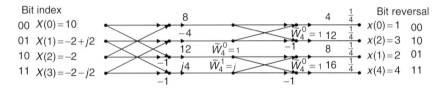

FIGURE 4.31 Four-point inverse FFT block diagram in Example 4.13.

4.5.2 Method of Decimation-in-Time

In this method, we split the input sequence $x(n)$ into the even indexed $x(2m)$ and $x(2m+1)$, each with N data points. Then Equation (4.33) becomes

$$X(k) = \sum_{m=0}^{(N/2)-1} x(2m) W_N^{2mk} + \sum_{m=0}^{(N/2)-1} x(2m+1) W_N^k W_N^{2mk},$$

for $k = 0, 1, \ldots, N - 1$. (4.46)

Using the relation $W_N^2 = W_{N/2}$, it follows that

$$X(k) = \sum_{m=0}^{(N/2)-1} x(2m) W_{N/2}^{mk} + W_N^k \sum_{m=0}^{(N/2)-1} x(2m+1) W_{N/2}^{mk}, \quad (4.47)$$

for $k = 0, 1, \ldots, N-1$.

Define new functions as

$$G(k) = \sum_{m=0}^{(N/2)-1} x(2m) W_{N/2}^{mk} = DFT\{x(2m) \text{ with } (N/2) \text{ points}\} \quad (4.48)$$

$$H(k) = \sum_{m=0}^{(N/2)-1} x(2m+1) W_{N/2}^{mk} = DFT\{x(2m+1) \text{ with } (N/2) \text{ points}\}. \quad (4.49)$$

Note that

$$G(k) = G\left(k + \frac{N}{2}\right), \text{ for } k = 0, 1, \ldots, \frac{N}{2} - 1 \quad (4.50)$$

$$H(k) = H\left(k + \frac{N}{2}\right), \text{ for } k = 0, 1, \ldots, \frac{N}{2} - 1. \quad (4.51)$$

Substituting Equations (4.50) and (4.51) into Equation (4.47) yields the first half frequency bins

$$X(k) = G(k) + W_N^k H(k), \text{ for } k = 0, 1, \ldots, \frac{N}{2} - 1. \quad (4.52)$$

Considering the following fact and using Equations (4.50) and (4.51),

$$W_N^{(N/2+k)} = -W_N^k. \quad (4.53)$$

Then the second half of frequency bins can be computed as follows:

$$X\left(\frac{N}{2} + k\right) = G(k) - W_N^k H(k), \text{ for } k = 0, 1, \ldots, \frac{N}{2} - 1. \quad (4.54)$$

If we perform backward iterations, we can obtain the FFT algorithm. Procedure using Equations (4.52) and (4.54) is illustrated in Figure 4.32, the block diagram for the eight-point FFT algorithm.

From a further iteration, we obtain Figure 4.33. Finally, after three recursions, we end up with the block diagram in Figure 4.34.

The index for each input sequence element can be achieved by bit reversal of the frequency index in a sequential order. Similar to the method of decimation-in-frequency, after we change W_N to $\tilde{W}_N$ in Figure 4.34 and multiply the output sequence by a factor of $1/N$, we derive the inverse FFT block diagram for the eight-point inverse FFT in Figure 4.35.

4.5 Fast Fourier Transform

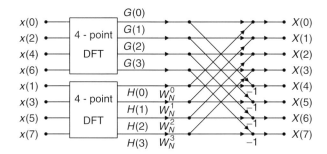

FIGURE 4.32 The first iteration.

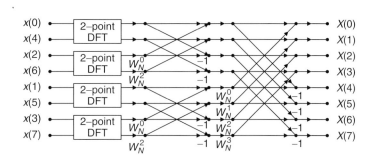

FIGURE 4.33 The second iteration.

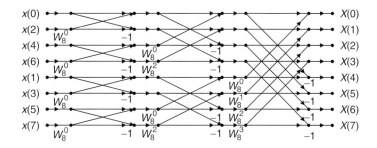

FIGURE 4.34 The eight-point FFT algorithm using decimation-in-time (twelve complex multiplications).

130   4 DISCRETE FOURIER TRANSFORM

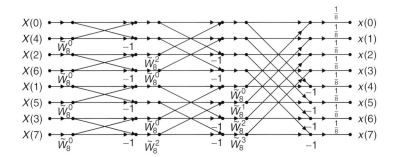

FIGURE 4.35   The eight-point IFFT using decimation-in-time.

Example 4.14.

Given a sequence $x(n)$ for $0 \leq n \leq 3$, where $x(0) = 1$, $x(1) = 2$, $x(2) = 3$, and $x(3) = 4$,

a. Evaluate its DFT $X(k)$ using the decimation-in-time FFT method.

Solution:

a. Using the block diagram in Figure 4.34 leads to

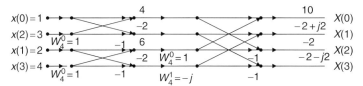

FIGURE 4.36   The four-point FFT using decimation in time.

Example 4.15.

Given the DFT sequence $X(k)$ for $0 \leq k \leq 3$ computed in Example 4.14,

a. Evaluate its inverse DFT $x(n)$ using the decimation-in-time FFT method.

Solution:

a. Using the block diagram in Figure 4.35 yields

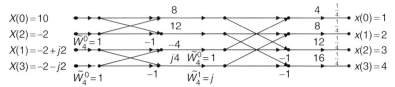

FIGURE 4.37   The four-point IFFT using decimation in time.

4.6 Summary

1. The Fourier series coefficients for a periodic digital signal can be used to develop the DFT.
2. The DFT transforms a time sequence to the complex DFT coefficients, while the inverse DFT transforms DFT coefficients back to the time sequence.
3. The *frequency bin number* is the same as the frequency index. *Frequency resolution* is the frequency spacing between two consecutive frequency indices (two consecutive spectrum components).
4. The DFT coefficients for a given digital signal are applied for computing the amplitude spectrum, power spectrum, or phase spectrum.
5. The spectrum calculated from all the DFT coefficients represents the signal frequency range from 0 Hz to the sampling rate. The spectrum beyond the folding frequency is equivalent to the negative-indexed spectrum from the negative folding frequency to 0 Hz. This two-sided spectrum can be converted into a one-sided spectrum by doubling alternating-current (AC) components from 0 Hz to the folding frequency and retaining the DC component as it is.
6. To reduce the burden of computing DFT coefficients, the FFT algorithm is used, which requires the data length to be a power of 2. Sometimes zero padding is employed to make up the data length. Zero padding actually does interpolation of the spectrum and does not carry any new information about the signal; even the calculated frequency resolution is smaller due to the zero padded longer length.
7. Applying the window function to the data sequence before DFT reduces the spectral leakage due to abrupt truncation of the data sequence when performing spectral calculation for a short sequence.
8. Two radix-2 FFT algorithms—decimation-in-frequency and decimation-in-time—are developed via the graphical illustrations.

4.7 Problems

4.1. Given a sequence $x(n)$ for $0 \leq n \leq 3$, where $x(0) = 1$, $x(1) = 1$, $x(2) = -1$, and $x(3) = 0$, compute its DFT $X(k)$.

4.2. Given a sequence $x(n)$ for $0 \leq n \leq 3$, where $x(0) = 4$, $x(1) = 3$, $x(2) = 2$, and $x(3) = 1$, evaluate its DFT $X(k)$.

4.3. Given the DFT sequence $X(k)$ for $0 \leq k \leq 3$ obtained in Problem 4.2, evaluate its inverse DFT $x(n)$.

4.4. Given a sequence $x(n)$, where $x(0) = 4$, $x(1) = 3$, $x(2) = 2$, and $x(3) = 1$ with the last two data zero-padded as $x(4) = 0$, and $x(5) = 0$, evaluate its DFT $X(k)$.

4.5. Using the DFT sequence $X(k)$ for $0 \leq k \leq 5$ computed in Problem 4.4, evaluate the inverse DFT $x(0)$ and $x(4)$.

4.6. Consider a digital sequence sampled at the rate of 20,000 Hz. If we use the 8,000-point DFT to compute the spectrum, determine

　a. the frequency resolution

　b. the folding frequency in the spectrum.

4.7. We use the DFT to compute the amplitude spectrum of a sampled data sequence with a sampling rate $f_s = 2{,}000$ Hz. It requires the frequency resolution to be less than 0.5 Hz. Determine the number of data points used by the FFT algorithm and actual frequency resolution in Hz, assuming that the data samples are available for selecting the number of data points.

4.8. Given the sequence in Figure 4.38

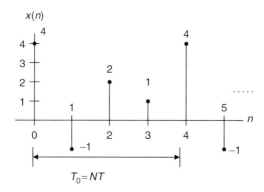

FIGURE 4.38 Data sequence in Problem 4.8.

and assuming that $f_s = 100$ Hz, compute the amplitude spectrum, phase spectrum, and power spectrum.

4.9. Compute the following window functions for a size of 8:

　a. Hamming window function.

　b. Hanning window function.

4.10. Given the following data sequence with a length of 6,

$x(0) = 0$, $x(1) = 1$, $x(2) = 0$, $x(3) = -1$, $x(4) = 0$, $x(5) = 1$

compute the windowed sequence $x_w(n)$ using the

a. triangular window function.

b. Hamming window function.

c. Hanning window function.

4.11. Given the sequence in Figure 4.39

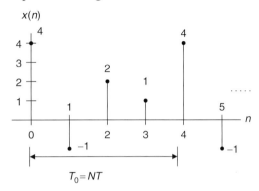

FIGURE 4.39 Data sequence in Problem 4.11.

where $f_s = 100$ Hz and $T = 0.01$ sec, compute the amplitude spectrum, phase spectrum, and power spectrum using the

a. triangular window.

b. Hamming window.

c. Hanning window.

4.12. Given the sinusoid

$$x(n) = 2 \cdot \sin\left(2000 \cdot 2\pi \cdot \frac{n}{8000}\right)$$

obtained by using the sampling rate at $f_s = 8{,}000$ Hz, we apply the DFT to compute the amplitude spectrum.

a. Determine the frequency resolution when the data length is 100 samples. Without using the window function, is there any spectral leakage in the computed spectrum? Explain.

b. Determine the frequency resolution when the data length is 73 samples. Without using the window function, is there any spectral leakage in the computed spectrum? Explain.

4.13. Given a sequence $x(n)$ for $0 \leq n \leq 3$, where $x(0) = 4$, $x(1) = 3$, $x(2) = 2$, and $x(3) = 1$, evaluate its DFT $X(k)$ using the decimation-in-frequency FFT method, and determine the number of complex multiplications.

4.14. Given the DFT sequence $X(k)$ for $0 \le k \le 3$ obtained in Problem 4.13, evaluate its inverse DFT $x(n)$ using the decimation-in-frequency FFT method.

4.15. Given a sequence $x(n)$ for $0 \le n \le 3$, where $x(0) = 4$, $x(1) = 3$, $x(2) = 2$, and $x(3) = 1$, evaluate its DFT $X(k)$ using the decimation-in-time FFT method, and determine the number of complex multiplications.

4.16. Given the DFT sequence $X(k)$ for $0 \le k \le 3$ computed in Problem 4.15, evaluate its inverse DFT $x(n)$ using the decimation-in-time FFT method.

4.17 Given three sinusoids with the following amplitude and phases:

$$x_1(t) = 5\cos(2\pi(500)t)$$
$$x_2(t) = 5\cos(2\pi(1200)t + 0.25\pi)$$
$$x_3(t) = 5\cos(2(1800)t + 0.5\pi)$$

a. Create a MATLAB program to sample each sinusoid and generate a sum of three sinusoids, that is, $x(n) = x_1(n) + x_2(n) + x_3(n)$, using a sampling rate of 8000 Hz, and plot the sum $x(n)$ over a range of time that will exhibit approximately 0.1 second.

b. Use the MATLAB function fft() to compute DFT coefficients, and plot and examine the spectrum of the signal $x(n)$.

4.18. Using the sum of sinusoids in Problem 4.17,

a. Generate the sum of sinusoids for 240 samples using a sampling rate of 8000 Hz.

b. Write a MATLAB program to compute and plot the amplitute spectrum of the signal $x(n)$ with the FFT and using each of the following window functions

(1) Rectangular window (no window)

(2) Triangular window

(3) Hamming window

c. Examine the effect of spectral leakage for each window used in (b).

References

Ahmed, N., and Natarajan, T. (1983). *Discrete-Time Signals and Systems*. Reston, VA: Reston Publishing Co.

Oppenheim, A. V., Schafer, R.W., and Buck, J. R. (1999). *Discrete-Time Signal Processing*, 2nd ed. Upper Saddle River, NJ: Prentice Hall.

Proakis, J. G., and Manolakis, D. G. (1996). *Digital Signal Processing: Principles, Algorithms, and Applications*, 3rd ed. Upper Saddle River, NJ: Prentice Hall.

Stearns, S. D., and Hush, D. R. (1990). *Digital Signal Analysis*, 2nd ed. Englewood Cliffs, NJ: Prentice Hall.

5

The z-Transform

Objectives:

This chapter introduces the z-transform and its properties; illustrates how to determine the inverse z-transform using partial fraction expansion; and applies the z-transform to solve linear difference equations.

5.1 Definition

The *z-transform* is a very important tool in describing and analyzing digital systems. It also offers the techniques for digital filter design and frequency analysis of digital signals. We begin with the definition of the z-transform.

The z-transform of a causal sequence $x(n)$, designated by $X(z)$ or $Z(x(n))$, is defined as

$$X(z) = Z(x(n)) = \sum_{n=0}^{\infty} x(n)z^{-n} \qquad (5.1)$$
$$= x(0)z^{-0} + x(1)z^{-1} + x(2)z^{-2} + \ldots$$

where z is the complex variable. Here, the summation taken from $n = 0$ to $n = \infty$ is according to the fact that for most situations, the digital signal $x(n)$ is the causal sequence, that is, $x(n) = 0$ for $n < 0$. Thus, the definition in Equation (5.1) is referred to as a *one-sided z-transform* or a *unilateral transform*. In Equation (5.1), all the values of z that make the summation to exist form a *region of convergence* in the z-transform domain, while all other values of z outside the region of convergence will cause the summation to diverge. The region of convergence is defined based on the particular sequence $x(n)$ being applied. Note that we deal with the unilateral z-transform in this book, and hence when performing inverse z-transform (which we shall study later), we are restricted to the causal sequence. Now let us study the following typical examples.

Example 5.1.

Given the sequence

$$x(n) = u(n),$$

a. Find the z-transform of $x(n)$.

Solution:

a. From the definition of Equation (5.1), the z-transform is given by

$$X(z) = \sum_{n=0}^{\infty} u(n) z^{-n} = \sum_{n=0}^{\infty} \left(z^{-1}\right)^n = 1 + \left(z^{-1}\right) + \left(z^{-1}\right)^2 + \ldots.$$

This is an infinite geometric series that converges to

$$X(z) = \frac{z}{z-1}$$

with a condition $\left|z^{-1}\right| < 1$. Note that for an infinite geometric series, we have $1 + r + r^2 + \ldots = \frac{1}{1-r}$ when $|r| < 1$. The region of convergence for all values of z is given as $|z| > 1$.

Example 5.2.

Considering the exponential sequence

$$x(n) = a^n u(n),$$

a. Find the z-transform of the sequence $x(n)$.

Solution:

a. From the definition of the z-transform in Equation (5.1), it follows that

$$X(z) = \sum_{n=0}^{\infty} a^n u(n) z^{-n} = \sum_{n=0}^{\infty} \left(az^{-1}\right)^n = 1 + \left(az^{-1}\right) + \left(az^{-1}\right)^2 + \ldots.$$

Since this is a geometric series which will converge for $\left|az^{-1}\right| < 1$, it is further expressed as

$$X(z) = \frac{z}{z-a}, \text{ for } |z| > |a|.$$

The z-transforms for common sequences are summarized in Table 5.1. Example 5.3 illustrates finding the z-transform using Table 5.1.

TABLE 5.1 **Table of z-transform pairs.**

Line No.	$x(n)$, $n \geq 0$	z-Transform $X(z)$	Region of Convergence
1	$x(n)$	$\sum_{n=0}^{\infty} x(n) z^{-n}$	
2	$\delta(n)$	1	$\|z\| > 0$
3	$au(n)$	$\dfrac{az}{z-1}$	$\|z\| > 1$
4	$nu(n)$	$\dfrac{z}{(z-1)^2}$	$\|z\| > 1$
5	$n^2 u(n)$	$\dfrac{z(z+1)}{(z-1)^3}$	$\|z\| > 1$
6	$a^n u(n)$	$\dfrac{z}{z-a}$	$\|z\| > \|a\|$
7	$e^{-na} u(n)$	$\dfrac{z}{(z-e^{-a})}$	$\|z\| > e^{-a}$
8	$na^n u(n)$	$\dfrac{az}{(z-a)^2}$	$\|z\| > \|a\|$
9	$\sin(an) u(n)$	$\dfrac{z \sin(a)}{z^2 - 2z \cos(a) + 1}$	$\|z\| > 1$
10	$\cos(an) u(n)$	$\dfrac{z[z - \cos(a)]}{z^2 - 2z \cos(a) + 1}$	$\|z\| > 1$
11	$a^n \sin(bn) u(n)$	$\dfrac{[a \sin(b)] z}{z^2 - [2a \cos(b)] z + a^2}$	$\|z\| > \|a\|$
12	$a^n \cos(bn) u(n)$	$\dfrac{z[z - a \cos(b)]}{z^2 - [2a \cos(b)] z + a^{-2}}$	$\|z\| > \|a\|$
13	$e^{-an} \sin(bn) u(n)$	$\dfrac{[e^{-a} \sin(b)] z}{z^2 - [2e^{-a} \cos(b)] z + e^{-2a}}$	$\|z\| > e^{-a}$
14	$e^{-an} \cos(bn) u(n)$	$\dfrac{z[z - e^{-a} \cos(b)]}{z^2 - [2e^{-a} \cos(b)] z + e^{-2a}}$	$\|z\| > e^{-a}$
15	$2\|A\|\|P\|^n \cos(n\theta + \phi) u(n)$ where P and A are complex constants defined by $P = \|P\| \angle \theta, A = \|A\| \angle \phi$	$\dfrac{Az}{z - P} + \dfrac{A^* z}{z - P^*}$	

Example 5.3.

Find the z-transform for each of the following sequences:

a. $x(n) = 10u(n)$

b. $x(n) = 10\sin(0.25\pi n)u(n)$

c. $x(n) = (0.5)^n u(n)$

d. $x(n) = (0.5)^n \sin(0.25\pi n)u(n)$

e. $x(n) = e^{-0.1n}\cos(0.25\pi n)u(n)$

Solution:

a. From Line 3 in Table 5.1, we get

$$X(z) = Z(10u(n)) = \frac{10z}{z-1}.$$

b. Line 9 in Table 5.1 leads to

$$X(z) = 10Z(\sin(0.2\pi n)u(n))$$

$$= \frac{10\sin(0.25\pi)z}{z^2 - 2z\cos(0.25\pi) + 1} = \frac{7.07z}{z^2 - 1.414z + 1}.$$

c. From Line 6 in Table 5.1, we yield

$$X(z) = Z((0.5)^n u(n)) = \frac{z}{z - 0.5}.$$

d. From Line 11 in Table 5.1, it follows that

$$X(z) = Z((0.5)^n \sin(0.25\pi n)u(n)) = \frac{0.5 \times \sin(0.25\pi)z}{z^2 - 2 \times 0.5\cos(0.25\pi)z + 0.5^2}$$

$$= \frac{0.3536z}{z^2 - 1.4142z + 0.25}.$$

e. From Line 14 in Table 5.1, it follows that

$$X(z) = Z(e^{-0.1n}\cos(0.25\pi n)u(n)) = \frac{z(z - e^{-0.1}\cos(0.25\pi))}{z^2 - 2e^{-0.1}\cos(0.25\pi)z + e^{-0.2}}$$

$$= \frac{z(z - 0.6397)}{z^2 - 1.2794z + 0.8187}.$$

5.2 Properties of the z-Transform

In this section, we study some important properties of the z-transform. These properties are widely used in deriving the z-transform functions of difference equations and solving the system output responses of linear digital systems with constant system coefficients, which will be discussed in the next chapter.

Linearity: The z-transform is a linear transformation, which implies

$$Z(ax_1(n) + bx_2(n)) = aZ(x_1(n)) + bZ(x_2(n)), \qquad (5.2)$$

where $x_1(n)$ and $x_2(n)$ denote the sampled sequences, while a and b are the arbitrary constants.

Example 5.4.

a. Find the z-transform of the sequence defined by

$$x(n) = u(n) - (0.5)^n u(n).$$

Solution:

a. Applying the linearity of the z-transform previously discussed, we have

$$X(z) = Z(x(n)) = Z(u(n)) - Z(0.5^n(n)).$$

Using Table 5.1 yields

$$Z(u(n)) = \frac{z}{z-1}$$

and $Z(0.5^n u(n)) = \dfrac{z}{z - 0.5}.$

Substituting these results into $X(z)$ leads to the final solution,

$$X(z) = \frac{z}{z-1} - \frac{z}{z-0.5}.$$

Shift theorem: Given $X(z)$, the z-transform of a sequence $x(n)$, the z-transform of $x(n-m)$, the time-shifted sequence, is given by

$$Z(x(n-m)) = z^{-m} X(z). \qquad (5.3)$$

Note that if $m \geq 0$, then $x(n-m)$ is obtained by right shifting $x(n)$ by m samples. Since the shift theorem plays a very important role in developing the transfer function from a difference equation, we verify the shift theorem for the causal sequence. Note that the shift thoerem also works for the noncausal sequence.

Verification: Applying the z-transform to the shifted causal signal $x(n-m)$ leads to

$$Z(x(n-m)) = \sum_{n=0}^{\infty} x(n-m)z^{-n}$$
$$= x(-m)z^{-0} + \ldots + x(-1)z^{-(m-1)} + x(0)z^{-m} + x(1)z^{-m-1} + \ldots$$

Since $x(n)$ is assumed to be a causal sequence, this means that

$$x(-m) = x(-m+1) = \ldots = x(-1) = 0.$$

Then we achieve

$$Z(x(n-m)) = x(0)z^{-m} + x(1)z^{-m-1} + x(2)z^{-m-2} + \ldots \quad (5.4)$$

Factoring z^{-m} from Equation (5.4) and applying the definition of z-transform of $X(z)$, we get

$$Z(x(n-m)) = z^{-m}\left(x(0) + x(1)z^{-1} + x(2)z^{-2} + \ldots\right) = z^{-m}X(z).$$

Example 5.5.

a. Determine the z-transform of the following sequence:

$$y(n) = (0.5)^{(n-5)} \cdot u(n-5),$$

where $u(n-5) = 1$ for $n \geq 5$ and $u(n-5) = 0$ for $n < 5$.

Solution:

a. We first use the shift theorem to have

$$Y(z) = Z\left[(0.5)^{n-5}u(n-5)\right] = z^{-5}Z[(0.5)^n u(n)].$$

Using Table 5.1 leads to

$$Y(z) = z^{-5} \cdot \frac{z}{z-0.5} = \frac{z^{-4}}{z-0.5}.$$

Convolution: Given two sequences $x_1(n)$ and $x_2(n)$, their convolution can be determined as follows:

$$x(n) = x_1(n) * x_2(n) = \sum_{k=0}^{\infty} x_1(n-k)x_2(k), \quad (5.5)$$

where $*$ designates the linear convolution. In z-transform domain, we have

$$X(z) = X_1(z)X_2(z). \quad (5.6)$$

Here, $X(z) = Z(x(n))$, $X_1(z) = Z(x_1(n))$, and $X_2(z) = Z(x_2(n))$.

Example 5.6.

a. Verify Equation (5.5) using causal sequences $x_1(n)$ and $x_2(n)$.

Solution:

a. Taking the z-transform of Equation (5.5) leads to

$$X(z) = \sum_{n=0}^{\infty} x(n)z^{-n} = \sum_{n=0}^{\infty}\sum_{k=0}^{\infty} x_1(n-k)x_2(k)z^{-n}.$$

This expression can be further modified to

$$X(z) = \sum_{n=0}^{\infty}\sum_{k=0}^{\infty} x_2(k)z^{-k}x_1(n-k)z^{-(n-k)}.$$

Now interchanging the order of the previous summation gives

$$X(z) = \sum_{k=0}^{\infty} x_2(k)z^{-k} \sum_{n=0}^{\infty} x_1(n-k)z^{-(n-k)}.$$

Now, let $m = n - k$:

$$X(z) = \sum_{k=0}^{\infty} x_2(k)z^{-k} \sum_{m=0}^{\infty} x_1(m)z^{-m}.$$

By the definition of Equation (5.1), it follows that

$$X(z) = X_2(z)X_1(z) = X_1(z)X_2(z).$$

Example 5.7.

Given two sequences,

$$x_1(n) = 3\delta(n) + 2\delta(n-1)$$
$$x_2(n) = 2\delta(n) - \delta(n-1),$$

a. Find the z-transform of their convolution:

$$X(z) = Z(x_1(n)*x_2(n)).$$

b. Determine the convolution sum using the z-transform:

$$x(n) = x_1(n)*x_2(n) = \sum_{k=0}^{\infty} x_1(k)x_2(n-k).$$

Solution:

a. Applying z-transform to $x_1(n)$ and $x_2(n)$, respectively, it follows that

$$X_1(z) = 3 + 2z^{-1}$$
$$X_2(z) = 2 - z^{-1}.$$

Using the convolution property, we have

$$X(z) = X_1(z)X_2(z) = (3 + 2z^{-1})(2 - z^{-1})$$
$$= 6 + z^{-1} - 2z^{-2}.$$

b. Applying the inverse z-transform and using the shift theorem and line 1 of Table 5.1 leads to

$$x(n) = Z^{-1}\left(6 + z^{-1} - 2z^{-2}\right) = 6\delta(n) + \delta(n-1) - 2\delta(n-2).$$

The properties of the z-transform discussed in this section are listed in Table 5.2.

5.3 Inverse z-Transform

The z-transform of the sequence $x(n)$ and the inverse z-transform of the function $X(z)$ are defined as, respectively,

$$X(z) = Z(x(n)) \tag{5.7}$$
$$\text{and } x(n) = Z^{-1}(X(z)), \tag{5.8}$$

where $Z(\ )$ is the z-transform operator, while $Z^{-1}(\ )$ is the inverse z-transform operator.

The inverse z-transform may be obtained by at least three methods:

1. Partial fraction expansion and look-up table
2. Power series expansion
3. Residue method.

TABLE 5.2 **Properties of z-transform.**

Property	Time Domain	z-Transform
Linearity	$ax_1(n) + bx_2(n)$	$aZ(x_1(n)) + bZ(x_2(n))$
Shift theorem	$x(n-m)$	$z^{-m}X(z)$
Linear convolution	$x_1(n)*x_2(n) = \sum_{k=0}^{\infty} x_1(n-k)x_2(k)$	$X_1(z)X_2(z)$

The first method is widely utilized, and it is assumed that the reader is well familiar with the partial fraction expansion method in learning Laplace transform. Therefore, we concentrate on the first method in this book. As for the power series expansion and residue methods, the interested reader is referred to the textbook by Oppenheim and Schafer (1975). The key idea of the partial fraction expansion is that if $X(z)$ is a proper rational function of z, we can expand it to a sum of the first-order factors or higher-order factors using the partial fraction expansion that could be inverted by inspecting the z-transform table. The partial fraction expansion method is illustrated via the following examples. (For simple z-transform functions, we can directly find the inverse z-transform using Table 5.1.)

Example 5.8.

Find the inverse z-transform for each of the following functions:

a. $X(z) = 2 + \dfrac{4z}{z-1} - \dfrac{z}{z-0.5}$

b. $X(z) = \dfrac{5z}{(z-1)^2} - \dfrac{2z}{(z-0.5)^2}$

c. $X(z) = \dfrac{10z}{z^2 - z + 1}$

d. $X(z) = \dfrac{z^{-4}}{z-1} + z^{-6} + \dfrac{z^{-3}}{z+0.5}$

Solution:

a. $x(n) = 2Z^{-1}(1) + 4Z^{-1}\left(\dfrac{z}{z-1}\right) - Z^{-1}\left(\dfrac{z}{z-0.5}\right).$

From Table 5.1, we have

$x(n) = 2\delta(n) + 4u(n) - (0.5)^n u(n).$

b. $x(n) = Z^{-1}\left(\dfrac{5z}{(z-1)^2}\right) - Z^{-1}\left(\dfrac{2z}{(z-0.5)^2}\right) = 5Z^{-1}\left(\dfrac{z}{(z-1)^2}\right) - \dfrac{2}{0.5}Z^{-1}\left(\dfrac{0.5z}{(z-0.5)^2}\right).$

Then $x(n) = 5nu(n) - 4n(0.5)^n u(n).$

c. Since $X(z) = \dfrac{10z}{z^2 - z + 1} = \left(\dfrac{10}{\sin(a)}\right)\dfrac{\sin(a)z}{z^2 - 2z\cos(a) + 1},$

by coefficient matching, we have

$$-2\cos(a) = -1.$$

Hence, $\cos(a) = 0.5$, and $a = 60°$. Substituting $a = 60°$ into the sine function leads to

$$\sin(a) = \sin(60°) = 0.866.$$

Finally, we have

$$x(n) = \frac{10}{\sin(a)} Z^{-1}\left(\frac{\sin(a)z}{z^2 - 2z\cos(a) + 1}\right) = \frac{10}{0.866}\sin(n \cdot 60°)$$

$$= 11.547 \sin(n \cdot 60°).$$

d. Since

$$x(n) = Z^{-1}\left(z^{-5}\frac{z}{z-1}\right) + Z^{-1}(z^{-6} \cdot 1) + Z^{-1}\left(z^{-4}\frac{z}{z+0.5}\right),$$

using Table 5.1 and the shift property, we get

$$x(n) = u(n-5) + \delta(n-6) + (-0.5)^{n-4}u(n-4).$$

Now, we are ready to deal with the inverse z-transform using the partial fraction expansion and look-up table. The general procedure is as follows:

1. Eliminate the negative powers of z for the z-transform function $X(z)$.

2. Determine the rational function $X(z)/z$ (assuming it is proper), and apply the partial fraction expansion to the determined rational function $X(z)/z$ using the formula in Table 5.3.

3. Multiply the expanded function $X(z)/z$ by z on both sides of the equation to obtain $X(z)$.

4. Apply the inverse z-transform using Table 5.1.

The partial fraction format and the formula for calculating the constants are listed in Table 5.3.

TABLE 5.3   Partial fraction(s) and formulas for constant(s).

Partial fraction with the first-order real pole:

$$\frac{R}{z-p} \qquad\qquad R = (z-p)\frac{X(z)}{z}\bigg|_{z=p}$$

Partial fraction with the first-order complex poles:

$$\frac{Az}{(z-P)} + \frac{A^*z}{(z-P^*)} \qquad\qquad A = (z-P)\frac{X(z)}{z}\bigg|_{z=P}$$

P^* = complex conjugate of P
A^* = complex conjugate of A

Partial fraction with mth-order real poles:

$$\frac{R_m}{(z-p)} + \frac{R_{m-1}}{(z-p)^2} + \cdots + \frac{R_1}{(z-p)^m} \qquad R_k = \frac{1}{(k-1)!}\frac{d^{k-1}}{dz^{k-1}}\left((z-p)^m\frac{X(z)}{z}\right)\bigg|_{z=p}$$

Example 5.9 considers the situation of the z-transform function having first-order poles.

Example 5.9.

a. Find the inverse of the following z-transform:
$$X(z) = \frac{1}{(1 - z^{-1})(1 - 0.5z^{-1})}.$$

Solution:

a. Eliminating the negative power of z by multiplying the numerator and denominator by z^2 yields

$$X(z) = \frac{z^2}{z^2(1 - z^{-1})(1 - 0.5z^{-1})}.$$

$$= \frac{z^2}{(z - 1)(z - 0.5)}.$$

Dividing both sides by z leads to

$$\frac{X(z)}{z} = \frac{z}{(z - 1)(z - 0.5)}.$$

Again, we write

$$\frac{X(z)}{z} = \frac{A}{(z - 1)} + \frac{B}{(z - 0.5)}.$$

Then A and B are constants found using the formula in Table 5.3, that is,

$$A = (z - 1)\frac{X(z)}{z}\bigg|_{z=1} = \frac{z}{(z - 0.5)}\bigg|_{z=1} = 2,$$

$$B = (z - 0.5)\frac{X(z)}{z}\bigg|_{z=0.5} = \frac{z}{(z - 1)}\bigg|_{z=0.5} = -1.$$

Thus

$$\frac{X(z)}{z} = \frac{2}{(z - 1)} + \frac{-1}{(z - 0.5)}.$$

Multiplying z on both sides gives

$$X(z) = \frac{2z}{(z - 1)} + \frac{-z}{(z - 0.5)}.$$

Using Table 5.1 of the z-transform pairs, it follows that

TABLE 5.4   Determined sequence in Example 5.9.

n	0	1	2	3	4	...	∞
$x(n)$	1.0	1.5	1.75	1.875	1.9375	...	2.0

$$x(n) = 2u(n) - (0.5)^n u(n).$$

Tabulating this solution in terms of integer values of n, we obtain the results in Table 5.4.

The following example considers the case where $X(z)$ has first-order complex poles.

Example 5.10.

a. Find $y(n)$ if $Y(z) = \dfrac{z^2(z+1)}{(z-1)(z^2-z+0.5)}$.

Solution:

a. Dividing $Y(z)$ by z, we have

$$\frac{Y(z)}{z} = \frac{z(z+1)}{(z-1)(z^2-z+0.5)}.$$

Applying the partial fraction expansion leads to

$$\frac{Y(z)}{z} = \frac{B}{z-1} + \frac{A}{(z-0.5-j0.5)} + \frac{A^*}{(z-0.5+j0.5)}.$$

We first find B:

$$B = (z-1)\frac{Y(z)}{z}\bigg|_{z=1} = \frac{z(z+1)}{(z^2-z+0.5)}\bigg|_{z=1} = \frac{1 \times (1+1)}{(1^2-1+0.5)} = 4.$$

Notice that A and A^* form a complex conjugate pair. We determine A as follows:

$$A = (z-0.5-j0.5)\frac{Y(z)}{z}\bigg|_{z=0.5+j0.5} = \frac{z(z+1)}{(z-1)(z-0.5+j0.5)}\bigg|_{z=0.5+j0.5}$$

$$A = \frac{(0.5+j0.5)(0.5+j0.5+1)}{(0.5+j0.5-1)(0.5+j0.5-0.5+j0.5)} = \frac{(0.5+j0.5)(1.5+j0.5)}{(-0.5+j0.5)j}.$$

Using the polar form, we get

$$A = \frac{(0.707\angle 45°)(1.58114\angle 18.43°)}{(0.707\angle 135°)(1\angle 90°)} = 1.58114\angle -161.57°$$

$$A^* = \bar{A} = 1.58114\angle 161.57°.$$

Assume that a first-order complex pole has the form
$P = 0.5 + 0.5j = |P|\angle\theta = 0.707\angle 45°$ and $P^* = |P|\angle-\theta = 0.707\angle-45°$.
We have

$$Y(z) = \frac{4z}{z-1} + \frac{Az}{(z-P)} + \frac{A^*z}{(z-P^*)}.$$

Applying the inverse z-transform from line 15 in Table 5.1 leads to

$$y(n) = 4Z^{-1}\left(\frac{z}{z-1}\right) + Z^{-1}\left(\frac{Az}{(z-P)} + \frac{A^*z}{(z-P^*)}\right).$$

Using the previous formula, the inversion and subsequent simplification yield

$$y(n) = 4u(n) + 2|A|(|P|)^n \cos(n\theta + \phi)u(n)$$
$$= 4u(n) + 3.1623(0.7071)^n \cos(45°n - 161.57°)u(n).$$

The situation dealing with the real repeated poles is presented in Example 5.11.

Example 5.11.

a. Find $x(n)$ if $X(z) = \dfrac{z^2}{(z-1)(z-0.5)^2}$.

Solution:

a. Dividing both sides of the previous z-transform by z yields

$$\frac{X(z)}{z} = \frac{z}{(z-1)(z-0.5)^2} = \frac{A}{z-1} + \frac{B}{z-0.5} + \frac{C}{(z-0.5)^2},$$

where $A = (z-1)\dfrac{X(z)}{z}\Big|_{z=1} = \dfrac{z}{(z-0.5)^2}\Big|_{z=1} = 4$.

Using the formulas for mth-order real poles in Table 5.3, where $m = 2$ and $p = 0.5$, to determine B and C yields

$$B = R_2 = \frac{1}{(2-1)!}\frac{d}{dz}\left\{(z-0.5)^2\frac{X(z)}{z}\right\}_{z=0.5}$$

$$= \frac{d}{dz}\left(\frac{z}{z-1}\right)\Big|_{z=0.5} = \frac{-1}{(z-1)^2}\Big|_{z=0.5} = -4$$

$$C = R_1 = \frac{1}{(1-1)!} \frac{d^0}{dz^0}\left\{(z-0.5)^2 \frac{X(z)}{z}\right\}_{z=0.5}$$

$$= \left.\frac{z}{z-1}\right|_{z=0.5} = -1.$$

Then $X(z) = \dfrac{4z}{z-1} + \dfrac{-4z}{z-0.5} + \dfrac{-1z}{(z-0.5)^2}.$ (5.9)

The inverse z-transform for each term on the right-hand side of Equation (5.9) can be achieved by the result listed in Table 5.1, that is,

$$Z^{-1}\left\{\frac{z}{z-1}\right\} = u(n),$$

$$Z^{-1}\left\{\frac{z}{z-0.5}\right\} = (0.5)^n u(n),$$

$$Z^{-1}\left\{\frac{z}{(z-0.5)^2}\right\} = 2n(0.5)^n u(n).$$

From these results, it follows that

$$x(n) = 4u(n) - 4(0.5)^n u(n) - 2n(0.5)^n u(n).$$

5.3.1 Partial Fraction Expansion Using MATLAB

The MATLAB function **residue( )** can be applied to perform the partial fraction expansion of a z-transform function $X(z)/z$. The syntax is given as

[R,P,K] = residue(B,A).

Here, B and A are the vectors consisting of coefficients for the numerator and denominator polynomials, $B(z)$ and $A(z)$, respectively. Notice that $B(z)$ and $A(z)$ are the polynomials with increasing positive powers of z.

$$\frac{B(z)}{A(z)} = \frac{b_0 z^M + b_1 z^{M-1} + b_2 z^{M-2} + \ldots + b_M}{z^N + a_1 z^{N-1} + a_2 z^{-2} + \ldots + a_N}.$$

The function returns the residues in vector R, corresponding poles in vector P, and polynomial coefficients (if any) in vector K. The expansion format is shown as

$$\frac{B(z)}{A(z)} = \frac{r_1}{z - p_1} + \frac{r_2}{z - p_2} + \ldots + k_0 + k_1 z^{-1} + \ldots.$$

For a pole p_j of multiplicity m, the partial fraction includes the following terms:

$$\frac{B(z)}{A(z)} = \ldots + \frac{r_j}{z - p_j} + \frac{r_{j+1}}{(z - p_j)^2} + \ldots + \frac{r_{j+m}}{(z - p_j)^m} + \ldots + k_0 + k_1 z^{-1} + \ldots.$$

Example 5.12.

Find the partial expansion for each of the following z-transform functions:

a. $X(z) = \dfrac{1}{(1 - z^{-1})(1 - 0.5 z^{-1})}$

b. $Y(z) = \dfrac{z^2(z + 1)}{(z - 1)(z^2 - z + 0.5)}$

c. $X(z) = \dfrac{z^2}{(z - 1)(z - 0.5)^2}$

Solution:

a. From MATLAB, we can show the denominator polynomial as
   ≫ **conv([1 −1], [1 −0.5])**
   D =
   1.0000 −1.5000 0.5000

   This leads to

   $$X(z) = \frac{1}{(1 - z^{-1})(1 - 0.5 z^{-1})} = \frac{1}{1 - 1.5 z^{-1} + 0.5^{-2}} = \frac{z^2}{z^2 - 1.5 z + 0.5}$$

   and $\dfrac{X(z)}{z} = \dfrac{z}{z^2 - 1.5 z + 0.5}$.

   From MATLAB, we have
   ≫ **[R,P,K] = residue([1 0], [1 −1.5 0.5])**
   R =
      2
      −1
   P =
      1.0000
      0.5000
   K =
      [ ]
   ≫

Then the expansion is written as

$$X(z) = \frac{2z}{z-1} - \frac{z}{z-0.5}.$$

b. From the MATLAB

≫ **N = conv([1 0 0], [1 1])**
N =
  1 1 0 0
≫ **D = conv([1 −1], [1 −1 0.5])**
D =
  1.0000 −2.0000 1.5000 −0.5000

we get

$$Y(z) = \frac{z^2(z+1)}{(z-1)(z^2-z+0.5)} = \frac{z^3+z^2}{z^3-2z^2+1.5z-0.5}$$

and $\dfrac{Y(z)}{z} = \dfrac{z^2+z}{z^3-2z^2+1.5z-0.5}.$

Using the MATLAB residue function yields

≫ **[R,P,K] = residue([1 1 0], [1 −2 1.5 −0.5])**
R =
  4.0000
  −1.5000 − 0.5000i
  −1.5000 + 0.5000i
P =
  1.0000
  0.5000 + 0.5000i
  0.5000 − 0.5000i
K =
  []
≫

Then the expansion is shown as:

$$X(z) = \frac{Bz}{z-p_1} + \frac{Az}{z-p} + \frac{A^*z}{z-p^*},$$

where $B = 4$,

$p_1 = 1$,
$A = -1.5 - 0.5j$,
$p = 0.5 + 0.5j$,
$A^* = -1.5 + 0.5j$, and
$p = 0.5 - 0.5j$.

c. Similarly,

≫ D = conv(conv([1 − 1], [1 − 0.5]), [1 − 0.5])
D =
  1.0000 − 2.0000 1.2500 − 0.2500

then $X(z) = \dfrac{z^2}{(z-1)(z-0.5)^2} = \dfrac{z^2}{z^3 - 2z^2 + 1.25z - 0.25}$ and

we yield $\dfrac{X(z)}{z} = \dfrac{z}{z^3 - 2z^2 + 1.25z - 0.25}$.

From MATLAB, we obtain
≫ [R,P,K] = residue([1 0], [1 − 2 1.25 − 0.25])
R =
  4.0000
  −4.0000
  −1.0000
P =
  1.0000
  0.5000
  0.5000
K =
  []
≫

Using the previous results leads to

$$X(z) = \dfrac{4z}{z-1} - \dfrac{4z}{z-0.5} - \dfrac{z}{(z-0.5)^2}.$$

5.4 Solution of Difference Equations Using the z-Transform

To solve a difference equation with initial conditions, we have to deal with time-shifted sequences such as $y(n-1)$, $y(n-2)$, ..., $y(n-m)$, and so on. Let us examine the z-transform of these terms. Using the definition of the z-transform, we have

$$Z(y(n-1)) = \sum_{n=0}^{\infty} y(n-1)z^{-n}$$
$$= y(-1) + y(0)z^{-1} + y(1)z^{-2} + \ldots$$
$$= y(-1) + z^{-1}\left(y(0) + y(1)z^{-1} + y(2)z^{-2} + \ldots\right)$$

It holds that

$$Z(y(n-1)) = y(-1) + z^{-1}Y(z). \tag{5.10}$$

Similarly, we can have

$$Z(y(n-2)) = \sum_{n=0}^{\infty} y(n-2)z^{-n}$$

$$= y(-2) + y(-1)z^{-1} + y(0)z^{-2} + y(1)z^{-3} + \ldots$$

$$= y(-2) + y(-1)z^{-1} + z^{-2}(y(0) + y(1)z^{-1} + y(2)z^{-2} + \ldots)$$

$$Z(y(n-2)) = y(-2) + y(-1)z^{-1} + z^{-2}Y(z) \tag{5.11}$$

$$Z(y(n-m)) = y(-m) + y(-m+1)z^{-1} + \ldots + y(-1)z^{-(m-1)}$$
$$+ z^{-m}Y(z), \tag{5.12}$$

where $y(-m), y(-m+1), \ldots, y(-1)$ are the initial conditions. If all initial conditions are considered to be zero, that is,

$$y(-m) = y(-m+1) = \ldots y(-1) = 0, \tag{5.13}$$

then Equation (5.12) becomes

$$Z(y(n-m)) = z^{-m}Y(z), \tag{5.14}$$

which is the same as the shift theorem in Equation (5.3).

The following two examples serve as illustrations of applying the z-transform to find the solutions of the difference equations. The procedure is:

1. Apply z-transform to the difference equation.
2. Substitute the initial conditions.
3. Solve for the difference equation in z-transform domain.
4. Find the solution in time domain by applying the inverse z-transform.

Example 5.13.

A digital signal processing (DSP) system is described by the difference equation

$$y(n) - 0.5y(n-1) = 5(0.2)^n u(n).$$

    a. Determine the solution when the initial condition is given by $y(-1) = 1$.

Solution:

    a. Applying the z-transform on both sides of the difference equation and using Equation (5.12), we have

$$Y(z) - 0.5(y(-1) + z^{-1}Y(z)) = 5Z(0.2^n u(n)).$$

Substituting the initial condition and $Z(0.2^n u(n)) = z/(z - 0.2)$, we achieve

$$Y(z) - 0.5(1 + z^{-1}Y(z)) = 5z/(z - 0.2).$$

Simplification yields

$$Y(z) - 0.5z^{-1}Y(z) = 0.5 + 5z/(z - 0.2).$$

Factoring out $Y(z)$ and combining the right-hand side of the equation, it follows that

$$Y(z)(1 - 0.5z^{-1}) = (5.5z - 0.1)/(z - 0.2).$$

Then we obtain

$$Y(z) = \frac{(5.5z - 0.1)}{(1 - 0.5z^{-1})(z - 0.2)} = \frac{z(5.5z - 0.1)}{(z - 0.5)(z - 0.2)}.$$

Using the partial fraction expansion method leads to

$$\frac{Y(z)}{z} = \frac{5.5z - 0.1}{(z - 0.5)(z - 0.2)} = \frac{A}{z - 0.5} + \frac{B}{z - 0.2},$$

where

$$A = (z - 0.5)\frac{Y(z)}{z}\bigg|_{z=0.5} = \frac{5.5z - 0.1}{z - 0.2}\bigg|_{z=0.5} = \frac{5.5 \times 0.5 - 0.1}{0.5 - 0.2} = 8.8333,$$

$$B = (z - 0.2)\frac{Y(z)}{z}\bigg|_{z=0.2} = \frac{5.5z - 0.1}{z - 0.5}\bigg|_{z=0.2} = \frac{5.5 \times 0.2 - 0.1}{0.2 - 0.5} = -3.3333.$$

Thus

$$Y(z) = \frac{8.8333z}{(z - 0.5)} + \frac{-3.3333z}{(z - 0.2)},$$

which gives the solution as

$$y(n) = 8.3333(0.5)^n u(n) - 3.3333(0.2)^n u(n).$$

Example 5.14.

A relaxed (zero initial conditions) DSP system is described by the difference equation

$$y(n) + 0.1y(n - 1) - 0.2y(n - 2) = x(n) + x(n - 1).$$

5 THE Z-TRANSFORM

a. Determine the impulse response $y(n)$ due to the impulse sequence $x(n) = \delta(n)$
b. Determine system response $y(n)$ due to the unit step function excitation, where $u(n) = 1$ for $n \geq 0$.

Solution:

a. Applying the z-transform on both sides of the difference equations and using Equation (5.3) or Equation (5.14), we yield

$$Y(z) + 0.1 Y(z)z^{-1} - 0.2 Y(z)z^{-2} = X(z) + X(z)z^{-1}. \qquad (5.15)$$

Factoring out $Y(z)$ on the left side and substituting $X(z) = Z(\delta(n)) = 1$ to the right side in Equation (5.15) achieves

$$Y(z)(1 + 0.1z^{-1} - 0.2z^{-2}) = 1(1 + z^{-1}).$$

Then $Y(z)$ can be expressed as

$$Y(z) = \frac{1 + z^{-1}}{1 + 0.1z^{-1} - 0.2z^{-2}}.$$

To obtain the impulse response, which is the inverse z-transform of the transfer function, we multiply the numerator and denominator by z^2. Thus

$$Y(z) = \frac{z^2 + z}{z^2 + 0.1z - 0.2} = \frac{z(z+1)}{(z-0.4)(z+0.5)}.$$

Using the partial fraction expansion method leads to

$$\frac{Y(z)}{z} = \frac{z+1}{(z-0.4)(z+0.5)} = \frac{A}{z-0.4} + \frac{B}{z+0.5},$$

where $A = (z - 0.4) \left.\frac{Y(z)}{z}\right|_{z=0.4} = \left.\frac{z+1}{z+0.5}\right|_{z=0.4} = \frac{0.4+1}{0.4+0.5} = 1.5556$

$B = (z + 0.5) \left.\frac{Y(z)}{z}\right|_{z=-0.5} = \left.\frac{z+1}{z-0.4}\right|_{z=-0.5} = \frac{-0.5+1}{-0.5-0.4} = -0.5556.$

Thus

$$Y(z) = \frac{1.5556z}{(z-0.4)} + \frac{-0.5556z}{(z+0.5)},$$

which gives the impulse response:

$$y(n) = 1.5556(0.4)^n u(n) - 0.5556(-0.5)^n u(n).$$

b. To obtain the response due to a unit step function, the input sequence is set to be

$$x(n) = u(n)$$

and the corresponding z-transform is given by

$$X(z) = \frac{z}{z-1},$$

and notice that

$$Y(z) + 0.1Y(z)z^{-1} - 0.2Y(z)z^{-2} = X(z) + X(z)z^{-1}.$$

Then the z-transform of the output sequence $y(n)$ can be yielded as

$$Y(z) = \left(\frac{z}{z-1}\right)\left(\frac{1+z^{-1}}{1+0.1z^{-1}-0.2z^{-2}}\right) = \frac{z^2(z+1)}{(z-1)(z-0.4)(z+0.5)}.$$

Using the partial fraction expansion method as before gives

$$Y(z) = \frac{2.2222z}{z-1} + \frac{-1.0370z}{z-0.4} + \frac{-0.1852z}{z+0.5},$$

and the system response is found by using Table 5.1:

$$y(n) = 2.2222u(n) - 1.0370(0.4)^n u(n) - 0.1852(-0.5)^n u(n).$$

5.5 Summary

1. The one-sided (unilateral) z-transform was defined, which can be used to transform any causal sequence to the z-transform domain.

2. The look-up table of the z-transform determines the z-transform for a simple causal sequence, or the causal sequence from a simple z-transform function.

3. The important properties of the z-transform, such as linearity, shift theorem, and convolution, were introduced. The shift theorem can be used to solve a difference equation. The z-transform of a digital convolution of two digital sequences is equal to the product of their z-transforms.

4. The method of the inverse z-transform, such as the partial fraction expansion, inverses the complicated z-transform function, which can have first-order real poles, multiple-order real poles, and first-order complex poles assuming that the z-transform function is proper. The MATLAB tool was introduced.

5. The application of the z-transform solves linear difference equations with nonzero initial conditions and zero initial conditions.

5.6 Problems

5.1. Find the z-transform for each of the following sequences:

a. $x(n) = 4u(n)$

b. $x(n) = (-0.7)^n u(n)$

c. $x(n) = 4e^{-2n} u(n)$

d. $x(n) = 4(0.8)^n \cos(0.1\pi n) u(n)$

e. $x(n) = 4e^{-3n} \sin(0.1\pi n) u(n)$.

5.2. Using the properties of the z-transform, find the z-transform for each of the following sequences:

a. $x(n) = u(n) + (0.5)^n u(n)$

b. $x(n) = e^{-3(n-4)} \cos(0.1\pi(n-4)) u(n-4)$,
where $u(n-4) = 1$ for $n \geq 4$ while $u(n-4) = 0$ for $n < 4$.

5.3. Given two sequences,

$$x_1(n) = 5\delta(n) - 2\delta(n-2) \text{ and}$$
$$x_2(n) = 3\delta(n-3),$$

a. determine the z-transform of convolution of the two sequences using the convolution property of z-transform

$$X(z) = X_1(z)X_2(z);$$

b. determine convolution by the inverse z-transform from the result in (a)

$$x(n) = Z^{-1}(X_1(z)X_2(z)).$$

5.4. Using Table 5.1 and z-transform properties, find the inverse z-transform for each of the following functions:

a. $X(z) = 4 - \dfrac{10z}{z-1} - \dfrac{z}{z+0.5}$

b. $X(z) = \dfrac{-5z}{(z-1)} + \dfrac{10z}{(z-1)^2} + \dfrac{2z}{(z-0.8)^2}$

c. $X(z) = \dfrac{z}{z^2 + 1.2z + 1}$

d. $X(z) = \dfrac{4z^{-4}}{z-1} + \dfrac{z^{-1}}{(z-1)^2} + z^{-8} + \dfrac{z^{-5}}{z - 0.5}$

5.5. Using the partial fraction expansion method, find the inverse of the following z-transforms:

a. $X(z) = \dfrac{1}{z^2 - 0.3z - 0.04}$

b. $X(z) = \dfrac{z}{(z - 0.2)(z + 0.4)}$

c. $X(z) = \dfrac{z}{(z + 0.2)(z^2 - z + 0.5)}$

d. $X(z) = \dfrac{z(z + 0.5)}{(z - 0.1)^2(z - 0.6)}$

5.6. A system is described by the difference equation

$$y(n) + 0.5y(n - 1) = 2(0.8)^n u(n).$$

Determine the solution when the initial condition is $y(-1) = 2$.

5.7. A system is described by the difference equation

$$y(n) - 0.5y(n - 1) + 0.06y(n - 2) = (0.4)^{n-1} u(n - 1).$$

Determine the solution when the initial conditions are $y(-1) = 1$ and $y(-2) = 2$.

5.8. Given the following difference equation with the input-output relationship of a certain initially relaxed system (all initial conditions are zero),

$$y(n) - 0.7y(n - 1) + 0.1y(n - 2) = x(n) + x(n - 1),$$

a. find the impulse response sequence $y(n)$ due to the impulse sequence $\delta(n)$;

b. find the output response of the system when the unit step function $u(n)$ is applied.

5.9. Given the following difference equation with the input-output relationship of a certain initially relaxed DSP system (all initial conditions are zero),

$$y(n) - 0.4y(n-1) + 0.29y(n-2) = x(n) + 0.5x(n-1),$$

a. find the impulse response sequence $y(n)$ due to an impulse sequence $\delta(n)$;

b. find the output response of the system when a unit step function $u(n)$ is applied.

Reference

Oppenheim, A. V., and Schafer, R. W. (1975). *Discrete-Time Signal Processing.* Englewood Cliffs, NJ: Prentice Hall.

6

Digital Signal Processing Systems, Basic Filtering Types, and Digital Filter Realizations

Objectives:

This chapter illustrates digital filtering operations for a given input sequence; derives transfer functions from the difference equations; analyzes stability of the linear systems using the z-plane pole-zero plot; and calculates the frequency responses of digital filters. Then the chapter further investigates realizations of the digital filters and examines spectral effects by filtering speech data using the digital filters.

6.1 The Difference Equation and Digital Filtering

In this chapter, we begin with developing the filtering concept of digital signal processing (DSP) systems. With the knowledge acquired in Chapter 5, dealing with the z-transform, we will learn how to describe and analyze linear time-invariant systems. We also will become familiar with digital filtering types and their realization structures. A DSP system (digital filter) is described in Figure 6.1.

Let $x(n)$ and $y(n)$ be a DSP system's input and output, respectively. We can express the relationship between the input and the output of a DSP system by the following *difference equation*:

$$y(n) = b_0 x(n) + b_1 x(n-1) + \cdots + b_M x(n-M) \\ - a_1 y(n-1) - \cdots - a_N y(n-N), \tag{6.1}$$

FIGURE 6.1 DSP system with input and output.

where b_i, $0 \le i \le M$ and a_j, $1 \le j \le N$, represent the coefficients of the system and n is the time index. Equation (6.1) can also be written as

$$y(n) = \sum_{i=0}^{M} b_i x(n-i) - \sum_{j=1}^{N} a_j y(n-j). \qquad (6.2)$$

From Equations (6.1) and (6.2), we observe that the DSP system output is the weighted summation of the current input value $x(n)$ and its past values: $x(n-1), \ldots, x(n-M)$, and past output sequence: $y(n-1), \ldots, y(n-N)$. The system can be verified as linear, time invariant, and causal. If the initial conditions are specified, we can compute system output (time response) $y(n)$ recursively. This process is referred to as *digital filtering*. We will illustrate filtering operations by Examples 6.1 and 6.2.

Example 6.1.

Compute the system output

$$y(n) = 0.5 y(n-2) + x(n-1)$$

for the first four samples using the following initial conditions:

a. initial conditions: $y(-2) = 1$, $y(-1) = 0$, $x(-1) = -1$, and input $x(n) = (0.5)^n u(n)$.

b. zero initial conditions: $y(-2) = 0$, $y(-1) = 0$, $x(-1) = 0$, and input $x(n) = (0.5)^n u(n)$.

Solution:

According to Equation (6.1), we identify the system coefficients as

$$N = 2,\ M = 1,\ a_1 = 0,\ a_2 = -0.5,\ b_0 = 0,\ \text{and } b_1 = 1.$$

a. Setting $n = 0$, and using initial conditions, we obtain the input and output as

$$x(0) = (0.5)^0 u(0) = 1$$
$$y(0) = 0.5 y(-2) + x(-1) = 0.5 \cdot 1 + (-1) = -0.5.$$

Setting $n = 1$ and using the initial condition $y(-1) = 0$, we achieve

$$x(1) = (0.5)^1 u(1) = 0.5$$
$$y(1) = 0.5y(-1) + x(0) = 0.5 \cdot 0 + 1 = 1.0.$$

Similarly, using the past output $y(0) = -0.5$, we get

$$x(2) = (0.5)^2 u(2) = 0.25$$
$$y(2) = 0.5y(0) + x(1) = 0.5 \cdot (-0.5) + 0.5 = 0.25$$

and with y(1)=1.0, we yield

$$x(3) = (0.5)^3 u(3) = 0.125$$
$$y(3) = 0.5y(1) + x(2) = 0.5 \cdot 1 + 0.25 = 0.75$$

. . . .

Clearly, $y(n)$ can be recursively computed for $n > 3$.

b. Setting $n = 0$, we obtain

$$x(0) = (0.5)^0 u(0) = 1$$
$$y(0) = 0.5y(-2) + x(-1) = 0 \cdot 1 + 0 = 0.$$

Setting $n = 1$, we achieve

$$x(1) = (0.5)^1 u(1) = 0.5$$
$$y(1) = 0.5y(-1) + x(0) = 0 \cdot 0 + 1 = 1$$

Similarly, with the past output $y(0) = 0$, we determine

$$x(2) = (0.5)^2 u(2) = 0.25$$
$$y(2) = 0.5y(0) + x(1) = 0.5 \cdot 0 + 0.5 = 0.5$$

and with $y(1) = 1$, we obtain

$$x(3) = (0.5)^3 u(3) = 0.125$$
$$y(3) = 0.5y(1) + x(2) = 0.5 \cdot 1 + 0.25 = 0.75$$

. . . .

Clearly, $y(n)$ can be recursively computed for $n > 3$.

Example 6.2.

Given the DSP system

$$y(n) = 2x(n) - 4x(n-1) - 0.5y(n-1) - y(n-2)$$

with initial conditions $y(-2) = 1$, $y(-1) = 0$, $x(-1) = -1$, and the input $x(n) = (0.8)^n u(n)$,

a. Compute the system response $y(n)$ for 20 samples using MATLAB.

Solution:

a. Program 6.1 on the next page lists the MATLAB program for computing the system response $y(n)$. The top plot in Figure 6.2 shows the input sequence. The middle plot displays the filtered output using the initial conditions, and the bottom plot shows the filtered output for zero initial conditions. As we can see, both system outputs are different at the beginning portion, while they approach the same value later.

MATLAB function **filter()**, developed using a direct-form II realization (which will be discussed in a later section), can be used to operate digital filtering, and the syntax is

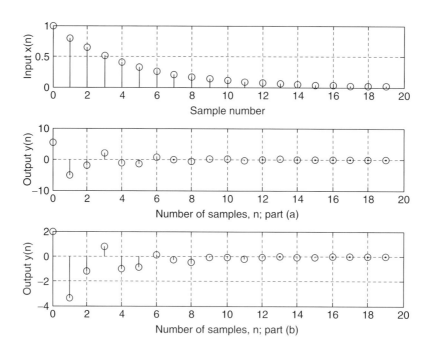

Part (a): response with initial conditions;

Part (b): response with zero initial conditions.

FIGURE 6.2 **Plots of the input and system outputs y(n) for Example 6.2.**

Program 6.1. MATLAB program for Example 6.2.

```
% Example 6.2
% Compute y(n) = 2x(n) − 4x(n − 1) − 0.5y(n − 1) − 0.5y(n − 2)
%Nonzero initial conditions:
% y(−2) = 1, y(−1) = 0, x(−1) = −1, and x(n) = (0.8)^n*u(n)
%
y = zeros(1,20);          %Set up a vector to store y(n)
y = [ 1 0 y];             %Set initial conditions of y(−2) and y(−1)
n = 0:1:19;               %Compute time indexes
x = (0.8).^n;             %Compute 20 input samples of x(n)
x = [0 −1 x];             %Set initial conditions of x(−2) = 0 and x(−1) = 1
for n = 1:20
y(n + 2) = 2*x(n + 2) − 4*x(n + 1) − 0.5*y(n + 1) − 0.5*y(n);%Compute 20 outputs
end
n= 0:1:19
subplot(3,1,1);stem(n,x(3:22));grid;ylabel('Input x(n)');
xlabel ('Sample number');
subplot(3,1,2); stem(n,y(3:22)),grid;
xlabel('Number of samples, n; part (a)');ylabel('Output y(n)');
y(3:22) %output y(n)
%Zero- initial conditions:
% y(−2) = 0, y(−1) = 0, x(−1) = 0, and x(n) = 1/(n + 1)
%
y = zeros(1,20);          %Set up a vector to store y(n)
y = [ 0 0 y];             %Set zero initial conditions of y(−2) and y(−1)
n=0:1:19;                 %Compute time indexes
x = (0.8).^n;             %Compute 20 input samples of x(n)
x = [0 0 x];              %Set zero initial conditions of x(−2) = 0 and x(−1) = 0
for n = 1:20
y(n + 2) = 2*x(n + 2) − 4*x(n + 1) − 0.5*y(n + 1) − 0.5*y(n);%Compute 20 outputs
end
n = 0:1:19
subplot(3,1,3);stem(n,y(3:22)),grid;
xlabel('Number of samples, n; part (b)');ylabel('Output y(n)');
y(3:22)%Output y(n)
```

$$\mathbf{Zi} = \mathbf{filtic(B, A, Yi, Xi)}$$
$$\mathbf{y} = \mathbf{filter(B, A, x, Zi)},$$

where B and A are vectors for the coefficients b_j and a_j, whose formats are

$$A = [1\ a_1\ a_2 \cdots a_N] \text{ and } B = [b_0\ b_1\ b_2\ \cdots\ b_M],$$

and x and y are the input data vector and the output data vector, respectively.

Note that the filter function **filtic()** is a MATLAB function used to obtain initial states required by the MATLAB filter function **filter()** (requirement by a direct-form II realization) from initial conditions in the difference equation. Hence, Z_i contains initial states required for operating MATLAB function **filter()**, that is,

$$Z_i = [w(-1)\ w(-2) \cdots],$$

which can be recovered by the MATLAB function, **filtic()**. X_i and Y_i are initial conditions with a length of the greater of M or N, given by

$$X_i = [x(-1)\ x(-2)\ \cdots\ ] \text{ and } Y_i = [y(-1)\ y(-2)\ \cdots\ ].$$

Especially for zero initial conditions, the syntax is reduced to

$$\mathbf{y} = \mathbf{filter(B, A, x)}.$$

Let us verify the filter operation results in Example 6.1 using the MATLAB functions. The MATLAB codes and results for Example 6.1 (a) with the non-zero initial conditions are listed as

```
>> B = [0   1]; A = [1   0  −0.5];
>> x = [1   0.5   0.25   0.125];
>> Xi = [−1   0]; Yi = [0   1];
>> Zi = filtic(B, A, Yi, Xi);
>> y = filter(B, A, x, Zi)
y =
 −0.5000   1.0000   0.2500   0.7500
>>
```

For the case of zero initial conditions in Example 6.1(b), the MATLAB codes and results are

```
>> B = [0 1];A = [1 0 -0.5];
>> x = [1 0.5 0.25 0.125];
>> y = filter(B, A, x)
y =
    0   1.0000   0.5000   0.7500
>>
```

As we expected, the filter outputs match those in Example 6.1.

6.2 Difference Equation and Transfer Function

To proceed in this section, Equation (6.1) is rewritten as

$$y(n) = b_0 x(n) + b_1 x(n-1) + \cdots + b_M x(n-M) \\ - a_1 y(n-1) - \cdots - a_N y(n-N).$$

With an assumption that all initial conditions of this system are zero, and with $X(z)$ and $Y(z)$ denoting the z-transforms of $x(n)$ and $y(n)$, respectively, taking the z-transform of Equation (6.1) yields

$$Y(z) = b_0 X(z) + b_1 X(z) z^{-1} + \cdots + b_M X(z) z^{-M} \\ - a_1 Y(z) z^{-1} - \cdots - a_N Y(z) z^{-N}. \tag{6.3}$$

Rearranging Equation (6.3), we yield

$$H(z) = \frac{Y(z)}{X(z)} = \frac{b_0 + b_1 z^{-1} + \cdots + b_M z^{-M}}{1 + a_1 z^{-1} + \cdots + a_N z^{-N}} = \frac{B(z)}{A(z)}, \tag{6.4}$$

where $H(z)$ is defined as the transfer function with its numerator and denominator polynomials defined below:

$$B(z) = b_0 + b_1 z^{-1} + \cdots + b_M z^{-M} \tag{6.5}$$
$$A(z) = 1 + a_1 z^{-1} + \cdots + a_N z^{-N}. \tag{6.6}$$

Clearly the z-transfer function is defined as

$$\text{ratio} = \frac{\text{z-transform of the output}}{\text{z-transform of the input}}.$$

In DSP applications, given the difference equation, we can develop the z-transfer function and represent the digital filter in the z-domain as shown in

FIGURE 6.3   **Digital filter transfer function.**

Figure 6.3. Then the stability and frequency response can be examined based on the developed transfer function.

Example 6.3.

A DSP system is described by the following difference equation:

$$y(n) = x(n) - x(n-2) - 1.3y(n-1) - 0.36y(n-2).$$

a. Find the transfer function $H(z)$, the denominator polynomial $A(z)$, and the numerator polynomial $B(z)$.

Solution:

a. Taking the z-transform on both sides of the previous difference equation, we achieve

$$Y(z) = X(z) - X(z)z^{-2} - 1.3Y(z)z^{-1} - 0.36Y(z)z^{-2}.$$

Moving the last two terms to the left side of the difference equation and factoring $Y(z)$ on the left side and $X(z)$ on the right side, we obtain

$$Y(z)(1 + 1.3z^{-1} + 0.36z^{-2}) = (1 - z^{-2})X(z).$$

Therefore, the transfer function, which is the ratio of $Y(z)$ to $X(z)$, can be found to be

$$H(z) = \frac{Y(z)}{X(z)} = \frac{1 - z^{-2}}{1 + 1.3z^{-1} + 0.36z^{-2}}.$$

From the derived transfer function $H(z)$, we can obtain the denominator polynomial and numerator polynomial as

$$A(z) = 1 + 1.3z^{-1} + 0.36z^{-2} \text{ and}$$
$$B(z) = 1 - z^{-2}.$$

The difference equation and its transfer function, as well as the stability issue of the linear time-invariant system, will be discussed in the following sections.

Example 6.4.

A digital system is described by the following difference equation:

$$y(n) = x(n) - 0.5x(n-1) + 0.36x(n-2).$$

a. Find the transfer function $H(z)$, the denominator polynomial $A(z)$, and the numerator polynomial $B(z)$.

Solution:

a. Taking the z-transform on both sides of the previous difference equation, we achieve

$$Y(z) = X(z) - 0.5X(z)z^{-1} + 0.36X(z)z^{-2}.$$

Therefore, the transfer function, that is, the ratio of $Y(z)$ to $X(z)$, can be found as

$$H(z) = \frac{Y(z)}{X(z)} = 1 - 0.5z^{-1} + 0.36z^{-2}.$$

From the derived transfer function $H(z)$, it follows that

$$A(z) = 1$$
$$B(z) = 1 - 0.5z^{-1} + 0.36z^{-2}.$$

In DSP applications, the given transfer function of a digital system can be converted into a difference equation for DSP implementation. The following example illustrates the procedure.

Example 6.5.

Convert each of the following transfer functions into its difference equation.

a. $H(z) = \dfrac{z^2 - 1}{z^2 + 1.3z + 0.36}$

b. $H(z) = \dfrac{z^2 - 0.5z + 0.36}{z^2}$

Solution:

a. Dividing the numerator and the denominator by z^2 to obtain the transfer function whose numerator and denominator polynomials have the negative power of z, it follows that

$$H(z) = \frac{(z^2 - 1)/z^2}{(z^2 + 1.3z + 0.36)/z^2} = \frac{1 - z^{-2}}{1 + 1.3z^{-1} + 0.36z^{-2}}.$$

We write the transfer function using the ratio of $Y(z)$ to $X(z)$:

$$\frac{Y(z)}{X(z)} = \frac{1 - z^{-2}}{1 + 1.3z^{-1} + 0.36z^{-2}}.$$

Then we have

$$Y(z)(1 + 1.3z^{-1} + 0.36z^{-2}) = X(z)(1 - z^{-2}).$$

By distributing $Y(z)$ and $X(z)$, we yield

$$Y(z) + 1.3z^{-1}Y(z) + 0.36z^{-2}Y(z) = X(z) - z^{-2}X(z).$$

Applying the inverse z-transform and using the shift property in Equation (5.3) of Chapter 5, we get

$$y(n) + 1.3y(n-1) + 0.36y(n-2) = x(n) - x(n-2).$$

Writing the output $y(n)$ in terms of inputs and past outputs leads to

$$y(n) = x(n) - x(n-2) - 1.3y(n-1) - 0.36y(n-2).$$

b. Similarly, dividing the numerator and denominator by z^2, we obtain

$$H(z) = \frac{Y(z)}{X(z)} = \frac{(z^2 - 0.5z + 0.36)/z^2}{z^2/z^2} = 1 - 0.5z^{-1} + 0.36z^{-2}.$$

Thus, $Y(z) = X(z)(1 - 0.5z^{-1} + 0.36z^{-2})$.

By distributing $X(z)$, we yield

$$Y(z) = X(z) - 0.5z^{-1}X(z) + 0.36z^{-2}X(z).$$

Applying the inverse z-transform while using the shift property in Equation (5.3), we obtain

$$y(n) = x(n) - 0.5x(n-1) + 0.36x(n-2).$$

The transfer function $H(z)$ can be factored into the *pole-zero form:*

$$H(z) = \frac{b_0(z - z_1)(z - z_2)\cdots(z - z_M)}{(z - p_1)(z - p_2)\cdots(z - p_N)}, \tag{6.7}$$

where the zeros z_i can be found by solving for the roots of the numerator polynomial, while the poles p_i can be solved for the roots of the denominator polynomial.

Example 6.6.

Given the following transfer function,

$$H(z) = \frac{1 - z^{-2}}{1 + 1.3z^{-1} + 0.36z^{-2}}$$

a. Convert it into the pole-zero form.

Solution:

a. We first multiply the numerator and denominator polynomials by z^2 to achieve its advanced form in which both numerator and denominator polynomials have positive powers of z, that is,

$$H(z) = \frac{(1 - z^{-2})z^2}{(1 + 1.3z^{-1} + 0.36z^{-2})z^2} = \frac{z^2 - 1}{z^2 + 1.3z + 0.36}.$$

Letting $z^2 - 1 = 0$, we get $z = 1$ and $z = -1$. Setting $z^2 + 1.3z + 0.36 = 0$ leads to $z = -0.4$ and $z = -0.9$. We then can write numerator and denominator polynomials in the factored form to obtain the pole-zero form:

$$H(z) = \frac{(z - 1)(z + 1)}{(z + 0.4)(z + 0.9)}.$$

6.2.1 Impulse Response, Step Response, and System Response

The impulse response $h(n)$ of the DSP system $H(z)$ can be obtained by solving its difference equation using a unit impulse input $\delta(n)$. With the help of the z-transform and noticing that $X(z) = Z\{\delta(n)\} = 1$, we yield

$$h(n) = Z^{-1}\{H(z)X(z)\} = Z^{-1}\{H(z)\}. \quad (6.8)$$

Similarly, for a step input, we can determine step response assuming the zero initial conditions. Letting

$$X(z) = Z[\,u(n)] = \frac{z}{z - 1},$$

the step response can be found as

$$y(n) = Z^{-1}\left\{H(z)\frac{z}{z - 1}\right\}. \quad (6.9)$$

Furthermore, the z-transform of the general system response is given by

$$Y(z) = H(z)X(z). \quad (6.10)$$

If we know the transfer function $H(z)$ and the z-transform of the input $X(z)$, we are able to determine the system response $y(n)$ by finding the inverse z-transform of the output $Y(z)$:

$$y(n) = Z^{-1}\{Y(z)\}. \quad (6.11)$$

Example 6.7.

Given a transfer function depicting a DSP system

$$H(z) = \frac{z+1}{z-0.5},$$

Determine

a. the impulse response $h(n)$,

b. the step response $y(n)$, and

c. the system response $y(n)$ if the input is given as $x(n) = (0.25)^n u(n)$.

Solution:

a. The transfer function can be rewritten as

$$\frac{H(z)}{z} = \frac{z+1}{z(z-0.5)} = \frac{A}{z} + \frac{B}{z-0.5},$$

where $A = \left.\frac{z+1}{(z-0.5)}\right|_{z=0} = -2,$ and $B = \left.\frac{z+1}{z}\right|_{z=0.5} = 3.$

Thus we have

$$\frac{H(z)}{z} = \frac{-2}{z} + \frac{3}{z-0.5} \quad \text{and}$$

$$H(z) = \left(-\frac{2}{z} + \frac{3}{z-0.5}\right)z = -2 + \frac{3z}{z-0.5}.$$

By taking the inverse z-transform as shown in Equation (6.8), we yield the impulse response

$$h(n) = -2\delta(n) + 3(0.5)^n u(n).$$

b. For the step input $x(n) = u(n)$ and its z-transform $X(z) = \frac{z}{z-1}$, we can determine the z-transform of the step response as

$$Y(z) = H(z)X(z) = \frac{z+1}{z-0.5} \cdot \frac{z}{z-1}.$$

Applying the partial fraction expansion leads to

$$\frac{Y(z)}{z} = \frac{z+1}{(z-0.5)(z-1)} = \frac{A}{z-0.5} + \frac{B}{z-1},$$

where

$$A = \left.\frac{z+1}{z-1}\right|_{z=0.5} = -3, \quad \text{and} \quad B = \left.\frac{z+1}{z-0.5}\right|_{z=1} = 4.$$

The z-transform step response is therefore

$$Y(z) = \frac{-3z}{z - 0.5} + \frac{4z}{z - 1}.$$

Applying the inverse z-transform yields the step response as

$$y(n) = -3(0.5)^n \, u(n) + 4 \, u(n).$$

c. To determine the system output response, we first find the z-transform of the input $x(n)$,

$$X(z) = Z\{(0.25)^n \, u(n)\} = \frac{z}{z - 0.25},$$

then $Y(z)$ can be yielded via Equation (6.10), that is,

$$Y(z) = H(z)X(z) = \frac{z+1}{z-0.5} \cdot \frac{z}{z-0.25} = \frac{z(z+1)}{(z-0.5)(z-0.25)}.$$

Using the partial fraction expansion, we have

$$\frac{Y(z)}{z} = \frac{(z+1)}{(z-0.5)(z-0.25)} = \left(\frac{A}{z-0.5} + \frac{B}{z-0.25}\right)$$

$$Y(z) = \left(\frac{6z}{z-0.5} + \frac{-5z}{z-0.25}\right).$$

Using Equation (6.11) and Table 5.1 in Chapter 5, we finally yield

$$y(n) = Z^{-1}\{Y(z)\} = 6(0.5)^n \, u(n) - 5(0.25)^n \, u(n).$$

The impulse response for (a), the step response for (b), and the system response for (c) are each plotted in Figure 6.4.

6.3 The z-Plane Pole-Zero Plot and Stability

A very useful tool to analyze digital systems is the z-plane pole-zero plot. This graphical technique allows us to investigate characteristics of the digital system shown in Figure 6.1, including the system stability. In general, a digital transfer function can be written in the pole-zero form as shown in Equation (6.7), and we can plot the poles and zeros on the z-plane. The z-plane is depicted in Figure 6.5 and has the following features:

1. The horizontal axis is the real part of the variable z, and the vertical axis represents the imaginary part of the variable z.

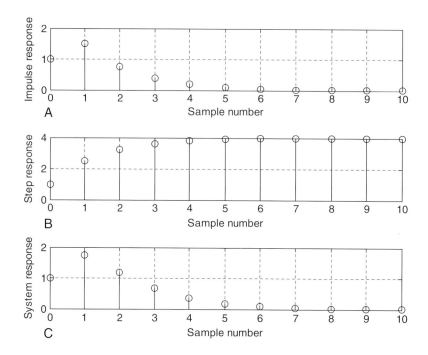

FIGURE 6.4 Impulse, step, and system responses in Example 6.7.

2. The z-plane is divided into two parts by a unit circle.
3. Each pole is marked on the z-plane using the cross symbol ×, while each zero is plotted using the small circle symbol ∘.

Let's investigate the z-plane pole-zero plot of a digital filter system via the following example.

Example 6.8.

Given the digital transfer function

$$H(z) = \frac{z^{-1} - 0.5z^{-2}}{1 + 1.2z^{-1} + 0.45z^{-2}},$$

a. Plot poles and zeros.

Solution:

a. Converting the transfer function to its advanced form by multiplying z^2 to both numerator and denominator, it follows that

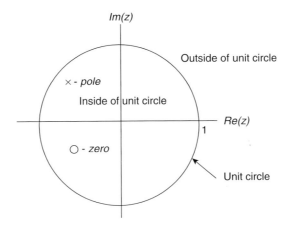

FIGURE 6.5 z-plane and pole-zero plot.

$$H(z) = \frac{(z^{-1} - 0.5z^{-2})z^2}{(1 + 1.2z^{-1} + 0.45z^{-2})z^2} = \frac{z - 0.5}{z^2 + 1.2z + 0.45}.$$

By setting $z^2 + 1.2z + 0.45 = 0$ and $z - 0.5 = 0$, we obtain two poles

$$p_1 = -0.6 + j0.3$$
$$p_2 = p_1^* = -0.6 - j0.3$$

and a zero $z_1 = 0.5$, which are plotted on the z-plane shown in Figure 6.6. According to the form of Equation (6.7), we also yield the pole-zero form as

$$H(z) = \frac{z^{-1} - 0.5z^{-2}}{1 + 1.2z^{-1} + 0.45z^{-2}} = \frac{(z - 0.5)}{(z + 0.6 - j0.3)(z + 0.6 + j0.3)}.$$

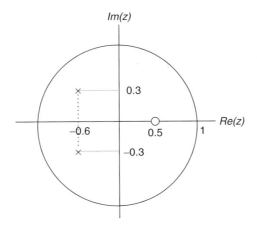

FIGURE 6.6 The z-plane pole-zero plot of Example 6.8.

Having zeros and poles plotted on the z-plane, we are able to study the system stability. We first establish the relationship between the s-plane in Laplace domain and the z-plane in z-transform domain, as illustrated in Figure 6.7.

As shown in Figure 6.7, the sampled signal, which is not quantized, with a sampling period of T is written as

$$x_s(t) = \sum_{n=0}^{\infty} x(nT)\delta(t - nT)$$
$$= x(0)\delta(t) + x(T)\delta(t - T) + x(2T)\delta(t - 2T) + \ldots. \quad (6.12)$$

Taking the Laplace transform and using the Laplace shift property as

$$L(\delta(t - nT)) = e^{-nTs} \quad (6.13)$$

leads to

$$X_s(s) = \sum_{n=0}^{\infty} x(nT)e^{-nTs} = x(0)e^{-0 \times Ts} + x(T)e^{-Ts} + x(2T)e^{-2Ts} + \ldots. \quad (6.14)$$

Comparing Equation (6.14) with the definition of a one-sided z-transform of the data sequence $x(n)$ from analog-to-digital conversion (ADC):

$$X(z) = Z(x(n)) = \sum_{n=0}^{\infty} x(n)z^{-n} = x(0)z^{-0} + x(1)z^{-1} + x(2)z^{-2} + \ldots. \quad (6.15)$$

Clearly, we see the relationship of the sampled system in Laplace domain and its digital system in z-transform domain by the following mapping:

$$z = e^{sT}. \quad (6.16)$$

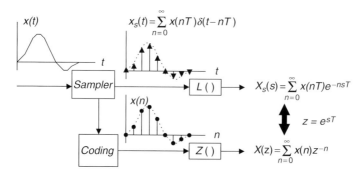

FIGURE 6.7 Relationship between Laplace transform and z-transform.

Substituting $s = -\alpha \pm j\omega$ into Equation (6.16), it follows that $z = e^{-\alpha T \pm j\omega T}$. In the polar form, we have

$$z = e^{-\alpha T} \angle \pm \omega T. \tag{6.17}$$

Equations (6.16) and (6.17) give the following important conclusions.

If $\alpha > 0$, this means $|z| = e^{-\alpha T} < 1$. Then the left-hand half plane (LHHP) of the s-plane is mapped to the inside of the unit circle of the z-plane. When $\alpha = 0$, this causes $|z| = e^{-\alpha T} = 1$. Thus the $j\omega$ axis of the s-plane is mapped on the unit circle of the z-plane, as shown in Figure 6.8. Obviously, the right-hand half plane (RHHP) of the s-plane is mapped to the outside of the unit circle in the z-plane. A stable system means that for a given bounded input, the system output must be bounded. Similar to the analog system, the digital system requires that all poles plotted on the z-plane must be inside the unit circle. We summarize the rules for determining the stability of a DSP system as follows:

1. If the outermost pole(s) of the z-transfer function $H(z)$ describing the DSP system is(are) inside the unit circle on the z-plane pole-zero plot, then the system is stable.

2. If the outermost pole(s) of the z-transfer function $H(z)$ is(are) outside the unit circle on the z-plane pole-zero plot, the system is unstable.

3. If the outermost pole(s) is(are) first-order pole(s) of the z-transfer function $H(z)$ and on the unit circle on the z-plane pole-zero plot, then the system is marginally stable.

4. If the outermost pole(s) is(are) multiple-order pole(s) of the z-transfer function $H(z)$ and on the unit circle on the z-plane pole-zero plot, then the system is unstable.

5. The zeros do not affect the system stability.

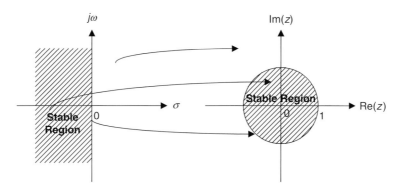

FIGURE 6.8   **Mapping between s-plane and z-plane.**

Notice that the following facts apply to a stable system (bounded-in/bounded-out [BIBO] stability discussed in Chapter 3):

1. If the input to the system is bounded, then the output of the system will also be bounded, or the impulse response of the system will go to zero in a finite number of steps.

2. An unstable system is one in which the output of the system will grow without bound due to any bounded input, initial condition, or noise, or its impulse response will grow without bound.

3. The impulse response of a marginally stable system stays at a constant level or oscillates between two finite values.

Examples illustrating these rules are shown in Figure 6.9.

Example 6.9.

The following transfer functions describe digital systems.

a. $H(z) = \dfrac{z + 0.5}{(z - 0.5)(z^2 + z + 0.5)}$

b. $H(z) = \dfrac{z^2 + 0.25}{(z - 0.5)(z^2 + 3z + 2.5)}$

c. $H(z) = \dfrac{z + 0.5}{(z - 0.5)(z^2 + 1.4141z + 1)}$

d. $H(z) = \dfrac{z^2 + z + 0.5}{(z - 1)^2(z + 1)(z - 0.6)}$

For each, sketch the z-plane pole-zero plot and determine the stability status for the digital system.

Solution:

a. A zero is found to be $z = -0.5$.

$$\text{Poles: } z = 0.5, \ |z| = 0.5 < 1; z = -0.5 \pm j0.5,$$

$$|z| = \sqrt{(-0.5)^2 + (\pm 0.5)^2} = 0.707 < 1.$$

The plot of poles and a zero is shown in Figure 6.10. Since the outermost poles are inside the unit circle, the system is stable.

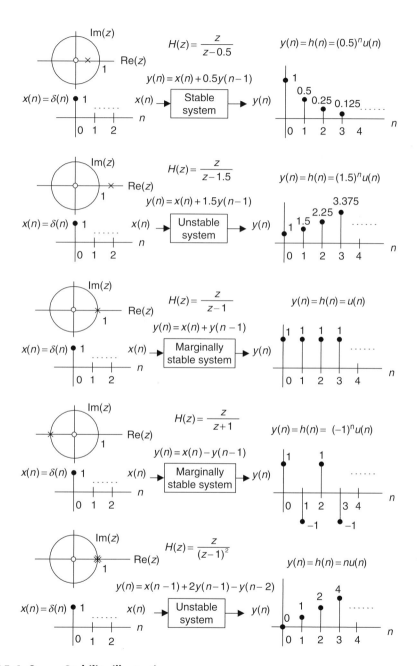

FIGURE 6.9   **Stability illustrations.**

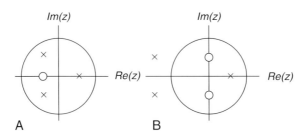

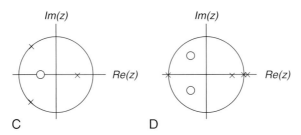

FIGURE 6.10 Pole-zero plots for Example 6.9.

b. Zeros are $z = \pm j0.5$.

$$\text{Poles:} z = 0.5, |z| = 0.5 < 1; z = -1.5 \pm j0.5,$$

$$|z| = \sqrt{(1.5)^2 + (\pm 0.5)^2} = 1.5811 > 1.$$

The plot of poles and zeros is shown in Figure 6.10. Since we have two poles at $z = -1.5 \pm j0.5$, which are outside the unit circle, the system is unstable.

c. A zero is found to be $z = -0.5$.

$$\text{Poles:} z = 0.5, |z| = 0.5 < 1; z = -0.707 \pm j0.707,$$

$$|z| = \sqrt{(0.707)^2 + (\pm 0.707)^2} = 1.$$

The zero and poles are plotted in Figure 6.10. Since the outermost poles are first order at $z = -0.707 \pm j0.707$ and are on the unit circle, the system is marginally stable.

d. Zeros are $z = -0.5 \pm j0.5$.

$$\text{Poles:} z = 1, |z| = 1; z = 1, |z| = 1; z = -1, |z| = 1; z = 0.6,$$

$$|z| = 0.6 < 1.$$

The zeros and poles are plotted in Figure 6.10. Since the outermost pole is multiple order (second order) at $z = 1$ and is on the unit circle, the system is unstable.

6.4 Digital Filter Frequency Response

From the Laplace transfer function, we can achieve the analog filter steady-state frequency response $H(j\omega)$ by substituting $s = j\omega$ into the transfer function $H(s)$. That is,

$$H(s)|_{s=j\omega} = H(j\omega).$$

Then we can study the magnitude frequency response $|H(j\omega)|$ and phase response $\angle H(j\omega)$. Similarly, in a DSP system, using the mapping Equation (6.16), we substitute $z = e^{sT}|_{s=j\omega} = e^{j\omega T}$ into the z-transfer function $H(z)$ to acquire the digital frequency response, which is converted into the magnitude frequency response $|H(e^{j\omega T})|$ and phase response $\angle H(e^{j\omega T})$. That is,

$$H(z)|_{z=e^{j\omega T}} = H(e^{j\omega T}) = |H(e^{j\omega T})| \angle H(e^{j\omega T}). \tag{6.18}$$

Let us introduce a normalized digital frequency in radians in digital domain

$$\Omega = \omega T. \tag{6.19}$$

Then the digital frequency response in Equation (6.18) would become

$$H(e^{j\Omega}) = H(z)|_{z=e^{j\Omega}} = |H(e^{j\Omega})| \angle H(e^{j\Omega}). \tag{6.20}$$

The formal derivation for Equation (6.20) can be found in Appendix D.

Now we verify the frequency response via the following simple digital filter. Consider a digital filter with a sinusoidal input of the amplitude K (Fig. 6.11):

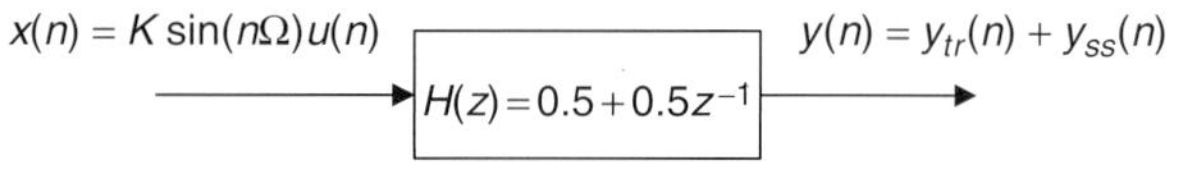

$x(n) = K \sin(n\Omega) u(n)$ → $H(z) = 0.5 + 0.5z^{-1}$ → $y(n) = y_{tr}(n) + y_{ss}(n)$

FIGURE 6.11 System transient and steady-state frequency responses.

We can determine the system output $y(n)$, which consists of the transient response $y_{tr}(n)$ and the steady-state response $y_{ss}(n)$. We find the z-transform output as

$$Y(z) = \left(\frac{0.5z + 0.5}{z}\right) \frac{Kz \sin \Omega}{z^2 - 2z \cos \Omega + 1}. \tag{6.21}$$

To perform the inverse z-transform to find the system output, we further rewrite Equation (6.21) as

$$\frac{Y(z)}{z} = \left(\frac{0.5z + 0.5}{z}\right) \frac{K \sin \Omega}{(z - e^{j\Omega})(z - e^{-j\Omega})} = \frac{A}{z} + \frac{B}{z - e^{j\Omega}} + \frac{B^*}{z - e^{-j\Omega}},$$

where A, B, and the complex conjugate B^* are the constants for the partial fractions. Applying the partial fraction expansion leads to

$$A = 0.5K \sin \Omega$$

$$B = \left.\frac{0.5z + 0.5}{z}\right|_{z=e^{j\Omega}} \frac{K}{2j} = |H(e^{j\Omega})| e^{j\angle H(e^{j\Omega})} \frac{K}{2j}.$$

Notice that the first part of constant B is a complex function, which is obtained by substituting $z = e^{j\Omega}$ into the filter z-transfer function. We can also express the complex function in terms of the polar form:

$$\left.\frac{0.5z + 0.5}{z}\right|_{z=e^{j\Omega}} = 0.5 + 0.5z^{-1}\big|_{z=e^{j\Omega}} = H(z)|_{z=e^{j\Omega}} = H(e^{j\Omega}) = |H(e^{j\Omega})| e^{j\angle H(e^{j\Omega})},$$

where $H(e^{j\Omega}) = 0.5 + 0.5e^{-j\Omega}$, and we call this complex function the steady-state frequency response. Based on the complex conjugate property, we get another residue as

$$B^* = |H(e^{j\Omega})| e^{-j\angle H(e^{j\Omega})} \frac{K}{-j2}.$$

The z-transform system output is then given by

$$Y(z) = A + \frac{Bz}{z - e^{j\Omega}} + \frac{B^* z}{z - e^{-j\Omega}}.$$

Taking the inverse z-transform, we achieve the following system transient and steady-state responses:

$$y(n) = \underbrace{0.5K \sin \Omega \delta(n)}_{y_{tr}(n)} + \underbrace{|H(e^{j\Omega})| e^{j\angle H(e^{j\Omega})} \frac{K}{j2} e^{jn\Omega} u(n) + |H(e^{j\Omega})| e^{-j\angle H(e^{j\Omega})} \frac{K}{-j2} e^{-jn\Omega} u(n)}_{y_{ss}(n)}.$$

Simplifying the response yields the form

$$y(n) = 0.5K \sin \Omega \delta(n) + |H(e^{j\Omega})| K \frac{e^{jn\Omega + j\angle H(e^{j\Omega})} u(n) - e^{-jn\Omega - j\angle H(e^{j\Omega})} u(n)}{j2}.$$

We can further combine the last term using Euler's formula to express the system response as

$$y(n) = \underbrace{0.5K \sin \Omega \delta(n)}_{y_{tr}(n) \text{ will decay to zero after the first sample}} + \underbrace{|H(e^{j\Omega})| K \sin(n\Omega + \angle H(e^{j\Omega})) u(n)}_{y_{ss}(n)}.$$

Finally, the steady-state response is identified as

$$y_{ss}(n) = K|H(e^{j\Omega})| \sin(n\Omega + \angle H(e^{j\Omega})) u(n).$$

For this particular filter, the transient response exists for only the first sample in the system response. By substituting $n = 0$ into $y(n)$ and after simplifying algebra, we achieve the response for the first output sample:

$$y(0) = y_{tr}(0) + y_{ss}(0) = 0.5K \sin(\Omega) - 0.5K \sin(\Omega) = 0.$$

Note that the first output sample of the transient response cancels the first output sample of the steady-state response, so the combined first output sample has a value of zero for this particular filter. The system response reaches the steady-state response after the first output sample. At this point, we can conclude:

$$\text{Steady-state magnitude frequency response}$$
$$= \frac{\text{Peak amplitude of steady-state response at } \Omega}{\text{Peak amplitude of sinusoidal input at } \Omega}$$
$$= \frac{|H(e^{j\Omega})|K}{K} = |H(e^{j\Omega})|$$

Steady-state phase frequency response = Phase difference = $\angle H(e^{j\Omega})$.

Figure 6.12 shows the system responses with sinusoidal inputs at $\Omega = 0.25\pi$, $\Omega = 0.5\pi$, and $\Omega = 0.75\pi$, respectively.

Next, we examine the properties of the filter frequency response $H(e^{j\Omega})$. From Euler's identity and trigonometric identity, we know that

$$e^{j(\Omega + k2\pi)} = \cos(\Omega + k2\pi) + j\sin(\Omega + k2\pi)$$
$$= \cos \Omega + j \sin \Omega = e^{j\Omega},$$

where k is an integer taking values of $k = 0, \pm 1, \pm 2, \ldots,$. Then the frequency response has the following property (assuming that all input sequences are real):

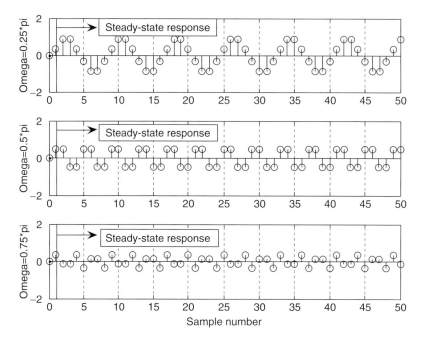

FIGURE 6.12 The digital filter responses to different input sinusoids.

1. Periodicity
   a. Frequency response: $H(e^{j\Omega}) = H(e^{j(\Omega+k2\pi)})$
   b. Magnitude frequency response: $|H(e^{j\Omega})| = |H(e^{j(\Omega+k2\pi)})|$
   c. Phase response: $\angle H(e^{j\Omega}) = \angle H(e^{j\Omega+k2\pi})$

The second property is given without proof (see proof in Appendix D) as shown:

2. Symmetry
   a. Magnitude frequency response: $|H(e^{-j\Omega})| = |H(e^{j\Omega})|$
   b. Phase response: $\angle H(e^{-j\Omega}) = -\angle H(e^{j\Omega})$

Since the maximum frequency in a DSP system is the folding frequency, $f_s/2$, where $f_s = 1/T$ and T designates the sampling period, the corresponding maximum normalized frequency can be calculated as

$$\Omega = \omega T = 2\pi \frac{f_s}{2} \times T = \pi \text{ radians.} \qquad (6.22)$$

6.4 Digital Filter Frequency Response

The frequency response $H(e^{j\Omega})$ for $|\Omega| > \pi$ consists of the image replicas of $H(e^{j\Omega})$ for $|\Omega| \le \pi$ and will be removed via the reconstruction filter later. Hence, we need to evaluate $H(e^{j\Omega})$ for only the positive normalized frequency range from $\Omega = 0$ to $\Omega = \pi$ radians. The frequency, in Hz, can be determined by

$$f = \frac{\Omega}{2\pi} f_s. \tag{6.23}$$

The magnitude frequency response is often expressed in decibels, defined as

$$\left| H(e^{j\Omega}) \right|_{dB} = 20 \log_{10} \left(\left| H(e^{j\Omega}) \right| \right). \tag{6.24}$$

The DSP system stability, magnitude response, and phase response are investigated via the following examples.

Example 6.10.

Given the following digital system with a sampling rate of 8,000 Hz,

$$y(n) = 0.5x(n) + 0.5x(n-1),$$

1. Determine the frequency response.

Solution:

1. Taking the z-transform on both sides of the difference equation leads to

$$Y(z) = 0.5X(z) + 0.5z^{-1}X(z).$$

Then the transfer function describing the system is easily found to be

$$H(z) = \frac{Y(z)}{X(z)} = 0.5 + 0.5z^{-1}.$$

Substituting $z = e^{j\Omega}$, we have the frequency response as

$$H(e^{j\Omega}) = 0.5 + 0.5e^{-j\Omega}$$
$$= 0.5 + 0.5\cos(\Omega) - j0.5\sin(\Omega).$$

Therefore, the magnitude frequency response and phase response are given by

$$\left| H(e^{j\Omega}) \right| = \sqrt{(0.5 + 0.5\cos(\Omega))^2 + (0.5\sin(\Omega))^2}$$

and

$$\angle H(e^{j\Omega}) = \tan^{-1}\left(\frac{-0.5\sin(\Omega)}{0.5 + 0.5\cos(\Omega)} \right).$$

Several points for the magnitude response and phase response are calculated and shown in Table 6.1.

According to data, we plot the magnitude frequency response and phase response of the DSP system as shown in Figure 6.13.

It is observed that when the frequency increases, the magnitude response decreases. The DSP system acts like a digital lowpass filter, and its phase response is linear.

TABLE 6.1 Frequency response calculations in Example 6.10.

Ω (radians)	$f = \frac{\Omega}{2\pi}f_s$ (Hz)	$\left|H(e^{j\Omega})\right|$	$\left|H(e^{j\Omega})\right|_{dB}$	$\angle H(e^{j\Omega})$
0	0	1.000	0 dB	0°
0.25π	1000	0.924	-0.687 dB	$-22.5°$
0.50π	2000	0.707	-3.012 dB	$-45.00°$
0.75π	3000	0.383	-8.336 dB	$-67.50°$
1.00π	4000	0.000	$-\infty$	$-90°$

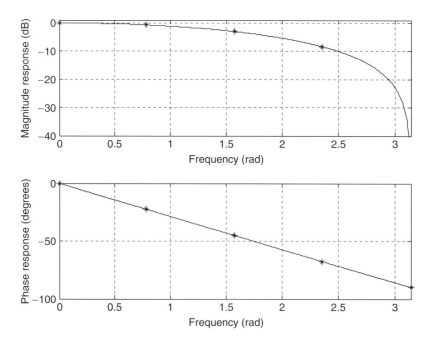

FIGURE 6.13 Frequency responses of the digital filter in Example 6.10.

6.4 Digital Filter Frequency Response

We can also verify the periodicity for $|H(e^{j\Omega})|$ and $\angle H(e^{j\Omega})$ when $\Omega = 0.25\pi + 2\pi$:

$$|H(e^{j(0.25\pi+2\pi)})| = \sqrt{(0.5 + 0.5\cos(0.25\pi + 2\pi))^2 + (0.5\sin(0.25\pi + 2\pi))^2}$$
$$= 0.924 = |H(e^{j0.25\pi})|$$

$$\angle H(e^{j(0.25\pi+2\pi)}) = \tan^{-1}\left(\frac{-0.5\sin(0.25\pi + 2\pi)}{0.5 + 0.5\cos(0.25\pi + 2\pi)}\right) = -22.5^0 = \angle H(e^{j0.25\pi}).$$

For $\Omega = -0.25\pi$, we can verify the symmetry property as

$$|H(e^{-j0.25\pi})| = \sqrt{(0.5 + 0.5\cos(-0.25\pi))^2 + (0.5\sin(-0.25\pi))^2}$$
$$= 0.924 = |H(e^{j0.25\pi})|$$

$$\angle H(e^{-j0.25\pi}) = \tan^{-1}\left(\frac{-0.5\sin(-0.25\pi)}{0.5 + 0.5\cos(-0.25\pi)}\right) = 22.5^0 = -\angle H(e^{j0.25\pi}).$$

The properties can be observed in Figure 6.14, where the frequency range is chosen from $\Omega = -2\pi$ to $\Omega = 4\pi$ radians. As shown in the figure, the magnitude and phase responses are periodic with a period of 2π. For a period

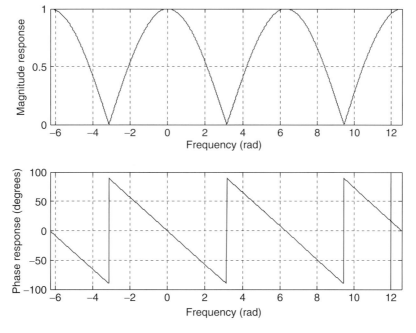

FIGURE 6.14 Periodicity of the magnitude response and phase response in Example 6.10.

between $\Omega = -\pi$ to $\Omega = \pi$, the magnitude responses for the portion $\Omega = -\pi$ to $\Omega = 0$ and the portion $\Omega = 0$ to $\Omega = \pi$ are the same, while the phase responses are opposite. The magnitude and phase responses calculated for the range from $\Omega = 0$ to $\Omega = \pi$ carry all the frequency response information, hence are required for generating only the frequency response plots.

Again, note that the phase plot shows a sawtooth shape instead of a linear straight line for this particular filter. This is due to the phase wrapping at $\Omega = 2\pi$ radians, since $e^{j(\Omega+k2\pi)} = e^{j\Omega}$ is used in the calculation. However, the phase plot shows that the phase is linear in the useful information range from $\Omega = 0$ to $\Omega = \pi$ radians.

Example 6.11.

Given a digital system with a sampling rate of 8,000 Hz,

$$y(n) = x(n) - 0.5y(n-1),$$

1. Determine the frequency response.

Solution:

1. Taking the z-transform on both sides of the difference equation leads to

$$Y(z) = X(z) - 0.5z^{-1}Y(z).$$

Then the transfer function describing the system is easily found to be

$$H(z) = \frac{Y(z)}{X(z)} = \frac{1}{1 + 0.5z^{-1}} = \frac{z}{z + 0.5}.$$

Substituting $z = e^{j\Omega}$, we have the frequency response as

$$H(e^{j\Omega}) = \frac{1}{1 + 0.5e^{-j\Omega}}$$

$$= \frac{1}{1 + 0.5\cos(\Omega) - j0.5\sin(\Omega)}.$$

Therefore, the magnitude frequency response and phase response are given by

$$|H(e^{j\Omega})| = \frac{1}{\sqrt{(1 + 0.5\cos(\Omega))^2 + (0.5\sin(\Omega))^2}}$$

and

$$\angle H(e^{j\Omega}) = -\tan^{-1}\left(\frac{-0.5\sin(\Omega)}{1 + 0.5\cos(\Omega)}\right), \text{ respectively.}$$

Several points for the magnitude response and phase response are calculated and shown in Table 6.2.

TABLE 6.2 Frequency response calculations in Example 6.11.

Ω (radians)	$f = \frac{\Omega}{2\pi} f_s$ (Hz)	$\left\lvert H(e^{j\Omega}) \right\rvert$	$\left\lvert H(e^{j\Omega}) \right\rvert_{dB}$	$\angle H(e^{j\Omega})$
0	0	0.670	−3.479 dB	0°
0.25π	1000	0.715	−2.914 dB	14.64°
0.50π	2000	0.894	−0.973 dB	26.57°
0.75π	3000	1.357	2.652 dB	28.68°
1.00π	4000	2.000	6.021 dB	0°

According to the achieved data, the magnitude response and phase response of the DSP system are roughly plotted in Figure 6.15.

From Table 6.2 and Figure 6.15, we can see that when the frequency increases, the magnitude response increases. The DSP system actually performs digital highpass filtering.

Notice that if all the coefficients a_i for $i = 0, 1, \ldots, M$ in Equation (6.1) are zeros, Equation (6.2) is reduced to

$$y(n) = \sum_{i=0}^{M} b_i x(n-i) \qquad (6.25)$$
$$= b_0 x(n) + b_1 x(n-1) + \cdots + b_K x(n-M).$$

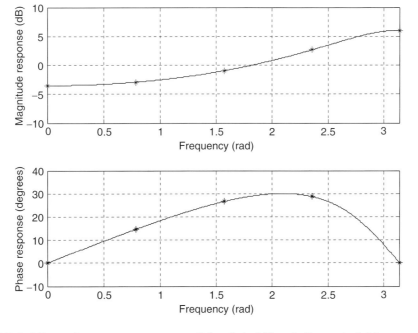

FIGURE 6.15 Frequency responses of the digital filter in Example 6.11.

Notice that b_i is the ith impulse response coefficient. Also, since M is a finite positive integer, b_i in this particular case is a finite set, $H(z) = B(z)$; note that the denominator $A(z) = 1$. Such systems are called *finite impulse response* (FIR) systems. If not all a_i in Equation (6.1) are zeros, the impulse response $h(i)$ will consist of an infinite number of coefficients. Such systems are called *infinite impulse response* (IIR) systems. The z-transform of the IIR $h(i)$, in general, is given by $H(z) = \frac{B(z)}{A(z)}$, where $A(z) \neq 1$.

6.5 Basic Types of Filtering

The basic filter types can be classified into four categories: *lowpass, highpass, bandpass*, and *bandstop*. Each of them finds a specific application in digital signal processing. One of the objectives in applications may involve the design of digital filters. In general, the filter is designed based on specifications primarily for the passband, stopband, and transition band of the filter frequency response. The filter passband is the frequency range with the amplitude gain of the filter response being approximately unity. The filter stopband is defined as the frequency range over which the filter magnitude response is attenuated to eliminate the input signal whose frequency components are within that range. The transition band denotes the frequency range between the passband and the stopband.

The design specifications of the lowpass filter are illustrated in Figure 6.16, where the low-frequency components are passed through the filter while the high-frequency components are attenuated. As shown in Figure 6.16, Ω_p and Ω_s are the passband cutoff frequency and the stopband cutoff frequency, respectively; δ_p is the design parameter to specify the ripple (fluctuation) of the frequency response in the passband, while δ_s specifies the ripple of the frequency response in the stopband.

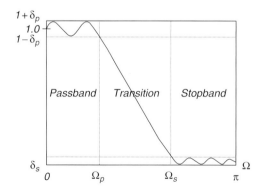

FIGURE 6.16 Magnitude response of the normalized lowpass filter.

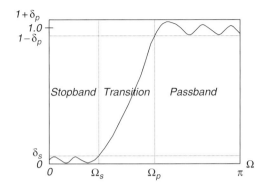

FIGURE 6.17 Magnitude response of the normalized highpass filter.

The highpass filter, remains high-frequency components and rejects low-frequency components. The magnitude frequency response for the highpass filter is demonstrated in Figure 6.17.

The bandpass filter attenuates both low- and high-frequency components while remaining the middle-frequency component, as shown in Figure 6.18.

As illustrated in Figure 6.18, Ω_{pL} and Ω_{sL} are the lower passband cutoff frequency and the lower stopband cutoff frequency, respectively. Ω_{pH} and Ω_{sH} are the upper passband cutoff frequency and the upper stopband cutoff frequency, respectively. δ_p is the design parameter to specify the ripple of the frequency response in the passband, while δ_s specifies the ripple of the frequency response in the stopbands.

Finally, the bandstop (band reject or notch) filter, shown in Figure 6.19, rejects the middle-frequency components and accepts both the low- and the high-frequency component.

As a matter of fact, all kinds of digital filters are implemented using FIR and IIR systems. Furthermore, the FIR and IIR systems can each be realized by

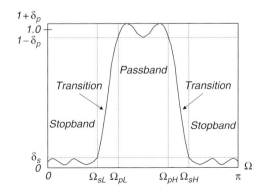

FIGURE 6.18 Magnitude response of the normalized bandpass filter.

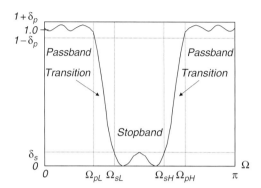

FIGURE 6.19 Magnitude of the normalized bandstop filter.

various filter configurations, such as direct forms, cascade forms, and parallel forms. Such topics will be included in the next section.

Given a transfer function, the MATLAB function **freqz()** can be used to determine the frequency response. The syntax is given by

$$[\mathbf{h}, \mathbf{w}] = \mathbf{freqz}(\mathbf{B}, \mathbf{A}, \mathbf{N}),$$

whose parameters are defined as:

h = an output vector containing frequency response
w = an output vector containing normalized frequency values distributed in the range from 0 to π radians.
B = an input vector for numerator coefficients
A = an input vector for denominator coefficients
N = the number of normalized frequency points used for calculating the frequency response

Let's consider Example 6.12.

Example 6.12.

Given each of the following digital transfer functions,

a. $H(z) = \dfrac{z}{z - 0.5}$

b. $H(z) = 1 - 0.5z^{-1}$

c. $H(z) = \dfrac{0.5z^2 - 0.32}{z^2 - 0.5z + 0.25}$

d. $H(z) = \dfrac{1 - 0.9z^{-1} + 0.81z^{-2}}{1 - 0.6z^{-1} + 0.36z^{-2}}$,

1. Plot the poles and zeros on the z-plane.

2. Use MATLAB function **freqz()** to plot the magnitude frequency response and phase response for each transfer function.

3. Identify the corresponding filter type, such as lowpass, highpass, bandpass, or bandstop.

Solution:

1. The pole-zero plot for each transfer function is demonstrated in Figure 6.20. The transfer functions of (a) and (c) need to be converted into the standard form (delay form) required by the MATLAB function **freqz()**, in which both numerator and denominator polynomials have negative powers of z. Hence, we obtain

$$H(z) = \frac{z}{z - 0.5} = \frac{1}{1 - 0.5z^{-1}}$$

$$H(z) = \frac{0.5z^2 - 0.32}{z^2 - 0.5z + 0.25} = \frac{0.5 - 0.32z^{-2}}{1 - 0.5z^{-1} + 0.25z^{-2}},$$

while the transfer functions of (b) and (d) are already in their standard forms (delay forms).

2. The MATLAB program for plotting the magnitude frequency response and the phase response for each case is listed in Program 6.2.

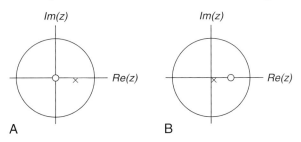

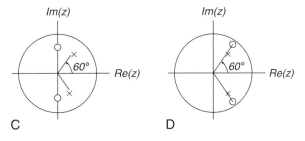

FIGURE 6.20 Pole-zero plots of Example 6.12.

Program 6.2. MATLAB program for Example 6.12.

```
% Example 6.12
% Plot the magnitude frequency response and phase response
% Case a
figure (1)
[h w] = freqz([1] [1 −0.5] 1024);           % Calculate the frequency response
phi = 180*unwrap(angle(h))/pi;
subplot(2,1,1),plot(w,abs(h)),grid;xlabel('Frequency (radians)'),
ylabel ('Magnitude')
subplot(2,1,2),plot(w,phi),grid;xlabel('Frequency (radians)'),
ylabel('Phase (degrees)')
% Case b
figure (2)
[h w] = freqz([1 −0.5] [1] 1024);           %Calculate the frequency response
phi = 180*unwrap(angle(h))/pi;
subplot(2,1,1),plot(w,abs(h)),grid;xlabel('Frequency (radians)'),
ylabel ('Magnitude')
subplot(2,1,2),plot(w,phi),grid;xlabel('Frequency (radians)'),
ylabel('Phase (degrees)')
% Case c
figure (3)
[h w] = freqz([0.5 0 -0.32],[1 -0.5 0.25],1024);%Calculate the frequency response
phi = 180*unwrap(angle(h))/pi;
subplot(2,1,1),plot(w,abs(h)),grid;xlabel('Frequency (radians)'),
ylabel('Magnitude')
subplot(2,1,2),plot(w,phi),grid;xlabel('Frequency (radians)'),
ylabel('Phase (degrees)')
% Case d
figure (4)
[h w] = freqz([1 -0.9 0.81],[1 -0.6 0.36],1024);%Calculate the frequency response
phi = 180*unwrap(angle(h))/pi;
subplot(2,1,1),plot(w,abs(h)),grid;xlabel('Frequency (radians)'),
ylabel ('Magnitude')
subplot(2,1,2),plot(w,phi),grid;xlabel('Frequency (radians)'),
ylabel('Phase (degrees)')
%
```

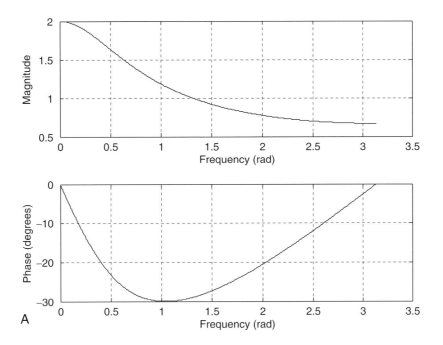

FIGURE 6.21A   Plots of frequency responses for Example 6.12 for (a).

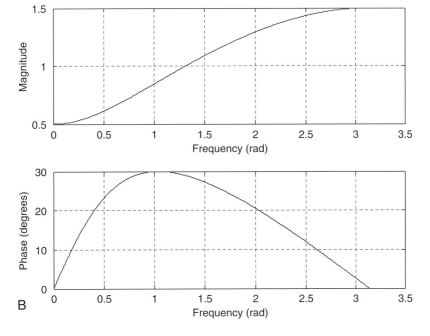

FIGURE 6.21B   Plots of frequency responses for Example 6.12 for (b).

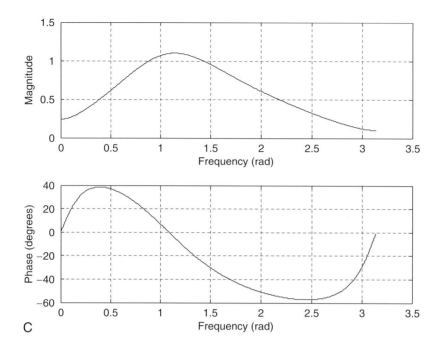

FIGURE 6.21C   Plots of frequency responses for Example 6.12 for (c).

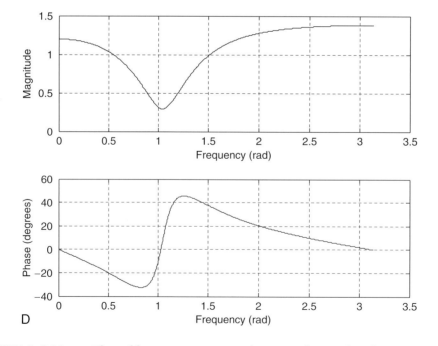

FIGURE 6.21D   Plots of frequency responses for Example 6.12 for (d).

3. From the plots in Figures 6.21a–6.21d of magnitude frequency responses for all cases, we can conclude that case (a) is a lowpass filter, (b) is a highpass filter, (c) is a bandpass filter, and (d) is a bandstop (band reject) filter.

6.6 Realization of Digital Filters

In this section, basic realization methods for digital filters are discussed. Digital filters described by the transfer function $H(z)$ may be generally realized in the following forms:

- Direct form I
- Direct form II
- Cascade
- Parallel.

(The reader can explore various lattice realizations in the textbook by Stearns and Hush [1990].)

6.6.1 Direct-Form I Realization

As we know, a digital filter transfer function, $H(z)$, is given by

$$H(z) = \frac{B(z)}{A(z)} = \frac{b_0 + b_1 z^{-1} + \cdots + b_M z^{-M}}{1 + a_1 z^{-1} + \cdots + a_N z^{-N}}. \quad (6.26)$$

Let $x(n)$ and $y(n)$ be the digital filter input and output, respectively. We can express the relationship in z-transform domain as

$$Y(z) = H(z)X(z), \quad (6.27)$$

where $X(z)$ and $Y(z)$ are the z-transforms of $x(n)$ and $y(n)$, respectively. If we substitute Equation (6.26) into $H(z)$ in Equation (6.27), we have

$$Y(z) = \left(\frac{b_0 + b_1 z^{-1} + \cdots + b_M z^{-M}}{1 + a_1 z^{-1} + \cdots + a_N z^{-N}} \right) X(z). \quad (6.28)$$

Taking the inverse of the z-transform of Equation (6.28), we yield the relationship between input $x(n)$ and output $y(n)$ in time domain, as follows:

$$\begin{aligned} y(n) = & b_0 x(n) + b_1 x(n-1) + \cdots + b_M x(n-M) \\ & - a_1 y(n-1) - a_2 y(n-2) - \cdots - a_N y(n-N). \end{aligned} \quad (6.29)$$

This difference equation thus can be implemented by a direct-form I realization shown in Figure 6.22(a). Figure 6.22(b) illustrates the realization of the second-order IIR filter ($M = N = 2$). Note that the notation used in Figures

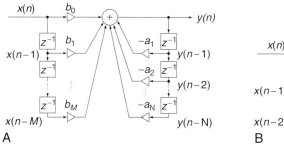
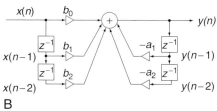

FIGURE 6.22 (a) Direct-form I realization. (b) Direct-form I realization with M = 2. (c) Notation.

6.22(a) and (b) are defined in Figure 6.22(c) and will be applied for discussion of other realizations.

Also, notice that any of the a_j and b_i can be zero, thus all the paths are not required to exist for the realization.

6.6.2 Direct-Form II Realization

Considering Equations (6.26) and (6.27) with $N = M$, we can express

$$Y(z) = H(z)X(z) = \frac{B(z)}{A(z)}X(z) = B(z)\left(\frac{X(z)}{A(z)}\right)$$

$$= (b_0 + b_1 z^{-1} + \cdots + b_M z^{-M})\underbrace{\left(\frac{X(z)}{1 + a_1 z^{-1} + \cdots + a_M z^{-M}}\right)}_{W(z)}. \quad (6.30)$$

Also, defining a new z-transform function as

$$W(z) = \frac{X(z)}{1 + a_1 z^{-1} + \cdots + a_M z^{-M}}, \quad (6.31)$$

we have

$$Y(z) = (b_0 + b_1 z^{-1} + \cdots + b_M z^{-M}) W(z). \tag{6.32}$$

The corresponding difference equations for Equations (6.31) and (6.32), respectively, become

$$w(n) = x(n) - a_1 w(n-1) - a_2 w(n-2) - \cdots - a_M w(n-M) \tag{6.33}$$

and

$$y(n) = b_0 w(n) + b_1 w(n-1) + \ldots + b_M w(n-M). \tag{6.34}$$

Realization of Equations (6.33) and (6.34) becomes another direct-form II realization, which is demonstrated in Figure 6.23(a). Again, the corresponding realization of the second-order IIR filter is described in Figure 6.23(b). Note that in Figure 6.23(a), the variables $w(n)$, $w(n-1)$, $w(n-2)$, ..., $w(n-M)$ are different from the filter inputs $x(n-1)$, $x(n-2)$, ..., $x(n-M)$.

6.6.3 Cascade (Series) Realization

An alternate way to filter realization is to cascade the factorized $H(z)$ in the following form:

$$H(z) = H_1(z) \cdot H_2(z) \cdots H_k(z), \tag{6.35}$$

where $H_k(z)$ is chosen to be the first- or second-order transfer function (section), which is defined by

$$H_k(z) = \frac{b_{k0} + b_{k1} z^{-1}}{1 + a_{k1} z^{-1}} \tag{6.36}$$

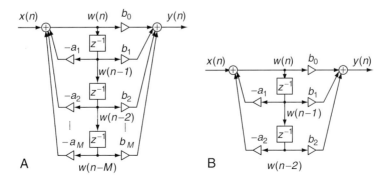

FIGURE 6.23 (A) Direct-form II realization. (B) Direct-form II realization with M = 2.

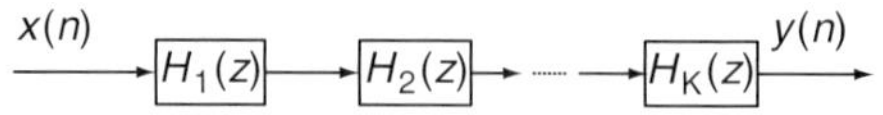

x(n) →[H₁(z)]→[H₂(z)]→ ⋯ →[H_K(z)]→ y(n)

FIGURE 6.24 Cascade realization.

or

$$H_k(z) = \frac{b_{k0} + b_{k1}z^{-1} + b_{k2}z^{-2}}{1 + a_{k1}z^{-1} + a_{k2}z^{-2}}, \quad (6.37)$$

respectively. The block diagram of the cascade, or series, realization is depicted in Figure 6.24.

6.6.4 Parallel Realization

Now we convert $H(z)$ into the following form:

$$H(z) = H_1(z) + H_2(z) + \cdots + H_k(z), \quad (6.38)$$

where $H_k(z)$ is defined as the first- or second-order transfer function (section) given by

$$H_k(z) = \frac{b_{k0}}{1 + a_{k1}z^{-1}} \quad (6.39)$$

or

$$H_k(z) = \frac{b_{k0} + b_{k1}z^{-1}}{1 + a_{k1}z^{-1} + a_{k2}z^{-2}}, \quad (6.40)$$

respectively. The resulting parallel realization is illustrated in the block diagram in Figure 6.25.

Example 6.13.

Given a second-order transfer function

$$H(z) = \frac{0.5(1 - z^{-2})}{1 + 1.3z^{-1} + 0.36z^{-2}},$$

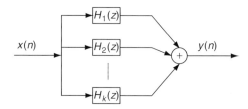

FIGURE 6.25 Parallel realization.

a. Perform the filter realizations and write the difference equations using the following realizations:

1. direct form I and direct form II
2. cascade form via the first-order sections
3. parallel form via the first-order sections.

Solution:

a. 1. To perform the filter realizations using the direct form I and direct form II, we rewrite the given second-order transfer function as

$$H(z) = \frac{0.5 - 0.5z^{-2}}{1 + 1.3z^{-1} + 0.36z^{-2}}$$

and identify that

$$a_1 = 1.3, \quad a_2 = 0.36, \quad b_0 = 0.5, \quad b_1 = 0, \quad \text{and } b_2 = -0.5.$$

Based on realizations in Figure 6.22, we sketch the direct-form I realization as Figure 6.26.

The difference equation for the direct-form I realization is given by

$$y(n) = 0.5x(n) - 0.5x(n-2) - 1.3y(n-1) - 0.36y(n-2).$$

Using the direct-form II realization shown in Figure 6.23, we have the realization in Figure 6.27.

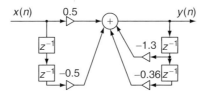

FIGURE 6.26   Direct-form I realization for Example 6.13.

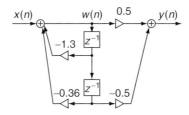

FIGURE 6.27   Direct-form II realization for Example 6.13.

The difference equations for the direct-form II realization are expressed as

$$w(n) = x(n) - 1.3w(n-1) - 0.36w(n-2)$$
$$y(n) = 0.5w(n) - 0.5w(n-2).$$

2. To achieve the cascade (series) form realization, we factor $H(z)$ into two first-order sections to yield

$$H(z) = \frac{0.5(1-z^{-2})}{1+1.3z^{-1}+0.36z^{-2}} = \frac{0.5-0.5z^{-1}}{1+0.4z^{-1}} \frac{1+z^{-1}}{1+0.9z^{-1}},$$

where $H_1(z)$ and $H_2(z)$ are chosen to be

$$H_1(z) = \frac{0.5-0.5z^{-1}}{1+0.4z^{-1}}$$
$$H_2(z) = \frac{1+z^{-1}}{1+0.9z^{-1}}.$$

Notice that the obtained $H_1(z)$ and $H_2(z)$ are not unique selections for realization. For example, there is another way of choosing $H_1(z) = \frac{0.5-0.5z^{-1}}{1+0.9z^{-1}}$ and $H_2(z) = \frac{1+z^{-1}}{1+0.4z^{-1}}$ to yield the same $H(z)$. Using the $H_1(z)$ and $H_2(z)$ we have obtained, and with the direct-form II realization, we achieve the cascade form depicted in Figure 6.28.

The difference equations for the direct-form II realization have two cascaded sections, expressed as

Section 1:

$$w_1(n) = x(n) - 0.4w(n-1)$$
$$y_1(n) = 0.5w_1(n) - 0.5w_1(n-1)$$

Section 2:

$$w_2(n) = y_1(n) - 0.9w_2(n-1)$$
$$y(n) = w_2(n) + w_2(n-1)$$

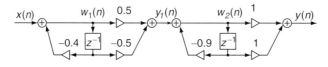

FIGURE 6.28 Cascade realization for Example 6.13.

3. In order to yield the parallel form of realization, we need to make use of the partial fraction expansion, and will first let

$$\frac{H(z)}{z} = \frac{0.5(z^2-1)}{z(z+0.4)(z+0.9)} = \frac{A}{z} + \frac{B}{z+0.4} + \frac{C}{z+0.9},$$

where

$$A = z\left(\frac{0.5(z^2-1)}{z(z+0.4)(z+0.9)}\right)\bigg|_{z=0} = \frac{0.5(z^2-1)}{(z+0.4)(z+0.9)}\bigg|_{z=0} = -1.39$$

$$B = (z+0.4)\left(\frac{0.5(z^2-1)}{z(z+0.4)(z+0.9)}\right)\bigg|_{z=-0.4} = \frac{0.5(z^2-1)}{z(z+0.9)}\bigg|_{z=-0.4} = -0.21$$

$$C = (z+0.9)\left(\frac{0.5(z^2-1)}{z(z+0.4)(z+0.9)}\right)\bigg|_{z=-0.9} = \frac{0.5(z^2-1)}{z(z+0.4)}\bigg|_{z=-0.9} = 0.21.$$

Therefore

$$H(z) = -1.39 + \frac{2.1z}{z+0.4} + \frac{-0.21z}{z+0.9} = -1.39 + \frac{2.1}{1+0.4z^{-1}} + \frac{-0.21}{1+0.9z^{-1}}.$$

Again, using the direct form II for each section, we obtain the parallel realization in Figure 6.29.

The difference equations for the direct-form II realization have three parallel sections, expressed as

$$y_1(n) = -1.39x(n)$$
$$w_2(n) = x(n) - 0.4w_2(n-1)$$
$$y_2(n) = 2.1w_2(n)$$
$$w_3(n) = x(n) - 0.9w_3(n-1)$$
$$y_3(n) = -0.21w_3(n)$$
$$y(n) = y_1(n) + y_2(n) + y_3(n).$$

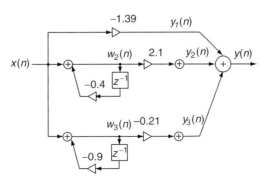

FIGURE 6.29 Parallel realization for Example 6.13.

In practice, the second-order filter module using the direct-form I or direct-form II is used. The high-order filter can be factored in the cascade form with the first- or second-order sections. In case the first-order filter is required, we can still modify the second-order filter module by setting the corresponding filter coefficients to be zero.

6.7 Application: Speech Enhancement and Filtering

This section investigates applications of speech enhancement using a pre-emphasis filter and speech filtering using a bandpass filter.

6.7.1 Pre-Emphasis of Speech

A speech signal may have frequency components that fall off at high frequencies. In some applications such as speech coding, to avoid overlooking the high frequencies, the high-frequency components are compensated using pre-emphasis filtering. A simple digital filter used for such compensation is given as:

$$y(n) = x(n) - \alpha x(n-1), \quad (6.41)$$

where α is the positive parameter to control the degree of pre-emphasis filtering and usually is chosen to be less than 1. The filter described in Equation (6.41) is essentially a highpass filter. Applying z-transform on both sides of Equation (6.41) and solving for the transfer function, we have

$$H(z) = 1 - \alpha z^{-1}. \quad (6.42)$$

The magnitude and phase responses adopting the pre-emphasis parameter $\alpha = 0.9$ and the sampling rate $f_s = 8,000\,\text{Hz}$ are plotted in Figure 6.30a using MATLAB.

Figure 6.30b compares the original speech waveform and the pre-emphasized speech using the filter in Equation (6.42). Again, we apply the fast Fourier transform (FFT) to estimate the spectrum of the original speech and the spectrum of the pre-emphasized speech. The plots are displayed in Figure 6.31.

From Figure 6.31, we can conclude that the filter does its job to boost the high-frequency components and attenuate the low-frequency components. We can also try this filter with different values of α to examine the degree of the pre-emphasis filtering of the digitally recorded speech. The MATLAB list is in Program 6.3.

6.7 Application: Speech Enhancement and Filtering 203

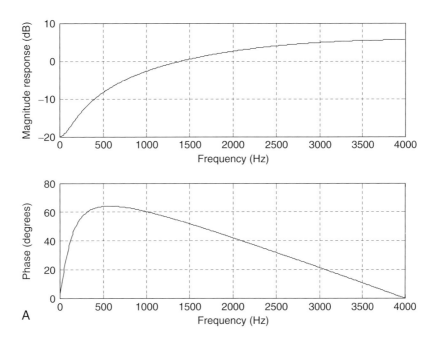

FIGURE 6.30A   Frequency responses of the pre-emphasis filter.

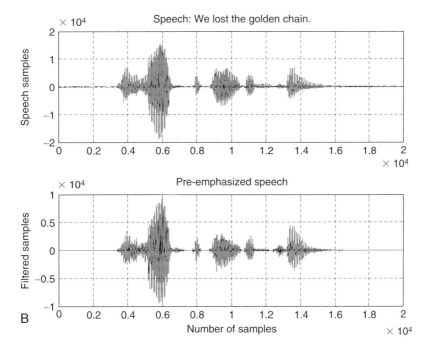

FIGURE 6.30B   Original speech and pre-emphasized speech waveforms.

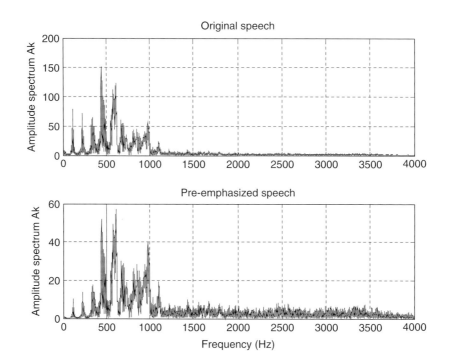

FIGURE 6.31 Amplitude spectral plots for the original speech and pre-emphasized speech.

Program 6.3. MATLAB program for pre-emphasis of speech.

```
% Matlab program for Figures 6.30 and 6.31
close all;clear all
fs = 8000;           % Sampling rate
alpha = 0.9;         % Degree of pre-emphasis
figure(1);
freqz([1-alpha],1,512,fs);    % Calculate and display frequency responses
load speech.dat
figure(2);
y = filter([1-alpha],1,speech);         % Filtering speech
subplot(2,1,1),plot(speech,'k');grid;
ylabel('Speech samples')
title('Speech: We lost the golden chain.')
subplot(2,1,2),plot(y,'k');grid
```

```
ylabel('Filtered samples')
xlabel('Number of samples');
title('Pre-emphasized speech.')
figure(3);
N = length(speech);          % Length of speech
Axk = abs(fft(speech.*hamming(N)'))/N;    % Two-sided spectrum of speech
Ayk = abs(fft(y.*hamming(N)'))/N;% Two-sided spectrum of pre-emphasized speech
f=[0:N/2]*fs/N;
Axk(2:N)=2*Axk(2:N);          % Get one-sided spectrum of speech
Ayk(2:N)= 2*Ayk(2:N);         % Get one-sided spectrum of filtered speech
subplot(2,1,1),plot(f,Axk(1:N/2+1),'k');grid
ylabel('Amplitude spectrum Ak')
title('Original speech');
subplot(2,1,2),plot(f,Ayk(1:N/2+1),'k');grid
ylabel('Amplitude spectrum Ak')
xlabel('Frequency (Hz)');
title('Preemphasized speech');
%
```

6.7.2 Bandpass Filtering of Speech

Bandpass filtering plays an important role in DSP applications. It can be used to pass the signals according to the specified frequency passband and reject the frequency other than the passband specification. Then the filtered signal can be further used for the signal feature extraction. Filtering can also be applied to perform applications such as noise reduction, frequency boosting, digital audio equalizing, and digital crossover, among others.

Let us consider the following digital fourth-order bandpass Butterworth filter with a lower cutoff frequency of 1,000 Hz, an upper cutoff frequency of 1,400 Hz (that is, the bandwidth is 400 Hz), and a sampling rate of 8,000 Hz:

$$H(z) = \frac{0.0201 - 0.0402z^{-2} + 0.0201z^{-4}}{1 - 2.1192z^{-1} + 2.6952z^{-2} - 1.6924z^{-3} + 0.6414z^{-4}}. \quad (6.43)$$

Converting the z-transfer function into the DSP difference equation yields

$$y(n) = 0.0201x(n) - 0.0402x(n-2) + 0.0201x(n-4) \\ + 2.1192y(n-1) - 2.6952y(n-2) + 1.6924y(n-3) - 0.6414y(n-4). \quad (6.44)$$

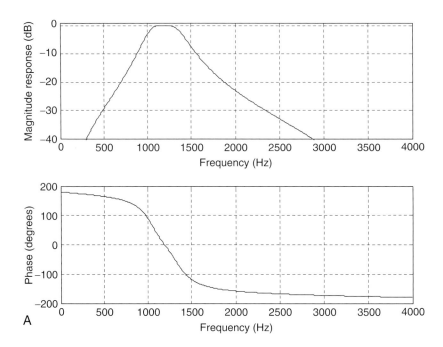

FIGURE 6.32A Frequency responses of the designed bandpass filter.

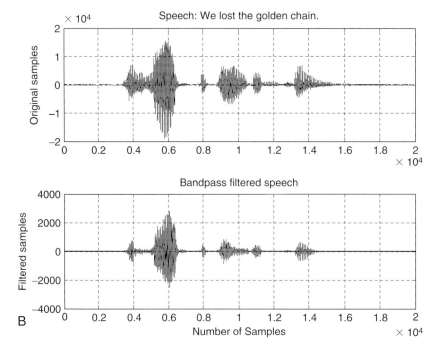

FIGURE 6.32B Plots of the original speech and filtered speech.

6.7 Application: Speech Enhancement and Filtering

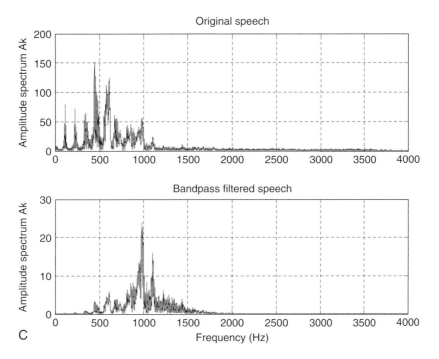

FIGURE 6.32C Amplitude spectra of the original speech and bandpass filtered speech.

The filter frequency responses are computed and plotted in Figure 6.32(a) with MATLAB. Figure 6.32(b) shows the original speech and filtered speech, while Figure 6.32(c) displays the spectral plots for the original speech and filtered speech.

As shown in Figure 6.32(c), the designed bandpass filter significantly reduces low-frequency components, which are less than 1,000 Hz, and high-frequency components, above 1,400 Hz, while letting the signals with the frequencies

Program 6.4. MATLAB program for bandpass filtering of speech.

```
fs=8000;         % Sampling rate
freqz[(0.0201 0.00 −0.0402 0 0.0201],[1 −2.1192 2.6952 −1.6924 0.6414],512,fs);
axis([0 fs/2 −40 1]);% Frequency responses of the bandpass filter
load speech.dat
y=filter([0.0201 0.00 −0.0402 0.0201],[1 −2.1192 2.6952 −1.6924 0.6414], speech);
subplot(2,1,1),plot(speech);grid;        % Filtering speech
ylabel('Origibal Samples')
```

(Continued)

```
title('Speech: We lost the golden chain.')
subplot(2,1,2),plot(y);grid
xlabel('Number of Samples');ylabel('Filtered Samples')
title('Bandpass filtered speech.')
figure
N=length(speech);
Axk=abs(fft(speech.*hamming(N)'))/N;      % One-sided spectrum of speech
Ayk=abs(fft(y.*hamming(N)'))/N;    % One-sided spectrum of filtered speech
f=[0:N/2]*fs/N;
Axk(2:N) = 2*Axk(2:N);Ayk(2:N) = 2*Ayk(2:N);            % One-sided spectra
subplot(2,1,1),plot(f,Axk(1:N/2+1));grid
ylabel('Amplitude spectrum Ak')
title('Original speech');
subplot(2,1,2),plot(f,Ayk(1:N/2+1),'w');grid
ylabel('Amplitude spectrum Ak');xlabel('Frequency (Hz)');
title('Bandpass filtered speech');
```

ranging from 1,000 to 1,400 Hz pass through the filter. Similarly, we can design and implement other types of filters, such as lowpass, highpass, and band reject to filter the signals and examine the performances of their designs. MATLAB implementation detail is given in Program 6.4.

6.8 Summary

1. The digital filter (DSP system) is represented by a difference equation, which is linear and time invariant.

2. The filter output depends on the filter current input, past input(s), and past output(s) in general. Given arbitrary inputs and nonzero or zero initial conditions, operating the difference equation can generate the filter output recursively.

3. System responses such as the impulse response and step response can be determined analytically using the z-transform.

4. The transfer function can be obtained by applying the z-transform to the difference equation to determine the ratio of the output z-transform over the input z-transform. A digital filter (DSP system) can be represented by its transfer function.

5. System stability can be studied using a very useful tool, a z-plane pole-zero plot.

6. The frequency responses of the DSP system were developed and illustrated to investigate magnitude and phase frequency responses. In addition, the FIR (finite impulse response) and IIR (infinite impulse response) systems were defined.

7. Digital filters and their specifications, such as lowpass, highpass, bandpass, and bandstop, were reviewed.

8. A digital filter can be realized using standard realization methods such as the direct form I; direct form II; cascade, or series form; and parallel form.

9. Digital processing of speech using the pre-emphasis filter and bandpass filter was investigated to study spectral effects of the processed digital speech. The pre-emphasis filter boosts the high-frequency components, while bandpass filtering keeps the midband frequency components and rejects other lower- and upper-band frequency components.

6.9 Problems

6.1. Given the difference equation

$$y(n) = x(n-1) - 0.75y(n-1) - 0.125y(n-2),$$

a. calculate the system response $y(n)$ for $n = 0, 1, 2, \ldots, 4$ with the input $x(n) = (0.5)^n u(n)$ and initial conditions: $x(-1) = -1$, $y(-2) = 2$, and $y(-1) = 1$;

b. calculate the system response $y(n)$ for $n = 0, 1, 2, \ldots, 4$ with the input $x(n) = (0.5)^n u(n)$ and zero initial conditions: $x(-1) = 0$, $y(-2) = 0$, and $y(-1) = 0$.

6.2. Given the following difference equation,

$$y(n) = 0.5x(n) + 0.5x(n-1),$$

a. find the $H(z)$;

b. determine the impulse response $y(n)$ if the input is $x(n) = 4\delta(n)$;

c. determine the step response $y(n)$ if the input is $x(n) = 10 \, u(n)$.

6.3. Given the following difference equation,
$$y(n) = x(n) - 0.5y(n-1),$$
a. find the $H(z)$;
b. determine the impulse response $y(n)$ if the input is $x(n) = \delta(n)$;
c. determine the step response $y(n)$ if the input is $x(n) = u(n)$.

6.4. A digital system is described by the following difference equation:
$$y(n) = x(n) - 0.25x(n-2) - 1.1y(n-1) - 0.28y(n-2).$$
a. Find the transfer function $H(z)$, the denominator polynomial $A(z)$, and the numerator polynomial $B(z)$.

6.5. A digital system is described by the following difference equation:
$$y(n) = x(n) - 0.3x(n-1) + 0.28x(n-2).$$
a. Find the transfer function $H(z)$, the denominator polynomial $A(z)$, and the numerator polynomial $B(z)$.

6.6. Convert each of the following transfer functions into its difference equation:
a. $H(z) = \dfrac{z^2 - 0.25}{z^2 + 1.1z + 0.18}$
b. $H(z) = \dfrac{z^2 - 0.1z + 0.3}{z^3}$

6.7. Convert the following transfer function into its pole-zero form:
a. $H(z) = \dfrac{1 - 0.16z^{-2}}{1 + 0.7z^{-1} + 0.1z^{-2}}$

6.8. A transfer function depicting a digital system is given by
$$H(z) = \dfrac{10(z+1)}{(z+0.75)}.$$
a. Determine the impulse response $h(n)$ and step response.
b. Determine the system response $y(n)$ if the input is $x(n) = (0.25)^n u(n)$.

6.9. Given each of the following transfer functions that describe digital system transfer functions, sketch the z-plane pole-zero plot and determine the stability for each digital system.
a. $H(z) = \dfrac{z - 0.5}{(z + 0.25)(z^2 + z + 0.8)}$

b. $H(z) = \dfrac{z^2 + 0.25}{(z - 0.5)(z^2 + 4z + 7)}$

c. $H(z) = \dfrac{z + 0.95}{(z + 0.2)(z^2 + 1.414z + 1)}$

d. $H(z) = \dfrac{z^2 + z + 0.25}{(z - 1)(z + 1)^2(z - 0.36)}$

6.10. Given the following digital system with a sampling rate of 8,000 Hz,

$$y(n) = 0.5x(n) + 0.5x(n - 2),$$

a. determine the frequency response;
b. calculate and plot the magnitude and phase frequency responses;
c. determine the filter type, based on the magnitude frequency response.

6.11. For the following digital system with a sampling rate of 8,000 Hz,

$$y(n) = x(n) - 0.5y(n - 2),$$

a. determine the frequency response;
b. calculate and plot the magnitude and phase frequency responses;
c. determine the filter type based on the magnitude frequency response.

6.12. Given the following difference equation,

$$y(n) = x(n) - 2 \cdot \cos(\alpha)x(n - 1) + x(n - 2) + 2\gamma \cdot \cos(\alpha) - \gamma^2,$$

where $\gamma = 0.8$ and $\alpha = 60°$,

a. find the transfer function $H(z)$;
b. plot the poles and zeros on the z-plane with the unit circle;
c. determine the stability of the system from the pole-zero plot;
d. calculate the amplitude (magnitude) response of $H(z)$;
e. calculate the phase response of $H(z)$.

6.13. For each of the following difference equations,

a. $y(n) = 0.5x(n) + 0.5x(n - 1)$

b. $y(n) = 0.5x(n) - 0.5x(n-1)$

c. $y(n) = 0.5x(n) + 0.5x(n-2)$

d. $y(n) = 0.5x(n) - 0.5x(n-2)$,

  1. find $H(z)$;
  2. calculate the magnitude response;
  3. specify the filter type based on the calculated magnitude response.

6.14. An IIR system is expressed as

$$y(n) = 0.5x(n) + 0.2y(n-1), \quad y(-1) = 0.$$

a. Find $H(z)$.

b. Find the system response $y(n)$ due to the input $x(n) = (0.5)^n u(n)$.

6.15. Given the following IIR system with zero initial conditions:

$$y(n) = 0.5x(n) - 0.7y(n-1) - 0.1y(n-2),$$

a. find $H(z)$;

b. find the unit step response.

6.16. Given the first-order IIR system

$$H(z) = \frac{1 + 2z^{-1}}{1 - 0.5z^{-1}},$$

realize $H(z)$ and develop the difference equations using the following forms:

a. direct-form I

b. direct-form II

6.17. Given the filter

$$H(z) = \frac{1 - 0.9z^{-1} - 0.1z^{-2}}{1 + 0.3z^{-1} - 0.04z^{-2}},$$

realize $H(z)$ and develop the difference equations using the following form:

a. direct-form I

b. direct-form II

c. cascade (series) form via the first-order sections

d. parallel form via the first-order sections

6.18. Given the following pre-emphasis filters:

$$H(z) = 1 - 0.5z^{-1}$$
$$H(z) = 1 - 0.7z^{-1}$$
$$H(z) = 1 - 0.9z^{-1},$$

a. write the difference equation for each;

b. determine which emphasizes high frequency components most.

MATLAB Problems

6.19. Given a filter

$$H(z) = \frac{1 + 2z^{-1} + z^{-2}}{1 - 0.5z^{-1} + 0.25z^{-2}},$$

a. use MATLAB to plot

1. its magnitude frequency response;
2. its phase response.

6.20. Given the difference equation

$$y(n) = x(n-1) - 0.75y(n-1) - 0.125y(n-2),$$

a. use the MATLAB functions **filter()** and **filtic()** to calculate the system response $y(n)$ for $n = 0, 1, 2, 3, \ldots, 4$ with the input of $x(n) = (0.5)^n u(n)$ and initial conditions: $x(-1) = -1$, $y(-2) = 2$, and $y(-1) = 1$;

b. use the MATLAB function **filter()** to calculate the system response $y(n)$ for $n = 0, 1, 2, 3, \ldots, 4$ with the input of $x(n) = (0.5)^n u(n)$ and zero initial conditions: $x(-1) = 0$, $y(-2) = 0$, and $y(-1) = 0$.

6.21. Given a filter

$$H(z) = \frac{1 - z^{-1} + z^{-2}}{1 - 0.9z^{-1} + 0.81z^{-2}},$$

a. plot the magnitude frequency response and phase response using MATLAB;

b. specify the type of filtering;

c. find the difference equation;

d. perform filtering, that is, calculate $y(n)$ for the first 1,000 samples for each of the following inputs and plot the filter outputs using MATLAB, assuming that all initial conditions are zeros and the sampling rate is 8,000 Hz:

1. $x(n) = \cos\left(\pi \cdot 10^3 \frac{n}{8000}\right)$
2. $x(n) = \cos\left(\frac{8}{3}\pi \cdot 10^3 \frac{n}{8000}\right)$
3. $x(n) = \cos\left(6\pi \cdot 10^3 \frac{n}{8000}\right)$;

e. repeat (d) using the MATLAB function **filter()**.

Reference

Stearns, S. D., and Hush, D. R. (1990). *Digital Signal Analysis*, 2nd ed. Englewood Cliffs, NJ: Prentice Hall.

7

Finite Impulse Response Filter Design

Objectives:

This chapter introduces principles of the finite impulse response (FIR) filter design and investigates the design methods such as the Fourier transform method, window method, frequency sampling method design, and optimal design method. Then the chapter illustrates how to apply the designed FIR filters to solve real-world problems such as noise reduction and digital crossover for audio applications. The major topics discussed in this chapter are included in the following outline.

7.1 Finite Impulse Response Filter Format

In this chapter, we describe techniques of designing *finite impulse response* (FIR) filters. An FIR filter is completely specified by the following input-output relationship:

$$y(n) = \sum_{i=0}^{K} b_i x(n-i) \qquad (7.1)$$
$$= b_0 x(n) + b_1 x(n-1) + b_2 x(n-2) + \cdots + b_K x(n-K)$$

where b_i represents FIR filter coefficients and $K + 1$ denotes the FIR filter length. Applying the z-transform on both sides of Equation (7.1) leads to

$$Y(z) = b_0 X(z) + b_1 z^{-1} X(z) + \cdots + b_K z^{-K} X(z). \qquad (7.2)$$

Factoring out $X(z)$ on the right-hand side of Equation (7.2) and then dividing $X(z)$ on both sides, we have the transfer function, which depicts the FIR filter, as

$$H(z) = \frac{Y(z)}{X(z)} = b_0 + b_1 z^{-1} + \cdots + b_K z^{-K}. \tag{7.3}$$

The following example serves to illustrate the notations used in Equations (7.1) and (7.3) numerically.

Example 7.1.

Given the following FIR filter:

$$y(n) = 0.1x(n) + 0.25x(n-1) + 0.2x(n-2),$$

a. Determine the transfer function, filter length, nonzero coefficients, and impulse response.

Solution:

a. Applying z-transform on both sides of the difference equation yields

$$Y(z) = 0.1X(z) + 0.25X(z)z^{-1} + 0.2X(z)z^{-2}.$$

Then the transfer function is found to be

$$H(z) = \frac{Y(z)}{X(z)} = 0.1 + 0.25z^{-1} + 0.2z^{-2}.$$

The filter length is $K + 1 = 3$, and the identified coefficients are

$$b_0 = 0.1 \ b_1 = 0.25 \text{ and } b_2 = 0.2.$$

Taking the inverse z-transform of the transfer function, we have

$$h(n) = 0.1\delta(n) + 0.25\delta(n-1) + 0.2\delta(n-2).$$

This FIR filter impulse response has only three terms.

The foregoing example is to help us understand the FIR filter format. We can conclude that

1. The transfer function in Equation (7.3) has a constant term, all the other terms each have a negative power of z, and all the poles are at the origin on the z-plane. Hence, the stability of filter is guaranteed. Its impulse response has only a finite number of terms.

2. The FIR filter operations involve only multiplying the filter inputs by their corresponding coefficients and accumulating them; the implementation of this filter type in real time is straightforward.

From the FIR filter format, the design objective can be to obtain the FIR filter b_i coefficients such that the magnitude frequency response of the FIR filter $H(z)$ will approximate the desired magnitude frequency response, such as that of a lowpass, highpass, bandpass, or bandstop filter. The following sections will introduce design methods to calculate the FIR filter coefficients.

7.2 Fourier Transform Design

We begin with an ideal lowpass filter with a normalized cutoff frequency Ω_c, whose magnitude frequency response in terms of the normalized digital frequency Ω is plotted in Figure 7.1 and is characterized by

$$H(e^{j\Omega}) = \begin{cases} 1, & 0 \le |\Omega| \le \Omega_c \\ 0, & \Omega_c \le |\Omega| \le \pi. \end{cases} \qquad (7.4)$$

Since the frequency response is periodic with a period of $\Omega = 2\pi$ radians, as we discussed in Chapter 6, we can extend the frequency response of the ideal filter $H(e^{j\Omega})$, as shown in Figure 7.2.

The periodic frequency response can be approximated using a complex Fourier series expansion (see this topic in Appendix B) in terms of the normalized digital frequency Ω, that is,

$$H(e^{j\Omega}) = \sum_{n=-\infty}^{\infty} c_n e^{-j\omega_0 n \Omega}, \qquad (7.5)$$

and the Fourier coefficients are given by

$$c_n = \frac{1}{2\pi} \int_{-\pi}^{\pi} H(e^{j\Omega}) e^{j\omega_0 n \Omega} d\Omega \text{ for } -\infty < n < \infty. \qquad (7.6)$$

Notice that we obtain Equations (7.5) and (7.6) simply by treating the Fourier series expansion in time domain with the time variable t replaced by the normalized digital frequency variable Ω. The fundamental frequency is easily found to be

$$\omega_0 = 2\pi/(period\ of\ waveform) = 2\pi/2\pi = 1. \qquad (7.7)$$

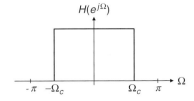

FIGURE 7.1 Frequency response of an ideal lowpass filter.

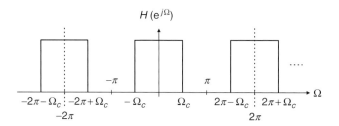

FIGURE 7.2 Periodicity of the ideal lowpass frequency response.

Substituting $\omega_0 = 1$ into Equation (7.6) and introducing $h(n) = c_n$, called the desired impulse response of the ideal filter, we obtain the Fourier transform design as

$$h(n) = \frac{1}{2\pi} \int_{-\pi}^{\pi} H(e^{j\Omega}) e^{j\Omega n} d\Omega \text{ for } -\infty < n < \infty. \quad (7.8)$$

Now, let us look at the possible z-transfer function. If we substitute $e^{j\Omega} = z$ and $\omega_0 = 1$ back into Equation (7.5), we yield a z-transfer function in the following format:

$$H(z) = \sum_{n=-\infty}^{\infty} h(n) z^{-n} \quad (7.9)$$
$$\cdots + h(-2)z^2 + h(-1)z^1 + h(0) + h(1)z^{-1} + h(2)z^{-2} + \cdots$$

This is a noncausal FIR filter. We will deal with this later in this section. Using the Fourier transform design shown in Equation (7.8), the desired impulse response approximation of the ideal lowpass filter is solved as

$$h(n) = \frac{1}{2\pi} \int_{-\pi}^{\pi} H(e^{j\Omega}) e^{j\Omega \times 0} d\Omega$$

For $n = 0$

$$= \frac{1}{2\pi} \int_{-\Omega_c}^{\Omega_c} 1 d\Omega = \frac{\Omega_c}{\pi}$$

$$h(n) = \frac{1}{2\pi} \int_{-\pi}^{\pi} H(e^{j\Omega}) e^{j\Omega n} d\Omega = \frac{1}{2\pi} \int_{-\Omega_c}^{\Omega_c} e^{j\Omega n} d\Omega$$

For $n \neq 0$

$$= \frac{e^{jn\Omega}}{2\pi j n} \bigg|_{-\Omega_c}^{\Omega_c} = \frac{1}{\pi n} \frac{e^{jn\Omega_c} - e^{-jn\Omega_c}}{2j} = \frac{\sin(\Omega_c n)}{\pi n}. \quad (7.10)$$

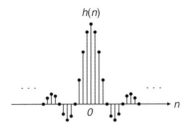

FIGURE 7.3 Impulse response of an ideal digital lowpass filter.

The desired impulse response $h(n)$ is plotted versus the sample number n in Figure 7.3.

Theoretically, $h(n)$ in Equation (7.10) exists for $-\infty < n < \infty$ and is symmetrical about $n = 0$; that is, $h(n) = h(-n)$. The amplitude of the impulse response sequence $h(n)$ becomes smaller when n increases in both directions. The FIR filter design must first be completed by truncating the infinite-length sequence $h(n)$ to achieve the $2M + 1$ dominant coefficients using the coefficient symmetry, that is,

$$H(z) = h(M)z^M + \cdots + h(1)z^1 + h(0) + h(1)z^{-1} + \cdots + h(M)z^{-M}.$$

The obtained filter is a noncausal z-transfer function of the FIR filter, since the filter transfer function contains terms with positive powers of z, which in turn means that the filter output depends on the future filter inputs. To remedy the noncausal z-transfer function, we delay the truncated impulse response $h(n)$ by M samples to yield the following causal FIR filter:

$$H(z) = b_0 + b_1 z^{-1} + \cdots + b_{2M}(2M)z^{-2M}, \tag{7.11}$$

where the delay operation is given by

$$b_n = h(n - M) \text{ for } n = 0, 1, \ldots, 2M. \tag{7.12}$$

Similarly, we can obtain the design equations for other types of FIR filters, such as highpass, bandpass, and bandstop, using their ideal frequency responses and Equation (7.8). The derivations are omitted here. Table 7.1 gives a summary of all the formulas for FIR filter coefficient calculations.

The following example illustrates the coefficient calculation for the lowpass FIR filter.

7 FINITE IMPULSE RESPONSE FILTER DESIGN

TABLE 7.1 Summary of ideal impulse responses for standard FIR filters.

Filter Type	Ideal Impulse Response $h(n)$ (noncausal FIR coefficients)
Lowpass:	$h(n) = \begin{cases} \frac{\Omega_c}{\pi} & n = 0 \\ \frac{\sin(\Omega_c n)}{n\pi} & \text{for } n \neq 0 \quad -M \leq n \leq M \end{cases}$
Highpass:	$h(n) = \begin{cases} \frac{\pi - \Omega_c}{\pi} & n = 0 \\ -\frac{\sin(\Omega_c n)}{n\pi} & \text{for } n \neq 0 \quad -M \leq n \leq M \end{cases}$
Bandpass:	$h(n) = \begin{cases} \frac{\Omega_H - \Omega_L}{\pi} & n = 0 \\ \frac{\sin(\Omega_H n)}{n\pi} - \frac{\sin(\Omega_L n)}{n\pi} & \text{for } n \neq 0 \quad -M \leq n \leq M \end{cases}$
Bandstop:	$h(n) = \begin{cases} \frac{\pi - \Omega_H + \Omega_L}{\pi} & n = 0 \\ -\frac{\sin(\Omega_H n)}{n\pi} + \frac{\sin(\Omega_L n)}{n\pi} & \text{for } n \neq 0 \quad -M \leq n \leq M \end{cases}$

Causal FIR filter coefficients: shifting $h(n)$ to the right by M samples.
Transfer function:
$$H(z) = b_0 + b_1 z^{-1} + b_2 z^{-2} + \cdots b_{2M} z^{-2M}$$
where $b_n = h(n - M)$, $n = 0, 1, \cdots, 2M$

Example 7.2.

a. Calculate the filter coefficients for a 3-tap FIR lowpass filter with a cutoff frequency of 800 Hz and a sampling rate of 8,000 Hz using the Fourier transform method.

b. Determine the transfer function and difference equation of the designed FIR system.

c. Compute and plot the magnitude frequency response for $\Omega = 0$, $\pi/4$, $\pi/2$, $3\pi/4$, and π radians.

Solution:

a. Calculating the normalized cutoff frequency leads to
$$\Omega_c = 2\pi f_c T_s = 2\pi \times 800/8000 = 0.2\pi \text{ radians.}$$

Since $2M + 1 = 3$ in this case, using the equation in Table 7.1 results in
$$h(0) = \frac{\Omega_c}{\pi} \quad \text{for } n = 0$$

$$h(n) = \frac{\sin(\Omega_c n)}{n\pi} = \frac{\sin(0.2\pi n)}{n\pi}, \quad \text{for } n \neq 1.$$

The computed filter coefficients via the previous expression are listed as:

$$h(0) = \frac{0.2\pi}{\pi} = 0.2$$

$$h(1) = \frac{\sin[0.2\pi \times 1]}{1 \times \pi} = 0.1871.$$

Using the symmetry leads to

$$h(-1) = h(1) = 0.1871.$$

Thus delaying $h(n)$ by $M = 1$ sample using Equation (7.12) gives

$$b_0 = h(0-1) = h(-1) = 0.1871$$
$$b_1 = h(1-1) = h(0) = 0.2$$

and $b_2 = h(2-1) = h(1) = 0.1871$.

b. The transfer function is achieved as

$$H(z) = 0.1871 + 0.2z^{-1} + 0.1871z^{-2}.$$

Using the technique described in Chapter 6, we have

$$\frac{Y(z)}{X(z)} = H(z) = 0.1871 + 0.2z^{-1} + 0.1817z^{-2}.$$

Multiplying $X(z)$ leads to

$$Y(z) = 0.1871X(z) + 0.2z^{-1}X(z) + 0.1871z^{-2}X(z).$$

Applying the inverse z-transform on both sides, the difference equation is yielded as

$$y(n) = 0.1871x(n) + 0.2x(n-1) + 0.1871x(n-2).$$

c. The magnitude frequency response and phase response can be obtained using the technique introduced in Chapter 6. Substituting $z = e^{j\Omega}$ into $H(z)$, it follows that

$$H(e^{j\Omega}) = 0.1871 + 0.2e^{-j\Omega} + 0.1871e^{-j2\Omega}.$$

Factoring the term $e^{-j\Omega}$ and using the Euler formula $e^{jx} + e^{-jx} = 2\cos(x)$, we achieve

$$H(e^{j\Omega}) = e^{-j\Omega}(0.1871e^{j\Omega} + 0.2 + 0.1871e^{-j\Omega})$$
$$= e^{-j\Omega}(0.2 + 0.3742\cos(\Omega))$$

Then the magnitude frequency response and phase response are found to be

$$|H(e^{j\Omega})| = |0.2 + 0.3472\cos\Omega|$$

$$\text{and } \angle H(e^{j\Omega}) = \begin{cases} -\Omega & \text{if } 0.2 + 0.3472\cos\Omega > 0 \\ -\Omega + \pi & \text{if } 0.2 + 0.3472\cos\Omega < 0. \end{cases}$$

Details of the magnitude calculations for several typical normalized frequencies are listed in Table 7.2.

Due to the symmetry of the coefficients, the obtained FIR filter has a linear phase response as shown in Figure 7.4. The sawtooth shape is produced by the contribution of the negative sign of the real magnitude term $0.2 + 0.3742\cos\Omega$ in the 3-tap filter frequency response, that is,

$$H(e^{j\Omega}) = e^{-j\Omega}(0.2 + 0.3742\cos\Omega).$$

In general, the FIR filter with symmetrical coefficients has a linear phase response (linear function of Ω) as follows:

$$\angle H(e^{j\Omega}) = -M\Omega + \text{possible phase of } 180°. \tag{7.13}$$

Next, we see that the 3-tap FIR filter does not give an acceptable magnitude frequency response. To explore this response further, Figure 7.5 displays the magnitude and phase responses of 3-tap ($M = 1$) and 17-tap ($M = 8$) FIR lowpass filters with a normalized cutoff frequency of $\Omega_c = 0.2\pi$ radian. The calculated coefficients for the 17-tap FIR lowpass filter are listed in Table 7.3.

We can make the following observations at this point:

1. The oscillations (ripples) exhibited in the passband (main lobe) and stopband (side lobes) of the magnitude frequency response constitute the *Gibbs effect*. Gibbs oscillatory behavior originates from the abrupt

TABLE 7.2 Frequency response calculation in Example 7.2.

Ω radians	$f = \Omega f_s/(2\pi)$ Hz	$0.2 + 0.3742\cos\Omega$	$|H(e^{j\Omega})|$	$|H(e^{j\Omega})|_{dB}$ dB	$\angle H(e^{j\Omega})$ degree
0	0	0.5742	0.5742	−4.82	0
$\pi/4$	1000	0.4646	0.4646	−6.66	−45
$\pi/2$	2000	0.2	0.2	−14.0	−90
$3\pi/4$	3000	−0.0646	0.0646	−23.8	45
π	4000	−0.1742	0.1742	−15.2	0

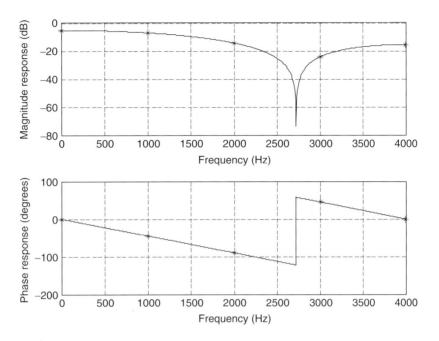

FIGURE 7.4 Magnitude frequency response in Example 7.2.

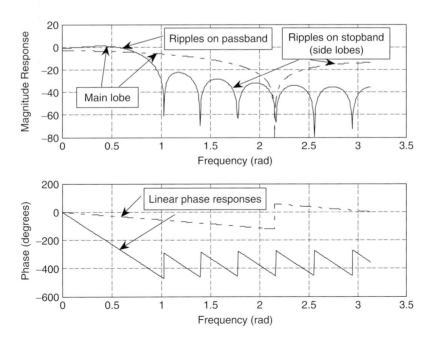

FIGURE 7.5 Magnitude and phase frequency responses of the lowpass FIR filters with 3 coefficients (dash-dotted line) and 17 coefficients (solid line).

TABLE 7.3 17-tap FIR lowpass filter coefficients in Example 7.2 (M = 8).

$b_0 = b_{16} = -0.0378$	$b_1 = b_{15} = -0.0432$	
$b_2 = b_{14} = -0.0312$	$b_3 = b_{13} = 0.0000$	
$b_4 = b_{12} = 0.0468$	$b_5 = b_{11} = 0.1009$	
$b_6 = b_{10} = 0.1514$	$b_7 = b_9 = 0.1871$	$b_8 = 0.2000$

truncation of the infinite impulse response in Equation (7.11). To remedy this problem, window functions will be used and will be discussed in the next section.

2. Using a larger number of the filter coefficients will produce the sharp roll-off characteristic of the transition band but may cause increased time delay and increased computational complexity for implementing the designed FIR filter.

3. The phase response is linear in the passband. This is consistent with Equation (7.13), which means that all frequency components of the filter input within the passband are subjected to the same amount of time delay at the filter output. This is a requirement for applications in audio and speech filtering, where phase distortion needs to be avoided. Note that we impose the following linear phase requirement, that is, the FIR coefficients are symmetrical about the middle coefficient, and the FIR filter order is an odd number. If the design method cannot produce the symmetric coefficients or can generate anti-symmetric coefficients, the resultant FIR filter does not have the linear phase property. (Linear phase even-order FIR filters and FIR filters using the anti-symmetry of coefficients are discussed in Proakis and Manolakis [1996].)

To further probe the linear phase property, we consider a sinusoidal sequence $x(n) = A \sin(n\Omega)$ as the FIR filter input, with the output expected to be

$$y(n) = A|H| \sin(n\Omega + \varphi),$$

where $\varphi = -M\Omega$. Substituting $\varphi = -M\Omega$ into $y(n)$ leads to

$$y(n) = A|H| \sin[\Omega(n - M)].$$

This clearly indicates that within the passband, all frequency components passing through the FIR filter will have the same constant delay at the output, which equals M samples. Hence, phase distortion is avoided.

Figure 7.6 verifies the linear phase property using the FIR filter with 17 taps. Two sinusoids of the normalized digital frequencies 0.05π and 0.15π radian, respectively, are used as inputs. These two input signals are within passband, so their magnitudes are not changed. As shown in Figure 7.6, the output

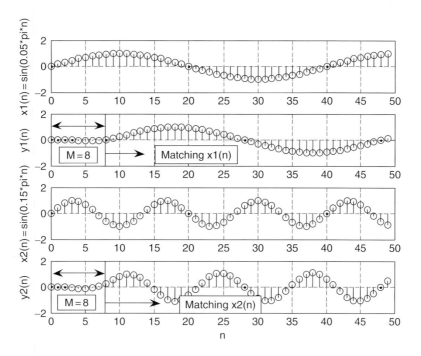

FIGURE 7.6 Illustration of FIR filter linear phase property (constant delay of 8 samples).

beginning the 9th sample matches the input, which is delayed by 8 samples for each case.

What would happen if the filter phase were nonlinear? This can be illustrated using the following combined sinusoids as the filter input:

$$x(n) = x_1(n) + x_2(n) = \sin(0.05\pi n)u(n) - \frac{1}{3}\sin(0.15\pi n)u(n).$$

The original $x(n)$ is the top plot shown in Figure 7.7. If the linear phase response of a filter is considered, such as $\varphi = -M\Omega_0$, where $M = 8$ in our illustration, we have the filtered output as

$$y_1(n) = \sin[0.05\pi(n-8)] - \frac{1}{3}\sin[0.15\pi(n-8)].$$

The linear phase effect is shown in the middle plot of Figure 7.7. We see that $y_1(n)$ is the 8-sample delayed version of $x(n)$. However, considering a unit gain filter with a phase delay of 90 degrees for all the frequency components, we have the filtered output as

$$y_2(n) = \sin(0.05\pi n - \pi/2) - \frac{1}{3}\sin(0.15\pi n - \pi/2),$$

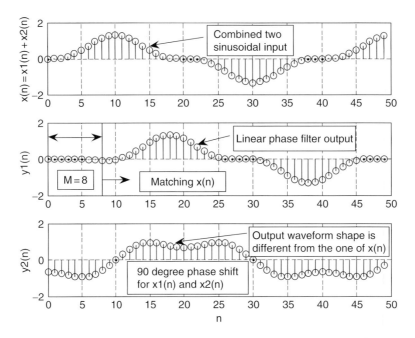

FIGURE 7.7 Comparison of linear and nonlinear phase responses.

where the first term has a phase shift of 10 samples (see $\sin[0.05\pi(n-10)]$), while the second term has a phase shift of 10/3 samples (see $\frac{1}{3}\sin[0.15\pi(n-\frac{10}{3})]$). Certainly, we do not have the linear phase feature. The signal $y_2(n)$ plotted in Figure 7.7 shows that the waveform shape is different from that of the original signal $x(n)$, hence has significant phase distortion. This phase distortion is audible for audio applications and can be avoided by using the FIR filter, which has the linear phase feature.

We now have finished discussing the coefficient calculation for the FIR lowpass filter, which has a good linear phase property. To explain the calculation of filter coefficients for the other types of filters and examine the Gibbs effect, we look at another simple example.

Example 7.3.

a. Calculate the filter coefficients for a 5-tap FIR bandpass filter with a lower cutoff frequency of 2,000 Hz and an upper cutoff frequency of 2,400 Hz at a sampling rate of 8,000 Hz.

b. Determine the transfer function and plot the frequency responses with MATLAB.

Solution:

a. Calculating the normalized cutoff frequencies leads to

$$\Omega_L = 2\pi f_L/f_s = 2\pi \times 2000/8000 = 0.5\pi \text{ radians}$$
$$\Omega_H = 2\pi f_H/f_s = 2\pi \times 2400/8000 = 0.6\pi \text{ radians}.$$

Since $2M + 1 = 5$ in this case, using the equation in Table 7.1 yields

$$h(n) = \begin{cases} \frac{\Omega_H - \Omega_L}{\pi} & n = 0 \\ \frac{\sin(\Omega_H n)}{n\pi} - \frac{\sin(\Omega_L n)}{n\pi} & n \neq 0 \quad -2 \leq n \leq 2 \end{cases} \quad (7.14)$$

Calculations for noncausal FIR coefficients are listed as

$$h(0) = \frac{\Omega_H - \Omega_L}{\pi} = \frac{0.6\pi - 0.5\pi}{\pi} = 0.1.$$

The other computed filter coefficients via Equation (7.14) are

$$h(1) = \frac{\sin[0.6\pi \times 1]}{1 \times \pi} - \frac{\sin[0.5\pi \times 1]}{1 \times \pi} = -0.01558$$

$$h(2) = \frac{\sin[0.6\pi \times 2]}{2 \times \pi} - \frac{\sin[0.5\pi \times 2]}{2 \times \pi} = -0.09355.$$

Using the symmetry leads to

$$h(-1) = h(1) = -0.01558$$
$$h(-2) = h(2) = -0.09355.$$

Thus, delaying $h(n)$ by $M = 2$ samples gives

$$b_0 = b_4 = -0.09355,$$
$$b_1 = b_3 = -0.01558, \text{ and } b_2 = 0.1.$$

b. The transfer function is achieved as

$$H(z) = -0.09355 - 0.01558z^{-1} + 0.1z^{-2} - 0.01558z^{-3} - 0.09355z^{-4}.$$

To complete Example 7.3, the magnitude frequency response plotted in terms of $|H(e^{j\Omega})|_{dB} = 20\log_{10}|H(e^{j\Omega})|$ using the MATLAB program 7.1 is displayed in Figure 7.8.

Program 7.1. MATLAB program for Example 7.3.

```
% Example 7.3
% MATLAB program to plot frequency responses
%
[hz, w] = freqz([-0.09355 -0.015580.1 -0.01558 -0.09355], [1], 512);
phi= 180*unwrap(angle(hz))/pi;
subplot(2,1,1), plot(w,20*log10(abs(hz))),grid;
xlabel('Frequency (radians)');
ylabel('Magnitude Response (dB)')
subplot(2,1,2), plot(w, phi);grid;
xlabel('Frequency (radians)');
ylabel('Phase (degrees)');
```

As a summary of Example 7.3, the magnitude frequency response demonstrates the Gibbs oscillatory behavior existing in the passband and stopband. The peak of the main lobe in the passband is dropped from 0 dB to approximately −10 dB, while for the stopband, the lower side lobe in the magnitude response plot swings approximately between −18 dB and −70 dB, and the upper side lobe swings between −25 dB and −68 dB. As we have pointed out, this is due to the abrupt truncation of the infinite impulse sequence $h(n)$.

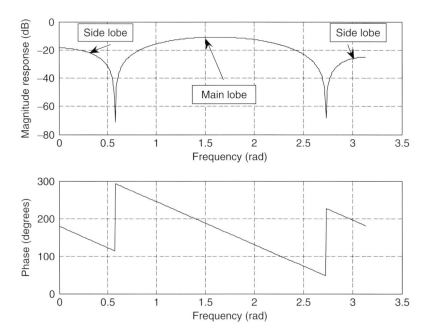

FIGURE 7.8 **Frequency responses for Example 7.3.**

The oscillations can be reduced by increasing the number of coefficients and using a window function, which will be studied next.

7.3 Window Method

In this section, the *window method* (Fourier transform design with window functions) is developed to remedy the undesirable Gibbs oscillations in the passband and stopband of the designed FIR filter. Recall that Gibbs oscillations originate from the abrupt truncation of the infinite-length coefficient sequence. Then it is natural to seek a window function, which is symmetrical and can gradually weight the designed FIR coefficients down to zeros at both ends for the range of $-M \le n \le M$. Applying the window sequence to the filter coefficients gives

$$h_w(n) = h(n) \cdot w(n),$$

where $w(n)$ designates the window function. Common window functions used in the FIR filter design are as follows:

1. Rectangular window:
$$w_{rec}(n) = 1, \quad -M \le n \le M. \tag{7.15}$$

2. Triangular (Bartlett) window:
$$w_{tri}(n) = 1 - \frac{|n|}{M}, \quad -M \le n \le M. \tag{7.16}$$

3. Hanning window:
$$w_{han}(n) = 0.5 + 0.5\cos\left(\frac{n\pi}{M}\right), \quad -M \le n \le M. \tag{7.17}$$

4. Hamming window:
$$w_{ham}(n) = 0.54 + 0.46\cos\left(\frac{n\pi}{M}\right), \quad -M \le n \le M. \tag{7.18}$$

5. Blackman window:
$$w_{black}(n) = 0.42 + 0.5\cos\left(\frac{n\pi}{M}\right) + 0.08\cos\left(\frac{2n\pi}{M}\right), \quad -M \le n \le M. \tag{7.19}$$

In addition, there is another popular window function, called the Kaiser window (its detailed information can be found in Oppenheim, Schafer, and Buck [1999]). As we expected, the rectangular window function has a constant value of 1 within the window, hence does only truncation. As a comparison, shapes of the other window functions from Equations (7.16) to (7.19) are plotted in Figure 7.9 for the case of $2M + 1 = 81$.

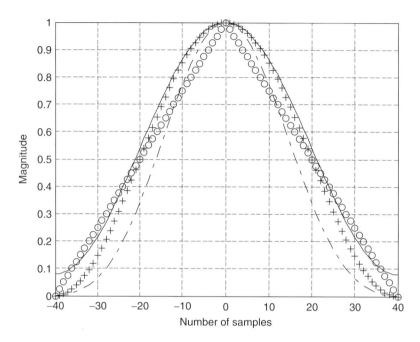

FIGURE 7.9 Shapes of window functions for the case of 2M + 1 = 81. ○ line, Triangular window; + line, Hanning window; Solid line, Hamming window; dashed line, Blackman window.

We apply the Hamming window function in Example 7.4.

Example 7.4.

Given the calculated filter coefficients

$h(0) = 0.25$, $h(-1) = h(1) = 0.22508$, $h(-2) = h(2) = 0.15915$, $h(-3) = h(3) = 0.07503$,

a. Apply the Hamming window function to obtain windowed coefficients $h_w(n)$.
b. Plot the impulse response $h(n)$ and windowed impulse response $h_w(n)$.

Solutions:

a. Since $M = 3$, applying Equation (7.18) leads to the window sequence

$$w_{ham}(-3) = 0.54 + 0.46 \cos\left(\frac{-3 \times \pi}{3}\right) = 0.08$$

$$w_{ham}(-2) = 0.54 + 0.46 \cos\left(\frac{-2 \times \pi}{3}\right) = 0.31$$

$$w_{ham}(-1) = 0.54 + 0.46\cos\left(\frac{-1\times\pi}{3}\right) = 0.77$$
$$w_{ham}(0) = 0.54 + 0.46\cos\left(\frac{0\times\pi}{3}\right) = 1$$
$$w_{ham}(1) = 0.54 + 0.46\cos\left(\frac{1\times\pi}{3}\right) = 0.77$$
$$w_{ham}(2) = 0.54 + 0.46\cos\left(\frac{2\times\pi}{3}\right) = 0.31$$
$$w_{ham}(3) = 0.54 + 0.46\cos\left(\frac{3\times\pi}{3}\right) = 0.08.$$

Applying the Hamming window function and its symmetrical property to the filter coefficients, we get

$$h_w(0) = h(0)\cdot w_{ham}(0) = 0.25\times 1 = 0.25$$
$$h_w(1) = h(1)\cdot w_{ham}(1) = 0.22508\times 0.77 = 0.17331 = h_w(-1)$$
$$h_w(2) = h(2)\cdot w_{ham}(2) = 0.15915\times 0.31 = 0.04934 = h_w(-2)$$
$$h_w(3) = h(3)\cdot w_{ham}(3) = 0.07503\times 0.08 = 0.00600 = h_w(-3).$$

b. Noncausal impulse responses $h(n)$ and $h_w(n)$ are plotted in Figure 7.10.

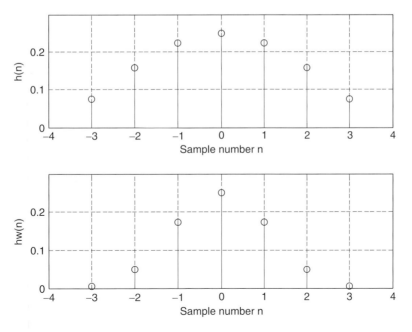

FIGURE 7.10 Plots of FIR non-causal coefficients and windowed FIR coefficients in Example 7.4.

We observe that the Hamming window does its job to weight the FIR filter coefficients to zero gradually at both ends. Hence, we can expect a reduced Gibbs effect in the magnitude frequency response.

Now the lowpass FIR filter design via the window method can be therefore achieved. The design procedure includes three steps. The first step is to obtain the truncated impulse response $h(n)$, where $-M \leq n \leq M$; then we multiply the obtained sequence $h(n)$ by the selected window data sequence to yield the windowed noncausal FIR filter coefficients $h_w(n)$; and the final step is to delay the windowed noncausal sequence $h_w(n)$ by M samples to achieve the causal FIR filter coefficients, $b_n = h_w(n - M)$. The design procedure of the FIR filter via windowing is summarized as follows:

1. Obtain the FIR filter coefficients $h(n)$ via the Fourier transform method (Table 7.1).

2. Multiply the generated FIR filter coefficients by the selected window sequence

$$h_w(n) = h(n)w(n), \quad n = -M, \ldots 0, 1, \ldots, M, \quad (7.20)$$

where $w(n)$ is chosen to be one of the window functions listed in Equations (7.15) to (7.19).

3. Delay the windowed impulse sequence $h_w(n)$ by M samples to get the windowed FIR filter coefficients:

$$b_n = h_w(n - M), \quad \text{for } n = 0, 1, \ldots, 2M. \quad (7.21)$$

Let us study the following design examples.

Example 7.5.

a. Design a 3-tap FIR lowpass filter with a cutoff frequency of 800 Hz and a sampling rate of 8,000 Hz using the Hamming window function.

b. Determine the transfer function and difference equation of the designed FIR system.

c. Compute and plot the magnitude frequency response for $\Omega = 0$, $\pi/4$, $\pi/2$, $3\pi/4$, and π radians.

Solution:

a. The normalized cutoff frequency is calculated as

$$\Omega_c = 2\pi f_c T_s = 2\pi \times 800/8000 = 0.2\pi \text{ radian}.$$

Since $2M+1=3$ in this case, FIR coefficients obtained by using the equation in Table 7.1 are listed as

$$h(0) = 0.2 \text{ and } h(-1) = h(1) = 0.1871$$

(see Example 7.2). Applying the Hamming window function defined in Equation (7.18), we have

$$w_{ham}(0) = 0.54 + 0.46\cos\left(\frac{0\pi}{1}\right) = 1$$

$$w_{ham}(1) = 0.54 + 0.46\cos\left(\frac{1\times\pi}{1}\right) = 0.08.$$

Using the symmetry of the window function gives

$$w_{ham}(-1) = w_{ham}(1) = 0.08.$$

The windowed impulse response is calculated as

$$h_w(0) = h(0)w_{ham}(0) = 0.2 \times 1 = 0.2$$
$$h_w(1) = h(1)w_{ham}(1) = 0.1871 \times 0.08 = 0.01497$$
$$h_w(-1) = h(-1)w_{ham}(-1) = 0.1871 \times 0.08 = 0.01497.$$

Thus, delaying $h_w(n)$ by $M=1$ sample gives

$$b_0 = b_2 = 0.01496 \text{ and } b_1 = 0.2.$$

b. The transfer function is achieved as

$$H(z) = 0.01497 + 0.2z^{-1} + 0.01497z^{-2}.$$

Using the technique described in Chapter 6, we have

$$\frac{Y(z)}{X(z)} = H(z) = 0.01497 + 0.2z^{-1} + 0.01497z^{-2}.$$

Multiplying $X(z)$ leads to

$$Y(z) = 0.01497X(z) + 0.2z^{-1}X(z) + 0.01497z^{-2}X(z).$$

Applying the inverse z-transform on both sides, the difference equation is yielded as

$$y(n) = 0.01497x(n) + 0.2x(n-1) + 0.01497x(n-2).$$

c. The magnitude frequency response and phase response can be obtained using the technique introduced in Chapter 6. Substituting $z = e^{j\Omega}$ into $H(z)$, it follows that

$$H(e^{j\Omega}) = 0.01497 + 0.2e^{-j\Omega} + 0.01497e^{-j2\Omega}$$
$$= e^{-j\Omega}\left(0.01497e^{j\Omega} + 0.2 + 0.01497e^{-j\Omega}\right).$$

Using Euler's formula leads to

$$H(e^{j\Omega}) = e^{-j\Omega}(0.2 + 0.02994\cos\Omega).$$

Then the magnitude frequency response and phase response are found to be

$$\left|H(e^{j\Omega})\right| = |0.2 + 0.2994\cos\Omega|$$

$$\text{and } \angle H(e^{j\Omega}) = \begin{cases} -\Omega & \text{if } 0.2 + 0.02994\cos\Omega > 0 \\ -\Omega + \pi & \text{if } 0.2 + 0.02994\cos\Omega < 0. \end{cases}$$

The calculation details of the magnitude response for several normalized frequency values are listed in Table 7.4. Figure 7.11 shows the plots of the frequency responses.

TABLE 7.4 Frequency response calculation in Example 7.5.

Ω radians	$f = \Omega f_s/(2\pi)$ Hz	$0.2 + 0.02994\cos\Omega$	$\left|H(e^{j\Omega})\right|$	$\left|H(e^{j\Omega})\right|_{dB}$ dB	$\angle H(e^{j\Omega})$ degree
0	0	0.2299	0.2299	−12.77	0
$\pi/4$	1000	0.1564	0.2212	−13.11	−45
$\pi/2$	2000	0.2000	0.2000	−13.98	−90
$3\pi/4$	3000	0.1788	0.1788	−14.95	−135
π	4000	0.1701	0.1701	−15.39	−180

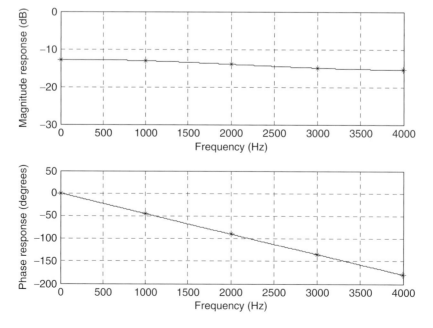

FIGURE 7.11 The frequency responses in Example 7.5.

Example 7.6.

a. Design a 5-tap FIR band reject filter with a lower cutoff frequency of 2,000 Hz, an upper cutoff frequency of 2,400 Hz, and a sampling rate of 8,000 Hz using the Hamming window method.

b. Determine the transfer function.

Solution:

a. Calculating the normalized cutoff frequencies leads to

$$\Omega_L = 2\pi f_L T = 2\pi \times 2000/8000 = 0.5\pi \text{ radians}$$
$$\Omega_H = 2\pi f_H T = 2\pi \times 2400/8000 = 0.6\pi \text{ radians}.$$

Since $2M + 1 = 5$ in this case, using the equation in Table 7.1 yields

$$h(n) = \begin{cases} \frac{\pi - \Omega_H + \Omega_L}{\pi} & n = 0 \\ -\frac{\sin(\Omega_H n)}{n\pi} + \frac{\sin(\Omega_L n)}{n\pi} & n \neq 0 \end{cases} \quad -2 \leq n \leq 2.$$

When $n = 0$, we have

$$h(0) = \frac{\pi - \Omega_H + \Omega_L}{\pi} = \frac{\pi - 0.6\pi + 0.5\pi}{\pi} = 0.9.$$

The other computed filter coefficients via the previous expression are listed as

$$h(1) = \frac{\sin[0.5\pi \times 1]}{1 \times \pi} - \frac{\sin[0.6\pi \times 1]}{1 \times \pi} = 0.01558$$

$$h(2) = \frac{\sin[0.5\pi \times 2]}{2 \times \pi} - \frac{\sin[0.6\pi \times 2]}{2 \times \pi} = 0.09355.$$

Using the symmetry leads to

$$h(-1) = h(1) = 0.01558$$
$$h(-2) = h(2) = 0.09355.$$

Applying the Hamming window function in Equation (7.18), we have

$$w_{ham}(0) = 0.54 + 0.46 \cos\left(\frac{0 \times \pi}{2}\right) = 1.0$$

$$w_{ham}(1) = 0.54 + 0.46 \cos\left(\frac{1 \times \pi}{2}\right) = 0.54$$

$$w_{ham}(2) = 0.54 + 0.46 \cos\left(\frac{2 \times \pi}{2}\right) = 0.08.$$

Using the symmetry of the window function gives

$$w_{ham}(-1) = w_{ham}(1) = 0.54$$
$$w_{ham}(-2) = w_{ham}(2) = 0.08.$$

The windowed impulse response is calculated as

$$h_w(0) = h(0)w_{ham}(0) = 0.9 \times 1 = 0.9$$
$$h_w(1) = h(1)w_{ham}(1) = 0.01558 \times 0.54 = 0.00841$$
$$h_w(2) = h(2)w_{ham}(2) = 0.09355 \times 0.08 = 0.00748$$
$$h_w(-1) = h(-1)w_{ham}(-1) = 0.00841$$
$$h_w(-2) = h(-2)w_{ham}(-2) = 0.00748.$$

Thus, delaying $h_w(n)$ by $M = 2$ samples gives

$$b_0 = b_4 = 0.00748, \ b_1 = b_3 = 0.00841, \text{ and } b_2 = 0.9.$$

b. The transfer function is achieved as

$$H(z) = 0.00748 + 0.00841z^{-1} + 0.9z^{-2} + 0.00841z^{-3} + 0.00748z^{-4}.$$

The following design examples are demonstrated using MATLAB programs. The MATLAB function **firwd(N, Ftype, WnL, WnH, Wtype)** is listed in the "MATLAB Programs" section at the end of this chapter. Table 7.5 lists comments to show the usage.

TABLE 7.5 Illustration of the MATLAB function for FIR filter design using the window methods.

```
function B = firwd (N, Ftype, WnL, WnH, Wtype)
% B = firwd(N, Ftype, WnL, WnH, Wtype)
% FIR filter design using the window function method.
% Input parameters:
% N: the number of the FIR filter taps.
% Note: It must be an odd number.
% Ftype: the filter type
% 1. Lowpass filter;
% 2. Highpass filter;
% 3. Bandpass filter;
% 4. Band reject filter;
% WnL: lower cutoff frequency in radians. Set WnL = 0 for the highpass filter.
% WnH: upper cutoff frequency in radians. Set WnH = 0 for the lowpass filter.
% Wtypw: window function type
% 1. Rectangular window;
% 2. Triangular window;
% 3. Hanning window;
% 4. Hamming window;
% 5. Blackman window;
```

Example 7.7.

a. Design a lowpass FIR filter with 25 taps using the MATLAB program listed in the "MATLAB Programs" section at the end of this chapter. The cutoff frequency of the filter is 2,000 Hz, assuming a sampling frequency of 8,000 Hz. The rectangular window and Hamming window functions are used for each design.

b. Plot the frequency responses along with those obtained using the rectangular window and Hamming window for comparison.

c. List FIR filter coefficients for each window design method.

Solution:

a. With a given sampling rate of 8,000 Hz, the normalized cutoff frequency can be found as

$$\Omega_c = \frac{2000 \times 2\pi}{8000} = 0.5\pi \text{ radians.}$$

Now we are ready to design FIR filters via the MATLAB program. The program, firwd(N,Ftype,WnL,WnH,Wtype), listed in the "MATLAB Programs" section at the end of this chapter, has five input parameters, which are described as follows:

- "N" is the number of specified filter coefficients (the number of filter taps).

- "Ftype" denotes the filter type, that is, input "1" for the lowpass filter design, input "2" for the highpass filter design, input "3" for the bandpass filter design, and input "4" for the band reject filter design.

- "WnL" and "WnH" are the lower and upper cutoff frequency inputs, respectively. Note that WnH = 0 when specifying WnL for the lowpass filter design, while WnL = 0 when specifying WnH for the highpass filter design.

- "Wtype" specifies the window data sequence to be used in the design, that is, input "1" for the rectangular window, input "2" for the triangular window, input "3" for the Hanning window, input "4" for the Hamming window, and input "5" for the Blackman window.

b. The following application program (Program 7.2) is used to generate FIR filter coefficients using the rectangular window. Its frequency responses will be plotted together with that obtained using the Hamming window for comparison, as shown in Program 7.3.

Program 7.2. MATLAB program for Example 7.7.

```
% Example 7.7
% MATLAB program to generate FIR coefficients
% using the rectangular window.
%
N=25; Ftype=1; WnL= 0.5*pi; WnH=0; Wtype=1;
B=firwd(N,Ftype,WnL,WnH,Wtype);
```

Results of the FIR filter design using the Hamming window are illustrated in Program 7.3.

Program 7.3. MATLAB program for Example 7.7.

```
% Figure 7.12
% MATLAB program to create Figure 7.12
%
N=25;Ftype=1;WnL= 0.5*pi;WnH=0;Wtype=1;fs=8000;
%Design using the rectangular window;
Brec=firwd(N,Ftype,WnL,WnH,Wtype);
N=25;Ftype=1;WnL= 0.5*pi;WnH=0;Wtype=4;
%Design using the Hamming window;
Bham=firwd(N,Ftype,WnL,WnH,Wtype);
[hrec,f]=freqz(Brec,1,512,fs);
[hham,f]=freqz(Bham,1,512,fs);
prec= 180*unwrap(angle(hrec))/pi;
pham= 180*unwrap(angle(hham))/pi
subplot(2,1,1);
plot(f,20*log10(abs(hrec)),'-.',f,20*log10(abs(hham)));grid
axis([0 4000 −100 10]);
xlabel('Frequency (Hz)');ylabel('Magnitude Response (dB)');
subplot(2,1,2);
plot(f,prec,'-.',f,pham);grid
xlabel('Frequency (Hz)');ylabel('Phase (degrees)');
```

As a comparison, the frequency responses achieved from the rectangular window and the Hamming window are plotted in Figure 7.12, where the dash-dotted line indicates the frequency response via the rectangular window, while the solid line indicates the frequency response via the Hamming window.

c. The FIR filter coefficients for both methods are listed in Table 7.6.

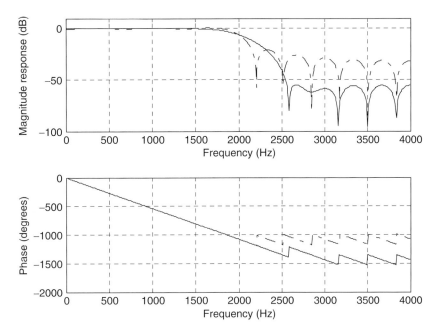

FIGURE 7.12 Frequency responses using the rectangular and Hamming windows.

TABLE 7.6 FIR filter coefficients in Example 7.7 (rectangular and Hamming windows).

B: FIR Filter Coefficients (rectangular window)	Bham: FIR Filter Coefficients (Hamming window)
$b_0 = b_{24} = 0.000000$	$b_0 = b_{24} = 0.000000$
$b_1 = b_{23} = -0.028937$	$b_1 = b_{23} = -0.002769$
$b_2 = b_{22} = 0.000000$	$b_2 = b_{22} = 0.000000$
$b_3 = b_{21} = 0.035368$	$b_3 = b_{21} = 0.007595$
$b_4 = b_{20} = 0.000000$	$b_4 = b_{20} = 0.000000$
$b_5 = b_{19} = -0.045473$	$b_5 = b_{19} = -0.019142$
$b_6 = b_{18} = 0.000000$	$b_6 = b_{18} = 0.000000$
$b_7 = b_{17} = 0.063662$	$b_7 = b_{17} = 0.041957$
$b_8 = b_{16} = 0.000000$	$b_8 = b_{16} = 0.000000$
$b_9 = b_{15} = -0.106103$	$b_9 = b_{15} = -0.091808$
$b_{10} = b_{14} = 0.000000$	$b_{10} = b_{14} = 0.000000$
$b_{11} = b_{13} = 0.318310$	$b_{11} = b_{13} = 0.313321$
$b_{12} = 0.500000$	$b_{12} = 0.500000$

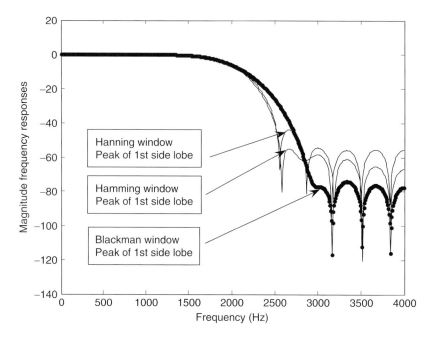

FIGURE 7.13 Comparisons of magnitude frequency responses for the Hanning, Hamming, and Blackman windows.

For comparison with other window functions, Figure 7.13 shows the magnitude frequency responses using the Hanning, Hamming, and Blackman windows, with 25 taps and a cutoff frequency of 2,000 Hz. The Blackman window offers the lowest side lobe, but with an increased width of the main lobe. The Hamming window and Hanning window have a similar narrow width of the main lobe, but the Hamming window accommodates a lower side lobe than the Hanning window. Next, we will study how to choose a window in practice.

Applying the window to remedy the Gibbs effect will change the characteristics of the magnitude frequency response of the FIR filter, where the width of the main lobe becomes wider, while more attenuation of side lobes is achieved.

Next, we illustrate the design for customer specifications in practice. Given the required stopband attenuation and passband ripple specifications shown in Figure 7.14, where the lowpass filter specifications are given for illustrative purposes, the appropriate window can be selected based on performances of the window functions listed in Table 7.7. For example, the Hamming window offers the passband ripple of 0.0194 dB and stopband attenuation of 53 dB. With the selected Hamming window and the calculated normalized transition band defined in Table 7.7,

$$\Delta f = |f_{stop} - f_{pass}|/f_s, \qquad (7.22)$$

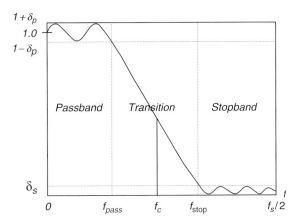

FIGURE 7.14 Lowpass filter frequency domain specifications.

the filter length using the Hamming window can be determined by

$$N = \frac{3.3}{\Delta f}. \qquad (7.23)$$

Note that the passband ripple is defined as

$$\delta_p \, dB = 20 \cdot \log_{10}(1 + \delta_p), \qquad (7.24)$$

while the stopband attenuation is defined as

$$\delta_s \, dB = -20 \log_{10}(\delta_s). \qquad (7.25)$$

The cutoff frequency used for the design will be chosen at the middle of the transition band, as illustrated for the lowpass filter shown in Figure 7.14.

As a rule of thumb, the cutoff frequency used for design is determined by

$$f_c = (f_{pass} + f_{stop})/2. \qquad (7.26)$$

TABLE 7.7 FIR filter length estimation using window functions (normalized transition width $\Delta f = |f_{stop} - f_{pass}|/f_s$).

Window Type	Window Function $w(n)$, $-M \leq n \leq M$	Window Length, N	Passband Ripple (dB)	Stopband Attenuation (dB)
Rectangular	1	$N = 0.9/\Delta f$	0.7416	21
Hanning	$0.5 + 0.5 \cos\left(\frac{\pi n}{M}\right)$	$N = 3.1/\Delta f$	0.0546	44
Hamming	$0.54 + 0.46 \cos\left(\frac{\pi n}{M}\right)$	$N = 3.3/\Delta f$	0.0194	53
Blackman	$0.42 + 0.5 \cos\left(\frac{n\pi}{M}\right) + 0.08 \cos\left(\frac{2n\pi}{M}\right)$	$N = 5.5/\Delta f$	0.0017	74

Note that Equation (7.23) and formulas for other window lengths in Table 7.7 are empirically derived based on the normalized spectral transition width of each window function. The spectrum of each window function appears to be a shape like the lowpass filter magnitude frequency response with ripples in the passband and side lobes in the stopband. The passband frequency edge of the spectrum is the frequency where the magnitude just begins to drop below the passband ripple and where the stop frequency edge is at the peak of the first side lobe in the spectrum. With the passband ripple and stopband attenuation specified for a particular window, the normalized transition width of the window is in inverse proportion to the window length N multiplied by a constant. For example, the normalized spectral transition Δf for the Hamming window is $3.3/N$. Hence, matching the FIR filter transition width with the transition width of the window spectrum gives the filter length estimation listed in Table 7.7.

The following examples illustrate the determination of each filter length and cutoff frequency/frequencies for the design of lowpass, highpass, bandpass, and bandstop filters. Application of each designed filter to the processing of speech data is included, along with an illustration of filtering effects in both time domain and frequency domain.

Example 7.8.

A lowpass FIR filter has the following specifications:

   Passband = 0 – 1,850 Hz
   Stopband = 2,150 – 4,000 Hz
   Stopband attenuation = 20 dB
   Passband ripple = 1 dB
   Sampling rate = 8,000 Hz

   a. Determine the FIR filter length and the cutoff frequency to be used in the design equation.

Solution:

   a. The normalized transition band as defined in Equation (7.22) and Table 7.7 is given by

$$\Delta f = |2150 - 1850|/8000 = 0.0375.$$

Again, based on Table 7.7, selecting the rectangular window will result in a passband ripple of 0.74 dB and a stopband attenuation of 21 dB. Thus, this window selection would satisfy the design requirement for the passband ripple of 1 dB and stopband attenuation of 20 dB. Next, we determine the length of the filter as

$$N = 0.9/\Delta f = 0.9/0.0375 = 24.$$

We choose the odd number $N = 25$. The cutoff frequency is determined by $(1850 + 2150)/2 = 2000$ Hz. Such a filter has been designed in Example 7.7, its filter coefficients are listed in Table 7.6, and its frequency responses can be found in Figure 7.12 (dashed lines).

Now we look at the time domain and frequency domain results from filtering a speech signal by using the lowpass filter we have just designed. Figure 7.15a shows the original speech and lowpass filtered speech. The spectral comparison is given in Figure 7.15b, where, as we can see, the frequency components beyond 2 kHz are filtered. The lowpass filtered speech would sound muffled.

We will continue to illustrate the determination of the filter length and cutoff frequency for other types of filters via the following examples.

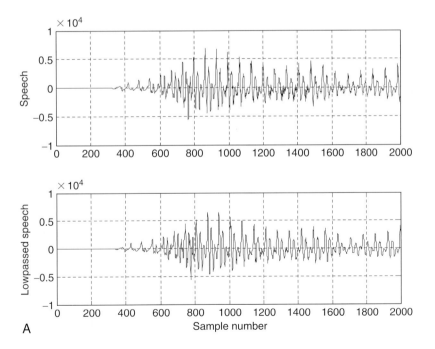

FIGURE 7.15A Original speech and processed speech using the lowpass filter.

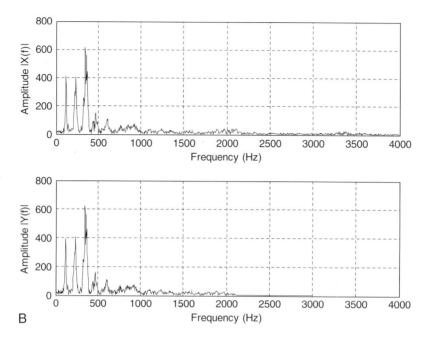

FIGURE 7.15B Spectral plots of the original speech and processed speech by the lowpass filter.

Example 7.9.

a. Design a highpass FIR filter with the following specifications:

Stopband = 0–1,500 Hz
Passband = 2,500–4,000 Hz
Stopband attenuation = 40 dB
Passband ripple = 0.1 dB
Sampling rate = 8,000 Hz

Solution:

a. Based on the specifications, the Hanning window will do the job, since it has a passband ripple of 0.0546 dB and a stopband attenuation of 44 dB. Then

$$\Delta f = |1500 - 2500|/8000 = 0.125$$
$$N = 3.1/\Delta f = 24.2. \text{ Choose } N = 25.$$

Hence, we choose 25 filter coefficients using the Hanning window method. The cutoff frequency is $(1500 + 2500)/2 = 2000$ Hz. The normalized cutoff frequency can be easily found as

$$\Omega_c = \frac{2000 \times 2\pi}{8000} = 0.5\pi \text{ radians.}$$

And notice that $2M + 1 = 25$. The application program and design results are listed in Program 7.4 and Table 7.8.

The corresponding frequency responses of the designed highpass FIR filter are displayed in Figure 7.16.

TABLE 7.8 FIR filter coefficients in Example 7.9 (Hanning window).

Bhan: FIR Filter Coefficients (Hanning window)

$b_0 = b_{24} = 0.000000$	$b_1 = b_{23} = 0.000493$
$b_2 = b_{22} = 0.000000$	$b_3 = b_{21} = -0.005179$
$b_4 = b_{20} = 0.000000$	$b_5 = b_{19} = 0.016852$
$b_6 = b_{18} = 0.000000$	$b_7 = b_{17} = -0.040069$
$b_8 = b_{16} = 0.0000000$	$b_9 = b_{15} = 0.090565$
$b_{10} = b_{14} = 0.000000$	$b_{11} = b_{13} = -0.312887$
$b_{12} = 0.500000$	

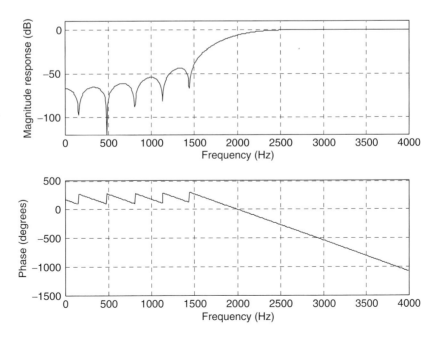

FIGURE 7.16 Frequency responses of the designed highpass filter using the Hanning window.

246 7 FINITE IMPULSE RESPONSE FILTER DESIGN

Program 7.4. MATLAB program for Example 7.9.

```
% Figure 7.16 (Example 7.9)
% MATLAB program to create Figure 7.16
%
N = 25;Ftype = 2;WnL = 0;WnH = 0.5*pi;Wtype = 3;fs = 8000;
Bhan=firwd(N, Ftype, WnL, WnH, Wtype);
freqz(Bhan,1,512,fs);
axis([0 fs/2 -120 10]);
```

Comparisons are given in Figure 7.17(a), where the original speech and processed speech using the highpass filter are plotted. The high-frequency components of speech generally contain small amounts of energy. Figure 7.17(b) displays the spectral plots, where clearly the frequency components lower than 1.5 kHz are filtered. The processed speech would sound crisp.

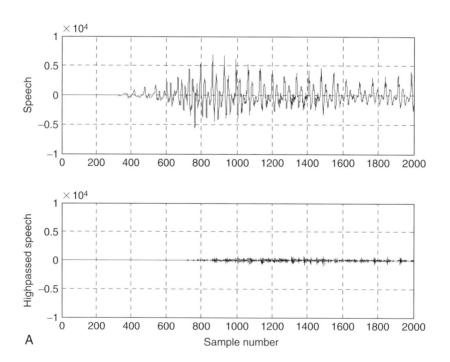

FIGURE 7.17A Original speech and processed speech using the highpass filter.

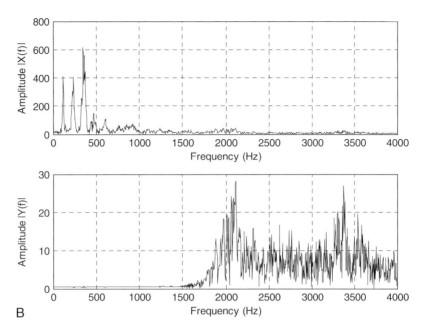

FIGURE 7.17B Spectral comparison of the original speech and processed speech using the highpass filter.

Example 7.10.

a. Design a bandpass FIR filter with the following specifications:

Lower stopband = 0–500 Hz
Passband = 1,600–2,300 Hz
Upper stopband = 3,500–4,000 Hz
Stopband attenuation = 50 dB
Passband ripple = 0.05 dB
Sampling rate = 8,000 Hz

Solution:

a. $\Delta f_1 = |1600 - 500|/8000 = 0.1375$ and $\Delta f_2 = |3500 - 2300|/8000 = 0.15$

$N_1 = 3.3/0.1375 = 24$ and $N_2 = 3.3/0.15 = 22$

Choosing $N = 25$ filter coefficients using the Hamming window method:

$f_1 = (1600 + 500)/2 = 1050 \text{ Hz}$ and $f_2 = (3500 + 2300)/2 = 2900 \text{ Hz}.$

The normalized lower and upper cutoff frequencies are calculated as:

$$\Omega_L = \frac{1050 \times 2\pi}{8000} = 0.2625\pi \text{ radians and}$$

$$\Omega_H = \frac{2900 \times 2\pi}{8000} = 0.725\pi \text{ radians,}$$

and $N = 2M + 1 = 25$. Using the MATLAB program, design results are achieved as shown in Program 7.5.

Figure 7.18 depicts the frequency responses of the designed bandpass FIR filter. Table 7.9 lists the designed FIR filter coefficients.

Program 7.5. MATLAB program for Example 7.10.

```
% Figure 7.18 (Example 7.10)
% MATLAB program to create Figure 7.18
%
N = 25; Ftype = 3; WnL = 0.2625*pi; WnH = 0.725*pi; Wtype = 4; fs = 8000;
Bham=firwd(N,Ftype,WnL,WnH,Wtype);
freqz(Bham,1,512,fs);
axis([0 fs/2 -130 10]);
```

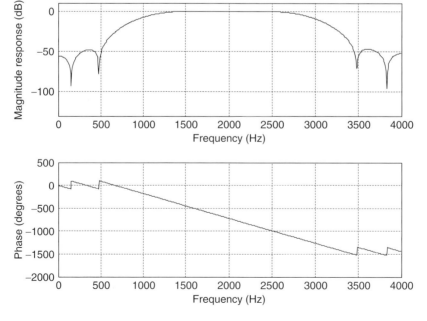

FIGURE 7.18 Frequency responses of the designed bandpass filter using the Hamming window.

7.3 Window Method

TABLE 7.9 FIR filter coefficients in Example 7.10 (Hamming window).

Bham: FIR Filter Coefficients (Hamming window)

$b_0 = b_{24} = 0.002680$	$b_1 = b_{23} = -0.001175$
$b_2 = b_{22} = -0.007353$	$b_3 = b_{21} = 0.000674$
$b_4 = b_{20} = -0.011063$	$b_5 = b_{19} = 0.004884$
$b_6 = b_{18} = 0.053382$	$b_7 = b_{17} = -0.003877$
$b_8 = b_{16} = 0.028520$	$b_9 = b_{15} = -0.008868$
$b_{10} = b_{14} = -0.296394$	$b_{11} = b_{13} = 0.008172$
$b_{12} = 0.462500$	

For comparison, the original speech and bandpass filtered speech are plotted in Figure 7.19a, where the bandpass frequency components contain a small portion of speech energy. Figure 7.19b shows a comparison indicating that the low frequency and high frequency are removed by the bandpass filter.

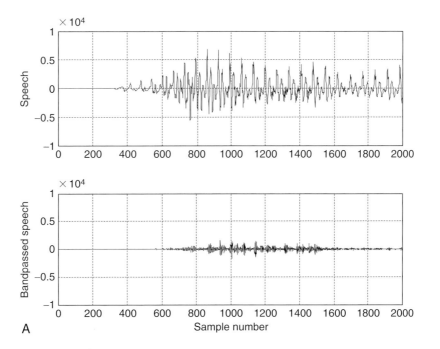

FIGURE 7.19A Original speech and processed speech using the bandpass filter.

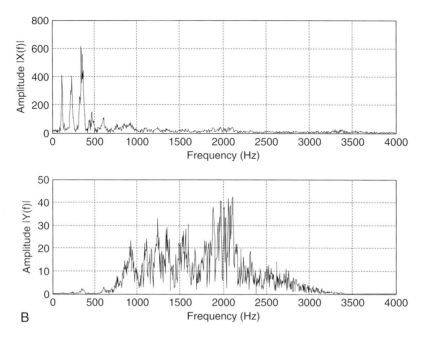

FIGURE 7.19B Spectral comparison of the original speech and processed speech using the bandpass filter.

Example 7.11.

a. Design a bandstop FIR filter with the following specifications:

   Lower cutoff frequency = 1,250 Hz
   Lower transition width = 1,500 Hz
   Upper cutoff frequency = 2,850 Hz
   Upper transition width = 1,300 Hz
   Stopband attenuation = 60 dB
   Passband ripple = 0.02 dB
   Sampling rate = 8,000 Hz

Solution:

a. We can directly compute the normalized transition widths:

$$\Delta f_1 = 1500/8000 = 0.1875, \text{ and } \Delta f_2 = 1300/8000 = 0.1625.$$

The filter lengths are determined using the Blackman window as:

$$N_1 = 5.5/0.1875 = 29.33, \text{ and } N_2 = 5.5/0.1625 = 33.8.$$

7.4 Applications: Noise Reduction and Two-Band Digital Crossover

We choose an odd number $N = 35$. The normalized lower and upper cutoff frequencies are calculated as:

$$\Omega_L = \frac{2\pi \times 1250}{8000} = 0.3125\pi \text{ radian and}$$

$$\Omega_H = \frac{2\pi \times 2850}{8000} = 0.7125\pi \text{ radians,}$$

and $N = 2M + 1 = 35$. Using MATLAB, the design results are demonstrated in Program 7.6.

Program 7.6. MATLAB program for Example 7.11.

```
% Figure 7.20 (Example 7.11)
% MATLAB program to create Figure 7.20
%
N = 35; Ftype = 4; WnL = 0.3125*pi; WnH = 0.7125*pi; Wtype = 5; fs = 8000;
Bblack = firwd(N, Ftype, WnL, WnH, Wtype);
freqz(Bblack, 1, 512, fs);
axis([0 fs/2 -120 10]);
```

Figure 7.20 shows the plot of the frequency responses of the designed bandstop filter. The designed filter coefficients are listed in Table 7.10.

Comparisons of filtering effects are illustrated in Figures 7.21a and 7.21b. In Figure 7.21a, the original speech and the processed speech by the bandstop filter are plotted. The processed speech contains most of the energy of the original speech because most of the energy of the speech signal exists in the low-frequency band. Figure 7.21b verifies the filtering frequency effects. The frequency components ranging from 2,000 to 2,200 Hz have been completely removed.

TABLE 7.10 FIR filter coefficients in Example 7.11 (Blackman window).

Black: FIR Filter Coefficients (Blackman window)

$b_0 = b_{34} = 0.000000$	$b_1 = b_{33} = 0.000059$
$b_2 = b_{32} = 0.000000$	$b_3 = b_{31} = 0.000696$
$b_4 = b_{30} = 0.001317$	$b_5 = b_{29} = -0.004351$
$b_6 = b_{28} = -0.002121$	$b_7 = b_{27} = 0.000000$
$b_8 = b_{26} = -0.004249$	$b_9 = b_{25} = 0.027891$
$b_{10} = b_{24} = 0.011476$	$b_{11} = b_{23} = -0.036062$
$b_{12} = b_{22} = 0.000000$	$b_{13} = b_{21} = -0.073630$
$b_{14} = b_{20} = -0.020893$	$b_{15} = b_{19} = 0.285306$
$b_{16} = b_{18} = 0.014486$	$b_{17} = 0.600000$

252 7 FINITE IMPULSE RESPONSE FILTER DESIGN

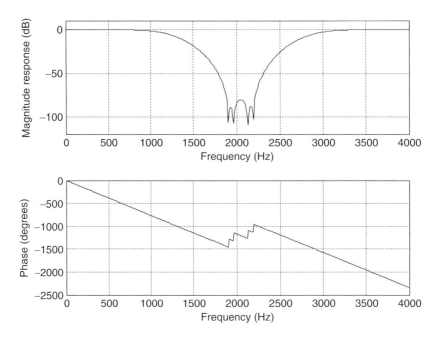

FIGURE 7.20 Frequency responses of the designed bandstop filter using the Blackman window.

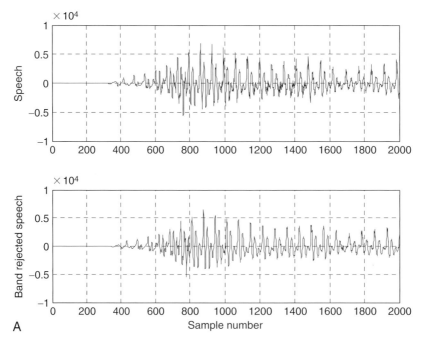

FIGURE 7.21A Original speech and processed speech using the bandstop filter.

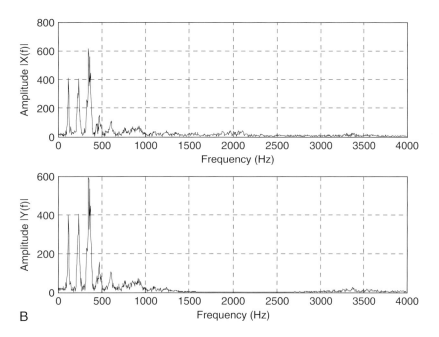

FIGURE 7.21B Spectral comparison of the original speech and the processed speech using the bandstop filter.

7.4 Applications: Noise Reduction and Two-Band Digital Crossover

In this section, we will investigate noise reduction and digital crossover design using the FIR filters.

7.4.1 Noise Reduction

One of the key digital signal processing (DSP) applications is noise reduction. In this application, a digital FIR filter removes noise in the signal that is contaminated by noise existing in the broad frequency range. For example, such noise often appears during the data acquisition process. In real-world applications, the desired signal usually occupies a certain frequency range. We can design a digital filter to remove frequency components other than the desired frequency range.

In a data acquisition system, we record a 500 Hz sine wave at a sampling rate of 8,000 Hz. The signal is corrupted by broadband noise $v(n)$:

$$x(n) = 1.4141 \cdot \sin(2\pi \cdot 500n/8000) + v(n).$$

The 500 Hz signal with noise and its spectrum are plotted in Figure 7.22, from which it is obvious that the digital sine wave contains noise. The spectrum is also displayed to give better understanding of the noise frequency level. We can see that noise is broadband, existing from 0 Hz to the folding frequency of 4,000 Hz. Assuming that the desired signal has a frequency range of only 0 to 800 Hz, we can filter noise from 800 Hz and beyond. A lowpass filter would complete such a task. Then we develop the filter specifications:

Passband frequency range: 0 Hz to 800 Hz with passband ripple less than 0.02 dB.
Stopband frequency range: 1 kHz to 4 kHz with 50 dB attenuation.

As we will see, lowpass filtering will remove the noise ranging from 1,000 to 4,000 Hz, and hence the signal-to-noise power ratio will be improved.

Based on the specifications, we design the FIR filter with a Hamming window, a cutoff frequency of 900 Hz, and an estimated filter length of 133 taps. The enhanced signal is depicted in Figure 7.23, where the clean signal can be observed. The amplitude spectrum for the enhanced signal is also plotted. As shown in the spectral plot, the noise level is almost neglected between 1 and 4 kHz. Notice that since we use the higher-order FIR filter, the signal experiences a linear phase delay of 66 samples, as is expected. We also see some transient response effects. However, the transient response effects will be

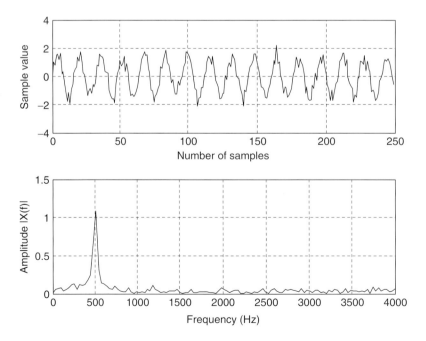

FIGURE 7.22 **Signal with noise and its spectrum.**

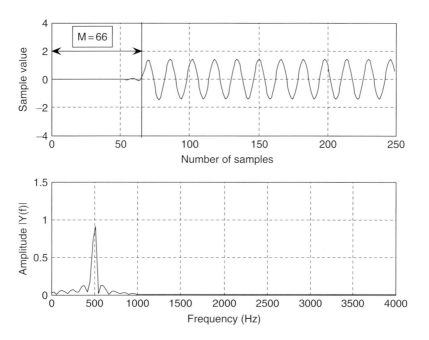

FIGURE 7.23 The noise-removed clean signal and spectrum.

ended totally after the first 132 samples due to the length of the FIR filter. MATLAB implementation is given in Program 7.7.

7.4.2 Speech Noise Reduction

In a speech recording system, we digitally record speech in a noisy environment at a sampling rate of 8,000 Hz. Assuming that the recorded speech contains information within 1,800 Hz, we can design a lowpass filter to remove the noise between 1,800 Hz and the Nyquist limit (the folding frequency of 4,000 Hz). Therefore, we have the following filter specifications:

Filter type = lowpass FIR
Passband frequency range = 0–1,800 Hz
Passband ripple = 0.02 dB
Stopband frequency range = 2,000–4,000 Hz
Stopband attenuation = 50 dB.

According to these specifications, we can determine the following parameters for filter design:

Window type = Hamming window
Number of filter taps = 133
Lowpass cutoff frequency = 1,900 Hz.

Program 7.7. MATLAB program for the application of noise filtering.

```
close all; clear all
fs = 8000;              % Sampling rate
T = 1/fs;               % Sampling period
v=sqrt(0.1)*randn(1,250);          % Generate the Gaussian random noise
n = 0:1:249;            % Indexes
x = sqrt(2)*sin(2*pi*500*n*T) +v; % Generate the 500-Hz sinusoid plus noise
subplot(2,1,1);plot(t,x);
xlabel('Number of samples');ylabel('Sample value');grid;
N=length(x);
f=[0:N/2]*fs/N;
Axk= 2*abs(fft(x))/N;Axk(1)=Axk(1)/2; % Calculate the single-sided spectrum
subplot(2,1,2);plot(f,Axk(1:N/2+1));
xlabel('Frequency (Hz)');ylabel('Amplitude (f)|');grid;
figure
Wnc= 2*pi*900/fs;       % Determine the normalized digital cutoff frequency
B=firwd(133,1,Wnc,0,4);          % Design the FIR filter
y=filter(B,1,x);        % Perform digital filtering
Ayk= 2*abs(fft(y))/N;Ayk(1)=Ayk(1)/2;% Single-sided spectrum of the filtered data
subplot(2,1,1);plot(t,y);
xlabel('Number of samples');ylabel('Sample value');grid;
subplot(2,1,2);plot(f,Ayk(1:N/2+1));axis([0 fs/2 0 1.5]);
xlabel('Frequency (Hz)');ylabel('Amplitude |Y(f)|');grid;
```

Figure 7.24(a) shows the plots of the recorded noisy speech and its spectrum. As we can see in the noisy spectrum, the noise level is high and broadband. After applying the designed lowpass filter, we plot the filtered speech and its spectrum shown in Figure 7.24(b), where the clean speech is clearly identified, while the spectrum shows that the noise components above 2 kHz have been completely removed.

7.4.3 Two-Band Digital Crossover

In audio systems, there is often a situation where the application requires the entire audible range of frequencies, but this is beyond the capability of any single speaker driver. So, we combine several drivers, such as the speaker cones and horns, each covering different frequency range, to reproduce the full audio frequency range.

A typical two-band digital crossover can be designed as shown in Figure 7.25. There are two speaker drivers. The woofer responds to low frequencies, and the

7.4 Applications: Noise Reduction and Two-Band Digital Crossover

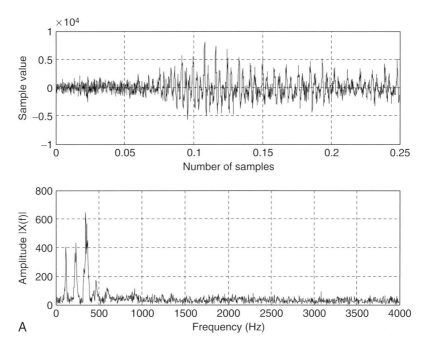

FIGURE 7.24A Noisy speech and its spectrum.

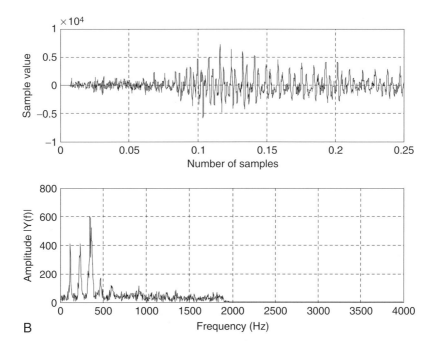

FIGURE 7.24B Enhanced speech and its spectrum.

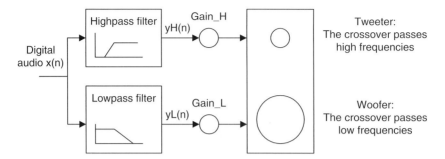

FIGURE 7.25 Two-band digital crossover.

tweeter responds to high frequencies. The incoming digital audio signal is split into two bands by using a lowpass filter and a highpass filter in parallel. We then amplify the separated audio signals and send them to their respective corresponding speaker drivers. Hence, the objective is to design the lowpass filter and the highpass filter so that their combined frequency response is flat, while keeping transition as sharp as possible to prevent audio signal distortion in the transition frequency range. Although traditional crossover systems are designed using active circuits (analog systems) or passive circuits, the digital crossover system provides a cost-effective solution with programmable ability, flexibility, and high quality.

A crossover system has the following specifications:

Sampling rate = 44,100 Hz
Crossover frequency = 1,000 Hz (cutoff frequency)
Transition band = 600 to 1,400 Hz
Lowpass filter = passband frequency range from 0 to 600 Hz with a ripple of 0.02 dB and stopband edge at 1,400 Hz with attenuation of 50 dB.
Highpass filter = passband frequency range from 1.4 to 44.1 kHz with ripple of 0.02 dB and stopband edge at 600 Hz with attenuation of 50 dB.

In the design of this crossover system, one possibility is to use an FIR filter, since it provides a linear phase for the audio system. However, an infinite impulse response (IIR) filter (which will be discussed in the next chapter) can be an alternative. Based on the transition band of 800 Hz and the passband ripple and stopband attenuation requirements, the Hamming window is chosen for both lowpass and highpass filters. We can determine the number of filter taps as 183, each with a cutoff frequency of 1,000 Hz.

The frequency responses for the designed lowpass filter and highpass filter are given in Figure 7.26(a), and for the lowpass filter, highpass filter, and combined responses in Figure 7.26(b). As we can see, the crossover frequency

7.4 Applications: Noise Reduction and Two-Band Digital Crossover

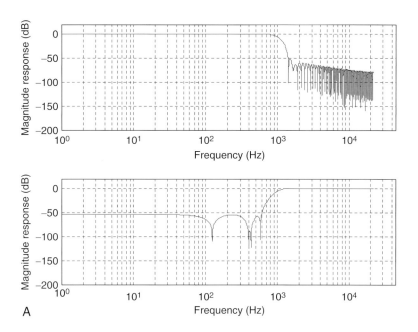

FIGURE 7.26A Magnitude frequency responses for lowpass filter and highpass filter.

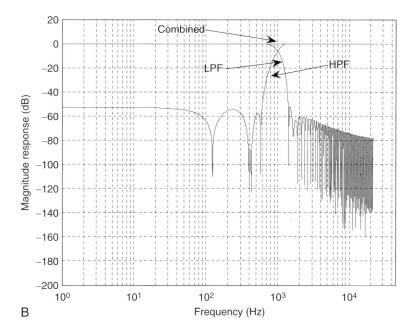

FIGURE 7.26B Magnitude frequency responses for both lowpass filter and highpass filter, and the combined magnitude frequency response for the digital audio crossover system.

260 7 FINITE IMPULSE RESPONSE FILTER DESIGN

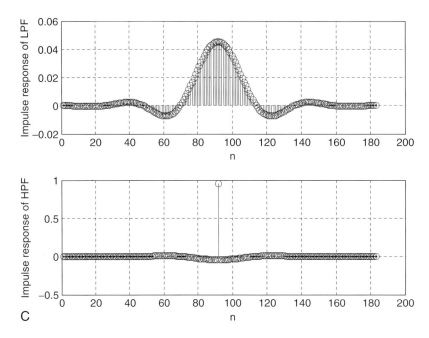

FIGURE 7.26C Impulse responses of both the FIR lowpass filter and the FIR highpass filter for the digital audio crossover system.

for both filters is at 1,000 Hz, and the combined frequency response is perfectly flat. The impulse responses (filter coefficients) for lowpass and highpass filters are plotted in Figure 7.26(c).

7.5 Frequency Sampling Design Method

In addition to methods of Fourier transform design and Fourier transform with windowing discussed in the previous section, *frequency sampling* is another alternative. The key feature of frequency sampling is that the filter coefficients can be calculated based on the specified magnitudes of the desired filter frequency response uniformly in frequency domain. Hence, it has design flexibility.

To begin with development, we let $h(n)$, for $n = 0, 1, \ldots, N-1$, be the causal impulse response (FIR filter coefficients) that approximates the FIR filter, and we let $H(k)$, for $k = 0, 1, \ldots, N-1$, represent the corresponding discrete Fourier transform (DFT) coefficients. We obtain $H(k)$ by sampling the desired frequency filter response $H(k) = H(e^{j\Omega})$ at equally spaced instants in frequency domain, as shown in Figure 7.27.

Then, according to the definition of the inverse DFT (IDFT), we can calculate the FIR coefficients:

$$h(n) = \frac{1}{N} \sum_{k=0}^{N-1} H(k) W_N^{-kn}, \text{ for } n = 0, 1, \ldots, N-1,$$

where

$$W_N = e^{-j\frac{2\pi}{N}} = \cos\left(\frac{2\pi}{N}\right) - j\sin\left(\frac{2\pi}{N}\right). \quad (7.27)$$

We assume that the FIR filter has linear phase and the number of taps $N = 2M + 1$. Equation (7.27) can be significantly simplified as

$$h(n) = \frac{1}{2M+1} \left\{ H_0 + 2 \sum_{k=1}^{M} H_k \cos\left(\frac{2\pi k(n-M)}{2M+1}\right) \right\}, \quad (7.28)$$

for $n = 0, 1, \ldots, 2M,$

where H_k, for $k = 0, 1, \ldots, 2M$, represents the magnitude values specifying the desired filter frequency response sampled at $\Omega_k = \frac{2\pi k}{(2M+1)}$. The derivation is

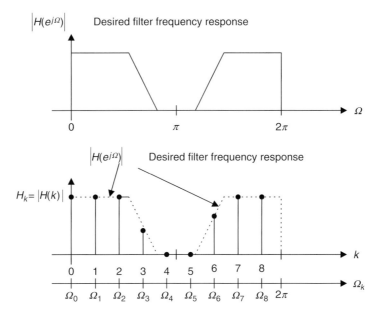

FIGURE 7.27 Desired filter frequency response and sampled frequency response.

detailed in Appendix E. The design procedure is therefore simply summarized as follows:

1. Given the filter length of $2M+1$, specify the magnitude frequency response for the normalized frequency range from 0 to π:

$$H_k \text{ at } \Omega_k = \frac{2\pi k}{(2M+1)} \quad \text{for } k = 0, 1, \ldots, M. \quad (7.29)$$

2. Calculate FIR filter coefficients:

$$h(n) = \frac{1}{2M+1}\left\{H_0 + 2\sum_{k=1}^{M} H_k \cos\left(\frac{2\pi k(n-M)}{2M+1}\right)\right\} \quad (7.30)$$

for $n = 0, 1, \ldots, M$.

3. Use the symmetry (linear phase requirement) to determine the rest of the coefficients:

$$h(n) = h(2M - n) \text{ for } n = M+1, \ldots, 2M. \quad (7.31)$$

Example 7.12 illustrates the design procedure.

Example 7.12.

a. Design a linear phase lowpass FIR filter with 7 taps and a cutoff frequency of $\Omega_c = 0.3\pi$ radian using the frequency sampling method.

Solution:

a. Since $N = 2M + 1 = 7$ and $M = 3$, the sampled frequencies are given by

$$\Omega_k = \frac{2\pi}{7}k \text{ radians}, k = 0, 1, 2, 3.$$

Next we specify the magnitude values H_k at the specified frequencies as follows:

$$\text{for } \Omega_0 = 0 \text{ radians}, \quad H_0 = 1.0$$

$$\text{for } \Omega_1 = \frac{2}{7}\pi \text{ radians}, \quad H_1 = 1.0$$

$$\text{for } \Omega_2 = \frac{4}{7}\pi \text{ radians}, \quad H_2 = 0.0$$

$$\text{for } \Omega_3 = \frac{6}{7}\pi \text{ radians}, \quad H_3 = 0.0.$$

Figure 7.28 shows the specifications.

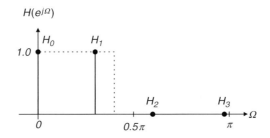

FIGURE 7.28 Sampled values of the frequency response in Example 7.12.

Using Equation (7.30), we achieve

$$h(n) = \frac{1}{7}\left\{1 + 2\sum_{k=1}^{3} H_k \cos[2\pi k(n-3)/7]\right\}, \quad n = 0, 1, \ldots, 3.$$

$$= \frac{1}{7}\{1 + 2\cos[2\pi(n-3)/7]\}$$

Thus, computing the FIR filter coefficients yields

$$h(0) = \frac{1}{7}\{1 + 2\cos(-6\pi/7)\} = -0.11456$$

$$h(1) = \frac{1}{7}\{1 + 2\cos(-4\pi/7)\} = 0.07928$$

$$h(2) = \frac{1}{7}\{1 + 2\cos(-2\pi/7)\} = 0.32100$$

$$h(3) = \frac{1}{7}\{1 + 2\cos(-0 \times \pi/7)\} = 0.42857.$$

By the symmetry, we obtain the rest of the coefficients as follows:

$$h(4) = h(2) = 0.32100$$
$$h(5) = h(1) = 0.07928$$
$$h(6) = h(0) = -0.11456.$$

The following two examples are devoted to illustrating the FIR filter design using the frequency sampling method. A MATLAB program, **firfs(N, Hk)**, is provided in the "MATLAB Programs" section at the end of this chapter (see its usage in Table 7.11) to implement the design in Equation (7.30) with input parameters of $N = 2M + 1$ (number of taps) and a vector H_k containing the specified magnitude values H_k, $k = 0, 1, \ldots, M$. Finally, the MATLAB function will return the calculated FIR filter coefficients.

TABLE 7.11 Illustrative usage for MATLAB function firfs(N, Hk).

```
function B=firfs(N, Hk)
% B=firls(N, Hk)
% FIR filter design using the frequency sampling method.
% Input parameters:
% N: the number of filter coefficients.
% Note: N must be an odd number.
% Hk: sampled frequency response for k = 0, 1, 2, ..., M = (N − 1)/2.
% Output:
% B: FIR filter coefficients.
```

Example 7.13.

a. Design a linear phase lowpass FIR filter with 25 coefficients using the frequency sampling method. Let the cutoff frequency be 2,000 Hz and assume a sampling frequency of 8,000 Hz.

b. Plot the frequency responses.

c. List the FIR filter coefficients.

Solution:

a. The normalized cutoff frequency for the lowpass filter is $\Omega = \omega T = 2\pi 2000/8000 = 0.5\pi$ radians, $N = 2M + 1 = 25$, and the specified values of the sampled magnitude frequency response are chosen to be

$$H_k = [1\ 1\ 1\ 1\ 1\ 1\ 1\ 0\ 0\ 0\ 0\ 0\ 0].$$

MATLAB Program 7.8 produces the design results.

b. The magnitude frequency response plotted using the dash-dotted line is displayed in Figure 7.29, where it is observed that oscillations (shown as the dash-dotted line) occur in the passband and stopband of the designed FIR filter. This is due to the abrupt change of the specification in the transition band (between the passband and the stopband). To reduce this ripple effect, the modified specification with a smooth transition band, H_k, $k = 0, 1, \ldots, 13$, is used:

$$H_k = [1\ 1\ 1\ 1\ 1\ 1\ 1\ 0.5\ 0\ 0\ 0\ 0\ 0].$$

Therefore the improved magnitude frequency response is shown in Figure 7.29 via the solid line.

c. The calculated FIR coefficients for both filters are listed in Table 7.12.

Program 7.8. MATLAB program for Example 7.13.

```
% Figure 7.29 (Example 7.13)
% MATLAB program to create Figure 7.29
fs = 8000;              % Sampling frequency
H1 = [1 1 1 1 1 1 1 0 0 0 0 0 0];       % Magnitude specifications
B1=firfs(25,H1);        % Design the filter
[h1,f]=freqz(B1,1,512,fs);   % Calculate the magnitude frequency response
H2 = [1 1 1 1 1 1 1 0.5 0 0 0 0 0];     % Magnitude specifications
B2=firfs(25,H2);        % Design the filter
[h2,f]=freqz(B2,1,512,fs);   % Calculate the magnitude frequency response
p1 = 180*unwrap(angle(h1))/pi;
p2 = 180*unwrap(angle(h2))/pi
subplot(2,1,1);plot(f,20*log10(abs(h1)),'-.',f,20*log10(abs(h2)));grid
axis([0 fs/2 -80 10]);
xlabel('Frequency (Hz)');ylabel('Magnitude Response (dB)');
subplot(2,1,2);plot(f,p1,'-.',f,p2);grid
xlabel('Frequency (Hz)');ylabel('Phase (degrees)');
```

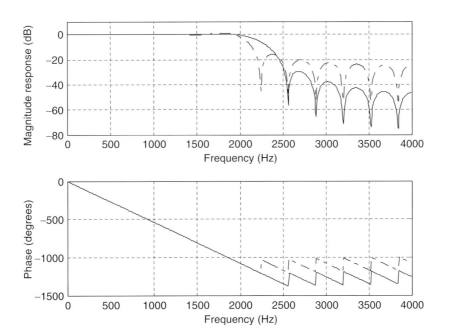

FIGURE 7.29 Frequency responses using the frequency sampling method in Example 7.13.

TABLE 7.12   FIR filter coefficients in Example 7.13 (frequency sampling method).

B1: FIR Filter Coefficients	B2: FIR Filter Coefficients
$b_0 = b_{24} = 0.027436$	$b_0 = b_{24} = 0.001939$
$b_1 = b_{23} = -0.031376$	$b_1 = b_{23} = 0.003676$
$b_2 = b_{22} = -0.024721$	$b_2 = b_{22} = -0.012361$
$b_3 = b_{21} = 0.037326$	$b_3 = b_{21} = -0.002359$
$b_4 = b_{20} = 0.022823$	$b_4 = b_{20} = 0.025335$
$b_5 = b_{19} = -0.046973$	$b_5 = b_{19} = -0.008229$
$b_6 = b_{18} = -0.021511$	$b_6 = b_{18} = -0.038542$
$b_7 = b_{17} = 0.064721$	$b_7 = b_{17} = 0.032361$
$b_8 = b_{16} = 0.020649$	$b_8 = b_{16} = 0.049808$
$b_9 = b_{15} = -0.106734$	$b_9 = b_{15} = -0.085301$
$b_{10} = b_{14} = -0.020159$	$b_{10} = b_{14} = -0.057350$
$b_{11} = b_{13} = 0.318519$	$b_{11} = b_{13} = 0.311024$
$b_{12} = 0.520000$	$b_{12} = 0.560000$

Example 7.14.

a. Design a linear phase bandpass FIR filter with 25 coefficients using the frequency sampling method. Let the lower and upper cutoff frequencies be 1,000 Hz and 3,000 Hz, respectively, and assume a sampling frequency of 8,000 Hz.

b. List the FIR filter coefficients.

c. Plot the frequency responses.

Solution:

a. First we calculate the normalized lower and upper cutoff frequencies for the bandpass filter; that is, $\Omega_L = 2\pi \times 1000/8000 = 0.25\pi$ radian and $\Omega_H = 2\pi \times 3000/8000 = 0.75\pi$ radians, respectively. The sampled values of the bandpass frequency response are specified by the following vector:

$$H_k = [0\ 0\ 0\ 0\ 1\ 1\ 1\ 1\ 0\ 0\ 0\ 0].$$

As a comparison, the second specification of H_k with a smooth transition band is used; that is,

$$H_k = [0\ 0\ 0\ 0.5\ 1\ 1\ 1\ 1\ 0.5\ 0\ 0\ 0].$$

b. The MATLAB list is shown in Program 7.9. The generated FIR coefficients are listed in Table 7.13.

Program 7.9. MATLAB program for Example 7.14.

```
% Figure 7.30 (Example 7.14)
% MATLAB program to create Figure 7.30
%
fs = 8000
H1 = [0 0 0 0 1 1 1 1 1 0 0 0 0];        % Magnitude specifications
B1=firfs(25,H1);             % Design the filter
[h1,f]=freqz(B1,1,512,fs);   % Calculate the magnitude frequency response
H2 = [0 0 0 0 0.5 1 1 1 1 0.5 0 0 0];    % Magnitude spectrum
B2=firfs(25,H2);             % Design the filter
[h2,f]=freqz(B2,1,512,fs);   % Calculate the magnitude frequency response
p1 = 180*unwrap(angle(h1)')/pi;
p2 = 180*unwrap(angle(h2)')/pi
subplot(2,1,1);plot(f,20*log10(abs(h1)),'-.',f,20*log10(abs(h2)));grid
axis([0 fs/2 -100 10]);
xlabel('Frequency (Hz)');ylabel('Magnitude Response (dB)');
subplot(2,1,2); plot(f,p1,'-.',f,p2);grid
xlabel('Frequency (Hz)');ylabel('Phase (degrees)');
```

c. Similar to the preceding example, Figure 7.30 shows the frequency responses. Focusing on the magnitude frequency responses depicted in Figure 7.30, the dash-dotted line indicates the magnitude frequency response obtained without specifying the smooth transition band, while the solid line indicates the magnitude frequency response achieved with the specification of the smooth transition band, hence resulting in the reduced ripple effect.

TABLE 7.13 FIR filter coefficients in Example 7.14 (frequency sampling method).

B1: FIR Filter Coefficients	B2: FIR Filter Coefficients
$b_0 = b_{24} = 0.055573$	$b_0 = b_{24} = 0.001351$
$b_1 = b_{23} = -0.030514$	$b_1 = b_{23} = -0.008802$
$b_2 = b_{22} = 0.000000$	$b_2 = b_{22} = -0.020000$
$b_3 = b_{21} = -0.027846$	$b_3 = b_{21} = 0.009718$
$b_4 = b_{20} = -0.078966$	$b_4 = b_{20} = -0.011064$
$b_5 = b_{19} = 0.042044$	$b_5 = b_{19} = 0.023792$
$b_6 = b_{18} = 0.063868$	$b_6 = b_{18} = 0.077806$
$b_7 = b_{17} = 0.000000$	$b_7 = b_{17} = -0.020000$
$b_8 = b_{16} = 0.094541$	$b_8 = b_{16} = 0.017665$
$b_9 = b_{15} = -0.038728$	$b_9 = b_{15} = -0.029173$
$b_{10} = b_{14} = -0.303529$	$b_{10} = b_{14} = -0.308513$
$b_{11} = b_{13} = 0.023558$	$b_{11} = b_{13} = 0.027220$
$b_{12} = 0.400000$	$b_{12} = 0.480000$

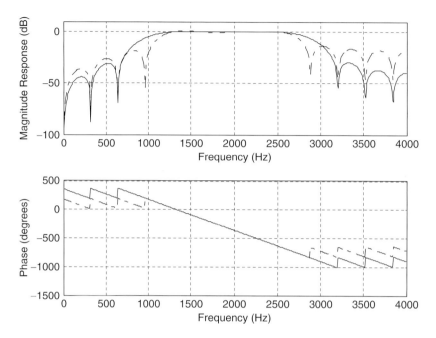

FIGURE 7.30 Frequency responses using the frequency sampling method in Example 7.14.

Observations can be made from examining Examples 7.13 and 7.14. First, the oscillations (Gibbs behavior) in the passband and stopband can be reduced at the expense of increasing the width of the main lobe. Second, we can modify the specification of the magnitude frequency response with a smooth transition band to reduce the oscillations and hence improve the performance of the FIR filter. Third, the magnitude values H_k, $k = 0, 1, \ldots, M$, in general can be arbitrarily specified. This indicates that the frequency sampling method is more flexible and can be used to design the FIR filter with an arbitrary specification of the magnitude frequency response.

7.6 Optimal Design Method

This section introduces Parks-McClellan algorithm, which is a most popular optimal design method used in industry due to its efficiency and flexibility. The FIR filter design using the Parks-McClellan algorithm is developed based on the idea of minimizing the maximum approximation error in a Chebyshev polynomial approximation to the desired filter magnitude frequency response. The details of this design development are beyond the scope of this text and can be found in Ambardar (1999) and Porat (1997). We will outline the design criteria and notation and then focus on the design procedure.

Given an ideal frequency response $H_d(e^{j\omega T})$, the approximation error $E(\omega)$ is defined as

$$E(\omega) = W(\omega)\left[H(e^{j\omega T}) - H_d(e^{j\omega T})\right], \quad (7.32)$$

where $H(e^{j\omega T})$ is the frequency response of the linear phase FIR filter to be designed, and $W(\omega)$ is the weight function for emphasizing certain frequency bands over others during the optimization process. This process is designed to minimize the error shown in Equation (7.33):

$$\min(\max|E(\omega)|) \quad (7.33)$$

over the set of FIR coefficients. With the help of Remez exchange algorithm, which is also beyond the scope of this book, we can obtain the best FIR filter whose magnitude response has an equiripple approximation to the ideal magnitude response. The achieved filters are optimal in the sense that the algorithms minimize the maximum error between the desired frequency response and the actual frequency response. These are often called *minimax filters*.

Next, we establish notations that will be used in the design procedure. Figure 7.31 shows the characteristics of the designed FIR filter by Parks-McClellan and Remez exchange algorithms. As illustrated in the top graph of Figure 7.31, the passband frequency response and stopband frequency response have equiripples. δ_p is used to specify the magnitude ripple in the passband, while δ_s specifies the stopband magnitude attenuation. In terms of dB value specification, we have $\delta_p\, dB = 20 \times \log_{10}(1 + \delta_p)$ and $\delta_s\, dB = -20 \times \log_{10} \delta_s$.

The middle graph in Figure 7.31 describes the error between the ideal frequency response and the actual frequency response. In general, the error magnitudes in the passband and stopband are different. This makes optimization unbalanced, since the optimization process involves an entire band. When the error magnitude in a band dominates the other(s), the optimization process may de-emphasize the contribution due to a small magnitude error. To make the error magnitudes balanced, a weight function can be introduced. The idea is to weight the band with the bigger magnitude error with a small weight factor and to weight the band with the smaller magnitude error with a big weight factor. We use a weight factor W_p for weighting the passband error and W_s for weighting the stopband error. The bottom graph in Figure 7.31 shows the weighted error, and clearly, the error magnitudes on both bands are at the same level. Selection of the weighting factors is further illustrated in the following design procedure.

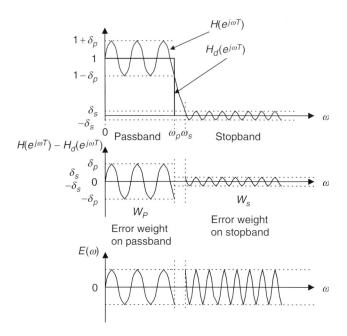

FIGURE 7.31 (*Top*) Magnitude frequency responses of an ideal lowpass filter and a typical lowpass filter designed using Parks-McClellan algorithm. (*Middle*) Error between the ideal and practical responses. (*Bottom*) Weighted error between the ideal and practical responses.

Optimal FIR Filter Design Procedure for Parks-McClellan Algorithm

1. Specify the band edge frequencies such as the passband and stopband frequencies, passband ripple, stopband attenuation, filter order, and sampling frequency of the DSP system.

2. Normalize band edge frequencies to the Nyquist limit (folding frequency $= f_s/2$) and specify the ideal magnitudes.

3. Calculate absolute values of the passband ripple and stopband attenuation if they are given in terms of dB values:

$$\delta_p = 10^{\left(\frac{\delta_p\ dB}{20}\right)} - 1 \qquad (7.34)$$

$$\delta_s = 10^{\left(\frac{-\delta_s\ dB}{20}\right)}. \qquad (7.35)$$

Then calculate the ratio and put it into a fraction form:

$$\frac{\delta_p}{\delta_s} = fraction\ form = \frac{numerator}{denominator} = \frac{W_s}{W_p}. \qquad (7.36)$$

7.6 Optimal Design Method

Next, set the error weight factors for passband and stopband, respectively:

$$W_p = denominator$$
$$W_s = numerator \qquad (7.37)$$

4. Apply the Remez algorithm to calculate filter coefficients.
5. If the specifications are not met, increase the filter order and repeat steps 1 to 4.

The following examples are given to illustrate the design procedure.

Example 7.15.

a. Design a lowpass filter with the following specifications:

DSP system sampling rate = 8,000 Hz
Passband = 0–800 Hz
Stopband = 1,000–4,000 Hz
Passband ripple = 1 dB
Stopband attenuation = 40 dB
Filter order = 53

Solution:

a. From the specifications, we have two bands: a lowpass band and a stopband. We perform normalization and specify ideal magnitudes as follows:

Folding frequency: $f_s/2 = 8000/2 = 4000$ Hz
For 0 Hz: $0/4000 = 0$, magnitude: 1
For 800 Hz: $800/4000 = 0.2$, magnitude: 1
For 1,000 Hz: $1000/4000 = 0.25$, magnitude: 0
For 4,000 Hz: $4000/4000 = 1$, magnitude: 0

Next, we determine the weights:

$$\delta_p = 10^{\left(\frac{1}{20}\right)} - 1 = 0.1220$$
$$\delta_s = 10^{\left(\frac{-40}{20}\right)} = 0.01.$$

Then, applying Equation (7.36) gives

$$\frac{\delta_p}{\delta_s} = 12.2 \approx \frac{12}{1} = \frac{W_s}{W_p}.$$

272 7 FINITE IMPULSE RESPONSE FILTER DESIGN

Hence, we have

$$W_s = 12 \text{ and } W_p = 1.$$

Applying **remez()** routine provided by MATLAB, we list MATLAB codes in Program 7.10. The filter coefficients are listed in Table 7.14.

Program 7.10. MATLAB program for Example 7.15.

```
% Figure 7.32 (Example 7.15)
% MATLAB program to create Figure 7.32
%
fs = 8000;
f = [0 0.2 0.25 1];        % Edge frequencies
m = [1 1 0 0];             % Ideal magnitudes
w = [1 12];                % Error weight factors
b = remez(53,f,m,w); % (53+1) Parks-McClellan algorithm and Remez exchange
format long
freqz(b,1,512,fs) % Plot the frequency responses
axis([0 fs/2 -80 10]);
```

Figure 7.32 shows the frequency responses.

Clearly, the stopband attenuation is satisfied. We plot the details for the filter passband in Figure 7.33.

TABLE 7.14 FIR filter coefficients in Example 7.15.

B: FIR Filter Coefficients (optimal design method)

$b_0 = b_{53} = -0.006075$	$b_1 = b_{52} = -0.00197$
$b_2 = b_{51} = 0.001277$	$b_3 = b_{50} = 0.006937$
$b_4 = b_{49} = 0.013488$	$b_5 = b_{48} = 0.018457$
$b_6 = b_{47} = 0.019347$	$b_7 = b_{46} = 0.014812$
$b_8 = b_{45} = 0.005568$	$b_9 = b_{44} = -0.005438$
$b_{10} = b_{43} = -0.013893$	$b_{11} = b_{42} = -0.015887$
$b_{12} = b_{41} = -0.009723$	$b_{13} = b_{40} = 0.002789$
$b_{14} = b_{39} = 0.016564$	$b_{15} = b_{38} = 0.024947$
$b_{16} = b_{37} = 0.022523$	$b_{17} = b_{36} = 0.007886$
$b_{18} = b_{35} = -0.014825$	$b_{19} = b_{34} = -0.036522$
$b_{20} = b_{33} = -0.045964$	$b_{21} = b_{32} = -0.033866$
$b_{22} = b_{31} = 0.003120$	$b_{23} = b_{30} = 0.060244$
$b_{24} = b_{29} = 0.125252$	$b_{25} = b_{28} = 0.181826$
$b_{26} = b_{27} = 0.214670$	

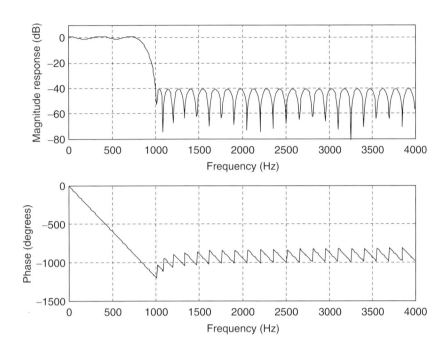

FIGURE 7.32 Frequency and phase responses for Example 7.15.

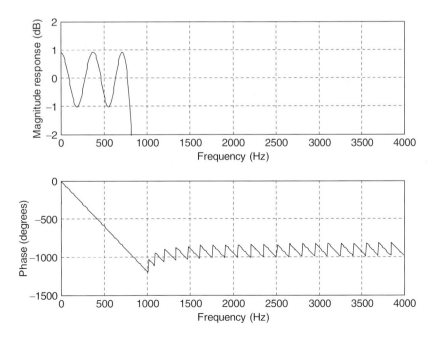

FIGURE 7.33 Frequency response details for passband in Example 7.15.

As shown in Figure 7.33, the ripples in the passband are between -1 and 1 dB. Hence, all the specifications are met. Note that if a specification is not satisfied, we will increase the order until the stopband attenuation and passband ripple are met.

Example 7.16.

This example illustrates the bandpass filter design.

a. Design a bandpass filter with the following specifications:

   DSP system sampling rate = 8,000 Hz
   Passband = 1,000–1,600 Hz
   Stopband = 0–600 Hz and 2,000–4,000 Hz
   Passband ripple = 1 dB
   Stopband attenuation = 30 dB
   Filter order = 25

Solution:

a. From the specifications, we have three bands: a passband, a lower stopband, and an upper stopband. We perform normalization and specify ideal magnitudes as follows:

   Folding frequency: $f_s/2 = 8000/2 = 4000$ Hz
   For 0 Hz: $0/4000 = 0$, magnitude: 0
   For 600 Hz: $600/4000 = 0.15$, magnitude: 0
   For 1,000 Hz: $1000/4000 = 0.25$, magnitude: 1
   For 1,600 Hz: $1600/4000 = 0.4$, magnitude: 1
   For 2,000 Hz: $2000/4000 = 0.5$, magnitude: 0
   For 4,000 Hz: $4000/4000 = 1$, magnitude: 0

Next, let us determine the weights:

$$\delta_p = 10^{\left(\frac{1}{20}\right)} - 1 = 0.1220$$

$$\delta_s = 10^{\left(\frac{-30}{20}\right)} = 0.0316.$$

Then, applying Equation (7.36), we get

$$\frac{\delta_p}{\delta_s} = 3.86 \approx \frac{39}{10} = \frac{W_s}{W_p}.$$

Hence, we have

$$W_s = 39 \text{ and } W_p = 10.$$

Applying the **remez()** routine provided by MATLAB and checking performance, we have Program 7.11. Table 7.15 lists the filter coefficients.

Program 7.11. MATLAB program for Example 7.16.

```
% Figure 7.34 (Example 7.16)
% MATLAB program to create Figure 7.34
%
fs = 8000;
f = [0 0.15 0.25 0.4 0.5 1];        % Edge frequencies
m = [0 0 1 1 0 0];          % Ideal magnitudes
w = [39 10 39];           % Error weight factors
format long
b = remez(25,f,m,w)% (25 + 1) taps Parks-McClellan algorithm and Remez exchange
freqz(b,1,512,fs);          % Plot the frequency responses
axis([0 fs/2 −80 10])
```

TABLE 7.15 FIR filter coefficients in Example 7.16.

B: FIR Filter Coefficients (optimal design method)

$b_0 = b_{25} = -0.022715$	$b_1 = b_{24} = -0.012753$
$b_2 = b_{23} = 0.005310$	$b_3 = b_{22} = 0.009627$
$b_4 = b_{21} = -0.004246$	$b_5 = b_{20} = 0.006211$
$b_6 = b_{19} = 0.057515$	$b_7 = b_{18} = 0.076593$
$b_8 = b_{17} = -0.015655$	$b_9 = b_{16} = -0.156828$
$b_{10} = b_{15} = -0.170369$	$b_{11} = b_{14} = 0.009447$
$b_{12} = b_{13} = 0.211453$	

The frequency responses are depicted in Figure 7.34.

Clearly, the stopband attenuation is satisfied. We also check the details for the passband as shown in Figure 7.35.

As shown in Figure 7.35, the ripples in the passband between 1,000 and 1,600 Hz are between −1 and 1 dB. Hence, all specifications are satisfied.

Example 7.17.

Now we show how the Remez exchange algorithm in Equation (7.32) is processed using a linear phase 3-tap FIR filter as

$$H(z) = b_0 + b_1 z^{-1} + b_0 z^{-2}.$$

The ideal frequency response specifications are shown in Figure 7.36(a), where the filter gain increases linearly from the gain of 0.5 at $\Omega = 0$ radian to the gain of 1 at $\Omega = \pi/4$ radian. The band between $\Omega = \pi/4$ radian and

276 7 FINITE IMPULSE RESPONSE FILTER DESIGN

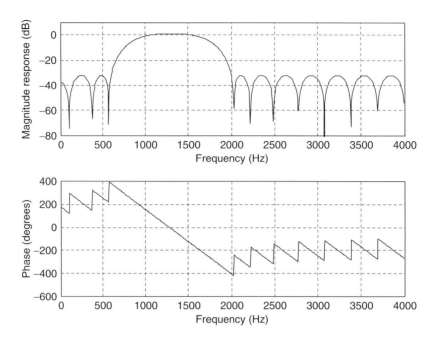

FIGURE 7.34 Frequency and phase responses for Example 7.16.

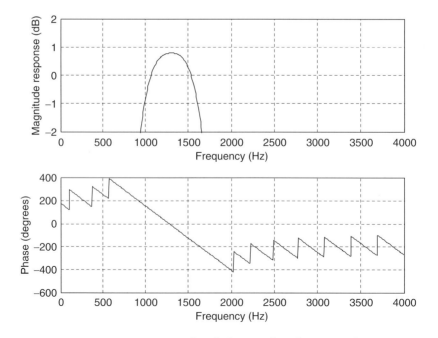

FIGURE 7.35 Frequency response details for passband in Example 7.16.

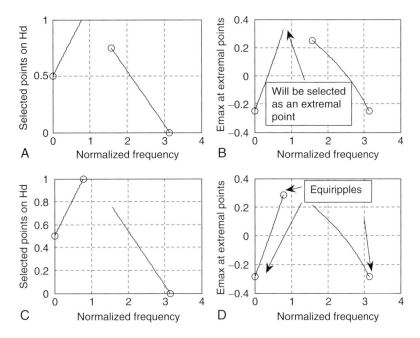

FIGURE 7.36 Determining the 3-tap FIR filter coefficients using the Remez algorithm in Example 7.17.

$\Omega = \pi/2$ radians is a transition band. Finally, the filter gain decreases linearly from the gain of 0.75 at $\Omega = \pi/2$ radians to the gain of 0 at $\Omega = \pi$ radians.

For simplicity, we use all the weight factors as 1, that is, $W(\Omega) = 1$. Equation (7.32) is simplified to be

$$E(\Omega) = H(e^{j\Omega}) - H_d(e^{j\Omega}).$$

Substituting $z = e^{j\Omega}$ to the transfer function $H(z)$ gives

$$H(e^{j\Omega}) = b_0 + b_1 e^{-j\Omega} + b_0 e^{-j2\Omega}.$$

After simplification using Euler's identity $e^{j\Omega} + e^{-j\Omega} = 2\cos\Omega$, the filter frequency response is given by

$$H(e^{j\Omega}) = e^{-j\Omega}(b_1 + 2b_0 \cos\Omega).$$

Regardless of the linear phase shift term $e^{-j\Omega}$ for the time being, we have a Chebyshev real magnitude function (there are a few other types as well) as

$$H(e^{j\Omega}) = b_1 + 2b_0 \cos\Omega.$$

The *alternation theorem* (Ambardar, 1999; Porat, 1997) must be used. The alternation theorem states that given Chebyshev polynomial $H(e^{j\Omega})$ to approximate the ideal magnitude response $H_d(e^{j\Omega})$, we can find at least $M+2$ (where $M=1$ for our case) frequencies $\Omega_0, \Omega_1, \ldots, \Omega_{M+1}$, called the extremal frequencies, so that signs of the error at the extremal frequencies alternate and the absolute error value at each extremal point reaches the maximum absolute error, that is,

$$E(\Omega_k) = -E(\Omega_{k+1}) \text{ for } \Omega_0, \Omega_1, \ldots \Omega_{M+1}$$

and $|E(\Omega_k)| = E_{\max}$.

But the alternation theorem does not tell us how to do the algorithm. The Remez exchange algorithm actually is employed to solve this problem. The equations and steps (Ambardar, 1999; Porat, 1997) are briefly summarized for our illustrative example:

1. Given the order of $N = 2M + 1$ choose initial extremal frequencies: $\Omega_0, \Omega_1, \ldots, \Omega_{M+1}$ (can be uniformly distributed first).

2. Solve the following equation to satisfy the alternation theorem:

   $$-(-1)^k E = W(\Omega_k)(H_d(e^{j\Omega_k}) - H(e^{j\Omega_k})) \text{ for } \Omega_0, \Omega_1, \ldots, \Omega_{M+1}.$$

   Note that since $H(e^{j\Omega}) = b_1 + 2b_0 \cos \Omega$, for example, the solution will include solving for three unknowns: b_0, b_1, and $E_{\max}$.

3. Determine the extremal points including band edges (can be more than $M + 2$ points), and retain $M + 2$ extremal points with the largest error values $E_{\max}$.

4. Output the coefficients if the extremal frequencies are not changed; otherwise, go to step 2 using the new set of extremal frequencies.

Now let us apply the Remez exchange algorithm.

First Iteration:

1. We use uniformly distributed extremal points: $\Omega_0 = 0$, $\Omega_1 = \pi/2$, $\Omega_2 = \pi$, whose ideal magnitudes are marked by the symbol "o" in Figure 7.36(a).

2. The alternation theorem requires:

   $$-(-1)^k E = H_d(e^{j\Omega}) - (b_1 + 2b_0 \cos \Omega).$$

   Applying extremal points yields the following three simultaneous equations with three unknowns, b_0, b_1, and E:

$$\begin{cases} -E = 0.5 - b_1 - 2b_0 \\ E = 0.75 - b_1 \\ -E = 0 - b_1 + 2b_0 \end{cases}.$$

We solve these three equations to get

$$b_0 = 0.125, \ b_1 = 0.5, \ E = 0.25, \ H(e^{j\Omega}) = 0.5 + 0.25 \cos \Omega.$$

3. We then determine the extremal points, including at the band edge, with their error values from Figure 7.36(b) using the following error function:

$$E(\Omega) = H_d(e^{j\Omega}) - 0.5 - 0.25 \cos \Omega.$$

These extremal points are marked by the symbol "o" and their error values are listed in Table 7.16.

4. Since the band edge at $\Omega = \pi/4$ has an error larger than others, it must be chosen as the extremal frequency. After deleting the extremal point at $\Omega = \pi/2$, a new set of extremal points are found according the largest error values as

$$\begin{aligned} \Omega_0 &= 0 \\ \Omega_1 &= \pi/4 \\ \Omega_2 &= \pi \end{aligned}.$$

The ideal magnitudes at these three extremal points are given in Figure 7.36(c), that is, 0.5, 1, 0. Now let us examine the second iteration.

Second Iteration:

Applying the alternation theorem at the new set of extremal points, we have

$$\begin{cases} -E = 0.5 - b_1 - 2b_0 \\ E = 1 - b_1 - 1.4142 b_0 \\ -E = 0 - b_1 + 2b_0 \end{cases}.$$

Solving these three simultaneous equations leads to

$$b_0 = 0.125, \ b_1 = 0.537, \ E = 0.287, \ \text{and} \ H(e^{j\Omega}) = 0.537 + 0.25 \cos \Omega.$$

TABLE 7.16 Extremal points and band edges with their error values for the first iteration.

Ω	0	$\pi/4$	$\pi/2$	π
$E_{\max}$	−0.25	0.323	0.25	−0.25

TABLE 7.17 Error values at extremal frequencies and band edge.

Ω	0	$\pi/4$	$\pi/2$	π
E_{max}	-0.287	0.287	0.213	-0.287

The determined extremal points and band edge with their error values are listed in Table 7.17 and shown in Figure 7.36(d), where the determined extremal points are marked by the symbol "o."

Since the extremal points have their same maximum error value of 0.287, they are found to be $\Omega_0 = 0$, $\Omega_1 = \pi/4$, and $\Omega_2 = \pi$, which are unchanged. Then we stop the iteration and output the filter transfer function as

$$H(z) = 0.125 + 0.537z^{-1} + 0.125z^{-2}.$$

As shown in Figure 7.35(d), we achieve the equiripples of error at the extemal points: $\Omega_0 = 0$, $\Omega_1 = \pi/4$, $\Omega_2 = \pi$; their signs are alternating, and the maximum absolute error of 0.287 is obtained at each point. It takes two iterations to determine the coefficients for this simplified example.

As we mentioned, the Parks-McClellan algorithm is one of the most popular filter design methods in industry due to its flexibility and performance. However, there are two disadvantages. The filter length has to be estimated by the empirical method. Once the frequency edges, magnitudes, and weighting factors are specified, applying the Remez exchange algorithm cannot control the actual ripple obtained from the design. We may often need to try a longer length of filter or different weight factors to remedy the situations where the ripple is unacceptable.

7.7 Realization Structures of Finite Impulse Response Filters

Using the direct form I (discussed in Chapter 6), we will get a special realization form, called the *transversal form*. Using the linear phase property will produce a linear phase realization structure.

7.7.1 Transversal Form

Given the transfer function of the FIR filter in Equation (7.38),

$$H(z) = b_0 + b_1 z^{-1} + \ldots + b_K z^{-K}, \qquad (7.38)$$

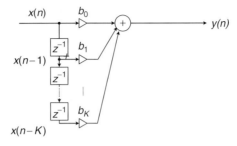

FIGURE 7.37 FIR filter realization (transversal form).

we obtain the difference equation as

$$y(n) = b_0 x(n) + b_1 x(n-1) + b_2 x(n-2) + \ldots + b_K x(n-K).$$

Realization of such a transfer function is the transversal form, displayed in Figure 7.37.

Example 7.18.

Given the FIR filter transfer function

$$H(z) = 1 + 1.2z^{-1} + 0.36z^{-2},$$

a. Perform the FIR filter realization.

Solution:

a. From the transfer function, we can identify that $b_0 = 1$, $b_1 = 1.2$, and $b_2 = 0.36$. Using Figure 7.37, we find the FIR realization to be as follows (Fig. 7.38):

We determine the DSP equation for implementation as

$$y(n) = x(n) + 1.2x(n-1) + 0.36x(n-2).$$

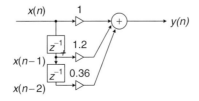

FIGURE 7.38 FIR filter realization for Example 7.18.

Program 7.12 (below) shows the MATLAB implementation.

Program 7.12. MATLAB program for Example 7.18.

```
%Sample MATLAB code
sample = 1:1:10;          %Input test array
x = [0 0 0];              %Input buffer [x(n) x(n − 1) ...]
y = [0];                  %Output buffer [y(n) y(n − 1) ...]
b = [1.0 1.2 0.36];       %FIR filter coefficients [b0 b1 ...]
KK = length(b);
for n = 1:1:length(sample) % Loop processing
for k = KK:−1:2% Shift the input by one sample
x(k) = x(k − 1);
end
x(1) = sample(n);         % Get new sample
y(1) = 0;                 % Perform FIR filtering
for k = 1:1:KK
y(1) = y(1) + b(k)*x(k);
end
out(n) = y(1);            %Send the filtered sample to the output array
end
out
```

7.7.2 Linear Phase Form

We illustrate the linear phase structure using the following simple example. Considering the transfer function with 5 taps obtained from the design as follows,

$$H(z) = b_0 + b_1 z^{-1} + b_2 z^{-2} + b_1 z^{-3} + b_0 z^{-4}, \qquad (7.39)$$

we can see that the coefficients are symmetrical and that the difference equation is

$$y(n) = b_0 x(n) + b_1 x(n-1) + b_2 x(n-2) + b_1 x(n-3) + b_0 x(n-4).$$

This DSP equation can further be combined to be

$$y(n) = b_0(x(n) + x(n-4)) + b_1(x(n-1) + x(n-3)) + b_2 x(n-2).$$

Then we obtain the realization structure in a linear phase form as follows (Fig. 7.39):

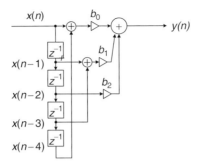

FIGURE 7.39 Linear phase FIR filter realization.

7.8 Coefficient Accuracy Effects on Finite Impulse Response Filters

In practical applications, the filter coefficients achieved through high-level software such as MATLAB must be quantized using finite word length. This may have two effects. First, the locations of zeros are changed; second, due to the location change of zeros, the filter frequency response will change correspondingly. In practice, there are two types of digital signal (DS) processors: *fixed-point* processors and *floating-point* processors. The fixed-point DS processor uses integer arithmetic, and the floating-point processor employs floating-point arithmetic. Such effects of filter coefficient quantization will be covered in Chapter 9.

In this section, we study effects of FIR filter coefficient quantization in general, since during practical filter realization, obtaining filter coefficients with infinite precision is impossible. Filter coefficients are usually truncated or rounded off for the application. Assume that the FIR filter transfer function with infinite precision is given by

$$H(z) = \sum_{n=0}^{K} b_n z^{-n} = b_0 + b_1 z^{-1} + \ldots + b_K z^{-K}, \quad (7.40)$$

where each filter coefficient b_n has infinite precision. Now let the quantized FIR filter transfer function be

$$H^q(z) = \sum_{n=0}^{K} b_n^q z^{-n} = b_0^q + b_1^q z^{-1} + \ldots + b_K^q z^{-K}, \quad (7.41)$$

where each filter coefficient b_n^q is quantized (round-off) using the specified number of bits. Then the error of the magnitude frequency response can be bounded as

$$\left|H(e^{j\Omega}) - H^q(e^{j\Omega})\right| = \sum_{n=0}^{K} \left|(b_n - b_n^q)e^{-jn\Omega}\right|$$
$$< \sum_{n=0}^{K} |b_n - b_n^q| < (K+1) \cdot 2^{-B-1} \qquad (7.42)$$

where B is the number of bits used to encode each magnitude of the filter coefficient. Look at Example 7.19.

Example 7.19.

In Example 7.7, a lowpass FIR filter with 25 taps using a Hamming window is designed, and FIR filter coefficients are listed for comparison in Table 7.18. One sign bit is used, and 7 bits are used for fractional parts, since all FIR filter coefficients are less than 1. We would multiply each filter coefficient by a scale factor of 2^7 and round off each scaled magnitude to an integer whose magnitude could be encoded using 7 bits. When the coefficient integer is scaled back, the coefficient with finite precision (quantized filter coefficient) using 8 bits, including the sign bit, will be achieved.

TABLE 7.18 FIR filter coefficients and their quantized filter coefficients in Example 7.19 (Hamming window).

Bham: FIR Filter Coefficients	BhamQ: FIR Filter Coefficients
$b_0 = b_{24} = 0.00000000000000$	$b_0 = b_{24} = 0.0000000$
$b_1 = b_{23} = -0.00276854711076$	$b_1 = b_{23} = -0.0000000$
$b_2 = b_{22} = 0.00000000000000$	$b_2 = b_{22} = 0.0000000$
$b_3 = b_{21} = 0.00759455135346$	$b_3 = b_{21} = 0.0078125$
$b_4 = b_{20} = 0.00000000000000$	$b_4 = b_{20} = 0.0000000$
$b_5 = b_{19} = -0.01914148493949$	$b_5 = b_{19} = -0.0156250$
$b_6 = b_{18} = 0.00000000000000$	$b_6 = b_{18} = 0.0000000$
$b_7 = b_{17} = 0.04195685650042$	$b_7 = b_{17} = 0.0390625$
$b_8 = b_{16} = 0.00000000000000$	$b_8 = b_{16} = 0.0000000$
$b_9 = b_{15} = -0.09180790496577$	$b_9 = b_{15} = -0.0859375$
$b_{10} = b_{14} = 0.00000000000000$	$b_{10} = b_{14} = 0.0000000$
$b_{11} = b_{13} = 0.31332065886015$	$b_{11} = b_{13} = 0.3125000$
$b_{12} = 0.50000000000000$	$b_{12} = 0.5000000$

7.8 Coefficient Accuracy Effects on Finite Impulse Response Filters

To see quantization, we take a look at one of the infinite precision coefficients, $Bham(3) = 0.00759455135346$, for illustration. The quantization using 7 magnitude bits is shown as:

$$0.00759455135346 \times 2^7 = 0.9721 = 1 \text{(rounded up to the integer)}.$$

Then the quantized filter coefficient is obtained as

$$BhamQ(3) = 1/2^7 = 0.0078125.$$

Since the poles for both FIR filters always reside at origin, we need to examine only their zeros. The z-plane zero plots for both FIR filters are shown in Figure 7.40a, where the circles are zeros from the FIR filter with infinite precision, while the crosses are zeros from the FIR filter with the quantized coefficients.

Most importantly, Figure 7.40b shows the difference of the frequency responses for both filters obtained using Program 7.13. In the figure, the solid line represents the frequency response with infinite filter coefficient precision, and the dot-dashed line indicates the frequency response with finite filter coefficients. It is observed that the stopband performance is degraded due to the filter coefficient quantization. The degradation in the passband is not severe.

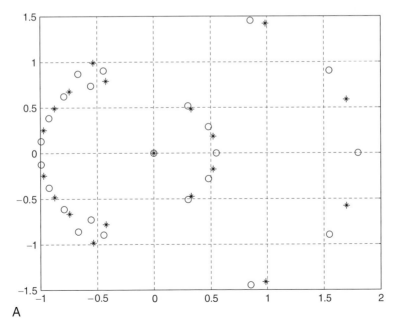

FIGURE 7.40A The z-plane zero plots for both FIR filters. The circles are zeros for infinite precision; the crosses are zeros for rounded-off coefficients.

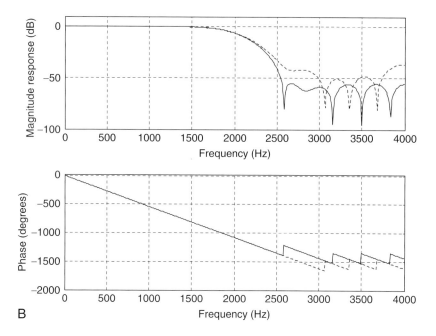

FIGURE 7.40B Frequency responses. The solid line indicates the FIR filter with infinite precision; the dashed line indicates the FIR filter with rounded-off coefficients.

Program 7.13. MATLAB program for Example 7.19.

```
fs=8000;
[hham,f]=freqz(Bham,1,512,fs);
[hhamQ,f]=freqz(BhamQ,1,512,fs);
p= 180*unwrap(angle(hham))/pi;
pQ= 180*unwrap(angle(hhamQ))/pi
subplot(2,1,1);plot(f,20*log10(abs(hham)),f,20*log10(abs(hhamQ)),':');grid
axis([0 fs/2 -100 10]);
xlabel('Frequency (Hz)');ylabel('Magnitude Response (dB)');
subplot(2,1,2); plot(f,p,f,pQ,':');grid
xlabel('Frequency (Hz)'); ylabel('Phase (degrees)');
```

Using Equation (7.42), the error of the magnitude frequency response due to quantization is bounded by

$$\left|H(e^{j\Omega}) - H^q(e^{j\Omega})\right| < 25/256 = 0.0977.$$

This can be easily verified at the stopband of the magnitude frequency response for the worst condition as follows:

$$\left|H(e^{j\Omega}) - H^q(e^{j\Omega})\right| = \left|10^{-100/20} - 10^{-30/20}\right| = 0.032 < 0.0977.$$

In practical situations, the same procedure can be used to analyze the effects of filter coefficient quantization to make sure that the designed filter meets the requirements.

7.9 Summary of Finite Impulse Response (FIR) Design Procedures and Selection of FIR Filter Design Methods in Practice

In this section, we first summarize the design procedures of the window design, frequency sampling design, and optimal design methods, and then discuss the selection of the particular filter for typical applications.

The window method (Fourier transform design using windows):

1. Given the filter frequency specifications, determine the filter order (odd number used in this book) and the cutoff frequency/frequencies using Table 7.7 and Equation (7.26).

2. Compute the impulse sequence $h(n)$ via the Fourier transform method using the appropriate equations (in Table 7.1).

3. Multiply the generated FIR filter coefficients $h(n)$ in (2) by the selected window sequence using Equation (7.20) to obtain the windowed impulse sequence $h_w(n)$.

4. Delay the windowed impulse sequence $h_w(n)$ by M samples to get the causal windowed FIR filter coefficients $b_n = h_w(n - M)$ using Equation (7.21).

5. Output the transfer function and plot the frequency responses.

6. If the frequency specifications are satisfied, output the difference equation. If the frequency specifications are not satisfied, increase the filter order and repeat beginning with step 2.

The frequency sampling method:

1. Given the filter frequency specifications, choose the filter order (odd number used in the book), and specify the equally spaced magnitudes of the frequency response for the normalized frequency range from 0 to π using Equation (7.29).

2. Calculate FIR filter coefficients using Equation (7.30).

3. Use the symmetry, in Equation (7.31), linear phase requirement, to determine the rest of the coefficients.

4. Output the transfer function and plot the frequency responses.

5. If the frequency specifications are satisfied, output the difference equation. If the frequency specifications are not satisfied, increase the filter order and repeat beginning with step 2.

The optimal design method (Parks-McClellan algorithm):

1. Given the band edge frequencies, choose the filter order, normalize each band edge frequency to the Nyquist limit (folding frequency $= f_s/2$), and specify the ideal magnitudes.

2. Calculate absolute values of the passband ripple and stopband attenuation, if they are given in terms of dB values, using Equations (7.34) and (7.35).

3. Determine the error weight factors for passband and stopband, respectively, using Equations (7.36) and (7.37).

4. Apply the Remez algorithm to calculate filter coefficients.

5. Output the transfer function and check the frequency responses.

6. If the frequency specifications are satisfied, output the difference equation. If the frequency specifications are not satisfied, increase the filter order and repeat beginning with step 4.

Table 7.19 shows the comparisons for the window, frequency sampling, and optimal methods. The table can be used as a selection guide for each design method in this book.

Example 7.20 describes the possible selection of the design method by a DSP engineer to solve a real-world problem.

Example 7.20.

a. Determine the appropriate FIR filter design method for each of the following DSP applications.

 1. A DSP engineer implements a digital two-band crossover system as described in the section in this book. He selects the FIR filters to satisfy the following specifications:

    Sampling rate $= 44,100$ Hz
    Crossover frequency $= 1,000$ Hz (cutoff frequency)
    Transition band $= 600$ to $1,400$ Hz
    Lowpass filter $=$ passband frequency range from 0 to 600 Hz with a ripple of 0.02 dB and stopband edge at 1,400 Hz with attenuation of 50 dB.

7.9 Summary of Finite Impulse Response (FIR)

TABLE 7.19 Comparisons of three design methods.

Design Method	Window Method	Frequency Sampling	Optimal Design
Filter type	1. Lowpass, highpass, bandpass, bandstop. 2. Formulas are not valid for arbitrary frequency selectivity.	1. Any type filter 2. The formula is valid for arbitrary frequency selectivity.	1. Any type filter 2. Valid for arbitrary frequency selectivity
Linear phase	Yes	Yes	Yes
Ripple and stopband specifications	Used for determining the filter order and cutoff frequency/-cies	Need to be checked after each design trial	Used in the algorithm; need to be checked after each design trial
Algorithm complexity for coefficients	Moderate: 1. Impulse sequence calculation 2. Window function weighting	Simple: Single equation	Complicated: 1. Parks-McClellan algorithm 2. Remez exchange algorithm
Minimal design tool	Calculator	Calculator	Software

290 7 FINITE IMPULSE RESPONSE FILTER DESIGN

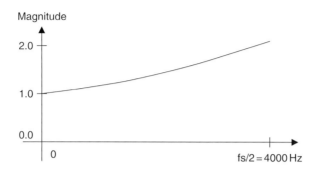

FIGURE 7.41 Magnitude frequency response in Example 7.20 (b).

Highpass filter = passband frequency range from 1.4 to 44.1 kHz with ripple of 0.02 dB and stopband edge at 600 Hz with attenuation of 50 dB.

The engineer does not have the software routine for the Remez algorithm.

2. An audio engineer tries to equalize the speech signal sampled at 8,000 Hz using a linear phase FIR filter based on the magnitude specifications in Figure 7.41. The engineer does not have the software routine for the Remez algorithm.

Solution:

a. 1. The window design method is the first choice, since the window design formula is in terms of the cutoff frequency (crossover frequency), the filter order is based on the transient band, and filter types are standard lowpass and highpass. The ripple and stopband specifications can be satisfied by selecting the Hamming window. The optimal design method will also do the job with a challenge to satisfy the combined unit gains at the crossover frequency of 1,000 Hz if the **remez()** algorithm is available.

2. Since the magnitude frequency response is not a standard filter type of lowpass, highpass, bandpass, or band reject, and the **remez()** algorithm is not available, the first choice should be the frequency sampling method.

7.10 Summary

1. The Fourier transform method is used to compute noncausal FIR filter coefficients, including those of lowpass, highpass, bandpass, and bandstop filters.

2. Converting noncausal FIR filter coefficients to causal FIR filter coefficients introduces only linear phase, which is a good property for audio applications. The linear phase filter output has the same amount of delay for all the input signals whose frequency components are within passband.

3. The causal FIR filter designed using the Fourier transform method generates ripple oscillations (Gibbs effect) in the passband and stopband in its filter magnitude frequency response due to abrupt truncation of the FIR filter coefficient sequence.

4. To reduce the oscillation effect, the window method is introduced to tap down the coefficient values toward both ends. A substantial improvement of the magnitude frequency response is achieved.

5. Real-life DSP applications such as noise reduction system and two-band digital audio crossover system were investigated.

6. Frequency sampling design is feasible for the FIR filter with an arbitrary magnitude response specification.

7. An optimal design method, Parks-McClellan algorithm using Remez exchange algorithm, offers the flexibility for filter specifications. The Remez exchange algorithm was explained using a simplified example.

8. Realization structures of FIR filters have special forms, such as the transversal form and the linear phase form.

9. The effect of quantizing FIR filter coefficients for implementation changes zero locations of the FIR filter. More effects on the stopband in the magnitude and phase responses are observed.

10. Guidelines for selecting an appropriate design method in practice were summarized considering the filter type, linear phase, ripple and stopband specifications, algorithm complexity, and design tools.

7.11 MATLAB Programs

Program 7.14 enables one to design FIR filters via the window method using functions such as the rectangular window, triangular (Bartlett) window, Hanning window, Hamming window, and Blackman window. Filter types of the design include lowpass, highpass, bandpass, and band reject.

Program 7.14. MATLAB function for FIR filter design using the window method.

```
function B = firwd(N, Ftype, WnL, WnH, Wtype)
% B = firwd(N,Ftype,WnL,WnH,Wtype)
% FIR filter design using the window function method.
% Input parameters:
% N: the number of the FIR filter taps.
% Note: It must be an odd number.
% Ftype: the filter type
%1. Lowpass filter;
%2. Highpass filter;
%3. Bandpass filter;
%4. Band reject filter;
% WnL: lower cutoff frequency in radians. Set WnL=0 for the highpass filter.
% WnH: upper cutoff frequency in radians. Set WnH=0 for the lowpass filter.
% Wtypw: window function type
%1. Rectangular window;
%2. Triangular window;
%3. Hanning window;
%4. Hamming window;
%5. Blackman window;
% Output:
% B: FIR filter coefficients.
  M = (N-1)/2;
  hH = sin(WnH*[-M:1:-1])./([-M:1:-1]*pi);
  hH(M+1) = WnH/pi;
  hH(M+2:1:N) = hH(M:-1:1);
  hL = sin(WnL*[-M:1:-1])./([-M:1:-1]*pi);
  hL(M+1) = WnL/pi;
  hL(M+2:1:N) = hL(M:-1:1);
  if Ftype == 1
  h(1:N) = hL(1:N);
  end
  if Ftype == 2
  h(1:N) = -hH(1:N);
  h(M+1) = 1+h(M+1);
  end
  if Ftype ==3
  h(1:N) = hH(1:N) - hL(1:N);
  end
  if Ftype == 4
  h(1:N) = hL(1:N) - hH(1:N);
  h(M+1) = 1+h(M+1);
  end
  % window functions;
  if Wtype ==1
  w(1:N)=ones(1,N);
```

```
end
if Wtype ==2
w = 1 - abs([-M:1:M])/M;
end
if Wtype ==3
w = 0.5 + 0.5*cos([-M:1:M]*pi/M);
end
if Wtype ==4
w = 0.54 + 0.46*cos([-M:1:M]*pi/M);
end
if Wtype ==5
w = 0.42 + 0.5*cos([-M:1:M]*pi/M) + 0.08*cos(2*[-M:1:M]*pi/M);
end
B = h.*w
```

Program 7.15. MATLAB function for FIR filter design using the frequency sampling method.

```
function B=firfs(N,Hk)
% B=firls(N,Hk)
% FIR filter design using the frequency sampling method.
% Input parameters:
% N: the number of filter coefficients.
% note: N must be an odd number.
% Hk: sampled frequency response for k = 0, 1, 2, ..., M = (N-1)/2.
% Output:
% B: FIR filter coefficients.
  M = (N-1)/2;
  for n = 1:1:N
  B(n) = (1/N)*(Hk(1) +...
  2*sum(Hk(2:1:M+1)...
  .*cos(2*pi*([1:1:M])*(n-1-M)/N)));
  end
```

Program 7.15 enables one to design FIR filters using the frequency sampling method. Note that values of the frequency response, which correspond to the equally spaced DFT frequency components, must be specified for design. Besides the lowpass, highpass, bandpass, and band reject filter designs, the method can be used to design FIR filters with an arbitrarily specified magnitude frequency response.

7.12 Problems

7.1. Design a 3-tap FIR lowpass filter with a cutoff frequency of 1,500 Hz and a sampling rate of 8,000 Hz using

  a. rectangular window function

  b. Hamming window function.

  Determine the transfer function and difference equation of the designed FIR system, and compute and plot the magnitude frequency response for $\Omega = 0$, $\pi/4$, $\pi/2$, $3\pi/4$, and π radians.

7.2. Design a 3-tap FIR highpass filter with a cutoff frequency of 1,600 Hz and a sampling rate of 8,000 Hz using

  a. rectangular window function

  b. Hamming window function.

  Determine the transfer function and difference equation of the designed FIR system, and compute and plot the magnitude frequency response for $\Omega = 0$, $\pi/4$, $\pi/2$, $3\pi/4$, and π radians.

7.3. Design a 5-tap FIR bandpass filter with a lower cutoff frequency of 1,600 Hz, an upper cutoff frequency of 1,800 Hz, and a sampling rate of 8,000 Hz using

  a. rectangular window function

  b. Hamming window function.

  Determine the transfer function and difference equation of the designed FIR system, and compute and plot the magnitude frequency response for $\Omega = 0$, $\pi/4$, $\pi/2$, $3\pi/4$, and π radians.

7.4. Design a 5-tap FIR band reject filter with a lower cutoff frequency of 1,600 Hz, an upper cutoff frequency of 1,800 Hz, and a sampling rate of 8,000 Hz using

  a. rectangular window function

  b. Hamming window function.

  Determine the transfer function and difference equation of the designed FIR system, and compute and plot the magnitude frequency response for $\Omega = 0$, $\pi/4$, $\pi/2$, $3\pi/4$, and π radians.

7.5. Given an FIR lowpass filter design with the following specifications:

Passband = 0–800 Hz
Stopband = 1,200–4,000 Hz
Passband ripple = 0.1 dB
Stopband attenuation = 40 dB
Sampling rate = 8,000 Hz,

determine the following:

a. window method

b. length of the FIR filter

c. cutoff frequency for the design equation.

7.6. Given an FIR highpass filter design with the following specifications:

Passband = 0–1,500 Hz
Stopband = 2,000–4,000 Hz
Passband ripple = 0.02 dB
Stopband attenuation = 60 dB
Sampling rate = 8,000 Hz,

determine the following:

a. window method

b. length of the FIR filter

c. cutoff frequency for the design equation.

7.7. Given an FIR bandpass filter design with the following specifications:

Lower cutoff frequency = 1,500 Hz
Lower transition width = 600 Hz
Upper cutoff frequency = 2,300 Hz
Upper transition width = 600 Hz
Passband ripple = 0.1 dB
Stopband attenuation = 50 dB
Sampling rate = 8,000 Hz,

determine the following:

a. window method

b. length of the FIR filter

c. cutoff frequencies for the design equation.

7.8. Given an FIR bandstop filter design with the following specifications:

Lower passband = 0–1,200 Hz
Stopband = 1,600–2,000 Hz
Upper passband = 2,400–4,000 Hz
Passband ripple = 0.05 dB
Stopband attenuation = 60 dB
Sampling rate = 8,000 Hz,

determine the following:

a. window method

b. length of the FIR filter

c. cutoff frequencies for the design equation.

7.9. Given an FIR system

$$H(z) = 0.25 - 0.5z^{-1} + 0.25z^{-2},$$

realize $H(z)$ using each of the following specified methods:

a. transversal form, and write the difference equation for implementation

b. linear phase form, and write the difference equation for implementation.

7.10. Given an FIR filter transfer function

$$H(z) = 0.2 + 0.5z^{-1} - 0.3z^{-2} + 0.5z^{-3} + 0.2z^{-4},$$

perform the linear phase FIR filter realization, and write the difference equation for implementation.

7.11. Determine the transfer function for a 5-tap FIR lowpass filter with a lower cutoff frequency of 2,000 Hz and a sampling rate of 8,000 Hz using the frequency sampling method.

7.12. Determine the transfer function for a 5-tap FIR highpass filter with a lower cutoff frequency of 3,000 Hz and a sampling rate of 8,000 Hz using the frequency sampling method.

7.13. Given the following specifications:

- a 7-tap FIR bandpass filter

- a lower cutoff frequency of 1,500 Hz and an upper cutoff frequency of 3,000 Hz

- a sampling rate of 8,000 Hz
- the frequency sampling design method,

determine the transfer function.

7.14. Given the following specifications:
- a 7-tap FIR band reject filter
- a lower cutoff frequency of 1,500 Hz and an upper cutoff frequency of 3,000 Hz
- a sampling rate of 8,000 Hz
- the frequency sampling design method,

determine the transfer function.

7.15. In a speech recording system with a sampling rate of 10,000 Hz, the speech is corrupted by broadband random noise. To remove the random noise while preserving speech information, the following specifications are given:

  Speech frequency range = 0–3,000 kHz
  Stopband range = 4,000–5,000 Hz
  Passband ripple = 0.1 dB
  Stopband attenuation = 45 dB
  FIR filter with Hamming window.

Determine the FIR filter length (number of taps) and the cutoff frequency; use MATLAB to design the filter; and plot the frequency response.

7.16. Given a speech equalizer shown in Figure 7.42 to compensate mid-range frequency loss of hearing:

  Sampling rate = 8,000 Hz
  Bandpass FIR filter with Hamming window
  Frequency range to be emphasized = 1,500–2,000 Hz
  Lower stopband = 0–1,000 Hz
  Upper stopband = 2,500–4,000 Hz

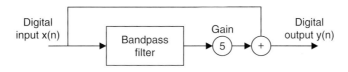

FIGURE 7.42   **Speech equalizer in Problem 7.16.**

Passband ripple = 0.1 dB
Stopband attenuation = 45 dB,

determine the filter length and the lower and upper cutoff frequencies.

7.17. A digital crossover can be designed as shown in Figure 7.43.

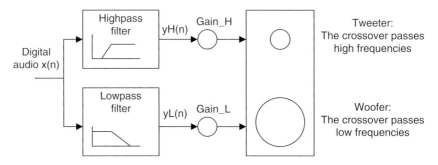

FIGURE 7.43   Two-band digital crossover in Problem 7.17.

Given the following audio specifications:

Sampling rate = 44,100 Hz
Crossover frequency = 2,000 Hz
Transition band range = 1,600 Hz
Passband ripple = 0.1 dB
Stopband attenuation = 50 dB
Filter type = FIR,

determine the following for each filter:

a. window function

b. filter length

c. cutoff frequency.

Use MATLAB to design and plot frequency responses for both filters.

Computer Problems with MATLAB

Use the MATLAB programs in Section 7.11 to design the following FIR filters.

7.18. Design a 41-tap lowpass FIR filter whose cutoff frequency is 1,600 Hz using the following window functions. Assume that the sampling frequency is 8,000 Hz.

a. rectangular window function

b. triangular window function

c. Hanning window function

d. Hamming window function

e. Blackman window function.

List the FIR filter coefficients and plot the frequency responses for each case.

7.19. Design a lowpass FIR filter whose cutoff frequency is 1,000 Hz using the Hamming window function for the following specified filter lengths. Assume that the sampling frequency is 8,000 Hz.

a. 21 filter coefficients

b. 31 filter coefficients

c. 41 filter coefficients.

List FIR filter coefficients for each design and compare the magnitude frequency responses.

7.20. Design a 31-tap highpass FIR filter whose cutoff frequency is 2,500 Hz using the following window functions. Assume that the sampling frequency is 8,000 Hz.

a. Hanning window function

b. Hamming window function

c. Blackman window function.

List the FIR filter coefficients and plot the frequency responses for each design.

7.21. Design a 41-tap bandpass FIR filter with the lower and upper cutoff frequencies being 2,500 Hz and 3,000 Hz, respectively, using the following window functions. Assume a sampling frequency of 8,000 Hz.

a. Hanning window function

b. Blackman window function.

List the FIR filter coefficients and plot the frequency responses for each design.

7.22. Design a 41-tap band reject FIR filter with frequencies 2,500 Hz and 3,000 Hz, respectively, using the Hamming window function. Assume a sampling frequency of 8,000 Hz. List the FIR filter coefficients and plot the frequency responses for each design.

7.23. Use the frequency sampling method to design a linear phase lowpass FIR filter with 17 coefficients. Let the cutoff frequency be 2,000 Hz and assume a sampling frequency of 8,000 Hz. List FIR filter coefficients and plot the frequency responses.

7.24. Use the frequency sampling method to design a linear phase bandpass FIR filter with 21 coefficients. Let the lower and upper cutoff frequencies be 2,000 Hz and 2,500 Hz, respectively, and assume a sampling frequency of 8,000 Hz. List the FIR filter coefficients and plot the frequency responses.

7.25. Given an input data sequence:

$$x(n) = 1.2 \cdot \sin(2\pi(1000)n/8000)) - 1.5 \cdot \cos(2\pi(2800)n/8000),$$

assuming a sampling frequency of 8,000 Hz, use the designed FIR filter with Hamming window in Problem 7.18 to filter 400 data points of $x(n)$, and plot the 400 samples of the input and output data.

7.26. Design a lowpass FIR filter with the following specifications:

> Design method: Parks-McClellan algorithm
> Sampling rate: 8000 Hz
> Passband: 0 – 1200 Hz
> Stopband 1500 – 4000 Hz
> Passband ripple: 1 dB
> Stopband attenuation: 40 dB

List the filter coefficients and plot the frequency responses.

7.27. Design a bandpass FIR filter with the following specifications:

> Design method: Parks-McClellan algorithm
> Sampling rate: 8000 Hz
> Passband: 1200 – 1600 Hz
> Lower stopband 0 – 800 Hz
> Upper stopband 2000 – 4000 Hz
> Passband ripple: 1 dB
> Stopband attenuation: 40 dB

List the filter coefficients and plot the frequency responses.

References

Ambardar, A. (1999). *Analog and Digital Signal Processing,* 2nd ed. Pacific Grove, CA: Brooks/Cole Publishing Company.

Oppenheim, A. V., Schafer, R. W., and Buck, J. R. (1999). *Discrete-Time Signal Processing*, 2nd ed. Upper Saddle River, NJ: Prentice Hall.

Porat, B. (1997). *A Course in Digital Signal Processing.* New York: John Wiley & Sons.

Proakis, J. G., and Manolakis, D. G. (1996). *Digital Signal Processing: Principles, Algorithms, and Applications*, 3rd ed. Upper Saddle River, NJ: Prentice Hall.

8

Infinite Impulse Response Filter Design

Objectives:

This chapter investigates a bilinear transformation method for infinite impulse response (IIR) filter design and develops a procedure to design digital Butterworth and Chebyshev filters. The chapter also investigates other IIR filter design methods, such as impulse invariant design and pole-zero placement design. Finally, the chapter illustrates how to apply the designed IIR filters to solve real-world problems such as digital audio equalization, 60-Hz interference cancellation in audio and electrocardiography signals, dual-tone multifrequency tone generation, and detection using the Goertzel algorithm.

8.1 Infinite Impulse Response Filter Format

In this chapter, we will study several methods for infinite impulse response (IIR) filter design. An IIR filter is described using the difference equation, as discussed in Chapter 6:

$$y(n) = b_0 x(n) + b_1 x(n-1) + \cdots + b_M x(n-M)$$
$$- a_1 y(n-1) - \cdots - a_N y(n-N).$$

Chapter 6 also gives the IIR filter transfer function as

$$H(z) = \frac{Y(z)}{X(z)} = \frac{b_0 + b_1 z^{-1} + \cdots + b_M z^{-M}}{1 + a_1 z^{-1} + \cdots + a_N z^{-N}},$$

where b_i and a_i are the $(M + 1)$ numerator and N denominator coefficients, respectively. $Y(z)$ and $X(z)$ are the z-transform functions of the filter input $x(n)$ and filter output $y(n)$. To become familiar with the form of the IIR filter, let us look at the following example.

Example 8.1.

Given the following IIR filter:

$$y(n) = 0.2x(n) + 0.4x(n-1) + 0.5y(n-1),$$

a. Determine the transfer function, nonzero coefficients, and impulse response.

Solution:

a. Applying the z-transform and solving for a ratio of the z-transform output over input, we have

$$H(z) = \frac{Y(z)}{X(z)} = \frac{0.2 + 0.4z^{-1}}{1 - 0.5z^{-1}}.$$

We also identify the nonzero numerator coefficients and denominator coefficient as

$$b_0 = 0.2, \ b_1 = 0.4, \text{ and } a_1 = -0.5.$$

To solve the impulse response, we rewrite the transfer function as

$$H(z) = \frac{0.2}{1 - 0.5z^{-1}} + \frac{0.4z^{-1}}{1 - 0.5z^{-1}}.$$

Using the inverse z-transform and shift theorem, we obtain the impulse response as

$$h(n) = 0.2(0.5)^n u(n) + 0.4(0.5)^{n-1} u(n-1).$$

The obtained impulse response has an infinite number of terms, where the first several terms are calculated as

$$h(0) = 0.2, \ h(1) = 0.7, \ h(2) = 0.25, \ldots.$$

At this point, we can make the following remarks:

1. The IIR filter output $y(n)$ depends not only on the current input $x(n)$ and past inputs $x(n-1), \ldots$, but also on the past output(s) $y(n-1), \ldots$ (recursive terms). Its transfer function is a ratio of the numerator polynomial over the denominator polynomial, and its impulse response has an infinite number of terms.

2. Since the transfer function has the denominator polynomial, the pole(s) of a designed IIR filter must be inside the unit circle on the z-plane to ensure its stability.

3. Compared with the finite impulse response (FIR) filter (see Chapter 7), the IIR filter offers a much smaller filter size. Hence, the filter operation requires a fewer number of computations, but the linear phase is not easily obtained. The IIR filter is thus preferred when a small filter size is called for but the application does not require a linear phase.

The objective of IIR filter design is to determine the filter numerator and denominator coefficients to satisfy filter specifications such as passband gain and stopband attenuation, as well as cutoff frequency/frequencies for the lowpass, highpass, bandpass, and bandstop filters.

We first focus on the bilinear transformation (BLT) design method. Then we introduce other design methods such as the impulse invariant design and the pole-zero placement design.

8.2 Bilinear Transformation Design Method

Figure 8.1 illustrates a flow chart of the BLT design used in this book. The design procedure includes the following steps: (1) transforming digital filter specifications into analog filter specifications, (2) performing analog filter design, and (3) applying bilinear transformation (which will be introduced in the next section) and verifying its frequency response.

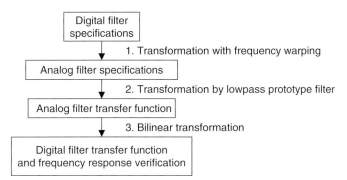

FIGURE 8.1   **General procedure for IIR filter design using bilinear transformation.**

8.2.1 Analog Filters Using Lowpass Prototype Transformation

Before we begin to develop the BLT design, let us review analog filter design using *lowpass prototype transformation.* This method converts the analog lowpass filter with a cutoff frequency of 1 radian per second, called the lowpass prototype, into practical analog lowpass, highpass, bandpass, and bandstop filters with their frequency specifications.

Letting $H_p(s)$ be a transfer function of the lowpass prototype, the transformation of the lowpass prototype into a lowpass filter is given in Figure 8.2.

As shown in Figure 8.2, $H_{LP}(s)$ designates the analog lowpass filter with a cutoff frequency of ω_c radians/second. The lowpass-prototype to lowpass-filter transformation substitutes s in the lowpass prototype function $H_P(s)$ with s/ω_c, where v is the normalized frequency of the lowpass prototype and ω_c is the cutoff frequency of the lowpass filter to be designed. Let us consider the following first-order lowpass prototype:

$$H_P(s) = \frac{1}{s+1}. \tag{8.1}$$

Its frequency response is obtained by substituting $s = jv$ into Equation (8.1), that is,

$$H_P(jv) = \frac{1}{jv+1}$$

with the magnitude gain given in Equation (8.2):

$$|H_P(jv)| = \frac{1}{\sqrt{1+v^2}}. \tag{8.2}$$

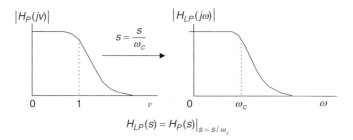

FIGURE 8.2 Analog lowpass prototype transformation into a lowpass filter.

We compute the gains at $v=0$, $v=1$, $v=100$, $v=10,000$ to obtain 1, $1/\sqrt{2}$, 0.0995, and 0.01, respectively. The cutoff frequency gain at $v=1$ equals $1/\sqrt{2}$, which is equivalent to -3 dB, and the direct-current (DC) gain is 1. The gain approaches zero when the frequency goes to $v=+\infty$. This verifies that the lowpass prototype is a normalized lowpass filter with a normalized cutoff frequency of 1. Applying the prototype transformation s/ω_c in Figure 8.2, we get an analog lowpass filter with a cutoff frequency of ω_c as

$$H(s) = \frac{1}{s/\omega_c + 1} = \frac{\omega_c}{s + \omega_c}. \tag{8.3}$$

We can obtain the analog frequency response by substituting $s = j\omega$ into Equation (8.3), that is,

$$H(j\omega) = \frac{1}{j\omega/\omega_c + 1}.$$

The magnitude response is determined by

$$|H(j\omega)| = \frac{1}{\sqrt{1 + \left(\frac{\omega}{\omega_c}\right)^2}}. \tag{8.4}$$

Similarly, we verify the gains at $\omega = 0$, $\omega = \omega_c$, $\omega = 100\omega_c$, $\omega = 10,000\omega_c$ to be 1, $1/\sqrt{2}$, 0.0995, and 0.01, respectively. The filter gain at the cutoff frequency ω_c equals $1/\sqrt{2}$, and the DC gain is 1. The gain approaches zero when $\omega = +\infty$. We notice that filter gains do not change but that the filter frequency is scaled up by a factor of ω_c. This verifies that the prototype transformation converts the lowpass prototype to the analog lowpass filter with the specified cutoff frequency of ω_c without an effect on the filter gain.

This first-order prototype function is used here for an illustrative purpose. We will obtain general functions for Butterworth and Chebyshev lowpass prototypes in a later section.

The highpass, bandpass, and bandstop filters using the specified lowpass prototype transformation can be easily verified. We review them in Figures 8.3, 8.4, and 8.5, respectively. The transformation from the lowpass prototype to the highpass filter $H_{HP}(s)$ with a cutoff frequency ω_c radians/second is given in Figure 8.3, where $s = \omega_c/s$ in the lowpass prototype transformation.

The transformation of the lowpass prototype function to a bandpass filter with a center frequency ω_0, a lower cutoff frequency ω_l, and an upper cutoff frequency ω_h in the passband is depicted in Figure 8.4, where $s = (s^2 + \omega_0^2)/(sW)$ is substituted into the lowpass prototype.

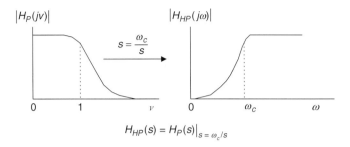

FIGURE 8.3 Analog lowpass prototype transformation to the highpass filter.

As shown in Figure 8.4, ω_0 is the geometric center frequency, which is defined as $\omega_0 = \sqrt{\omega_l \omega_h}$, while the passband bandwidth is given by $W = \omega_h - \omega_l$. Similarly, the transformation from the lowpass prototype to a bandstop (band reject) filter is illustrated in Figure 8.5, with $s = sW/(s^2 + \omega_0^2)$ substituted into the lowpass prototype.

Finally, the lowpass prototype transformations are summarized in Table 8.1.

MATLAB function **freqs()** can be used to plot analog filter frequency responses for verification with the following syntax:

H = freqs(B, A, W)
B = the vector containing the numerator coefficients
A = the vector containing the denominator coefficients
W = the vector containing the specified analog frequency points (radians per second)
H = the vector containing the frequency response.

The following example verifies the lowpass prototype transformation.

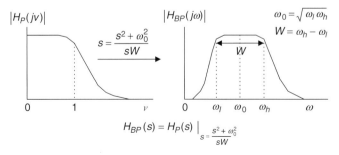

FIGURE 8.4 Analog lowpass prototype transformation to the bandpass filter.

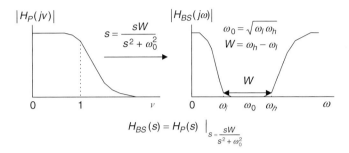

FIGURE 8.5 Analog lowpass prototype transformation to a bandstop filter.

Example 8.2.

Given a lowpass prototype

$$H_P(s) = \frac{1}{s+1},$$

a. Determine each of the following analog filters and plot their magnitude responses from 0 to 200 radians per second.

1. The highpass filter with a cutoff frequency of 40 radians per second.
2. The bandpass filter with a center frequency of 100 radians per second and bandwidth of 20 radians per second.

Solution:

a. 1. Applying the lowpass prototype transformation by substituting $s = 40/s$ into the lowpass prototype, we have an analog highpass filter as

$$H_{HP}(s) = \frac{1}{\frac{40}{s}+1} = \frac{s}{s+40}.$$

TABLE 8.1 Analog lowpass prototype transformations.

Filter Type	Prototype Transformation
Lowpass	$\frac{s}{\omega_c}$, ω_c is the cutoff frequency
Highpass	$\frac{\omega_c}{s}$, ω_c is the cutoff frequency
Bandpass	$\frac{s^2+\omega_0^2}{sW}$, $\omega_0 = \sqrt{\omega_l \omega_h}$, $W = \omega_h - \omega_l$
Bandstop	$\frac{sW}{s^2+\omega_0^2}$, $\omega_0 = \sqrt{\omega_l \omega_h}$, $W = \omega_h - \omega_l$

2. Similarly, substituting the lowpass-to-bandpass transformation $s = (s^2 + 100)/(20s)$ into the lowpass prototype leads to

$$H_{BP}(s) = \frac{1}{\frac{s^2+100}{20s} + 1} = \frac{20s}{s^2 + 20s + 100}.$$

The program for plotting the magnitude responses for highpass and bandpass filters is shown in Program 8.1, and Figure 8.6 displays the magnitude responses for the highpass filter and bandpass filter, respectively.

Program 8.1. MATLAB program in Example 8.2.

```
W = 0:1:200;        %Analog frequency points for computing the filter gains
Ha = freqs([1 0],[1 40],W);    % Frequency response for the highpass filter
Hb = freqs([20 0],[1 20 100],W); % Frequency response for the bandpass filter
subplot(2,1,1);plot(W,abs(Ha),'k');grid % Filter gain plot for the highpass filter
xlabel('(a) Frequency (radians per second)')
ylabel('Absolute filter gain');
subplot(2,1,2);plot(W,abs(Hb),'k');grid % Filter gain plot for the bandpass filter
xlabel('(b) Frequency (radians per second)')
ylabel('Absolute filter gain');
```

Figure 8.6 confirms the lowpass prototype transformation into a highpass filter and a bandpass filter, respectively. To obtain the transfer function of an analog filter, we always begin with a lowpass prototype and apply the corresponding lowpass prototype transformation. To transfer from a lowpass prototype to a bandpass or bandstop filter, the resultant order of the analog filter is twice that of the lowpass prototype order.

8.2.2 Bilinear Transformation and Frequency Warping

In this subsection, we develop the BLT, which converts an analog filter into a digital filter. We begin by finding the area under a curve using the integration of calculus and the numerical recursive method. The area under the curve is a common problem in early calculus courses. As shown in Figure 8.7, the area under the curve can be determined using the following integration:

$$y(t) = \int_0^t x(t)dt, \tag{8.5}$$

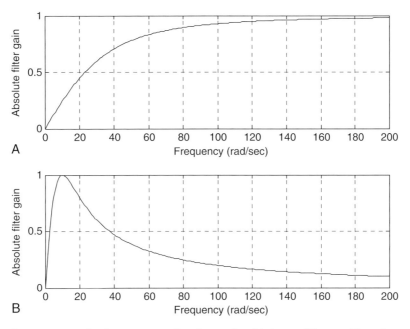

FIGURE 8.6 Magnitude responses for the analog highpass filter and bandpass filter in Example 8.2.

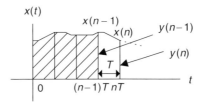

FIGURE 8.7 Digital integration method to calculate the area under the curve.

where $y(t)$ (area under the curve) and $x(t)$ (curve function) are the output and input of the analog integrator, respectively, and t is the upper limit of the integration.

Applying Laplace transform on Equation (8.5), we have

$$Y(s) = \frac{X(s)}{s} \tag{8.6}$$

and find the Laplace transfer function as

8 INFINITE IMPULSE RESPONSE FILTER DESIGN

$$G(s) = \frac{Y(s)}{X(s)} = \frac{1}{s}. \tag{8.7}$$

Now we examine the numerical integration method shown in Figure 8.7 to approximate the integration of Equation (8.5) using the following difference equation:

$$y(n) = y(n-1) + \frac{x(n) + x(n-1)}{2} T, \tag{8.8}$$

where T denotes the sampling period. $y(n) = y(nT)$ is the output sample that is the whole area under the curve, while $y(n-1) = y(nT-T)$ is the previous output sample from the integrator indicating the previously computed area under the curve (the shaded area in Figure 8.7). Notice that $x(n) = x(nT)$ and $x(n-1) = x(nT-T)$, sample amplitudes from the curve, are the current input sample and the previous input sample in Equation (8.8). Applying the z-transform on both sides of Equation (8.8) leads to

$$Y(z) = z^{-1}Y(z) + \frac{T}{2}\left(X(z) + z^{-1}X(z)\right).$$

Solving for the ratio $Y(z)/X(z)$, we achieve the z-transfer function as

$$H(z) = \frac{Y(z)}{X(z)} = \frac{T}{2}\frac{1+z^{-1}}{1-z^{-1}}. \tag{8.9}$$

Next, comparing Equation (8.9) with Equation (8.7), it follows that

$$\frac{1}{s} = \frac{T}{2}\frac{1+z^{-1}}{1-z^{-1}} = \frac{T}{2}\frac{z+1}{z-1}. \tag{8.10}$$

Solving for s in Equation (8.10) gives the bilinear transformation

$$s = \frac{2}{T}\frac{z-1}{z+1}. \tag{8.11}$$

The BLT method is a mapping or transformation of points from the s-plane to the z-plane. Equation (8.11) can be alternatively written as

$$z = \frac{1+sT/2}{1-sT/2}. \tag{8.12}$$

The general mapping properties are summarized as following:

1. The left-half s-plane is mapped onto the inside of the unit circle of the z-plane.

2. The right-half s-plane is mapped onto the outside of the unit circle of the z-plane.

3. The positive $j\omega$ axis portion in the s-plane is mapped onto the positive half circle (the dashed-line arrow in Figure 8.8) on the unit circle, while the negative $j\omega$ axis is mapped onto the negative half circle (the dotted-line arrow in Figure 8.8) on the unit circle.

To verify these features, let us look at the following illustrative example:

Example 8.3.

Assuming that $T = 2$ seconds in Equation (8.12), and given the following points:

1. $s = -1 + j$, on the left half of the s-plane
2. $s = 1 - j$, on the right half of the s-plane
3. $s = j$, on the positive $j\omega$ on the s-plane
4. $s = -j$, on the negative $j\omega$ on the s-plane,

a. Convert each of the points in the s-plane to the z-plane, and verify the mapping properties (1) to (3).

Solution:

a. Substituting $T = 2$ into Equation (8.12) leads to

$$z = \frac{1+s}{1-s}.$$

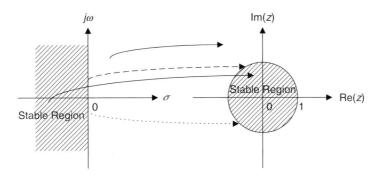

FIGURE 8.8 Mapping between the s-plane and the z-plane by the bilinear transformation.

314   8 INFINITE IMPULSE RESPONSE FILTER DESIGN

We can carry out mapping for each point as follows:

1. $z = \dfrac{1 + (-1 + j)}{1 - (-1 + j)} = \dfrac{j}{2 - j} = \dfrac{1\angle 90°}{\sqrt{5}\angle - 26.57°} = 0.4472\angle 116.57°,$

   since $|z| = 0.4472 < 1$, which is inside the unit circle on the z-plane.

2. $z = \dfrac{1 + (1 - j)}{1 - (1 - j)} = \dfrac{2 - j}{j} = \dfrac{\sqrt{5}\angle - 26.57°}{1\angle 90°} = 2.2361\angle - 116.57°,$

   since $|z| = 2.2361 > 1$, which is outside the unit circle on the z-plane.

3. $z = \dfrac{1 + j}{1 - j} = \dfrac{\sqrt{2}\angle 45°}{\sqrt{2}\angle - 45°} = 1\angle 90°,$

   since $|z| = 1$ and $\theta = 90°$, which is on the positive half circle on the unit circle on the z-plane.

4. $z = \dfrac{1 - j}{1 - (-j)} = \dfrac{1 - j}{1 + j} = \dfrac{\sqrt{2}\angle - 45°}{\sqrt{2}\angle 45°} = 1\angle - 90°,$

   since $|z| = 1$ and $\theta = -90°$, which is on the negative half circle on the unit circle on the z-plane.

As shown in Example 8.3, the BLT offers conversion of an analog transfer function to a digital transfer function. Example 8.4 shows how to perform the BLT.

Example 8.4.

Given an analog filter whose transfer function is

$$H(s) = \dfrac{10}{s + 10},$$

a. Convert it to the digital filter transfer function and difference equation, respectively, when a sampling period is given as $T = 0.01$ second.

Solution:

a. Applying the BLT, we have

$$H(z) = H(s)\Big|_{s=\frac{2}{T}\frac{z-1}{z+1}} = \dfrac{10}{s + 10}\Bigg|_{s=\frac{2}{T}\frac{z-1}{z+1}}.$$

Substituting $T = 0.01$, it follows that

$$H(z) = \frac{10}{\frac{200(z-1)}{z+1} + 10} = \frac{0.05}{\frac{z-1}{z+1} + 0.05} = \frac{0.05(z+1)}{z - 1 + 0.05(z+1)} = \frac{0.05z + 0.05}{1.05z - 0.95}.$$

Finally, we get

$$H(z) = \frac{(0.05z + 0.05)/(1.05z)}{(1.05z - 0.95)/(1.05z)} = \frac{0.0476 + 0.0476z^{-1}}{1 - 0.9048z^{-1}}.$$

Applying the technique in Chapter 6, we yield the difference equation as

$$y(n) = 0.0476x(n) + 0.0476x(n-1) + 0.9048y(n-1).$$

Next, we examine frequency mapping between the s-plane and the z-plane. As illustrated in Figure 8.9, the analog frequency ω_a is marked on the $j\omega$ axis on the s-plane, whereas ω_d is the digital frequency labeled on the unit circle in the z-plane.

We substitute $s = j\omega_a$ and $z = e^{j\omega_d T}$ into the BLT in Equation (8.11) to get

$$j\omega_a = \frac{2}{T} \frac{e^{j\omega_d T} - 1}{e^{j\omega_d T} + 1}. \tag{8.13}$$

Simplifying Equation (8.13) leads to

$$\omega_a = \frac{2}{T} \tan\left(\frac{\omega_d T}{2}\right). \tag{8.14}$$

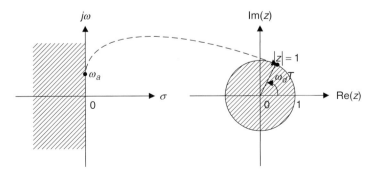

FIGURE 8.9 Frequency mapping from the analog domain to the digital domain.

Equation (8.14) explores the relation between the analog frequency on the $j\omega$ axis and the corresponding digital frequency ω_d on the unit circle. We can also write its inverse as

$$\omega_d = \frac{2}{T}\tan^{-1}\left(\frac{\omega_a T}{2}\right). \tag{8.15}$$

The range of the digital frequency ω_d is from 0 radian per second to the folding frequency $\omega_s/2$ radians per second, where ω_s is the sampling frequency in radians per second. We make a plot of Equation (8.14) in Figure 8.10.

From Figure 8.10 when the digital frequency range $0 \le \omega_d \le 0.25\omega_s$ is mapped to the analog frequency range $0 \le \omega_a \le 0.32\omega_s$, the transformation appears to be linear; however, when the digital frequency range $0.25\omega_s \le \omega_d \le 0.5\omega_s$ is mapped to the analog frequency range for $\omega_a > 0.32\omega_s$, the transformation is nonlinear. The analog frequency range for $\omega_a > 0.32\omega_s$ is compressed into the digital frequency range $0.25\omega_s \le \omega_d \le 0.5\omega_s$. This nonlinear frequency mapping effect is called *frequency warping*. We must incorporate the frequency warping into the IIR filter design. The following example will illustrate the frequency warping effect in the BLT.

Example 8.5.

Assume the following analog frequencies:
$\omega_a = 10$ radians per second
$\omega_a = \omega_s/4 = 50\pi = 157$ radians per second
$\omega_a = \omega_s/2 = 100\pi = 314$ radians per second.

a. Find their digital frequencies using the BLT with a sampling period of 0.01 second, given the analog filter in Example 8.4 and the developed digital filter.

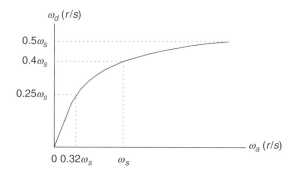

FIGURE 8.10 **Frequency warping from bilinear transformation.**

Solution:

a. From Equation (8.15), we can calculate digital frequency ω_d as follows: When $\omega_a = 10$ radians/sec and $T = 0.01$ second,

$$\omega_d = \frac{2}{T}\tan^{-1}\left(\frac{\omega_a T}{2}\right) = \frac{2}{0.01}\tan^{-1}\left(\frac{10 \times 0.01}{2}\right) = 9.99 \text{ rad/sec},$$

which is close to the analog frequency of 10 radians per second. When $\omega_a = 157$ rad/sec and $T = 0.01$ second,

$$\omega_d = \frac{2}{0.01}\tan^{-1}\left(\frac{157 \times 0.01}{2}\right) = 133.11 \text{ rad/sec},$$

which has an error as compared with the desired value 157. When $\omega_a = 314$ rad/sec and $T = 0.01$ second,

$$\omega_d = \frac{2}{0.01}\tan^{-1}\left(\frac{314 \times 0.01}{2}\right) = 252.5 \text{ rad/sec},$$

which gives a bigger error compared with the digital folding frequency of 314 radians per second.

Figure 8.11 shows how to correct the frequency warping error. First, given the digital frequency specification, we prewarp the digital frequency specification to the analog frequency specification by Equation (8.14).

Second, we obtain the analog lowpass filter $H(s)$ using the prewarped analog frequency ω_a and the lowpass prototype. For the lowpass analog filter, we have

$$H(s) = H_P(s)|_{s=\frac{s}{\omega_a}} = H_P\left(\frac{s}{\omega_a}\right). \tag{8.16}$$

Finally, substituting BLT Equation (8.11) into Equation (8.16) yields the digital filter as

$$H(z) = H(s)|_{s=\frac{2}{T}\frac{z-1}{z+1}}. \tag{8.17}$$

This approach can be extended to the other type of filter design similarly.

8.2.3 Bilinear Transformation Design Procedure

Now we can summarize the BLT design procedure.

1. Given the digital filter frequency specifications, prewarp the digital frequency specifications to the analog frequency specifications.

8 INFINITE IMPULSE RESPONSE FILTER DESIGN

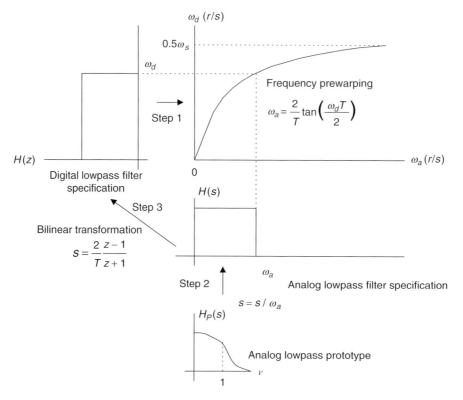

FIGURE 8.11 Graphical representation of IIR filter design using the bilinear transformation.

For the lowpass filter and highpass filter:

$$\omega_a = \frac{2}{T} \tan\left(\frac{\omega_d T}{2}\right). \tag{8.18}$$

For the bandpass filter and bandstop filter:

$$\omega_{al} = \frac{2}{T} \tan\left(\frac{\omega_l T}{2}\right), \quad \omega_{ah} = \frac{2}{T} \tan\left(\frac{\omega_h T}{2}\right), \tag{8.19}$$

where

$$\omega_0 = \sqrt{\omega_{al}\omega_{ah}}, \quad W = \omega_{ah} - \omega_{al}$$

2. Perform the prototype transformation using the lowpass prototype $H_p(s)$.

From lowpass to lowpass: $H(s) = H_P(s)\big|_{s=\frac{s}{\omega_a}}$ \qquad (8.20)

From lowpass to highpass: $H(s) = H_P(s)\big|_{s=\frac{\omega_a}{s}}$ (8.21)

From lowpass to bandpass: $H(s) = H_P(s)\big|_{s=\frac{s^2+\omega_0^2}{sW}}$ (8.22)

From lowpass to bandstop: $H(s) = H_P(s)\big|_{s=\frac{sW}{s^2+\omega_0^2}}$ (8.23)

3. Substitute the BLT to obtain the digital filter

$$H(z) = H(s)\big|_{s=\frac{2}{T}\frac{z-1}{z+1}}.$$ (8.24)

Table 8.2 lists MATLAB functions for the BLT design.

We illustrate the lowpass filter design procedure in Example 8.6. Other types of filter, such as highpass, bandpass, and bandstop, will be illustrated in the next section.

TABLE 8.2 MATLAB functions for the bilinear transformation design.

Lowpass to lowpass: $H(s) = H_P(s)\big|_{s=\frac{s}{\omega_a}}$
≫[B,A] = lp2lp(Bp,Ap,wa)
Lowpass to highpass: $H(s) = H_P(s)\big|_{s=\frac{\omega_a}{s}}$
≫[B,A] = lp2hp(Bp,Ap,wa)
Lowpass to bandpass: $H(s) = H_P(s)\big|_{s=\frac{s^2+\omega_0^2}{sW}}$
≫[B,A] = lp2bp(Bp,Ap,w0,W)
Lowpass to bandstop: $H(s) = H_P(s)\big|_{s=\frac{sW}{s^2+\omega_0^2}}$
≫[B,A] = lp2bs(Bp,Ap,w0,W)
Bilinear transformation to achieve the digital filter:
≫[b, a] = bilinear(B,A,fs)
Plot of the magnitude and phase frequency responses of the digital filter:
≫freqz(b,a,512,fs)
Definitions of design parameters:
Bp = vector containing the numerator coefficients of the lowpass prototype.
Ap = vector containing the denominator coefficients of the lowpass prototype.
wa = cutoff frequency for the lowpass or highpass analog filter (rad/sec).
w0 = center frequency for the bandpass or bandstop analog filter (rad/sec).
W = bandwidth for the bandpass or bandstop analog filter (rad/sec).
B = vector containing the numerator coefficients of the analog filter.
A = vector containing the denominator coefficients of the analog filter.
b = vector containing the numerator coefficients of the digital filter.
a = vector containing the denominator coefficients of the digital filter.
fs = sampling rate (samples/sec).

Example 8.6.

The normalized lowpass filter with a cutoff frequency of 1 rad/sec is given as:

$$H_P(s) = \frac{1}{s+1}.$$

a. Use the given $H_p(s)$ and the BLT to design a corresponding digital IIR lowpass filter with a cutoff frequency of 15 Hz and a sampling rate of 90 Hz.

b. Use MATLAB to plot the magnitude response and phase response of $H(z)$.

Solution:

a. First, we obtain the digital frequency as

$$\omega_d = 2\pi f = 2\pi(15) = 30\pi \text{ rad/sec}, \text{ and } T = 1/f_s = 1/90 \text{ sec}.$$

We then follow the design procedure:

1. First calculate the prewarped analog frequency as

$$\omega_a = \frac{2}{T} \tan\left(\frac{\omega_d T}{2}\right) = \frac{2}{1/90} \tan\left(\frac{30\pi/90}{2}\right),$$

that is, $\omega_a = 180 \times \tan(\pi/6) = 180 \times \tan(30°) = 103.92$ rad/sec.

2. Then perform the prototype transformation (lowpass to lowpass) as follows:

$$H(s) = H_P(s)\big|_{s=\frac{s}{\omega_a}} = \frac{1}{\frac{s}{\omega_a}+1} = \frac{\omega_a}{s+\omega_a},$$

which yields an analog filter:

$$H(s) = \frac{103.92}{s+103.92}.$$

3. Apply the BLT, which yields

$$H(z) = \frac{103.92}{s+103.92}\bigg|_{s=\frac{2}{T}\frac{z-1}{z+1}}.$$

We simplify the algebra by dividing both the numerator and the denominator by 180:

$$H(z) = \frac{103.92}{180 \times \frac{z-1}{z+1} + 103.92} = \frac{103.92/180}{\frac{z-1}{z+1} + 103.92/180} = \frac{0.5773}{\frac{z-1}{z+1} + 0.5773}.$$

Then we multiply both numerator and denominator by $(z+1)$ to obtain

$$H(z) = \frac{0.5773(z+1)}{\left(\frac{z-1}{z+1} + 0.5773\right)(z+1)} = \frac{0.5773z + 0.5773}{(z-1) + 0.5773(z+1)}$$

$$= \frac{0.5773z + 0.5773}{1.5773z - 0.4227}.$$

Finally, we divide both numerator and denominator by $1.5773z$ to get the transfer function in the standard format:

$$H(z) = \frac{(0.5773z + 0.5773)/(1.5773z)}{(1.5773z - 0.4227)/(1.5773z)} = \frac{0.3660 + 0.3660z^{-1}}{1 - 0.2679z^{-1}}.$$

b. The corresponding MATLAB design is listed in Program 8.2. Figure 8.12 shows the magnitude and phase frequency responses.

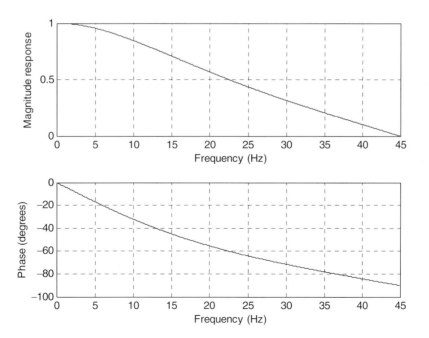

FIGURE 8.12 Frequency responses of the designed digital filter for Example 8.6.

Program 8.2. MATLAB program for Example 8.6.

```
%Example 8.6
% Plot the magnitude and phase responses
fs = 90;% Sampling rate (Hz)
[B, A] = lp2lp([1],[1 1],103.92);
[b, a] = bilinear(B, A, fs)
%b = [0.3660 0.3660] numerator coefficients of the digital filter from MATLAB
%a = [1 −0.2679]denominator coefficients of the digital filter from MATLAB
[hz, f] = freqz([0.3660 0.3660],[1 −0.2679],512,fs);%the frequency response
phi = 180*unwrap(angle(hz))/pi;
subplot(2,1,1), plot(f, abs(hz)),grid;
axis([0 fs/2 0 1]);
xlabel('Frequency (Hz)'); ylabel('Magnitude Response')
subplot(2,1,2), plot(f, phi); grid;
axis([0 fs/2 −100 0]);
xlabel('Frequency (Hz)'); ylabel('Phase (degrees)')
```

8.3 Digital Butterworth and Chebyshev Filter Designs

In this section, we design various types of digital Butterworth and Chebyshev filters using the BLT design method developed in the previous section.

8.3.1 Lowpass Prototype Function and Its Order

As described in the Section 8.2.3 (Bilinear Transformation Design Procedure), BLT design requires obtaining the analog filter with prewarped frequency specifications. These analog filter design requirements include the ripple specification at the passband frequency edge, the attenuation specification at the stopband frequency edge, and the type of lowpass prototype (which we shall discuss) and its order.

Table 8.3 lists the Butterworth prototype functions with 3 dB passband ripple specification. Tables 8.4 and 8.5 contain the Chebyshev prototype functions (type I) with 1 dB and 0.5 dB passband ripple specifications, respectively. Other lowpass prototypes with different ripple specifications and order can be computed using the methods described in Appendix C.

In this section, we will focus on the Chebyshev type I filter. The Chebyshev type II filter design can be found in Proakis and Manolakis (1996) and Porat (1997).

TABLE 8.3 3 dB Butterworth lowpass prototype transfer functions ($\varepsilon = 1$)

n	$H_P(s)$
1	$\frac{1}{s+1}$
2	$\frac{1}{s^2+1.4142s+1}$
3	$\frac{1}{s^3+2s^2+2s+1}$
4	$\frac{1}{s^4+2.6131s^3+3.4142s^2+2.6131s+1}$
5	$\frac{1}{s^5+3.2361s^4+5.2361s^3+5.2361s^2+3.2361s+1}$
6	$\frac{1}{s^6+3.8637s^5+7.4641s^4+9.1416s^3+7.4641s^2+3.8637s+1}$

TABLE 8.4 Chebyshev lowpass prototype transfer functions with 0.5 dB ripple ($\varepsilon = 0.3493$)

n	$H_P(s)$
1	$\frac{2.8628}{s+2.8628}$
2	$\frac{1.4314}{s^2+1.4256s+1.5162}$
3	$\frac{0.7157}{s^3+1.2529s^2+1.5349s+0.7157}$
4	$\frac{0.3579}{s^4+1.1974s^3+1.7169s^2+1.0255s+0.3791}$
5	$\frac{0.1789}{s^5+1.1725s^4+1.9374s^3+1.3096s^2+0.7525s+0.1789}$
6	$\frac{0.0895}{s^6+1.1592s^5+2.1718s^4+1.5898s^3+1.1719s^2+0.4324s+0.0948}$

TABLE 8.5 Chebyshev lowpass prototype transfer functions with 1 dB ripple ($\varepsilon = 0.5088$)

n	$H_P(s)$
1	$\frac{1.9652}{s+1.9652}$
2	$\frac{0.9826}{s^2+1.0977s+1.1025}$
3	$\frac{0.4913}{s^3+0.9883s^2+1.2384s+0.4913}$
4	$\frac{0.2456}{s^4+0.9528s^3+1.4539s^2+0.7426s+0.2756}$
5	$\frac{0.1228}{s^5+0.9368s^4+1.6888s^3+0.9744s^2+0.5805s+0.1228}$
6	$\frac{0.0614}{s^6+0.9283s^5+1.9308s^4+1.20121s^3+0.9393s^2+0.3071s+0.0689}$

The magnitude response function of the Butterworth lowpass prototype with an order of n is shown in Figure 8.13, where the magnitude response $|H_p(v)|$ versus the normalized frequency v is given by Equation (8.25):

$$|H_P(v)| = \frac{1}{\sqrt{1+\varepsilon^2 v^{2n}}} \quad (8.25)$$

With the given passband ripple A_p dB at the normalized passband frequency edge $v_p = 1$, and the stopband attenuation A_s dB at the normalized stopband

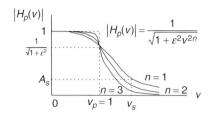

FIGURE 8.13 Normalized Butterworth magnitude response function.

frequency edge v_s, the following two equations must be satisfied to determine the prototype filter order:

$$A_P \, dB = -20 \cdot \log_{10}\left(\frac{1}{\sqrt{1+\varepsilon^2}}\right) \tag{8.26}$$

$$A_s \, dB = -20 \cdot \log_{10}\left(\frac{1}{\sqrt{1+\varepsilon^2 v_s^{2n}}}\right). \tag{8.27}$$

Solving Equations (8.26) and (8.27), we determine the lowpass prototype order as

$$\varepsilon^2 = 10^{0.1 A_p} - 1 \tag{8.28}$$

$$n \geq \frac{\log_{10}\left(\frac{10^{0.1 A_s}-1}{\varepsilon^2}\right)}{[2 \cdot \log_{10}(v_s)]}, \tag{8.29}$$

where ε is the absolute ripple specification.

The magnitude response function of the Chebyshev lowpass prototype with an order of n is shown in Figure 8.14, where the magnitude response $|H_p(v)|$ versus the normalized frequency v is given by

$$|H_P(v)| = \frac{1}{\sqrt{1+\varepsilon^2 C_n^2(v)}} \tag{8.30}$$

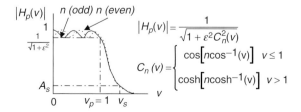

FIGURE 8.14 Normalized Chebyshev magnitude response function.

8.3 Digital Butterworth and Chebyshev Filter Designs

where

$$C_n(v_s) = \cosh\left[n \cosh^{-1}(v_s)\right] \quad (8.31)$$

$$\cosh^{-1}(v_s) = \ln\left(v_s + \sqrt{v_s^2 - 1}\right) \quad (8.32)$$

As shown in Figure 8.14, the magnitude response for the Chebyshev lowpass prototype with the order of an odd number begins with the filter DC gain of 1. In the case of a Chebyshev lowpass prototype with the order of an even number, the magnitude starts at the filter DC gain of $1/\sqrt{1+\varepsilon^2}$. For both cases, the filter gain at the normalized cutoff frequency $v_p = 1$ is $1/\sqrt{1+\varepsilon^2}$.

Similarly, Equations (8.33) and (8.34) must be satisfied:

$$A_P \, dB = -20 \cdot \log_{10}\left(\frac{1}{\sqrt{1+\varepsilon^2}}\right) \quad (8.33)$$

$$A_s \, dB = -20 \cdot \log_{10}\left(\frac{1}{\sqrt{1+\varepsilon^2 C_n^2(v_s)}}\right). \quad (8.34)$$

The lowpass prototype order can be solved in Equation (8.35b):

$$\varepsilon^2 = 10^{0.1 A_p} - 1 \quad (8.35a)$$

$$n \geq \frac{\cosh^{-1}\left[\left(\frac{10^{0.1 A_s} - 1}{\varepsilon^2}\right)^{0.5}\right]}{\cosh^{-1}(v_s)}, \quad (8.35b)$$

where $\cosh^{-1}(x) = \ln(x + \sqrt{x^2 - 1})$, ε is the absolute ripple parameter.

The normalized stopband frequency v_s can be determined from the frequency specifications of an analog filter in Table 8.6. Then the order of the lowpass

TABLE 8.6 Conversion from analog filter specifications to lowpass prototype specifications.

Analog Filter Specifications	Lowpass Prototype Specifications
Lowpass: ω_{ap}, ω_{as}	$v_p = 1$, $v_s = \omega_{as}/\omega_{ap}$
Highpass: ω_{ap}, ω_{as}	$v_p = 1$, $v_s = \omega_{ap}/\omega_{as}$
Bandpass: ω_{apl}, ω_{aph}, ω_{asl}, ω_{ash}	$v_p = 1, v_s = \frac{\omega_{ash} - \omega_{asl}}{\omega_{aph} - \omega_{apl}}$
$\omega_0 = \sqrt{\omega_{apl}\omega_{aph}}$, $\omega_0 = \sqrt{\omega_{asl}\omega_{ash}}$	
Bandstop: ω_{apl}, ω_{aph}, ω_{asl}, ω_{ash}	$v_p = 1, v_s = \frac{\omega_{aph} - \omega_{apl}}{\omega_{ash} - \omega_{asl}}$
$\omega_0 = \sqrt{\omega_{apl}\omega_{aph}}$, $\omega_0 = \sqrt{\omega_{asl}\omega_{ash}}$	

ω_{ap}, passband frequency edge; ω_{as}, stopband frequency edge; ω_{apl}, lower cutoff frequency in passband; ω_{aph}, upper cutoff frequency in passband; ω_{asl}, lower cutoff frequency in stopband; ω_{ash}, upper cutoff frequency in stopband; ω_0, geometric center frequency.

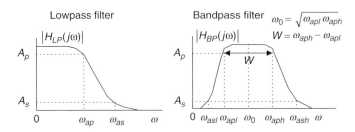

FIGURE 8.15 Specifications for analog lowpass and bandpass filters.

prototype can be determined by Equation (8.29) for the Butterworth function and Equation (8.35b) for the Chebyshev function. Figure 8.15 gives frequency edge notations for analog lowpass and bandpass filters. The notations for analog highpass and bandstop filters can be defined correspondingly.

8.3.2 Lowpass and Highpass Filter Design Examples

The following examples illustrate various designs for the Butterworth and Chebyshev lowpass and highpass filters.

Example 8.7.

a. Design a digital lowpass Butterworth filter with the following specifications:

1. 3 dB attenuation at the passband frequency of 1.5 kHz
2. 10 dB stopband attenuation at the frequency of 3 kHz
3. Sampling frequency of 8,000 Hz.

b. Use MATLAB to plot the magnitude and phase responses.

Solution:

a. First, we obtain the digital frequencies in radians per second:

$$\omega_{dp} = 2\pi f = 2\pi(1500) = 3000\pi \text{ rad/sec}$$
$$\omega_{ds} = 2\pi f = 2\pi(3000) = 6000\pi \text{ rad/sec}$$
$$T = 1/f_s = 1/8000 \text{ sec}$$

Following the steps of the design procedure,

1. We apply the warping equation as

$$\omega_{ap} = \frac{2}{T}\tan\left(\frac{\omega_d T}{2}\right) = 16000 \times \tan\left(\frac{3000\pi/8000}{2}\right) = 1.0691 \times 10^4 \text{ rad/sec}.$$

$$\omega_{as} = \frac{2}{T}\tan\left(\frac{\omega_d T}{2}\right) = 16000 \times \tan\left(\frac{6000\pi/8000}{2}\right) = 3.8627 \times 10^4 \text{ rad/sec}.$$

We then find the lowpass prototype specifications using Table 8.6 as follows:

$$v_s = \omega_{as}/\omega_{ap} = 3.8627 \times 10^4/(1.0691 \times 10^4)$$
$$= 3.6130 \text{ rad/sec and } A_s = 10 \text{ dB}.$$

The filter order is computed as

$$\varepsilon^2 = 10^{0.1 \times 3} - 1 = 1$$

$$n = \frac{\log_{10}(10^{0.1 \times 10} - 1)}{2 \cdot \log_{10}(3.6130)} = 0.8553.$$

2. Rounding n up, we choose $n = 1$ for the lowpass prototype. From Table 8.3, we have

$$H_P(s) = \frac{1}{s+1}.$$

Applying the prototype transformation (lowpass to lowpass) yields the analog filter

$$H(s) = H_P(s)\Big|_{\frac{s}{\omega_{ap}}} = \frac{1}{\frac{s}{\omega_{ap}}+1} = \frac{\omega_{ap}}{s+\omega_{ap}} = \frac{1.0691 \times 10^4}{s + 1.0691 \times 10^4}.$$

3. Finally, using the BLT, we have

$$H(z) = \frac{1.0691 \times 10^4}{s + 1.0691 \times 10^4}\Big|_{s=16000(z-1)/(z+1)}.$$

Substituting the BLT leads to

$$H(z) = \frac{1.0691 \times 10^4}{\left(16000\frac{z-1}{z+1}\right) + 1.0691 \times 10^4}.$$

To simplify the algebra, we divide both numerator and denominator by 16000 to get

$$H(z) = \frac{0.6682}{\left(\frac{z-1}{z+1}\right) + 0.6682}.$$

Then multiplying $(z + 1)$ to both numerator and denominator leads to

$$H(z) = \frac{0.6682(z + 1)}{(z - 1) + 0.6682(z + 1)} = \frac{0.6682z + 0.6682}{1.6682z - 0.3318}.$$

Dividing both numerator and denominator by $(1.6682 \cdot z)$ leads to

$$H(z) = \frac{0.4006 + 0.4006z^{-1}}{1 - 0.1989z^{-1}}.$$

b. Steps 2 and 3 can be carried out using MATLAB Program 8.3, as shown in the first three lines of the MATLAB codes. Figure 8.16 describes the filter frequency responses.

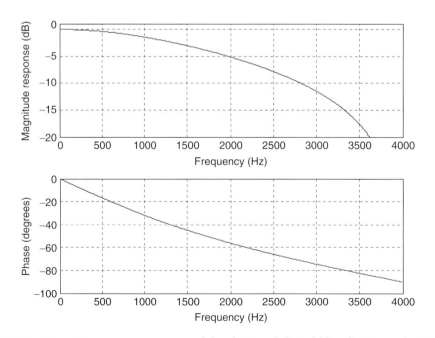

FIGURE 8.16 Frequency responses of the designed digital filter for Example 8.7.

Program 8.3. MATLAB program for Example 8.7.

```
%Example 8.7
% Design of the digital lowpass Butterworth filter
format long
fs=8000;% Sampling rate
[B A]=lp2lp([1],[1 1], 1.0691*10^4)% Complete step 2
[b a]=bilinear(B,A,fs) % Complete step 3
% Plot the magnitude and phase responses
%b=[0.4005 0.4005]; numerator coefficients from MATLAB
%a= [1 -0.1989]; denominator coefficients from MATLAB
freqz(b,a,512,fs);
axis([0 fs/2 -20 1])
```

Example 8.8.

a. Design a first-order digital highpass Chebyshev filter with a cutoff frequency of 3 kHz and 1 dB ripple on passband using a sampling frequency of 8,000 Hz.

b. Use MATLAB to plot the magnitude and phase responses.

Solution:

a. First, we obtain the digital frequency in radians per second:

$$\omega_d = 2\pi f = 2\pi(3000) = 6000\pi \text{ rad/sec, and } T = 1/f_s = 1/8000 \text{ sec.}$$

Following the steps of the design procedure, we have

1. $\omega_a = \frac{2}{T} \tan\left(\frac{\omega_d T}{2}\right) = 16000 \times \tan\left(\frac{6000\pi/8000}{2}\right) = 3.8627 \times 10^4$ rad/sec.

2. Since the filter order is given as 1, we select the first-order lowpass prototype from Table 8.5 as

$$H_P(s) = \frac{1.9652}{s + 1.9625}.$$

Applying the prototype transformation (lowpass to highpass), we yield

$$H(s) = H_P(s)\big|_{\frac{\omega_a}{s}} = \frac{1.9652}{\frac{\omega_a}{s} + 1.9652} = \frac{1.9652s}{1.9652s + 3.8627 \times 10^4}.$$

Dividing both numerator and denominator by 1.9652 gives

$$H(s) = \frac{s}{s + 1.9656 \times 10^4}.$$

3. Using the BLT, we have

$$H(z) = \left. \frac{s}{s + 1.9656 \times 10^4} \right|_{s=16000(z-1)/(z+1)}.$$

Algebra work is demonstrated as follows:

$$H(z) = \frac{16000 \frac{z-1}{z+1}}{16000 \frac{z-1}{z+1} + 1.9656 \times 10^4}.$$

Simplifying the transfer function yields

$$H(z) = \frac{0.4487 - 0.4487 z^{-1}}{1 + 0.1025 z^{-1}}.$$

b. Steps 2 and 3 and the frequency response plots shown in Figure 8.17 can be carried out using MATLAB Program 8.4.

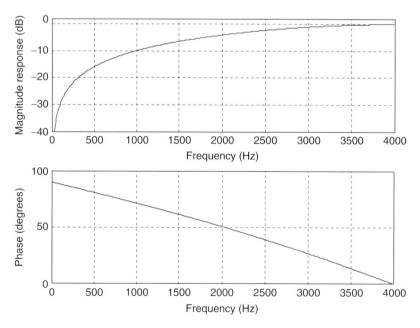

FIGURE 8.17 Frequency responses of the designed digital filter for Example 8.8.

Program 8.4. MATLAB program for Example 8.8

```
%Example 8.8
% Design of the digital highpass Butterworth filter
format long
fs=8000;           % Sampling rate
[B A]=lp2hp([1.9652],[1 1.9652], 3.8627*10^4) % Complete step 2
[b a]=bilinear(B,A,fs)% Complete step 3
% Plot the magnitude and phase responses
%b = [0.4487 -0.4487]; numerator coefficients from MATLAB
%a=[1 0.1025]; denominator coefficients from MATLAB
freqz(b,a,512,fs);
axis([0 fs/2 -40 2])
```

Example 8.9.

a. Design a second-order digital lowpass Butterworth filter with a cutoff frequency of 3.4 kHz at a sampling frequency of 8,000 Hz.

b. Use MATLAB to plot the magnitude and phase responses.

Solution:

a. First, we obtain the digital frequency in radians per second:

$\omega_d = 2\pi f = 2\pi(3400) = 6800\pi$ rad/sec, and $T = 1/f_s = 1/8000$ sec.

Following the steps of the design procedure, we compute the prewarped analog frequency as

1. $\omega_a = \frac{2}{T}\tan\left(\frac{\omega_d T}{2}\right) = 16000 \times \tan\left(\frac{6800\pi/8000}{2}\right) = 6.6645 \times 10^4$ rad/sec.

2. Since the order of 2 is given in the specification, we directly pick the second-order lowpass prototype from Table 8.3:

$$H_P(s) = \frac{1}{s^2 + 1.4142s + 1}.$$

After applying the prototype transformation (lowpass to lowpass), we have

$$H(s) = H_P(s)\Big|_{\frac{s}{\omega_a}} = \frac{4.4416 \times 10^9}{s^2 + 9.4249 \times 10^4 s + 4.4416 \times 10^9}.$$

3. Carrying out the BLT yields

$$H(z) = \frac{4.4416 \times 10^9}{s^2 + 9.4249 \times 10^4 s + 4.4416 \times 10^9}\Bigg|_{s=16000(z-1)/(z+1)}.$$

Let us work on algebra:

$$H(z) = \frac{4.4416 \times 10^9}{\left(16000\frac{z-1}{z+1}\right)^2 + 9.4249 \times 10^4\left(16000\frac{z-1}{z+1}\right) + 4.4416 \times 10^9}.$$

To simplify, we divide both numerator and denominator by $(16000)^2$ to get

$$H(z) = \frac{17.35}{\left(\frac{z-1}{z+1}\right)^2 + 5.8906\left(\frac{z-1}{z+1}\right) + 17.35}.$$

Then multiplying both numerator and denominator by $(z+1)^2$ leads to

$$H(z) = \frac{17.35(z+1)^2}{(z-1)^2 + 5.8906(z-1)(z+1) + 17.35(z+1)^2}.$$

Using identities, we have

$$H(z) = \frac{17.35(z^2 + 2z + 1)}{(z^2 - 2z + 1) + 5.8906(z^2 - 1) + 17.35(z^2 + 2z + 1)}$$

$$= \frac{17.35z^2 + 34.7z + 17.35}{24.2406z^2 + 32.7z + 12.4594}.$$

Dividing both numerator and denominator by $(24.2406z^2)$ leads to

$$H(z) = \frac{0.7157 + 1.4314z^{-1} + 0.7151z^{-2}}{1 + 1.3490z^{-1} + 0.5140z^{-2}}.$$

b. Steps 2 and 3 require a certain amount of algebra work and can be verified using MATLAB Program 8.5, as shown in the first three lines. Figure 8.18 plots the filter magnitude and phase frequency responses.

Program 8.5. MATLAB program for Example 8.9.

```
%Example 8.9
% Design of the digital lowpass Butterworth filter
format long
fs = 8000;           % Sampling rate
[B A] = lp2lp([1],[1 1.4142 1], 6.6645*10^4) % Complete step 2
[b a] = bilinear(B,A,fs)% Complete step 3
% Plot the magnitude and phase responses
%b = [0.7157 1.4315 0.7157]; numerator coefficients from MATLAB
%a = [1 1.3490 0.5140]; denominator coefficients from MATLAB
freqz(b,a,512,fs);
axis([0 fs/2 -40 10])
```

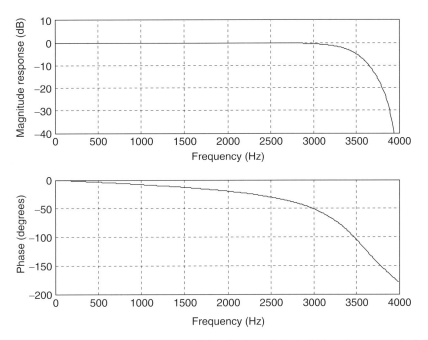

FIGURE 8.18 Frequency responses of the designed digital filter for Example 8.9.

Example 8.10.

a. Design a digital highpass Chebyshev filter with the following specifications:

1. 0.5 dB ripple on passband at the frequency of 3,000 Hz
2. 25 dB attenuation at the frequency of 1,000 Hz
3. Sampling frequency of 8,000 Hz.

b. Use MATLAB to plot the magnitude and phase responses.

Solution:

a. From the specifications, the digital frequencies are

$\omega_{dp} = 2\pi f = 2\pi(3000) = 6000\pi$ rad/sec

$\omega_{ds} = 2\pi f = 2\pi(1000) = 2000\pi$ rad/sec

$T = 1/f_s = 1/8000$ sec.

Using the design procedure, it follows that

$$\omega_{ap} = \frac{2}{T}\tan\left(\frac{\omega_{dp}T}{2}\right) = 16000 \times \tan\left(\frac{6000\pi/8000}{2}\right) = 3.8627 \times 10^4 \text{ rad/sec}$$

$$\omega_{as} = 16000 \times \tan\left(\frac{\omega_{ds}T}{2}\right) = 16000 \times \tan\left(\frac{2000\pi/8000}{2}\right)$$

$$= 6.6274 \times 10^3 \text{ rad/sec}$$

We find the lowpass prototype specifications using Table 8.6 as follows:

$v_s = \omega_{ps}/\omega_{sp} = 3.8627 \times 10^4/6.6274 \times 10^3 = 5.8284 \text{ rad/s}$ and $A_s = 25 \text{ dB}$, then the filter order is computed as

$$\varepsilon^2 = 10^{0.1 \times 0.5} - 1 = 0.1220$$

$$(10^{0.1 \times 25} - 1)/0.1220 = 2583.8341$$

$$n = \frac{\cosh^{-1}\left[(2583.8341)^{0.5}\right]}{\cosh^{-1}(5.8284)} = \frac{\ln(50.8314 + \sqrt{50.8314^2 - 1})}{\ln(5.8284 + \sqrt{5.8284^2 - 1})} = 1.8875.$$

We select $n = 2$ for the lowpass prototype function. Following the steps of the design procedure, it follows that

1. $\omega_p = 3.8627 \times 10^4 \text{ rad/sec}$.

2. Performing the prototype transformation (lowpass to lowpass) using the prototype filter in Table 8.4, we have

$$H_P(s) = \frac{1.4314}{s^2 + 1.4256s + 1.5162} \text{ and}$$

$$H(s) = H_P(s)\big|_{\frac{s}{\omega_a}} = \frac{1.4314}{\left(\frac{\omega_p}{s}\right)^2 + 1.4256\left(\frac{\omega_p}{s}\right) + 1.5162}$$

$$= \frac{0.9441s^2}{s^2 + 3.6319 \times 10^4 s + 9.8407 \times 10^8}.$$

3. Hence, applying the BLT, we convert the analog filter to a digital filter as

$$H(z) = \frac{0.9441s^2}{s^2 + 3.6319 \times 10^4 s + 9.8407 \times 10^8}\bigg|_{s=16000(z-1)/(z+1)}.$$

After algebra simplification, it follows that

$$H(z) = \frac{0.1327 - 0.2654z^{-1} + 0.1327z^{-2}}{1 + 0.7996z^{-1} + 0.3618z^{-2}}.$$

b. MATLAB Program 8.6 is listed for this example, and the frequency responses are given in Figure 8.19.

Program 8.6. MATLAB program for Example 8.10.

```
%Example 8.10
% Design of the digital lowpass Chebyshev filter
format long
fs = 8000;           % Sampling rate
% BLT design
[B A] = lp2hp([1.4314],[1 1.4256 1.5162], 3.8627*10 ∧ 4)   % Complete step 2
[b a] = bilinear(B,A,fs)% Complete step 3
% Plot the magnitude and phase responses
%b = [0.1327 −0.2654 0.1327]; numerator coefficients from MATLAB
%a = [1 0.7996 0.3618];denominator coefficients from MATLAB
freqz(b,a,512,fs);
axis([0 fs/2 −40 10])
```

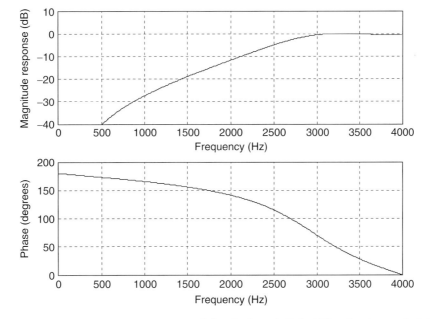

FIGURE 8.19 Frequency responses of the designed digital filter for Example 8.10.

8.33 Bandpass and Bandstop Filter Design Examples

Example 8.11.

a. Design a second-order digital bandpass Butterworth filter with the following specifications:
   - an upper cutoff frequency of 2.6 kHz and
   - a lower cutoff frequency of 2.4 kHz,
   - a sampling frequency of 8,000 Hz.

b. Use MATLAB to plot the magnitude and phase responses.

Solution:

a. Let us find the digital frequencies in radians per second:
   - $\omega_h = 2\pi f_h = 2\pi(2600) = 5200\pi$ rad/sec
   - $\omega_l = 2\pi f_l = 2\pi(2400) = 4800\pi$ rad/sec, and $T = 1/f_s = 1/8000$ sec.

Following the steps of the design procedure, we have the following:

1. $\omega_{ah} = \frac{2}{T}\tan\left(\frac{\omega_h T}{2}\right) = 16000 \times \tan\left(\frac{5200\pi/8000}{2}\right) = 2.6110 \times 10^4$ rad/sec

   $\omega_{al} = 16000 \times \tan\left(\frac{\omega_l T}{2}\right) = 16000 \times \tan(0.3\pi) = 2.2022 \times 10^4$ rad/sec

   $W = \omega_{ah} - \omega_{al} = 26110 - 22022 = 4088$ rad/sec

   $\omega_0^2 = \omega_{ah} \times \omega_{al} = 5.7499 \times 10^8$

2. We perform the prototype transformation (lowpass to bandpass) to obtain $H(s)$. From Table 8.3, we pick the lowpass prototype with the order of 1 to produce the bandpass filter with the order of 2, as

   $$H_P(s) = \frac{1}{s+1},$$

   and applying the lowpass-to-bandpass transformation, it follows that

   $$H(s) = H_P(s)\Big|_{\frac{s^2+\omega_0^2}{sW}} = \frac{Ws}{s^2 + Ws + \omega_0^2} = \frac{4088s}{s^2 + 4088s + 5.7499 \times 10^8}.$$

3. Hence we apply the BLT to yield

   $$H(z) = \frac{4088s}{s^2 + 4088s + 5.7499 \times 10^8}\Bigg|_{s=16000(z-1)/(z+1)}.$$

Via algebra work, we obtain the digital filter as

$$H(z) = \frac{0.0730 - 0.0730z^{-2}}{1 + 0.7117z^{-1} + 0.8541z^{-2}}.$$

b. MATLAB Program 8.7 is given for this example, and the corresponding frequency response plots are illustrated in Figure 8.20.

Program 8.7. MATLAB program for Example 8.11.

```
%Example 8.11
% Design of the digital bandpass Butterworth filter
format long
fs = 8000;
[B A] = lp2bp([1],[1 1],sqrt(5.7499*10^8),4088) % Complete step 2
[b a] = bilinear(B,A,fs) % Complete step 3
% Plot the magnitude and phase responses
%b = [0.0730 −0.0730]; numerator coefficients from MATLAB
%a = [1 0.7117 0.8541]; denominator coefficients from MATLAB
freqz(b, a, 512, fs);
axis([0 fs/2 −40 10])
```

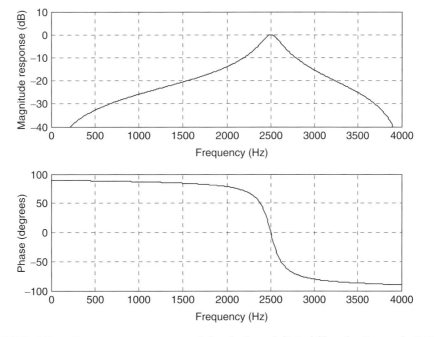

FIGURE 8.20 Frequency responses of the designed digital filter for Example 8.11.

Example 8.12.

Now let us examine the bandstop Butterworth filter design.

a. Design a digital bandstop Butterworth filter with the following specifications:

- Center frequency of 2.5 kHz
- Passband width of 200 Hz and ripple of 3 dB
- Stopband width of 50 Hz and attenuation of 10 dB
- Sampling frequency of 8,000 Hz.

b. Use MATLAB to plot the magnitude and phase responses.

Solution:

a. The digital frequencies of the digital filter are

$$\omega_h = 2\pi f_h = 2\pi(2600) = 5200\pi \text{ rad/sec},$$
$$\omega_l = 2\pi f_l = 2\pi(2400) = 4800\pi \text{ rad/sec},$$
$$\omega_{d0} = 2\pi f_0 = 2\pi(2500) = 5000\pi \text{ rad/sec, and } T = 1/fs = 1/8000 \text{ sec}.$$

Applying the three steps of the IIR fitter design approach, it follows that

1. $\omega_{ah} = \frac{2}{T}\tan\left(\frac{\omega_h T}{2}\right) = 16000 \times \tan\left(\frac{5200\pi/8000}{2}\right) = 2.6110 \times 10^4$ rad/sec

    $\omega_{al} = 16000 \times \tan\left(\frac{\omega_l T}{2}\right) = 16000 \times \tan(0.3\pi) = 2.2022 \times 10^4$ rad/sec

    $\omega_0 = 16000 \times \tan\left(\frac{\omega_{d0} T}{2}\right) = 16000 \times \tan(0.3125\pi) = 2.3946 \times 10^4$ rad/sec

    $\omega_{sh} = \frac{2}{T}\tan\left(\frac{2525 \times 2\pi/8000}{2}\right) = 16000 \times \tan(56.8125^0) = 2.4462 \times 10^4$ rad/sec

    $\omega_{sl} = 16000 \times \tan\left(\frac{2475 \times 2\pi/8000}{2}\right) = 16000 \times \tan(55.6875^0) = 2.3444 \times 10^4$ rad/sec.

To adjust the unit passband gain at the center frequency of 2,500 Hz, we perform the following:

Fixing $\omega_{al} = 2.2022 \times 10^4$, we compute $\omega_{ah} = \omega_0^2/\omega_{al} = \frac{(2.3946 \times 10^4)^2}{2.2022 \times 10^4} = 2.6037 \times 10^4$ and the passband bandwidth: $W = \omega_{ah} - \omega_{al} = 4015$

fixing $\omega_{sl} = 2.3444 \times 10^4$, $\omega_{sh} = \omega_0^2/\omega_{sl} = \frac{(2.3946 \times 10^4)^2}{2.3444 \times 10^4} = 2.4459 \times 10^4$ and the stopband bandwidth: $W_s = \omega_{sh} - \omega_{sl} = 1015$

Again, fixing $\omega_{ah} = 2.6110 \times 10^4$, we got $\omega_{al} = \omega_0^2/\omega_{ah} = \frac{(2.3946 \times 10^4)^2}{2.6110 \times 10^4} = 2.1961 \times 10^4$ and the passband bandwidth: $W = \omega_{ah} - \omega_{al} = 4149$

Fixing $\omega_{sh} = 2.4462 \times 10^4$, $\omega_{sl} = \omega_0^2/\omega_{sh} = \frac{(2.3946 \times 10^4)^2}{2.4462 \times 10^4} = 2.3441 \times 10^4$
and the stopband bandwidth: $W_s = \omega_{sh} - \omega_{sl} = 1021$

For an aggressive bandstop design, we choose $\omega_{al} = 2.6110 \times 10^4$, $\omega_{ah} = 2.1961 \times 10^4$, $\omega_{sl} = 2.3441 \times 10^4$, $\omega_{sh} = 2.4462 \times 10^4$, and $\omega_0 = 2.3946 \times 10^4$ to satisfy a larger bandwidth.

Thus, we develop the prototype specification

$$v_s = (26110 - 21916)/(24462 - 23441) = 4.0177$$

$$n = \left(\frac{\log_{10}(10^{0.1 \times 10} - 1)}{2 \cdot \log_{10}(4.0177)}\right) = 0.7899, \text{ choose } n = 1.$$

$W = \omega_{ah} - \omega_{al} = 26110 - 21961 = 4149 \text{ rad/sec}$, $\omega_0^2 = 5.7341 \times 10^8$.

2. Then, carrying out the prototype transformation (lowpass to bandstop) using the first-order lowpass prototype filter given by

$$H_P(s) = \frac{1}{s+1},$$

it follows that

$$H(s) = H_P(s)\Big|_{\frac{sW}{s^2+\omega_0^2}} = \frac{(s^2 + \omega_0^2)}{s^2 + Ws + \omega_0^2}.$$

Substituting the values of ω_0^2 and W yields

$$H(s) = \frac{s^2 + 5.7341 \times 10^8}{s^2 + 4149s + 5.7341 \times 10^8}.$$

3. Hence, applying the BLT leads to

$$H(z) = \frac{s^2 + 5.7341 \times 10^8}{s^2 + 4149s + 5.73411 \times 10^8}\Big|_{s=16000(z-1)/(z+1)}.$$

After algebra, we get

$$H(z) = \frac{0.9259 + 0.7078z^{-1} + 0.9259z^{-2}}{1 + 0.7078z^{-1} + 0.8518z^{-2}}.$$

b. MATLAB Program 8.8 includes the design steps. Figure 8.21 shows the filter frequency responses.

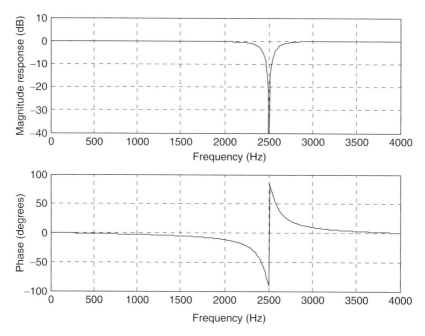

FIGURE 8.21 Frequency responses of the designed digital filter for Example 8.12.

Program 8.8. MATLAB program for Example 8.12.

```
%Example 8.12
% Design of the digital bandstop Butterworth filter
format long
fs = 8000;           % Sampling rate
[B A] = lp2bs([1],[1 1],sqrt(5.7341*10^8),4149)% Complete step 2
[b a] = bilinear(B,A,fs)% Complete step 3
% Plot the magnitude and phase responses
%b = [0.9259 0.7078 0.9259]; numerator coefficients from MATLAB
%a = [1 0.7078 0.8518]; denominator coefficients from MATLAB
freqz(b,a,512,fs);
axis([0 fs/2 -40 10])
```

Example 8.13.

a. Design a digital bandpass Chebyshev filter with the following specifications:
- Center frequency of 2.5 kHz
- Passband bandwidth of 200 Hz, 0.5 dB ripple on passband
- Lower stop frequency of 1.5 kHz, upper stop frequency of 3.5 kHz

- Stopband attenuation of 10 dB
- Sampling frequency of 8,000 Hz.

b. Use MATLAB to plot the magnitude and phase responses.

Solution:

a. The digital frequencies are given as:

$\omega_{dph} = 2\pi f_{dph} = 2\pi(2600) = 5200\pi$ rad/sec,

$\omega_{dpl} = 2\pi f_{dpl} = 2\pi(2400) = 4800\pi$ rad/sec,

$\omega_{d0} = 2\pi f_0 = 2\pi(2500) = 5000\pi$ rad/sec, and $T = 1/f_s = 1/8000$ sec.

Applying the frequency prewarping equation, it follows that

$\omega_{aph} = \dfrac{2}{T}\tan\left(\dfrac{\omega_{dph}T}{2}\right) = 16000 \times \tan\left(\dfrac{5200\pi/8000}{2}\right) = 2.6110 \times 10^4$ rad/sec

$\omega_{apl} = 16000 \times \tan\left(\dfrac{\omega_{dpl}T}{2}\right) = 16000 \times \tan(0.3\pi) = 2.2022 \times 10^4$ rad/sec

$\omega_0 = 16000 \times \tan\left(\dfrac{\omega_{d0}T}{2}\right) = 16000 \times \tan(0.3125\pi) = 2.3946 \times 10^4$ rad/sec

$\omega_{ash} = 16000 \times \tan\left(\dfrac{3500 \times 2\pi/8000}{2}\right) = 16000 \times \tan(78.75^0) = 8.0437 \times 10^4$ rad/sec

$\omega_{asl} = 16000 \times \tan\left(\dfrac{1500 \times 2\pi/8000}{2}\right) = 1.0691 \times 10^4$ rad/sec.

Now, adjusting the unit gain for the center frequency of 2,500 Hz leads to the following:

Fixing $\omega_{apl} = 2.2022 \times 10^4$, we have $\omega_{aph} = \dfrac{\omega_0^2}{\omega_{apl}} = \dfrac{(2.3946\times 10^4)^2}{2.2022\times 10^4} = 2.6038 \times 10^4$ and the passband bandwidth: $W = \omega_{aph} - \omega_{apl} = 4016$

Fixing $\omega_{asl} = 1.0691 \times 10^4$, $\omega_{ash} = \dfrac{\omega_0^2}{\omega_{asl}} = \dfrac{(2.3946\times 10^4)^2}{2.10691\times 10^4} = 5.3635 \times 10^4$ and the stopband bandwidth: $W_s = \omega_{ash} - \omega_{asl} = 42944$

Again, fixing $\omega_{aph} = 2.6110 \times 10^4$, we have $\omega_{apl} = \dfrac{\omega_0^2}{\omega_{aph}} = \dfrac{(2.3946\times 10^4)^2}{2.6110\times 10^4} = 2.1961 \times 10^4$ and the passband bandwidth: $W = \omega_{aph} - \omega_{apl} = 4149$

Fixing $\omega_{ash} = 8.0437 \times 10^4$, $\omega_{asl} = \dfrac{\omega_0^2}{\omega_{ash}} = \dfrac{(2.3946\times 10^4)^2}{8.0437\times 10^4} = 0.7137 \times 10^4$ and the stopband bandwidth: $W_s = \omega_{ash} - \omega_{asl} = 73300$

For an aggressive bandpass design, we select $\omega_{apl} = 2.2022 \times 10^4$, $\omega_{aph} = 2.6038 \times 10^4$, $\omega_{asl} = 1.0691 \times 10^4$, $\omega_{ash} = 5.3635 \times 10^4$, and for a smaller bandwidth for passband.

Thus, we obtain the prototype specifications:

$v_s = (53635 - 10691)/(26038 - 22022) = 10.6932$

$\varepsilon^2 = 10^{0.1\times 0.5} - 1 = 0.1220$

$$(10^{0.1\times 10} - 1)/0.1220 = 73.7705$$

$$n = \frac{\cosh^{-1}\left[(73.7705)^{0.5}\right]}{\cosh^{-1}(10.6932)} = \frac{\ln(8.5890 + \sqrt{8.5890^2 - 1})}{\ln(10.6932 + \sqrt{10.6932^2 - 1})} = 0.9280;$$

rounding up n leads to $n = 1$.

Applying the design steps leads to:

1. $\omega_{aph} = 2.6038 \times 10^4$ rad/sec, $\omega_{apl} = 2.2022 \times 10^4$ rad/sec, $W = 4016$ rad/sec, $\omega_0^2 = 5.7341 \times 10^8$
2. Performing the prototype transformation (lowpass to bandpass), we obtain

$$H_P(s) = \frac{2.8628}{s + 2.8628}$$

and

$$H(s) = H_P(s)\Big|_{s=\frac{s^2+\omega_0^2}{sW}} = \frac{2.8628Ws}{s^2 + 2.8628Ws + \omega_0^2}$$

$$= \frac{1.1497 \times 10^4 s}{s^2 + 1.1497 \times 10^4 s + 5.7341 \times 10^8}.$$

3. Applying the BLT, the analog filter is converted into a digital filter as follows:

$$H(z) = \frac{1.1497 \times 10^4 s}{s^2 + 1.1497 \times 10^4 s + 5.7341 \times 10^8}\Big|_{s=16000(z-1)/(z+1)},$$

which is simplified and arranged to be

$$H(z) = \frac{0.1815 - 0.1815z^{-2}}{1 + 0.6264z^{-1} + 0.6369z^{-2}}.$$

b. Program 8.9 lists the MATLAB details. Figure 8.22 displays the frequency responses.

Program 8.9. MATLAB program for Example 8.13.

```
%Example 8.13
% Design of the digital bandpass Chebyshev filter
format long
fs = 8000;
[B A] = lp2bp([2.8628],[1 2.8628],sqrt(5.7341*10^8),4016) % Complete step 2
[b a] = bilinear(B,A,fs) % Complete step 3
% Plot the magnitude and phase responses
% b = [0.1815 0.0 -0.1815]; numerator coefficients from MATLAB
% a = [1 0.6264 0.6369]; denominator coefficients from MATLAB
freqz(b,a,512,fs);
axis([0 fs/2 -40 10])
```

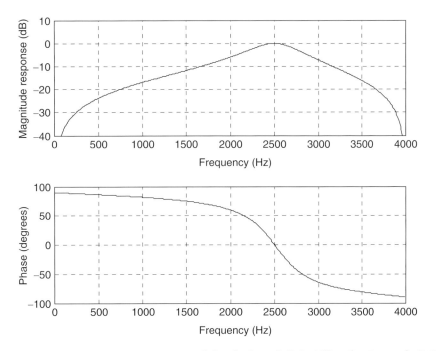

FIGURE 8.22 Frequency responses of the designed digital filter for Example 8.13.

8.4 Higher-Order Infinite Impulse Response Filter Design Using the Cascade Method

For the higher-order IIR filter design, use of a cascade transfer function is preferred. The factored forms for the lowpass prototype transfer functions for Butterworth and Chebyshev filters are given in Tables 8.7, 8.8, and 8.9. The Butterworth filter design example will be provided next. A similar procedure can be adopted for the Chebyshev filters.

Example 8.14.

a. Design a fourth-order digital lowpass Butterworth filter with a cutoff frequency of 2.5 kHz at a sampling frequency of 8,000 Hz.

b. Use MATLAB to plot the magnitude and phase responses.

Solution:

a. First, we obtain the digital frequency in radians per second:

$$\omega_d = 2\pi f = 2\pi(2500) = 5000\pi \text{ rad/sec, and } T = 1/f_s = 1/8000 \text{ sec.}$$

TABLE 8.7 3 dB Butterworth prototype functions in the cascade form.

n	$H_P(s)$
3	$\dfrac{1}{(s+1)(s^2+s+1)}$
4	$\dfrac{1}{(s^2+0.7654s+1)(s^2+1.8478s+1)}$
5	$\dfrac{1}{(s+1)(s^2+0.6180s+1)(s^2+1.6180s+1)}$
6	$\dfrac{1}{(s^2+0.5176s+1)(s^2+1.4142s+1)(s^2+1.9319s+1)}$

TABLE 8.8 Chebyshev prototype functions in the cascade form with 0.5 dB ripple ($\varepsilon = 0.3493$)

n	$H_P(s)$ 0.5 dB Ripple ($\varepsilon = 0.3493$)
3	$\dfrac{0.7157}{(s+0.6265)(s^2+0.6265s+1.1425)}$
4	$\dfrac{0.3579}{(s^2+0.3507s+1.0635)(s^2+0.8467s+0.3564)}$
5	$\dfrac{0.1789}{(s+0.3623)(s^2+0.2239s+1.0358)(s^2+0.5862s+0.4768)}$
6	$\dfrac{0.0895}{(s^2+0.1553s+1.0230)(s^2+0.4243s+0.5900)(s^2+0.5796s+0.1570)}$

TABLE 8.9 Chebyshev prototype functions in the cascade form with 1 dB ripple ($\varepsilon = 0.5088$).

n	$H_P(s)$ 1 dB Ripple ($\varepsilon = 0.5088$)
3	$\dfrac{0.4913}{(s+0.4942)(s^2+0.4942s+0.9942)}$
4	$\dfrac{0.2456}{(s^2+0.2791s+0.9865)(s^2+0.6737s+0.2794)}$
5	$\dfrac{0.1228}{(s+0.2895)(s^2+0.1789s+0.9883)(s^2+0.4684s+0.4293)}$
6	$\dfrac{0.0614}{(s^2+0.1244s+0.9907)(s^2+0.3398s+0.5577)(s^2+0.4641s+0.1247)}$

Following the design steps, we compute the specifications for the analog filter.

1. $\omega_a = \frac{2}{T}\tan\left(\frac{\omega_d T}{2}\right) = 16000 \times \tan\left(\frac{5000\pi/8000}{2}\right) = 2.3946 \times 10^4$ rad/sec.

2. From Table 8.7, we have the fourth-order factored prototype transfer function as

$$H_P(s) = \dfrac{1}{(s^2+0.7654s+1)(s^2+1.8478s+1)}.$$

Applying the prototype transformation, we yield

$$H(s) = H_P(s)\big|_{\frac{s}{\omega_a}} = \dfrac{\omega_a^2 \times \omega_a^2}{(s^2+0.7654\omega_a s+\omega_a^2)(s^2+1.8478\omega_a s+\omega_a^2)}.$$

8.4 Higher-Order Infinite Impulse Response Filter Design Using the Cascade Method

Substituting $\omega_a = 2.3946 \times 10^4$ rad/sec yields

$$H(s) = \frac{(5.7340 \times 10^8) \times (5.7340 \times 10^8)}{(s^2 + 1.8328s + 5.7340 \times 10^8)(s^2 + 4.4247 \times 10^4 s + 5.7340 \times 10^8)}.$$

3. Hence, after applying BLT, we have

$$H(z) = \frac{(5.7340 \times 10^8) \times (5.7340 \times 10^8)}{(s^2 + 1.8328s + 5.7340 \times 10^8)(s^2 + 4.4247 \times 10^4 s + 5.7340 \times 10^8)}\bigg|_{s=16000(z-1)/(z+1)}.$$

Simplifying algebra, we have the digital filter as

$$H(z) = \frac{0.5108 + 1.0215z^{-1} + 0.5108z^{-2}}{1 + 0.5654z^{-1} + 0.4776z^{-2}} \times \frac{0.3730 + 0.7460z^{-1} + 0.3730z^{-2}}{1 + 0.4129z^{-1} + 0.0790z^{-2}}.$$

b. A MATLAB program is better to use to carry out algebra and is listed in Program 8.10. Figure 8.23 shows the filter magnitude and phase frequency responses.

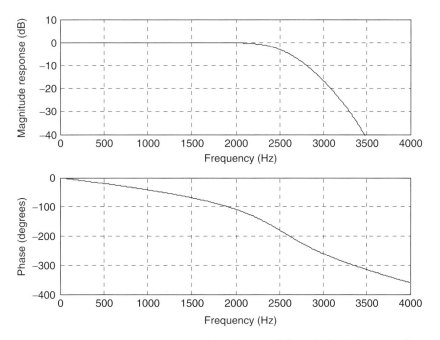

FIGURE 8.23 Frequency responses of the designed digital filter for Example 8.14.

Program 8.10. MATLAB program for Example 8.14.

```
%Example 8.14
% Design of the fourth-order digital lowpass Butterworth filter
% in the cascade form
format long
fs = 8000;           % Sampling rate
[B1 A1] = lp2lp([1],[1 0.7654 1], 2.3946*10^4)% Complete step 2
[b1 a1] = bilinear(B1,A1,fs)% Complete step 3
[B2 A2] = lp2lp([1],[1 1.8478 1], 2.3946*10^4)% Complete step 2
[b2 a2] = bilinear(B2,A2,fs)% Complete step 3
% Plot the magnitude and phase responses
% b1 = [0.5108 1.0215 0.5108];a1 = [1 0.5654 0.4776]; coefficients from MATLAB
% b2 = [0.3730 0.7460 0.3730];a2 = [1 0.4129 0.0790]; coefficients from MATLAB
freqz(conv(b1,b2),conv(a1,a2),512,fs);% Combined filter responses
axis([0 fs/2 −40 10]);
```

The higher-order bandpass, highpass, and bandstop filters using the cascade form can be designed similarly.

8.5 Application: Digital Audio Equalizer

In this section, the design of a digital audio equalizer is introduced. For an audio application such as the CD player, the digital audio equalizer is used to make the sound as one desires by changing filter gains for different audio frequency bands. Other applications include adjusting the sound source to take room acoustics into account, removing undesired noise, and boosting the desired signal in the specified passband. The simulation is based on the consumer digital audio processor—such as a CD player—handling the 16-bit digital samples with a sampling rate of 44.1 kHz and an audio signal bandwidth at 22.05 kHz. A block diagram of the digital audio equalizer is depicted in Figure 8.24.

A seven-band audio equalizer is adopted for discussion. The center frequencies are listed in Table 8.10. The 3 dB bandwidth for each bandpass filter is chosen to be 50% of the center frequency. As shown in Figure 8.24, g_0 through g_6 are the digital gains for each bandpass filter output and can be adjusted to make sound effects, while $y_0(n)$ through $y_6(n)$ are the digital amplified bandpass filter outputs. Finally, the equalized signal is the sum of the amplified bandpass filter outputs and itself. By changing the digital gains of the equalizer, many sound effects can be produced.

To complete the design and simulation, second-order IIR bandpass Butterworth filters are chosen for the audio equalizer. The coefficients are achieved using the BLT method, and are given in Table 8.11.

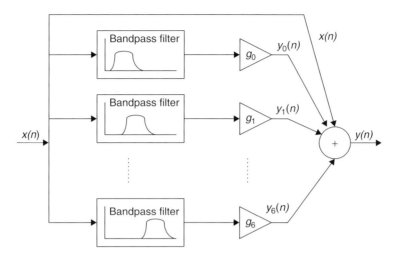

FIGURE 8.24 Simplified block diagram of the audio equalizer.

TABLE 8.10 Specifications for an audio equalizer to be designed.

Center frequency (Hz)	100	200	400	1000	2500	6000	15000
Bandwidth (Hz)	50	100	200	500	1250	3000	7500

TABLE 8.11 Designed filter banks.

Filter Banks	Coefficients for the Numerator	Coefficients for the Denominator
Bandpass filter 0	0.0031954934, 0, −0.0031954934	1, −1.9934066716, 0.9936090132
Bandpass filter 1	0.0063708102, 0, −0.0063708102	1, −1.9864516324, 0.9872583796
Bandpass filter 2	0.0126623878, 0, −0.0126623878	1, −1.9714693192, 0.9746752244
Bandpass filter 3	0.0310900413, 0, −0.0310900413	1, −1.9181849043, 0.9378199174
Bandpass filter 4	0.0746111954, 0, −0.0746111954	1, −1.7346085867, 0.8507776092
Bandpass filter 5	0.1663862883, 0, −0.1663862884	1, −1.0942477187, 0.6672274233
Bandpass filter 6	0.3354404899, 0, −0.3354404899	1, 0.7131366534, 0.3291190202

The magnitude response for each filter bank is plotted in Figure 8.25 for design verification. As shown in the figure, after careful examination, the magnitude response of each filter band meets the design specification. We will perform simulation next.

Simulation in the MATLAB environment is based on the following setting. The audio test signal having frequency components of 100 Hz, 200 Hz, 400 Hz, 1,000 Hz, 2,500 Hz, 6,000 Hz, and 15,000 Hz is generated from Equation (8.36):

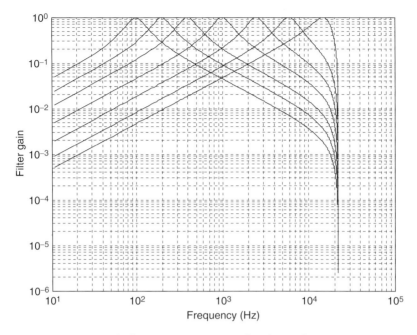

FIGURE 8.25 Magnitude frequency responses for the audio equalizer.

$$\begin{aligned}x(n) = {} & \sin(200\pi n/44100) + \sin(400\pi n/44100 + \pi/14) \\ & + \sin(800\pi n/44100 + \pi/7) + \sin(2000\pi n/44100 + 3\pi/14) \\ & + \sin(5000\pi n/44100 + 2\pi/7) + \sin(12000\pi n/44100 + 5\pi/14) \\ & + \sin(30000\pi n/44100 + 3\pi/7) \end{aligned} \qquad (8.36)$$

The gains set for the filter banks are:

$$g_0 = 10; \quad g_1 = 10; \quad g_2 = 0; \quad g_3 = 0; \quad g_4 = 0; \quad g_5 = 10; \quad g_6 = 10.$$

After simulation, we notice that the frequency components at 100 Hz, 200 Hz, 6,000 Hz, and 15,000 Hz will be boosted by $20 \bullet \log_{10} 10 = 20$ dB. The top plot in Figure 8.26 shows the spectrum for the audio test signal, while the bottom plot depicts the spectrum for the equalized audio test signal. As shown in the plots, before audio digital equalization, the spectral peaks at all bands are at the same level; after audio digital equalization, the frequency components at bank 0, bank 1, bank 5, and bank 6 are amplified. Therefore, as we expected, the operation of the digital equalizer boosts the low frequency components and the high frequency components. The MATLAB list for the simulation is shown in Program 8.11.

8.5 Application: Digital Audio Equalizer

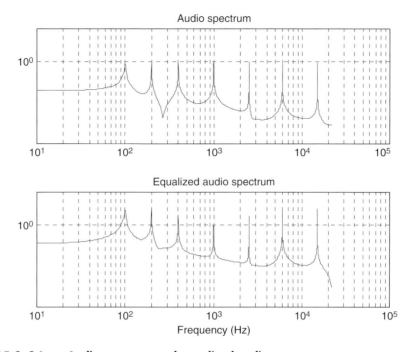

FIGURE 8.26 Audio spectrum and equalized audio spectrum.

Program 8.11. MATLAB program for the digital audio equalizer.

```
close all; clear all
% Filter coefficients (Butterworth type designed using the BLT)
B0 = [0.0031954934 0 −0.0031954934]; A0 = [1.0000000000 −1.9934066716 0.9936090132];
B1 = [0.0063708102 0 −0.0063708102]; A1 = [1.0000000000 −1.9864516324 0.9872583796];
B2 = [0.0126623878 0 −0.0126623878]; A2 = [1.0000000000 −1.9714693192 0.9746752244];
B3 = [0.0310900413 0 −0.0310900413]; A3 = [1.0000000000 −1.9181849043 0.9378199174];
B4 = [ 0.0746111954 0.000000000 −0.0746111954];
A4 = [1.0000000000 −1.7346085867 0.8507776092];
B5 = [0.1663862883 0.0000000000 −0.1663862884];
A5 = [1.0000000000 −1.0942477187 0.6672274233];
B6 = [0.3354404899 0.0000000000 −0.3354404899];
A6 = [1.0000000000 0.7131366534 0.3291190202];
[h0,f] = freqz(B0,A0,2048,44100);
[h1,f] = freqz(B1,A1,2048,44100);
[h2,f] = freqz(B2,A2,2048,44100);
[h3,f] = freqz(B3,A3,2048,44100);
[h4,f] = freqz(B4,A4,2048,44100);
```

(*Continued*)

```
[h5,f] = freqz(B5,A5,2048,44100);
[h6,f] = freqz(B6,A6,2048,44100);
loglog(f,abs(h0),f,abs(h1),f,abs(h2), ...
f,abs(h3),f,abs(h4),f,abs(h5),f,abs(h6));
xlabel('Frequency (Hz)');
ylabel('Filter Gain');grid
axis([10 10^5 10^(-6) 1]);
figure(2)
g0 = 10;g1 = 10;g2 = 0;g3 = 0;g4 = 0;g5 = 10;g6 = 10;
p0 = 0;p1 = pi/14;p2 = 2*p1;p3 = 3*p1;p4 = 4*p1;p5 = 5*p1;p6 = 6*p1;
n = 0:1:20480;           % Indices of samples
fs = 44100;              % Sampling rate
x = sin(2*pi*100*n/fs) + sin(2*pi*200*n/fs + p1) + ...
sin(2*pi*400*n/fs + p2) + sin(2*pi*1000*n/fs + p3) + ...
sin(2*pi*2500*n/fs + p4) + sin(2*pi*6000*n/fs + p5) + ...
sin(2*pi*15000*n/fs + p6);        % Generate test audio signals
y0 = filter(B0,A0,x);             % Bandpass filter 0
y1 = filter(B1,A1,x);             % Bandpass filter 1
y2 = filter(B2,A2,x);             % Bandpass filter 2
y3 = filter(B3,A3,x);             % Bandpass filter 3
y4 = filter(B4,A4,x);             % Bandpass filter 4
y5 = filter(B5,A5,x);             % Bandpass filter 5
y6 = filter(B6,A6,x);             % Bandpass filter 6
y = g0.*y0 + g1.*y1 + g2.*y2 + g3.*y3 + g4.*y4 + g5.*y5 + g6.*y6 + x; % Equalizer output
N = length(x);
Axk = 2*abs(fft(x))/N;Axk(1) = Axk(1)/2; % One-sided amplitude spectrum of the input
f = [0:N/2]*fs/N;
subplot(2,1,1);loglog(f,Axk(1:N/2 + 1));
title('Audio spectrum');
axis([10 100000 0.00001 100]);grid;
Ayk = 2*abs(fft(y))/N; Ayk(1) = Ayk(1)/2; % One-sided amplitude spectrum of the output
subplot(2,1,2);loglog(f,Ayk(1:N/2+1));
xlabel('Frequency (Hz)');
title('Equalized audio spectrum');
axis([10 100000 0.00001 100]);grid;
```

8.6 Impulse Invariant Design Method

We illustrate the concept of the impulse invariant design in Figure 8.27. Given the transfer function of a designed analog filter, an analog impulse response can be easily found by the inverse Laplace transform of the transfer function. To replace the analog filter by the equivalent digital filter, we apply an approximation in time domain in which the digital impulse response must be equivalent to the analog impulse response. Therefore, we can sample the analog impulse

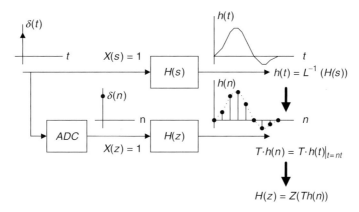

FIGURE 8.27 Impulse invariant design method.

response to get the digital impulse response and take the z-transform of the sampled analog impulse response to obtain the transfer function of the digital filter.

The analog impulse response can be achieved by taking the inverse Laplace transform of the analog filter $H(s)$, that is,

$$h(t) = L^{-1}(H(s)). \tag{8.37}$$

Now, if we sample the analog impulse response with a sampling interval of T and use T as a scale factor, it follows that

$$T \cdot h(n) = T \cdot h(t)|_{t=nT}, \; n \geq 0. \tag{8.38}$$

Taking the z-transform on both sides of Equation (8.38) yields the digital filter as

$$H(z) = Z[T \cdot h(n)]. \tag{8.39}$$

The effect of the scale factor T in Equation (8.38) can be explained as follows. We approximate the area under the curve specified by the analog impulse function $h(t)$ using a digital sum given by

$$\text{area} = \int_0^\infty h(t)dt \approx T \cdot h(0) + T \cdot h(1) + T \cdot h(2) + \cdots \tag{8.40}$$

Note that the area under the curve indicates the DC gain of the analog filter while the digital sum in Equation (8.40) is the DC gain of the digital filter.

The rectangular approximation is used, since each sample amplitude is multiplied by the sampling interval T. Due to the interval size for approximation in practice, we cannot guarantee that the digital sum has exactly the same value as the one from the integration unless the sampling interval T in Equation (8.40) approaches zero. This means that the higher the sampling rate—that is, the smaller the sampling interval—the more accurately the digital filter gain matches the analog filter gain. Hence, in practice, we need to further apply gain scaling for adjustment if it is a requirement. We look at the following examples.

Example 8.15.

Consider the following Laplace transfer function:

$$H(s) = \frac{2}{s+2}.$$

a. Determine $H(z)$ using the impulse invariant method if the sampling rate $f_s = 10\,\text{Hz}$.

b. Use MATLAB to plot

1. the magnitude response $|H(f)|$ and the phase response $\varphi(f)$ with respect to $H(s)$ for the frequency range from 0 to $f_s/2$ Hz.

2. the magnitude response $|H(e^{j\Omega})| = |H(e^{j2\pi f/T})|$ and the phase response $\varphi(f)$ with respect to $H(z)$ for the frequency range from 0 to $f_s/2$ Hz.

Solution:

a. Taking the inverse Laplace transform of the analog transfer function, the impulse response is found to be

$$h(t) = L^{-1}\left[\frac{2}{s+2}\right] = 2e^{-2t}u(t).$$

Sampling the impulse response $h(t)$ with $T = 1/f_s = 0.1$ second, we have

$$Th(n) = T2e^{-2nT}u(n) = 0.2e^{-0.2n}u(n).$$

Using the z-transform table in Chapter 5, we yield

$$Z[e^{-an}u(n)] = \frac{z}{z - e^{-a}}.$$

And noting that $e^{-a} = e^{-0.2} = 0.8187$, the digital filter transfer function $H(z)$ is finally given by

$$H(z) = \frac{0.2z}{z - 0.8187} = \frac{0.2}{1 - 0.8187z^{-1}}.$$

b. The MATLAB list is Program 8.12. The first and third plots in Figure 8.28 show comparisons of the magnitude and phase frequency responses. The shape of the magnitude response (first plot) closely matches that of the analog filter, while the phase response (third plot) differs from the analog phase response in this example.

Program 8.12. MATLAB program for Example 8.15.

```
%Example 8.15.
% Plot the magnitude responses |H(s)| and |H(z)|
% For the Laplace transfer function H(s)
f = 0:0.1:5=0.1%Frequency range and sampling interval
w = 2*pi*f;            %Frequency range in rad/sec
hs = freqs([2], [1 2],w);       % Analog frequency response
phis = 180*angle(hs)/pi;
% For the z-transfer function H(z)
hz = freqz([0.2],[1 -0.8187],length(w)); % Digital frequency response
hzscale = freqz([0.1813],[1 -0.8187],length(w)); % Scaled digital mag. response
phiz = 180*angle(hz)/pi;
%Plot magnitude and phase responses.
subplot(3,1,1), plot(f,abs(hs), 'kx',f, abs(hz), 'k-'),grid;axis([0 5 0 1.2]);
xlabel('Frequency (Hz)');ylabel('Mag. Responses')
subplot(3,1,2), plot(f,abs(hs),'kx',f, abs(hz_scale),'k-'),grid;axis([0 5 0 1.2]);
xlabel('Frequency (Hz)');ylabel('Scaled Mag. Responses')
subplot(3,1,3), plot(f,phis,'kx',f, phiz, 'k-');grid;
xlabel('Frequency (Hz)');ylabel('Phases (deg.)');
```

The filter DC gain is given by

$$H(e^{j\Omega})|_{\Omega=0} = H(1) = 1.1031.$$

We can further scale the filter to have a unit gain of

$$H(z) = \frac{1}{1.1031} \frac{0.2}{1-0.8187z^{-1}} = \frac{0.1813}{1-0.8187z^{-1}}.$$

The scaled magnitude frequency response is shown in the middle plot along with that of the analog filter in Figure 8.28, where the magnitudes are matched very well below 1.8 Hz.

Example 8.15 demonstrates the design procedure using the impulse invariant design. The filter performance depends on the sampling interval (Lynn and Fuerst, 1999). As shown in Figure 8.27, the analog impulse response $h(t)$ is not a band-limited signal whose frequency components generally are larger than the Nyquist limit (folding frequency); hence, sampling $h(t)$ could cause aliasing.

354   **8** INFINITE IMPULSE RESPONSE FILTER DESIGN

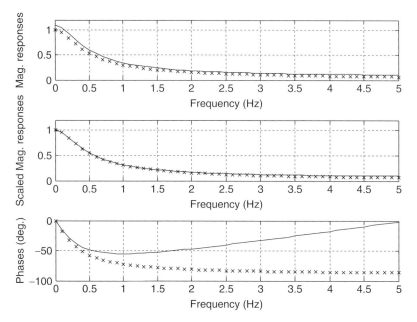

FIGURE 8.28   Frequency responses. Line of x's, frequency responses of the analog filter; solid line, frequency responses of the designed digital filter.

Figure 8.29(a) shows the analog impulse response $Th(t)$ in Example 8.15 and its sampled version $Th(nT)$, where the sampling interval is 0.125 second. The analog filter and digital filter magnitude responses are plotted in Figure 8.29(b). Aliasing occurs, since the impulse response contains the frequency components beyond the Nyquist limit, that is, 4 Hz, in this case. Furthermore, using the lower sampling rate of 8 Hz causes less accuracy in the digital filter magnitude response, so that more aliasing develops.

Figure 8.29(c) shows the analog impulse response and its sampled version using a higher sampling rate of 16 Hz. Figure 8.29(d) displays the more accurate magnitude response of the digital filter. Hence, we can obtain the reduced aliasing level. Note that aliasing cannot be avoided, due to sampling of the analog impulse response. The only way to reduce the aliasing is to use a higher sampling frequency or design a filter with a very low cutoff frequency to reduce the aliasing to a minimum level.

Investigation of the sampling interval effect leads us to the following conclusions. Note that the analog impulse response for the highpass filter or band reject filter contains frequency components at the maximum level at the Nyquist limit (folding frequency), even assuming that the sampling rate is much higher than the cutoff frequency of a highpass filter or the upper cutoff frequency of a

8.6 Impulse Invariant Design Method

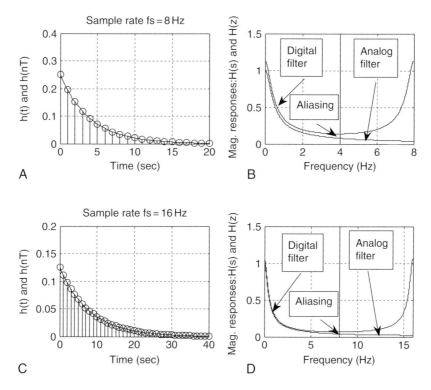

FIGURE 8.29 Sampling interval effect in the impulse invariant IIR filter design.

band reject filter. Hence, sampling the analog impulse response always produces the maximum aliasing level. Without using an additional anti-aliasing filter, the impulse invariant method alone cannot be used for designing the highpass filter or band reject filter.

Instead, in practice, we should apply the BLT design method. The impulse invariant design method is only appropriate for designing a lowpass filter or bandpass filter with a sampling rate much larger than the cutoff frequency of the lowpass filter or the upper cutoff frequency of the bandpass filter.

Next, let us focus on the second-order filter design via Example 8.16.

Example 8.16.

Consider the following Laplace transfer function:

$$H(s) = \frac{s}{s^2 + 2s + 5}.$$

a. Determine $H(z)$ using the impulse invariant method if the sampling rate $f_s = 10\,\text{Hz}$.

b. Use MATLAB to plot:

1. the magnitude response $|H(f)|$ and the phase response $\varphi(f)$ with respect to $H(s)$ for the frequency range from 0 to $f_s/2\,\text{Hz}$.

2. the magnitude response $|H(e^{j\Omega})| = |H(e^{j2\pi fT})|$ and the phase response $\varphi(f)$ with respect to $H(z)$ for the frequency range from 0 to $f_s/2\,\text{Hz}$.

Solution:

a. Since $H(s)$ has complex poles located at $s = -1 \pm 2j$, we can write it in a quadratic form as

$$H(s) = \frac{s}{s^2 + 2s + 5} = \frac{s}{(s+1)^2 + 2^2}.$$

We can further write the transfer function as

$$H(s) = \frac{(s+1) - 1}{(s+1)^2 + 2^2} = \frac{(s+1)}{(s+1)^2 + 2^2} - 0.5 \times \frac{2}{(s+1)^2 + 2^2}.$$

From the Laplace transform table (Appendix B), the analog impulse response can easily be found as

$$h(t) = e^{-t}\cos(2t)u(t) - 0.5e^{-t}\sin(2t)u(t).$$

Sampling the impulse response $h(t)$ using a sampling interval $T = 0.1$ and using the scale factor of $T = 0.1$, we have

$$Th(n) = Th(t)|_{t=nT} = 0.1e^{-0.1n}\cos(0.2n)u(n) - 0.05e^{-0.1n}\sin(0.2n)u(n).$$

Applying the z-transform (Chapter 5) leads to

$$H(z) = Z\left[0.1e^{-0.1n}\cos(0.2n)u(n) - 0.05e^{-0.1n}\sin(0.2n)u(n)\right]$$

$$= \frac{0.1z(z - e^{-0.1}\cos(0.2))}{z^2 - 2e^{-0.1}\cos(0.2)z + e^{-0.2}} - \frac{0.05e^{-0.1}\sin(0.2)z}{z^2 - 2e^{-0.1}\cos(0.2)z + e^{-0.2}}.$$

After algebra simplification, we obtain the second-order digital filter as

$$H(z) = \frac{0.1 - 0.09767z^{-1}}{1 - 1.7735z^{-1} + 0.8187z^{-2}}.$$

b. The magnitude and phase frequency responses are shown in Figure 8.30, and MATLAB Program 8.13 is given. The passband gain of the digital filter is higher than that of the analog filter, but their shapes are the same.

Program 8.13. MATLAB program for Example 8.16.

```
%Example 8.16
% Plot the magnitude responses |H(s)| and |H(z)|
% For the Laplace transfer function H(s)
f = 0:0.1:5;T = 0.1; % Initialize analog frequency range in Hz and sampling interval
w = 2*pi*f;          % Convert the frequency range to radians/second
hs = freqs([1 0], [1 2 5],w);        % Calculate analog filter frequency responses
phis = 180*angle(hs)/pi;
% For the z-transfer function H(z)
% Calculate digital filter frequency responses
hz = freqz([0.1 -0.09766],[1 -1.7735 0.8187],length(w));
phiz = 180*angle(hz)/pi;
% Plot magnitude and phase responses
subplot(2,1,1), plot(f,abs(hs),'x',f, abs(hz),'-'),grid;
xlabel('Frequency (Hz)');ylabel('Magnitude Responses')
subplot(2,1,2), plot(f,phis, 'x',f, phiz,'-');grid;
xlabel('Frequency (Hz)');ylabel('Phases (degrees)')
```

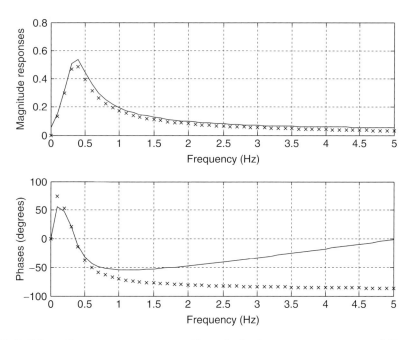

FIGURE 8.30 Frequency responses. Line of x's, frequency responses of the analog filter; solid line, frequency responses of the designed digital filter.

8.7 Pole-Zero Placement Method for Simple Infinite Impulse Response Filters

This section introduces a pole-zero placement method for a simple IIR filter design. Let us first examine effects of the pole-zero placement on the magnitude response in the z-plane shown in Figure 8.31.

In the z-plane, when we place a pair of complex conjugate zeros at a given point on the unit circle with an angle θ (usually we do), we will have a numerator factor of $(z - e^{j\theta})(z - e^{-j\theta})$ in the transfer function. Its magnitude contribution to the frequency response at $z = e^{j\Omega}$ is $(e^{j\Omega} - e^{j\theta})(e^{j\Omega} - e^{-j\theta})$. When $\Omega = \theta$, the magnitude will reach zero, since the first factor $(e^{j\theta} - e^{j\theta}) = 0$ gives a zero magnitude. When a pair of complex conjugate poles are placed at a given point within the unit circle, we have a denominator factor of $(z - re^{j\theta})(z - re^{-j\theta})$, where r is the radius chosen to be less than and close to 1 to place the poles inside the unit circle. The magnitude contribution to the frequency response at $\Omega = \theta$ will rise to a large magnitude, since the first factor $(e^{j\theta} - re^{j\theta}) = (1 - r)e^{j\theta}$ gives a small magnitude of $1 - r$, which is the length between the pole and the unit circle at the angle $\Omega = \theta$. Note that the magnitude of $e^{j\theta}$ is 1.

Therefore, we can reduce the magnitude response using zero placement, while we increase the magnitude response using pole placement. Placing a combination of poles and zeros will result in different frequency responses,

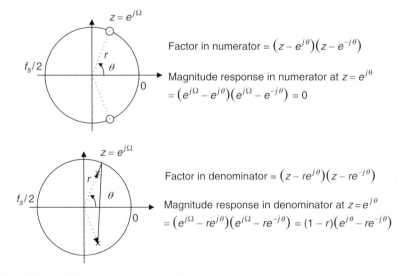

FIGURE 8.31 Effects of pole-zero placement on the magnitude response.

such as lowpass, highpass, bandpass, and bandstop. The method is intuitive and approximate. However, it is easy to compute filter coefficients for simple IIR filters. Here, we describe the design procedures for second-order bandpass and bandstop filters, as well as first-order lowpass and highpass filters. (For details of derivations, readers are referred to Lynn and Fuerst [1999]). Practically, the pole-zero placement method has good performance when the bandpass and bandstop filters have very narrow bandwidth requirements and the lowpass and highpass filters have either very low cutoff frequencies close to the DC or very high cutoff frequencies close to the folding frequency (the Nyquist limit).

8.7.1 Second-Order Bandpass Filter Design

Typical pairs of poles and zeros for a bandpass filter are placed in Figure 8.32. Poles are complex conjugate, with the magnitude r controlling the bandwidth and the angle θ controlling the center frequency. The zeros are placed at $z = 1$, corresponding to DC, and $z = -1$, corresponding to the folding frequency.

The poles will raise the magnitude response at the center frequency while the zeros will cause zero gains at DC (zero frequency) and at the folding frequency.

The following equations give the bandpass filter design formulas using pole-zero placement:

$$r \approx 1 - (BW_{3dB}/f_s) \times \pi, \text{ good for } 0.9 \leq r < 1 \tag{8.41}$$

$$\theta = \left(\frac{f_0}{f_s}\right) \times 360^0 \tag{8.42}$$

$$H(z) = \frac{K(z-1)(z+1)}{(z-re^{j\theta})(z-re^{-j\theta})} = \frac{K(z^2-1)}{(z^2-2rz\cos\theta+r^2)}, \tag{8.43}$$

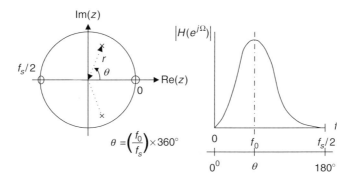

FIGURE 8.32 Pole-zero placement for a second-order narrow bandpass filter.

where K is a scale factor to adjust the bandpass filter to have a unit passband gain given by

$$K = \frac{(1-r)\sqrt{1-2r\cos 2\theta + r^2}}{2|\sin\theta|}. \tag{8.44}$$

Example 8.17.

A second-order bandpass filter is required to satisfy the following specifications:

- Sampling rate = 8,000 Hz
- 3 dB bandwidth: $BW = 200$ Hz
- Narrow passband centered at $f_0 = 1,000$ Hz
- Zero gain at 0 Hz and 4,000 Hz.

a. Find the transfer function using the pole-zero placement method.

Solution:

a. First, we calculate the required magnitude of the poles.

$$r = 1 - (200/8000)\pi = 0.9215,$$

which is a good approximation. Use the center frequency to obtain the angle of the pole location:

$$\theta = \left(\frac{1000}{8000}\right) \times 360 = 45^0.$$

Compute the unit-gain scale factor as

$$K = \frac{(1-0.9215)\sqrt{1-2\times 0.9215 \times \cos 2 \times 45^0 + 0.9215^2}}{2|\sin 45^0|} = 0.0755.$$

Finally, the transfer function is given by

$$H(z) = \frac{0.0755(z^2-1)}{(z^2 - 2\times 0.9215z\cos 45^0 + 0.9215^2)} = \frac{0.0755 - 0.0755z^{-2}}{1 - 1.3031z^{-1} + 0.8491z^{-2}}.$$

8.7.2 Second-Order Bandstop (Notch) Filter Design

For this type of filter, the pole placement is the same as the bandpass filter (Fig. 8.33). The zeros are placed on the unit circle with the same angles with respect to the poles. This will improve passband performance. The magnitude

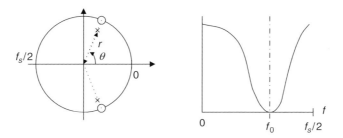

FIGURE 8.33 Pole-zero placement for a second-order notch filter.

and the angle of the complex conjugate poles determine the 3 dB bandwidth and the center frequency, respectively.

Design formulas for bandstop filters are given in the following equations:

$$r \approx 1 - (BW_{3dB}/f_s) \times \pi, \text{ good for } 0.9 \leq r < 1 \quad (8.45)$$

$$\theta = \left(\frac{f_0}{f_s}\right) \times 360^0 \quad (8.46)$$

$$H(z) = \frac{K(z - e^{j\theta})(z + e^{-j\theta})}{(z - re^{j\theta})(z - re^{-j\theta})} = \frac{K(z^2 - 2z\cos\theta + 1)}{(z^2 - 2rz\cos\theta + r^2)}. \quad (8.47)$$

The scale factor to adjust the bandstop filter to have a unit passband gain is given by

$$K = \frac{(1 - 2r\cos\theta + r^2)}{(2 - 2\cos\theta)}. \quad (8.48)$$

Example 8.18.

A second-order notch filter is required to satisfy the following specifications:

- Sampling rate = 8,000 Hz
- 3 dB bandwidth: $BW = 100$ Hz
- Narrow passband centered at $f_0 = 1,500$ Hz.

a. Find the transfer function using the pole-zero placement approach.

Solution:

a. We first calculate the required magnitude of the poles:

$$r \approx 1 - (100/8000) \times \pi = 0.9607,$$

which is a good approximation. We use the center frequency to obtain the angle of the pole location:

$$\theta = \left(\frac{1500}{8000}\right) \times 360^0 = 67.5^0.$$

The unit-gain scale factor is calculated as

$$K = \frac{(1 - 2 \times 0.9607 \cos 67.5^0 + 0.9607^2)}{(2 - 2\cos 67.5^0)} = 0.9620.$$

Finally, we obtain the transfer function:

$$H(z) = \frac{0.9620(z^2 - 2z\cos 67.5^0 + 1)}{(z^2 - 2 \times 0.9607z \cos 67.5^0 + 0.9607^2)}$$

$$= \frac{0.9620 - 0.7363z^{-1} + 0.9620z^{-2}}{1 - 0.7353z^{-1} + 0.9229}.$$

8.7.3 First-Order Lowpass Filter Design

The first-order pole-zero placement can be operated in two cases. The first situation is when the cutoff frequency is less than $f_s/4$. Then the pole-zero placement is shown in Figure 8.34.

As shown in Figure 8.34, the pole $z = \alpha$ is placed in the real axis. The zero is placed at $z = -1$ to ensure zero gain at the folding frequency (Nyquist limit). When the cutoff frequency is above $f_s/4$, the pole-zero placement is adopted as shown in Figure 8.35.

Design formulas for lowpass filters using the pole-zero placement are given in the following equations:

When $f_c < f_s/4$, $\alpha \approx 1 - 2 \times (f_c/f_s) \times \pi$, good for $0.9 \leq r < 1$. (8.49)

When $f_c > f_s/4$, $\alpha \approx -(1 - \pi + 2 \times (f_c/f_s) \times \pi)$, good for $-1 < r \leq -0.9$. (8.50)

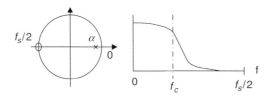

FIGURE 8.34 Pole-zero placement for the first-order lowpass filter with $f_c < f_s/4$.

8.7 Pole-Zero Placement Method for Simple Infinite Impulse Response Filters

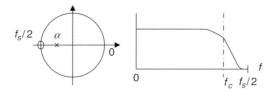

FIGURE 8.35 Pole-zero placement for the first-order lowpass filter with $f_c > f_s/4$.

The transfer function is

$$H(z) = \frac{K(z+1)}{(z-\alpha)}, \qquad (8.51)$$

and the unit passband gain scale factor is given by

$$K = \frac{(1-\alpha)}{2}. \qquad (8.52)$$

Example 8.19.

A first-order lowpass filter is required to satisfy the following specifications:

- Sampling rate = 8,000 Hz
- 3 dB cutoff frequency: $f_c = 100$ Hz
- Zero gain at 4,000 Hz.

   a. Find the transfer function using the pole-zero placement method.

Solution:

   a. Since the cutoff frequency of 100 Hz is much less than $f_s/4 = 2,000$ Hz, we determine the pole as

$$\alpha \approx 1 - 2 \times (100/8000) \times \pi = 0.9215,$$

which is above 0.9. Hence, we have a good approximation. The unit-gain scale factor is calculated by

$$K = \frac{(1 - 0.9215)}{2} = 0.03925.$$

Last, we can develop the transfer function as

$$H(z) = \frac{0.03925(z+1)}{(z-0.9215)} = \frac{0.03925 + 0.03925z^{-1}}{1 - 0.9215z^{-1}}.$$

Note that we can also determine the unit-gain factor K by substituting $z = e^{j0} = 1$ to the transfer function $H(z) = \frac{(z+1)}{(z-\alpha)}$, then find a DC gain. Set the scale factor to be a reciprocal of the DC gain. This can be easily done, that is,

$$DC\ gain = \left.\frac{z+1}{z-0.9215}\right|_{z=1} = \frac{1+1}{1-0.9215} = 25.4777.$$

Hence, $K = 1/25.4777 = 0.03925$.

8.7.4 First-Order Highpass Filter Design

Similar to the lowpass filter design, the pole-zero placements for first-order highpass filters in two cases are shown in Figures 8.36a and 8.36b.

Formulas for designing highpass filters using the pole-zero placement are listed in the following equations:

When $f_c < f_s/4$, $\alpha \approx 1 - 2 \times (f_c/f_s) \times \pi$, good for $0.9 \leq r < 1$. (8.53)

When $f_c > f_s/4$, $\alpha \approx -(1 - \pi + 2 \times (f_c/f_s) \times \pi)$, good for $-1 < r \leq -0.9$ (8.54)

$$H(z) = \frac{K(z-1)}{(z-\alpha)} \tag{8.55}$$

$$K = \frac{(1+\alpha)}{2}. \tag{8.56}$$

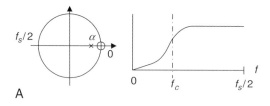

FIGURE 8.36A Pole-zero placement for the first-order highpass filter with $f_c < f_s/4$.

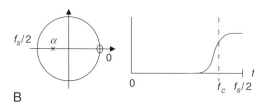

FIGURE 8.36B Pole-zero placement for the first-order highpass filter with $f_c > f_s/4$.

Example 8.20.

A first-order highpass filter is required to satisfy the following specifications:

- Sampling rate = 8,000 Hz
- 3 dB cutoff frequency: $f_c = 3,800$ Hz
- Zero gain at 0 Hz.

a. Find the transfer function using the pole-zero placement method.

Solution:

a. Since the cutoff frequency of 3,800 Hz is much larger than $f_s/4 = 2,000$ Hz, we determine the pole as

$$\alpha \approx -(1 - \pi + 2 \times (3800/8000) \times \pi) = -0.8429,$$

The unit-gain scale factor and transfer functions are obtained as

$$K = \frac{(1 - 0.8429)}{2} = 0.07854$$

$$H(z) = \frac{0.07854(z - 1)}{(z + 0.8429)} = \frac{0.07854 - 0.07854z^{-1}}{1 + 0.8429z^{-1}}.$$

Note that we can also determine the unit-gain scale factor K by substituting $z = e^{j180°} = -1$ into the transfer function $H(z) = \frac{(z-1)}{(z-\alpha)}$, finding a passband gain at the Nyquist limit $f_s/2 = 4,000$ Hz. We then set the scale factor to be a reciprocal of the passband gain. That is,

$$passband\ gain = \left.\frac{z - 1}{z + 0.8429}\right|_{z=1} = \frac{-1 - 1}{-1 + 0.8429} = 12.7307.$$

Hence, $K = 1/12.7307 = 0.07854$.

8.8 Realization Structures of Infinite Impulse Response Filters

In this section, we will realize the designed IIR filter using direct form I and direct form II. We will then realize a higher-order IIR filter using a cascade form.

8.8.1 Realization of Infinite Impulse Response Filters in Direct-Form I and Direct-Form II

Example 8.21.

a. Realize the first-order digital highpass Butterworth filter

$$H(z) = \frac{0.1936 - 0.1936z^{-1}}{1 + 0.6128z^{-1}}$$

using a direct form I.

Solution:

a. From the transfer function, we can identify

$$b_0 = 0.1936, \ b_1 = -0.1936, \text{ and } a_1 = 0.6128.$$

Applying the direct form I developed in Chapter 6 results in the diagram in Figure 8.37.

The digital signal processing (DSP) equation for implementation is then given by

$$y(n) = -0.6128y(n-1) + 0.1936x(n) - 0.1936x(n-1).$$

Program 8.14 lists the MATLAB implementation.

Program 8.14. *m*-File for Example 8.21.

```
%Sample MATLAB code
sample = 2:2:20;            %Input test array
x = [0 0];          % Input buffer [x(n) x(n − 1)...]
y = [0 0];          % Output buffer [y(n) y(n − 1)...]
b = [0.1936 −0.1936];       % Numerator coefficients [b0 b1...]
a = [1 0.6128];     % Denominator coefficients [1 a0 a1...]
for n = 1:1:length(sample)% Processing loop
for k = 2:−1:2
x(k) = x(k − 1);            % Shift the input by one sample
y(k) = y(k − 1);            % Shift the output by one sample
end
x(1) = sample(n);           % Get new sample
y(1) = 0;           % Digital filtering
for k = 1:1:2
y(1) = y(1) + x(k)*b(k);
end
for k = 2:2
y(1) = y(1) − a(k)*y(k);
end
out(n) = y(1);              %Output the filtered sample to the output array
end
out
```

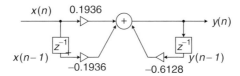

FIGURE 8.37 Realization of IIR filter in Example 8.21 in direct form I.

Example 8.22.

a. Realize the following digital filter using a direct form II:

$$H(z) = \frac{0.7157 + 1.4314z^{-1} + 0.7151z^{-2}}{1 + 1.3490z^{-1} + 0.5140z^{-2}}.$$

Solution:

a. First, we can identify

$$b_0 = 0.7157, \ b_1 = 1.4314, \ b_2 = 0.7151$$
$$\text{and } a_1 = 1.3490, \ a_2 = 0.5140.$$

Applying the direct form II developed in Chapter 6 leads to Figure 8.38. There are two difference equations required for implementation:

$$w(n) = x(n) - 1.3490w(n-1) - 0.5140w(n-2) \text{ and}$$
$$y(n) = 0.7157w(n) + 1.4314w(n-1) + 0.7157w(n-2).$$

The MATLAB implementation is listed in Program 8.15.

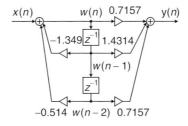

FIGURE 8.38 Realization of IIR filter in Example 8.22 in direct form II.

Program 8.15. *m*-File for Example 8.22.

```
%Sample MATLAB code
sample = 2:2:20;          % Input test array
x = [0];                  %Input buffer [x(n)]
y = [0];                  %Output buffer [y(n)]
w = [0 0 0];              % Buffer for w(n) [w(n)w(n − 1)...]
b = [0.7157 1.4314 0.7157];        % Numerator coefficients [b0 b1...]
a = [1 1.3490 0.5140];             % Denominator coefficients [1 a1 a2...]
for n = 1:1:length(sample)% Processing loop
for k = 3:−1:2
w(k) = w(k − 1);          %Shift w(n) by one sample
end
x(1) = sample(n);         % Get new sample
w(1) = x(1);              % Perform IIR filtering
for k = 2:1:3
w(1) = w(1) − a(k)*w(k);
end
y(1) = 0;                 % Perform FIR filtering
for k = 1:1:3
y(1) = y(1) + b(k)*w(k);
end
out(n) = y(1);            % Send the filtered sample to the output array
end
out
```

8.8.2 Realization of Higher-Order Infinite Impulse Response Filters via the Cascade Form

Example 8.23.

Given a fourth-order filter transfer function designed as

$$H(z) = \frac{0.5108z^2 + 1.0215z + 0.5108}{z^2 + 0.5654z + 0.4776} \times \frac{0.3730z^2 + 0.7460z + 0.3730}{z^2 + 0.4129z + 0.0790},$$

a. Realize the digital filter using the cascade (series) form via second-order sections.

Solution:

a. Since the filter is designed using the cascade form, we have two sections of the second-order filters, whose transfers are

$$H_1(z) = \frac{0.5108z^2 + 1.0215z + 0.5108}{z^2 + 0.5654z + 0.4776} = \frac{0.5180 + 1.0215z^{-1} + 0.5108z^{-2}}{1 + 0.5654z^{-1} + 0.4776z^{-2}}$$

and

$$H_2(z) = \frac{0.3730z^2 + 0.7460z + 0.3730}{z^2 + 0.4129z + 0.0790} = \frac{0.3730 + 0.7460z^{-1} + 0.3730z^{-2}}{1 + 0.4129z^{-1} + 0.0790z^{-2}}.$$

Each filter section is developed using the direct form I, shown in Figure 8.39.

There are two sets of DSP equations for implementation of the first and second sections, respectively.

First section:

$$y_1(n) = -0.5654y_1(n-1) - 0.4776y_1(n-2) \\ + 0.5108x(n) + 1.0215x(n-1) + 0.5108x(n-2)$$

Second section:

$$y(n) = -0.4129y(n-1) - 0.0790y(n-2) \\ + 0.3730y_1(n) + 0.7460y_1(n-1) + 0.3730y_1(n-2).$$

Again, after we use the direct form II for realizing each second-order filter, the realization shown in Figure 8.40 is developed.

The difference equations for the implementation of the first section are:

$$w_1(n) = x(n) - 0.5654w_1(n-1) - 0.4776w_1(n-2) \\ y_1(n) = 0.5108w_1(n) + 1.0215w_1(n-1) + 0.5108w_1(n-2).$$

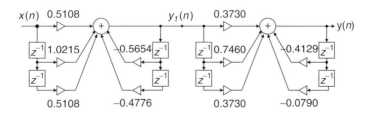

FIGURE 8.39 Cascade realization of IIR filter in Example 8.23 in direct form I.

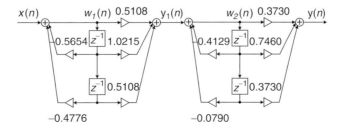

FIGURE 8.40 Cascade realization of IIR filter in Example 8.23 in direct form II.

The difference equations for the implementation of the second section are:

$$w_2(n) = y_1(n) - 0.4129 w_2(n-1) - 0.0790 w_2(n-2)$$
$$y(n) = 0.3730 w_2(n) + 0.7460 w_2(n-1) + 0.3730 w_2(n-2).$$

Note that for both direct form I and direct form II, the output from the first filter section becomes the input for the second filter section.

8.9 Application: 60-Hz Hum Eliminator and Heart Rate Detection Using Electrocardiography

Hum noise created by poor power supplies, transformers, or electromagnetic interference sourced by a main power supply is characterized by a frequency of 60 Hz and its harmonics. If this noise interferes with a desired audio or biomedical signal (e.g., in electrocardiography [ECG]), the desired signal could be corrupted. The corrupted signal is useless without signal processing. It is sufficient to eliminate the 60-Hz hum frequency with its second and third harmonics in most practical applications. We can complete this by cascading with notch filters having notch frequencies of 60 Hz, 120 Hz, and 180 Hz, respectively. Figure 8.41 depicts the functional block diagram.

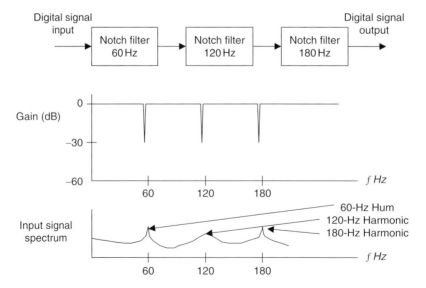

FIGURE 8.41 (*Top*) 60-Hz hum eliminator; (*middle*) the filter frequency response of the eliminator; (*bottom*) the input signal spectrum corrupted by the 60-Hz hum and its second and third harmonics.

8.9 Application: 60-Hz Hum Eliminator and Heart Rate Detection Using Electrocardiography

Now let us apply the 60-Hz hum eliminator to an ECG recording system. ECG is a small electrical signal captured from an ECG sensor. The ECG signal is produced by the activity of the human heart, thus can be used for heart rate detection, fetal monitoring, and diagnostic purposes. The single pulse of the ECG is depicted in Figure 8.42, which shows that the ECG signal is characterized by five peaks and valleys, labeled P, Q, R, S, and T. The highest positive wave is the R wave. Shortly before and after the R wave are negative waves called Q wave and S wave. The P wave comes before the Q wave, while the T wave comes after the S wave. The Q, R, and S waves together are called the QRS complex.

The properties of the QRS complex, with its rate of occurrence and times, highs, and widths, provide information to cardiologists concerning various pathological conditions of the heart. The reciprocal of the time period between R wave peaks (in milliseconds) multiplied by 60,000 gives the instantaneous heart rate in beats per minute. On a modern ECG monitor, the acquired ECG signal is displayed for diagnostic purposes.

However, a major source of frequent interference is the electric-power system. Such interference appears on the recorded ECG data due to electric-field coupling between the power lines and the electrocardiograph or patient, which is the cause of the electrical field surrounding mains power lines. Another cause is magnetic induction in the power line, whereby current in the power line

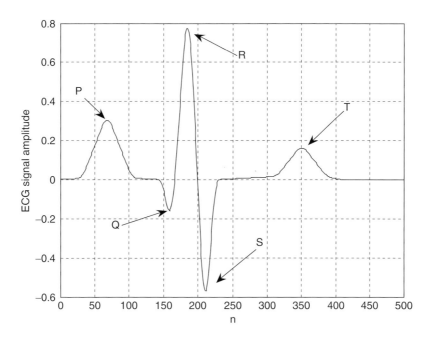

FIGURE 8.42 The characteristics of the ECG pulse.

generates a magnetic field around the line. Sometimes, the harmonics of 60-Hz hum exist due to nonlinear sensor and signal amplifier effects. If such interference is severe, the recorded ECG data become useless.

In this application, we focus on ECG enhancement for heart rate detection. To significantly reduce 60-Hz interference, we apply signal enhancement to the ECG recording system, as shown in Figure 8.43.

The 60-Hz hum eliminator removes the 60-Hz interference and has the capability to reduce its second harmonic of 120 Hz and its third harmonic of 180 Hz. The next objective is to detect the heart rate using the enhanced ECG signal. We need to remove DC drift and to filter muscle noise, which may occur at approximately 40 Hz or more. If we consider the lowest heart rate as 30 beats per minute, the corresponding frequency is $30/60 = 0.5$ Hz. Choosing the lower cutoff frequency of 0.25 Hz should be reasonable.

Thus, a bandpass filter with a passband from 0.25 to 40 Hz (range 0.67–40 Hz, discussed in Webster [1998]), either FIR or IIR type, can be designed to reduce such effects. The resultant ECG signal is valid only for the detection of heart rate. Notice that the ECG signal after bandpass filtering with a passband from 0.25 to 40 Hz is no longer valid for general ECG applications, since the original ECG signal occupies the frequency range from 0.01 to 250 Hz (diagnostic-quality ECG), as discussed in Carr and Brown (2001) and Webster (1998). The enhanced ECG signal from the 60-Hz hum eliminator can serve for general ECG signal analysis (which is beyond the scope of this book). We summarize the design specifications for the heart rate detection application as:

System outputs: Enhanced ECG signal with 60-Hz elimination
Processed ECG signal for heart rate detection
60 Hz eliminator:
Harmonics to be removed: 60 Hz (fundamental)
120 Hz (second harmonic)
180 Hz (third harmonic)
3 dB bandwidth for each filter: 4 Hz
Sampling rate: 600 Hz
Notch filter type: Second-order IIR
Design method: Pole-zero placement
Bandpass filter:

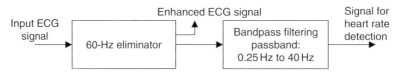

FIGURE 8.43 **ECG signal enhancement system.**

8.9 Application: 60-Hz Hum Eliminator and Heart Rate Detection Using Electrocardiography

Passband frequency range: 0.25–40 Hz
Passband ripple: 0.5 dB
Filter type: Chebyshev fourth order
Design method: Bilinear transformation
DSP sampling rate: 600 Hz

Let us carry out the 60-Hz eliminator design and determine the transfer function and difference equation for each notch filter and bandpass filter. For the first section with the notch frequency of 60 Hz, applying Equations (8.45) to (8.48) leads to

$$r = 1 - (4/600) \times \pi = 0.9791$$

$$\theta = \left(\frac{60}{600}\right) \times 360^\circ = 36^\circ.$$

We calculate $2\cos(36^\circ) = 1.6180$, $2r\cos(36^\circ) = 1.5842$, and

$$K = \frac{(1 - 2r\cos\theta + r^2)}{(2 - 2\cos\theta)} = 0.9803.$$

Hence it follows that

$$H_1(z) = \frac{0.9803 - 1.5862z^{-1} + 0.9803z^{-2}}{1 - 1.5842z^{-1} + 0.9586z^{-2}}$$

$$y_1(n) = 0.9803x(n) - 1.5862x(n-1) + 0.9802x(n-2)$$
$$+ 1.5842y_1(n-1) - 0.9586y_1(n-2).$$

Similarly, we yield the transfer functions and difference equations for the second and third sections as:

Second section:

$$H_2(z) = \frac{0.9794 - 0.6053z^{-1} + 0.9794z^{-2}}{1 - 0.6051z^{-1} + 0.9586z^{-2}}$$

$$y_2(n) = 0.9794y_1(n) - 0.6053y_1(n-1) + 0.9794y_1(n-2)$$
$$+ 0.6051y_2(n-1) - 0.9586y_2(n-2)$$

Third section:

$$H_3(z) = \frac{0.9793 + 0.6052z^{-1} + 0.9793z^{-2}}{1 + 0.6051z^{-1} + 0.9586z^{-2}}$$

$$y_3(n) = 0.9793y_2(n) + 0.6052y_2(n-1) + 0.9793y_2(n-2) - 0.6051y_3(n-1) - 0.9586y_3(n-2).$$

The cascaded frequency responses are plotted in Figure 8.44. As we can see, the rejection for each notch frequency is below 50 dB.

The second-stage design using the BLT gives the bandpass filter transfer function and difference equation

$$H_4(z) = \frac{0.0464 - 0.0927z^{-2} + 0.0464z^{-4}}{1 - 3.3523z^{-1} + 4.2557z^{-2} - 2.4540z^{-3} + 0.5506z^{-4}}$$

$$y_4(n) = 0.046361 y_3(n) - 0.092722 y_3(n-2) + 0.046361 y_3(n-4)$$
$$+ 03.352292 y_4(n-1) - 4.255671 y_4(n-2)$$
$$+ 2.453965 y_4(n-3) - 0.550587 y_4(n-4).$$

Figure 8.45 depicts the processed results at each stage. In Figure 8.45, plot (a) shows the initial corrupted ECG data, which include 60 Hz interference and its 120 and 180 Hz harmonics, along with muscle noise. Plot (b) shows that the interferences of 60 Hz and its harmonics of 120 and 180 Hz have been removed. Finally, plot (c) displays the result after the bandpass filter. As we expected, the muscle noise has been removed; and the enhanced ECG signal is observed. The MATLAB simulation is listed in Program 8.16.

With the processed ECG signal, a simple zero-cross algorithm can be designed to detect the heart rate. Based on plot (c) in Figure 8.45, we use a

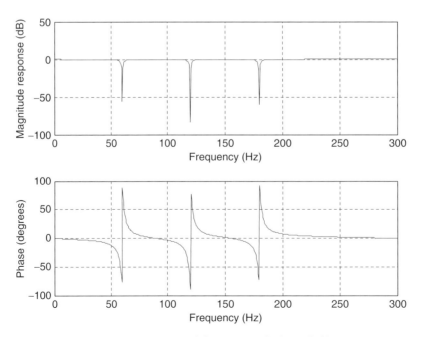

FIGURE 8.44 Frequency responses of three cascaded notch filters.

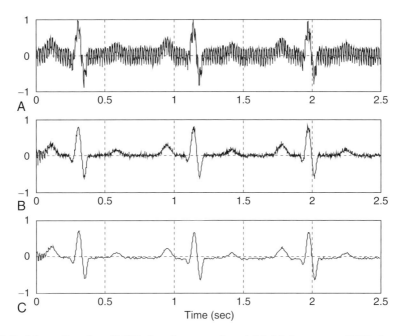

FIGURE 8.45 Results of ECG signal processing. (a) Initial corrupted ECG data; (b) ECG data enhanced by removing 60 Hz; (c) ECG data with DC blocking and noise removal for heart rate detection.

threshold value of 0.5 and continuously compare each of two consecutive samples with the threshold. If both results are opposite, then a zero crossing is detected. Each zero-crossing measure is given by

$$\text{zero crossing} = \frac{|cur\_sign - pre\_sign|}{2},$$

where cur_sign and pre_sign are determined based on the current input $x(n)$, the past input $x(n-1)$, and the threshold value, given as

if $x(n) \geq threshold$ $cur\_sign = 1$ else $cur\_sign = -1$
if $x(n-1) \geq threshold$ $pre\_sign = 1$ else $pre\_sign = -1$.

Figure 8.46 summarizes the algorithm.

After detecting the total number of zero crossings, the number of the peaks will be half the number of the zero crossings. The heart rate in terms of pulses per minute can be determined by

$$\text{Heart rate} = \frac{60}{\left(\frac{\text{Number of enhanced ECG data}}{f_s}\right)} \times \left(\frac{\text{zero-crossing number}}{2}\right).$$

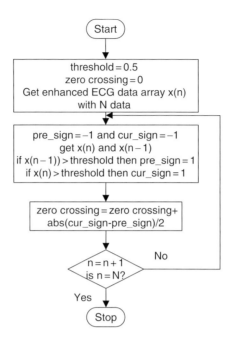

FIGURE 8.46 A simple zero-cross algorithm.

In our simulation, we have detected 6 zero-crossing points using 1,500 captured data at a sampling rate of 600 samples per second. Hence,

$$\text{Heart rate} = \frac{60}{\left(\frac{1500}{600}\right)} \times \left(\frac{6}{2}\right) = 72 \text{ pulses per minute.}$$

The MATLAB implementation of the zero-crossing detection can be found in the last part in Program 8.16.

Program 8.16. MATLAB program for heart rate detection using an ECG signal.

```
load ecgbn.dat;          % Load noisy ECG recording
b1 = [0.9803 -1.5862 0.9803]; %Notch filter with a notch frequency of 60 Hz
a1 = [1 -1.5842 0.9586];
b2 = [0.9794 -0.6053 0.9794]; % Notch filter with a notch frequency of 120 Hz
a2 = [1 -0.6051 0.9586];
b3 = [0.9793 0.6052 0.9793]; % Notch filter with a notch frequency of 180 Hz
a3 = [1 0.6051 0.9586];
y1 = filter(b1,a1,ecgbn);          % First section filtering
y2 = filter(b2,a2,y1);             % Second section filtering
y3 = filter(b3,a3,y2);             % Third section filtering
%bandpass filter
fs = 600;            % Sampling rate
```

```
T = 1/600;           % Sampling interval
% BLT design
wd1 = 2*pi*0.25;
wd2 = 2*pi*40;
wa1 = (2/T)*tan(wd1*T/2);
wa2 = (2/T)*tan(wd2*T/2);
[B,A] = lp2bp([1.4314], [1 1.4652 1.5162],sqrt(wa1*wa2),wa2-wa1);
[b,a] = bilinear(B,A,fs);
%b = [ 0.046361 0 -0.092722 0 0.046361] numerator coefficients from MATLAB
%a = [1 -3.352292 4.255671 -2.453965 0.550587] denominator coefficients from MATLAB
y4 = filter(b,a,y3);          %Bandpass filtering
t = 0:T:1499*T;          % Recover time
subplot(3,1,1);plot(t,ecgbn);grid;ylabel('(a)');
subplot(3,1,2);plot(t,y3);grid;ylabel('(b)');
subplot(3,1,3);plot(t,y4);grid;ylabel('(c)');
xlabel('Time (sec.)');
%Zero crossing algorithm
zcross = 0.0; threshold = 0.5
for n = 2:length(y4)
  pre_sign = -1; cur_sign = -1;
  if y4(n-1) >threshold
  pre_sign = 1;
  end
if y4(n) >threshold
  cur_sign = 1;
end
zcross = zcross+abs(cur_sign-pre_sign)/2;
end
zcross % Output the number of zero crossings
rate = 60*zcross/(2*length(y4)/600) % Output the heart rate
```

8.10 Coefficient Accuracy Effects on Infinite Impulse Response Filters

In practical applications, the IIR filter coefficients with infinite precision may be quantized due to the finite word length. Quantization of infinite precision filter coefficients changes the locations of the zeros and poles of the designed filter transfer function, hence changes the filter frequency responses. Since analysis of filter coefficient quantization for the IIR filter is very complicated and beyond the scope of this textbook, we pick only a couple of simple cases for discussion. Filter coefficient quantization for specific processors such as the fixed-point DSP processor and floating-point processor will be included in Chapter 9. To illustrate this effect, we look at the following first-order IIR filter transfer function having filter coefficients with infinite precision,

$$H(z) = \frac{b_0 + b_1 z^{-1}}{1 + a_1 z^{-1}}. \quad (8.57)$$

After filter coefficient quantization, we have the quantized digital IIR filter transfer function

$$H^q(z) = \frac{b_0^q + b_1^q z^{-1}}{1 + a_1^q z^{-1}}. \quad (8.58)$$

Solving for pole and zero, we achieve

$$p_1 = -a_1^q \quad (8.59)$$

$$z_1 = -\frac{b_1^q}{b_0^q}. \quad (8.60)$$

Now considering a second-order IIR filter transfer function as

$$H(z) = \frac{b_0 + b_1 z^{-1} + b_2 z^{-2}}{1 + a_1 z^{-1} + a_2 z^{-2}}, \quad (8.61)$$

and its quantized IIR filter transfer function

$$H^q(z) = \frac{b_0^q + b_1^q z^{-1} + b_2^q z^{-2}}{1 + a_1^q z^{-1} + a_2^q z^{-2}}, \quad (8.62)$$

solving for poles and zeros finds:

$$p_{1,2} = -0.5 \cdot a_1^q \pm j\left(a_2^q - 0.25 \cdot \left(a_1^q\right)^2\right)^{\frac{1}{2}} \quad (8.63)$$

$$z_{1,2} = -0.5 \cdot \frac{b_1^q}{b_0^q} \pm j\left(\frac{b_2^q}{b_0^q} - 0.25 \cdot \left(\frac{b_1^q}{b_0^q}\right)^2\right)^{\frac{1}{2}}. \quad (8.64)$$

With the developed Equations (8.59) and (8.60) for the first-order IIR filter, and Equations (8.63) and (8.64) for the second-order IIR filter, we can study the effects of the location changes of the poles and zeros, and the frequency responses due to filter coefficient quantization.

Example 8.24.

Given the following first-order IIR filter,

$$H(z) = \frac{1.2341 + 0.2126 z^{-1}}{1 - 0.5126 z^{-1}},$$

and assuming that we use 1 sign bit and 6 bits for encoding the magnitude of the filter coefficients,

a. Find the quantized transfer function and pole-zero locations.

8.10 Coefficient Accuracy Effects on Infinite Impulse Response Filters

Solution:

a. Let us find the pole and zero for infinite precision filter coefficients:

Solving $1.2341z + 0.2126 = 0$ leads to a zero location $z_1 = -0.17227$.
Solving $z - 0.5126 = 0$ gives a pole location $p_1 = 0.5126$.

Now let us quantize the filter coefficients. Quantizing 1.2341 can be illustrated as

$$1.2341 \times 2^5 = 39.4912 = 39 \text{ (rounded to integer)}.$$

Since the maximum magnitude of the filter coefficients is 1.2341, which is between 1 and 2, we scale all coefficient magnitudes by a factor of 2^5 and round off each value to an integer whose magnitude is encoded using 6 bits. As shown in the quantization, 6 bits are required to encode the integer 39. When the coefficient integer is scaled back by the same scale factor, the corresponding quantized coefficient with finite precision (7 bits, including the sign bit) is found to be

$$b_0^q = 39/2^5 = 1.21875.$$

Following the same procedure, we can obtain

$$b_1^q = 0.1875$$

and

$$a_1^q = -0.5.$$

Thus we achieve the quantized transfer function

$$H^q(z) = \frac{1.21875 + 0.1875z^{-1}}{1 - 0.5z^{-1}}.$$

Solving for pole and zero leads to

$$p_1 = 0.5$$

and

$$z_1 = -0.1538.$$

It is clear that the pole and zero locations change after the filter coefficients are quantized. This effect can change the frequency responses of the designed filter as well. In Example 8.25, we study quantization of the filter coefficients for the second-order IIR filter and examine the pole/zero location changes and magnitude/phase frequency responses.

Example 8.25.

A second-order digital lowpass Chebyshev filter with a cutoff frequency of 3.4 kHz and 0.5 dB ripple on passband at a sampling frequency at 8,000 Hz is designed. Assume that we use 1 sign bit and 7 bits for encoding the magnitude of each filter coefficient. The z-transfer function is given by

$$H(z) = \frac{0.7434 + 1.4865z^{-1} + 0.7434z^{-2}}{1 + 1.5149z^{-1} + 0.6346z^{-2}}.$$

a. Find the quantized transfer function and pole and zero locations.

b. Plot the magnitude and phase responses, respectively.

Solution:

a. Since the maximum magnitude of the filter coefficients is between 1 and 2, the scale factor for quantization is chosen to be 2^6, so that the coefficient integer can be encoded using 7 bits.

After performing filter coefficient encoding, we have

$$H^q(z) = \frac{0.7500 + 1.484375z^{-1} + 0.7500z^{-2}}{1 + 1.515625z^{-1} + 0.640625z^{-2}}.$$

For comparison, the uncoded zeros and encoded zeros of the transfer function $H(z)$ are

Uncoded zeros: $-1, -1$;
Coded zeros: $-0.9896 + 0.1440\,i, -0.9896 - 0.1440\,i$.

Similarly, the uncoded poles and coded poles of the transfer function $H^q(z)$ are

Uncoded poles: $-0.7574 + 0.2467\,i, -0.7574 - 0.2467\,i$;
Coded poles: $-0.7578 + 0.2569\,i, -0.7578 - 0.2569\,i$.

b. The comparisons for the magnitude responses and phase responses are listed in Program 8.17 and plotted in Figure 8.47.

Program 8.17. MATLAB *m*-file for Example 8.25.

```
% Example 8.25
% Plot the magnitude and phase responses
fs = 8000;          % Sampling rate
B = [0.7434 1.4868 0.7434];
A = [1 1.5149 0.6346];
[hz,f] = freqz(B,A,512,fs); % Calculate reponses without coefficient quantization
phi = 180*unwrap(angle(hz))/pi;
Bq = [0.750 1.4834375 0.75000];
Aq = [1 1.515625 0.640625];
[hzq,f] = freqz(Bq,Aq,512,fs); % Calculate responses with coefficient quantization
phiq = 180*unwrap(angle(hzq))/pi;
subplot(2,1,1),plot(f,20*log10(abs(hz)),f,20*log10(abs(hzq)),'-.');grid;
```

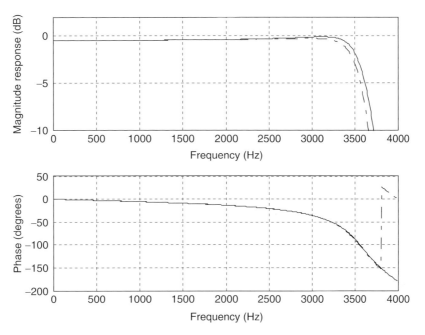

FIGURE 8.47 Frequency responses (dash-dotted line, quantized coefficients; solid line, unquantized coefficients).

```
axis([0 fs/2 -10 2])
xlabel('Frequency (Hz)');
ylabel('Magnitude Response (dB)');
subplot(2,1,2), plot(f, phi, f, phiq,'-.');grid;
xlabel('Frequency (Hz)');
ylabel('Phase (degrees)');
```

From Figure 8.47, we observe that the quantization of IIR filter coefficients has more effect on magnitude response and less effect on the phase response in the passband. In practice, one needs to verify this effect to make sure that the magnitude frequency response meets the filter specifications.

8.11 Application: Generation and Detection of Dual-Tone Multifrequency Tones Using the Goertzel Algorithm

In this section, we study an application of the digital filters to the generation and detection of dual-tone multifrequency (DTMF) signals used for telephone touch

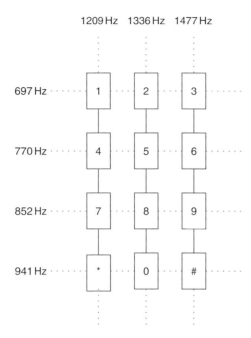

FIGURE 8.48 DTMF tone specifications.

keypads. In our daily life, DTMF touch tones produced by telephone keypads on handsets are applied to dial telephone numbers routed to telephone companies, where the DTMF tones are digitized and processed and the detected dialed telephone digits are used for the telephone switching system to ring the party being called. A telephone touch keypad is shown in Figure 8.48, where each key is represented by two tones with their specified frequencies. For example, if the key "7" is pressed, the DTMF signal with the designated frequencies of 852 Hz and 1,209 Hz is generated, which is sent to the central office at the telephone company for processing. At the central office, the received DTMF tones are detected through the digital filters and some logic operations to decode the dialed signal consisting of 852 Hz and 1,209 Hz to be key "7." The frequencies defined for each key are in Figure 8.48.

8.11.1 Single-Tone Generator

Now, let us look at a digital tone generator whose transfer function is obtained from the z-transform function of a sinusoidal sequence $\sin(n\Omega_0)$ as

8.11 Generation and Detection of Dual-Tone Multifrequency Tones Using the Goertzel Algorithm

$$H(z) = \frac{z \sin \Omega_0}{z^2 - 2z \cos \Omega_0 + 1} = \frac{z^{-1} \sin \Omega_0}{1 - 2z^{-1} \cos \Omega_0 + z^{-2}}, \quad (8.65)$$

where Ω_0 is the normalized digital frequency. Given the sampling rate of the DSP system and the frequency of the tone to be generated, we have the relationship

$$\Omega_0 = 2\pi f_0 / f_s. \quad (8.66)$$

Applying the inverse z-transform to the transfer function leads to the difference equation

$$y(n) = \sin \Omega_0 x(n-1) + 2 \cos \Omega_0 y(n-1) - y(n-2), \quad (8.67)$$

since

$$Z^{-1}(H(z)) = Z^{-1}\left(\frac{z \sin \Omega_0}{z^2 - 2z \cos \Omega_0 + 1}\right) = \sin(\Omega_0 n) = \sin(2\pi f_0 n / f_s),$$

which is the impulse response. Hence, to generate a pure tone with the amplitude of A, an impulse function $x(n) = A\delta(n)$ must be used as an input to the digital filter, as illustrated in Figure 8.49.

Now, we illustrate implementation. Assuming that the sampling rate of the DSP system is 8,000 Hz, we need to generate a digital tone of 1 kHz. Then we compute

$$\Omega_0 = 2\pi \times 1000/8000 = \pi/4, \sin \Omega_0 = 0.707107, \text{ and } 2\cos \Omega_0 = 1.414214.$$

The required filter transfer function is determined as

$$H(z) = \frac{0.707107 z^{-1}}{1 - 1.414214 z^{-1} + z^{-2}}.$$

The MATLAB simulation using the input $x(n) = \delta(n)$ is displayed in Figure 8.50, where the top plot is the generated 1 kHz tone, and the bottom plot shows its spectrum. The corresponding MATLAB list is in Program 8.18.

Note that if we replace the filter $H(z)$ with the z-transform of other sequences such as a cosine function and use the impulse sequence as the filter input, the filter will generate the corresponding digital wave such as the digital cosine wave.

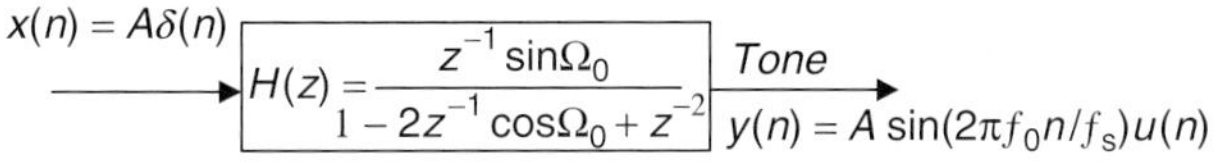

FIGURE 8.49 Single-tone generator.

384 8 INFINITE IMPULSE RESPONSE FILTER DESIGN

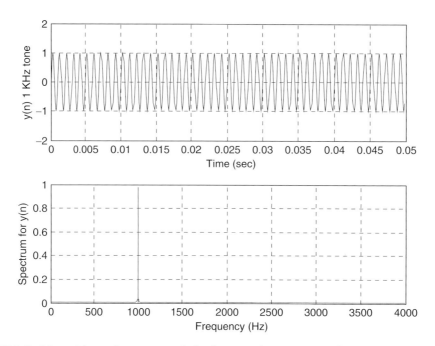

FIGURE 8.50 Plots of a generated single tone of 1,000 Hz and its spectrum.

Program 8.18. MATLAB program for generating a sinusoid.

```
fs = 8000;           % Sampling rate
t = 0:1/fs:1;        % Time vector for 1 second
x = zeros(1,length(t));       % Initialize input to be zero
x(1) = 1;            % Set up impulse function
y = filter([0 0.707107],[1 -1.414214 1],x);        % Perform filtering
subplot(2,1,1);plot(t(1:400),y(1:400));grid
ylabel('y(n) 1 kHz tone');xlabel('time (second)')
Ak = 2*abs(fft(y))/length(y);Ak(1) = Ak(1)/2; % One-sided amplitude spectrum
f = [0:1:(length(y)-1)/2]*fs/length(y); % Indices to frequencies (Hz) for plot
subplot(2,1,2);plot(f,Ak(1:(length(y)+1)/2));grid
ylabel('Spectrum for y(n)');xlabel('frequency (Hz)')
```

8.11.2 Dual-Tone Multifrequency Tone Generator

Now that the principle of a single-tone generator has been illustrated, we can extend it to develop the DTMF tone generator using two digital filters in parallel. The DTMF tone generator for key "7" is depicted in Figure 8.51.

8.11 Generation and Detection of Dual-Tone Multifrequency Tones Using the Goertzel Algorithm

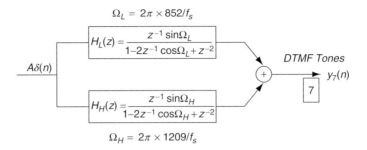

FIGURE 8.51 Digital DTMF tone generator for the keypad digit "7."

Here we generate the DTMF tone for key "7" for a duration of one second, assuming the sampling rate of 8,000 Hz. The generated tone and its spectrum are plotted in Figure 8.52 for verification, while the MATLAB implementation is given in Program 8.19.

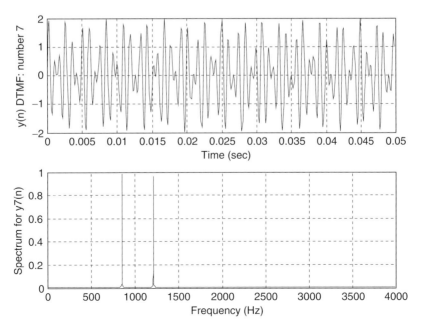

FIGURE 8.52 Plots of the generated DTMF tone "7" and its spectrum.

Program 8.19. MATLAB program for DTMF tone generation.

```
close all; clear all
fs = 8000;            % Sampling rate
t = 0:1/fs:1;         % 1-second time vector
x = zeros(1,length(t));      % Initialize the input to be zero
x(1) = 1;             % Set up the impulse function
% generate the 852-Hz tone
y852 = filter([0 sin(2*pi*852/fs)],[1 −2*cos(2*pi*852/fs)1],x);
% generate the 1209-Hz tone
y1209 = filter([0 sin(2*pi*1209/fs)],[1 − 2*cos(2*pi*1209/fs) 1],x);     % Filtering
y7 = y852 + y1209;          % Generate the DTMF tone
subplot(2,1,1);plot(t(1:400),y7(1:400));grid
ylabel('y(n) DTMF: number 7');
xlabel('time (second)')
Ak = 2*abs(fft(y7))/length(y7);Ak(1)=Ak(1)/2; % One-sided amplitude spectrum
f = [0:1:(length(y7) − 1)/2]*fs/length(y7); % Map indices to frequencies (Hz)
subplot(2,1,2);plot(f,Ak(1:(length(y7)+1)/2));grid
ylabel('Spectrum for y7(n)');
xlabel('frequency (Hz)');
```

8.11.3 Goertzel Algorithm

In practice, the DTMF tone detector is designed to use the Goertzel algorithm. This is a special and powerful algorithm used for computing discrete Fourier transform (DFT) coefficients and signal spectra using a digital filtering method. The modified Goertzel algorithm can be used for computing signal spectra without involving complex algebra like the DFT algorithm.

Specifically, the Goertzel algorithm is a filtering method for computing the DFT coefficient $X(k)$ at the specified frequency bin k with the given N digital data $x(0), x(1), \ldots, x(N-1)$. We can begin to illustrate the Goertzel algorithm using the second-order IIR digital Goertzel filter, whose transfer function is given by

$$H_k(z) = \frac{Y_k(z)}{X(z)} = \frac{1 - W_N^k z^{-1}}{1 - 2\cos\left(\frac{2\pi k}{N}\right)z^{-1} + z^{-2}}, \qquad (8.68)$$

with the input data $x(n)$ for $n = 0, 1, \ldots, N - 1$, and the last element set to be $x(N) = 0$. Notice that $W_N^k = e^{-\frac{2\pi k}{N}}$. We will process the data sequence $N + 1$ times to achieve the filter output as $y_k(n)$ for $n = 0, 1, \ldots, N$, where k is the frequency index (bin number) of interest. The DFT coefficient $X(k)$ is the last datum from the Goertzel filter, that is,

8.11 Generation and Detection of Dual-Tone Multifrequency Tones Using the Goertzel Algorithm

$$X(k) = y_k(N). \tag{8.69}$$

The implementation of the Goertzel filter is presented by direct-form II realization in Figure 8.53.

According to the direct-form II realization, we can write the Goertzel algorithm as

$$x(N) = 0, \tag{8.70}$$
$$\text{for } n = 0, 1, \ldots, N$$

$$v_k(n) = 2\cos\left(\frac{2\pi k}{N}\right) v_k(n-1) - v_k(n-2) + x(n) \tag{8.71}$$

$$y_k(n) = v_k(n) - W_N^k v_k(n-1) \tag{8.72}$$

with initial conditions: $v_k(-2) = 0$, $v_k(-1) = 0$.

Then the DFT coefficient $X(k)$ is given as

$$X(k) = y_k(N). \tag{8.73}$$

The squared magnitude of $x(k)$ is computed as

$$|X(k)|^2 = v_k^2(N) + v_k^2(N-1) - 2\cos\left(\frac{2\pi k}{N}\right) v_k(N) v_k(N-1). \tag{8.74}$$

We show the derivation of Equation (8.74) as follows. Note that Equation (8.72) involves complex algebra, since the equation contains only one complex number, a factor

$$W_N^k = e^{-j\frac{2\pi k}{N}} = \cos\left(\frac{2\pi k}{N}\right) - j\sin\left(\frac{2\pi k}{N}\right)$$

discussed in Chapter 4. If our objective is to compute the spectral value, we can substitute $n = N$ into Equation (8.72) to obtain $X(k)$ and multiply $X(k)$ by its

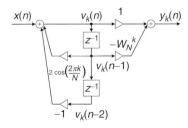

FIGURE 8.53 Second-order Goertzel IIR filter.

complex conjugate $X^*(k)$ to achieve the squared magnitude of the DFT coefficient. It follows (Ifeachor and Jervis, 2002) that

$$|X(k)|^2 = X(k)X^*(k)$$

Since

$$X(k) = v_k(N) - W_N^k v_k(N-1)$$
$$X^*(k) = v_k(N) - W_N^{-k} v_k(N-1)$$

then

$$|X(k)|^2 = [v_k(N) - W_N^k v_k(N-1)][v_k(N) - W_N^{-k} v_k(N-1)]$$
$$= v_k^2(N) + v_k^2(N-1) - (W_N^k + W_N^{-k})v_k(N)v_k(N-1). \quad (8.75)$$

Using Euler's identity yields

$$W_N^k + W_N^{-k} = e^{-j\frac{2\pi}{N}k} + e^{j\frac{2\pi}{N}k} = 2\cos\left(\frac{2\pi k}{N}\right). \quad (8.76)$$

Substituting Equation (8.76) into Equation (8.75) leads to Equation (8.74).

We can see that the DSP equation for $v_k(k)$ and computation of the squared magnitude of the DFT coefficient $|X(k)|^2$ do not involve any complex algebra. Hence, we will use this advantage for later development. To illustrate the algorithm, let us consider Example 8.26.

Example 8.26.

Given the digital data sequence of length 4 as $x(0) = 1$, $x(1) = 2$, $x(2) = 3$, and $x(3) = 4$,

   a. Use the Goertzel algorithm to compute DFT coefficient $X(1)$ and the corresponding spectral amplitude at the frequency bin $k = 1$.

Solution:

   a. We have $k = 1$, $N = 4$, $x(0) = 1$, $x(1) = 2$, $x(2) = 3$, and $x(3) = 4$. Note that

$$2\cos\left(\frac{2\pi}{4}\right) = 0 \text{ and } W_4^1 = e^{-j\frac{2\pi \times 1}{4}} = \cos\left(\frac{\pi}{2}\right) - j\sin\left(\frac{\pi}{2}\right) = -j.$$

We first write the simplified difference equations:

$$x(4) = 0$$
$$\text{for } n = 0, 1, \ldots, 4$$
$$v_1(n) = -v_1(n-2) + x(n)$$
$$y_1(n) = v_1(n) + jv_1(n-1)$$

then
$$X(1) = y_1(4)$$
$$|X(1)|^2 = v_1^2(4) + v_1^2(3).$$

The digital filter process is demonstrated in the following:
$$v_1(0) = -v_1(-2) + x(0) = 0 + 1 = 1$$
$$y_1(0) = v_1(0) + jv_1(-1) = 1 + j \times 0 = 1$$
$$v_1(1) = -v_1(-1) + x(1) = 0 + 2 = 2$$
$$y_1(1) = v_1(1) + jv_1(0) = 2 + j \times 1 = 2 + j$$
$$v_1(2) = -v_1(0) + x(2) = -1 + 3 = 2$$
$$y_1(2) = v_1(2) + jv_1(1) = 2 + j \times 2 = 2 + j2$$
$$v_1(3) = -v_1(1) + x(3) = -2 + 4 = 2$$
$$y_1(3) = v_1(3) + jv_1(2) = 2 + j \times 2 = 2 + j2$$
$$v_1(4) = -v_1(2) + x(4) = -2 + 0 = -2$$
$$y_1(4) = v_1(4) + jv_1(3) = -2 + j \times 2 = -2 + j2.$$

Then the DFT coefficient and its squared magnitude are determined as
$$X(1) = y_1(4) = -2 + j2$$
$$|X(1)|^2 = v_1^2(4) + v_1^2(3) = (-2)^2 + (2)^2 = 8.$$

Thus, the two-sided amplitude spectrum is computed as
$$A_1 = \frac{1}{4}\sqrt{\left(|X(1)|^2\right)} = 0.7071$$

and the corresponding single-sided amplitude spectrum is $A_1 = 2 \times 0.707 = 1.4141$.

From this simple illustrative example, we see that the Goertzel algorithm has the following advantages:

1. We can apply the algorithm for computing the DFT coefficient $X(k)$ for a specified frequency bin k; unlike the fast Fourier transform (FFT) algorithm, all the DFT coefficients are computed once it is applied.

2. If we want to compute the spectrum at frequency bin k, that is, $|X(k)|$, Equation (8.71) shows that we need to process only $v_k(n)$ for $N + 1$ times and then compute $|X(k)|^2$. The operations avoid complex algebra.

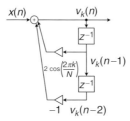

FIGURE 8.54 Modified second-order Goertzel IIR filter.

If we use the modified Goertzel filter in Figure 8.54, then the corresponding transfer function is given by

$$G_k(z) = \frac{V_k(z)}{X(z)} = \frac{1}{1 - 2\cos\left(\frac{2\pi k}{N}\right)z^{-1} + z^{-2}}. \quad (8.77)$$

The modified Goertzel algorithm becomes
$x(N) = 0$
for $n = 0, 1, \ldots, N$
$v_k(n) = 2\cos\left(\frac{2\pi k}{N}\right)v_k(n-1) - v_k(n-2) + x(n)$
with initial conditions: $v_k(-2) = 0$, and $v_k(-1) = 0$
then the squared magnitude of the DFT coefficient is given by

$$|X(k)|^2 = v_k^2(N) + v_k^2(N-1) - 2\cos\left(\frac{2\pi k}{N}\right)v_k(N)v_k(N-1)$$

Example 8.27.

Given the digital data sequence of length 4 as $x(0) = 1$, $x(1) = 2$, $x(2) = 3$, and $x(3) = 4$,

  a. Use the Goertzel algorithm to compute the spectral amplitude at the frequency bin $k = 0$.

Solution:

$k = 0$, $N = 4$, $x(0) = 1$, $x(1) = 2$, $x(2) = 3$, and $x(3) = 4$.

  a. Using the modified Goertzel algorithm and noting that $2 \cdot \cos\left(\frac{2\pi}{4} \times 0\right) = 2$, we get the simplified difference equations as:

$$x(4) = 0$$
for $n = 0, 1, \ldots, 4$
$$v_0(n) = 2v_0(n-1) - v_0(n-2) + x(n)$$
then $|X(0)|^2 = v_0^2(4) + v_0^2(3) - 2v_0(4)v_0(3)$.

The digital filtering is performed as:
$$v_0(0) = 2v_0(-1) - v_0(-2) + x(0) = 0 + 0 + 1 = 1$$
$$v_0(1) = 2v_0(0) - v_0(-1) + x(1) = 2 \times 1 + 0 + 2 = 4$$
$$v_0(2) = 2v_0(1) - v_0(0) + x(2) = 2 \times 4 - 1 + 3 = 10$$
$$v_0(3) = 2v_0(2) - v_0(1) + x(3) = 2 \times 10 - 4 + 4 = 20$$
$$v_0(4) = 2v_0(3) - v_0(2) + x(4) = 2 \times 20 - 10 + 0 = 30.$$

Then the squared magnitude is determined by
$$|X(0)|^2 = v_0^2(4) + v_0^2(3) - 2v_0(4)v_0(3) = (30)^2 + (20)^2 - 2 \times 30 \times 20 = 100.$$

Thus, the amplitude spectrum is computed as
$$A_0 = \frac{1}{4}\sqrt{\left(|X(0)|^2\right)} = 2.5.$$

8.11.4 Dual-Tone Multifrequency Tone Detection Using the Modified Goertzel Algorithm

Based on the specified frequencies of each DTMF tone shown in Figure 8.48 and the modified Goertzel algorithm, we can develop the following design principles for DTMF tone detection:

1. When the digitized DTMF tone $x(n)$ is received, it has two nonzero frequency components from the following seven: 679, 770, 852, 941, 1209, 1336, and 1477 Hz.

2. We can apply the modified Goertzel algorithm to compute seven spectral values, which correspond to the seven frequencies in (1). The single-sided amplitude spectrum is computed as

$$A_k = \frac{2}{N}\sqrt{|X(k)|^2}. \tag{8.78}$$

3. Since the modified Goertzel algorithm is used, there is no complex algebra involved. Ideally, there are two nonzero spectral components. We will use these two nonzero spectral components to determine which key is pressed.

4. The frequency bin number (frequency index) can be determined based on the sampling rate f_s, and the data size of N via the following relation:

$$k = \frac{f}{f_s} \times N \text{ (rounded off to an integer).} \tag{8.79}$$

Given the key frequency specification in Table 8.12, we can determine the frequency bin k for each DTMF frequency with $f_s = 8,000$ Hz and $N = 205$. The results are summarized in Table 8.12.

The DTMF detector block diagram is shown in Figure 8.55.

5. The threshold value can be the sum of all seven spectral values divided by a factor of 4. Note that there are only two nonzero spectral values, hence the threshold value should ideally be half of the individual nonzero spectral value. If the spectral value is larger than the threshold value, then the logic operation outputs logic 1; otherwise, it outputs logic 0. Finally, the logic operation at the last stage decodes the key information based on the 7-bit binary pattern.

TABLE 8.12 DTMF frequencies and their frequency bins.

DTMF Frequency (Hz)	Frequency Bin: $k = \frac{f}{f_s} \times N$
697	18
770	20
852	22
941	24
1209	31
1336	34
1477	38

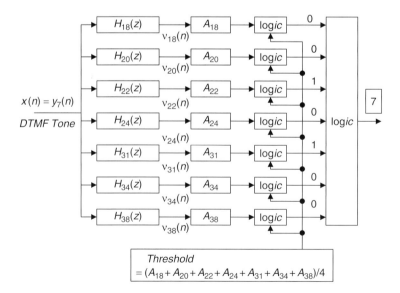

FIGURE 8.55 DTMF detector using the Goertzel algorithm.

8.11 Generation and Detection of Dual-Tone Multifrequency Tones Using the Goertzel Algorithm

Example 8.28.

Given a DSP system with $f_s = 8,000$ Hz and data size $N = 205$, seven Goertzel IIR filters are implemented for DTMF tone detection.

a. Determine the following for the frequencies corresponding to key 7:
   1. frequency bin numbers
   2. the Goertzel filter transfer functions and DSP equations
   3. equations for calculating amplitude spectral values.

Solution:

a. For key 7, we have $f_L = 852$ Hz and $f_H = 1,209$ Hz.

   1. Using Equation (8.79), we get
   $$k_L = \frac{852}{8000} \times 205 \approx 22, \text{ and } k_H = \frac{1209}{8000} \times 205 \approx 31.$$

   2. Since $2\cos\left(\frac{2\pi \times 22}{205}\right) = 1.5623$, and $2\cos\left(\frac{2\pi \times 31}{205}\right) = 1.1631$, it follows that
   $$H_{22}(z) = \frac{1}{1 - 1.5623z^{-1} + z^{-2}} \text{ and}$$
   $$H_{31}(z) = \frac{1}{1 - 1.1631z^{-1} + z^{-2}}.$$

   The DSP equations are therefore given by:

   $v_{22}(n) = 1.5623v_{22}(n-1) - v_{22}(n-2) + x(n)$ with $x(205) = 0$, for $n = 0, 1, \ldots, 205$
   $v_{31}(n) = 1.1631v_{31}(n-1) - v_{31}(n-2) + x(n)$ with $x(205) = 0$, for $n = 0, 1, \ldots, 205$.

   3. The amplitude spectral values are determined by
   $$|X(22)|^2 = (v_{22}(205))^2 + (v_{22}(204))^2 - 1.5623(v_{22}(205)) \times (v_{22}(204))$$
   $$A_{22} = \frac{2\sqrt{|X(22)|^2}}{205}$$
   and
   $$|X(31)|^2 = (v_{31}(205))^2 + (v_{31}(204))^2 - 1.1631(v_{31}(205)) \times (v_{31}(204))$$
   $$A_{31} = \frac{2\sqrt{|X(31)|^2}}{205}.$$

The MATLAB simulation for decoding the key 7 is shown in Program 8.20. Figure 8.56(a) shows the frequency responses of the second-order Goertzel bandpass filters.

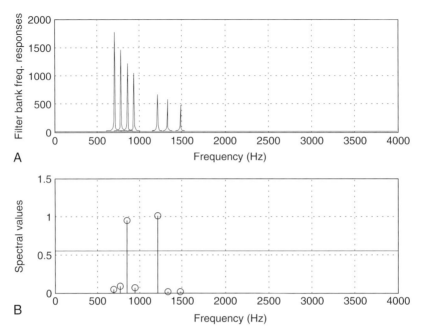

FIGURE 8.56 (a) Goertzel filter bank frequency responses; (b) display of spectral values and threshold for key 7.

The input is generated as shown in Figure 8.52. After filtering, the calculated spectral values and threshold value for decoding key 7 are displayed in Figure 8.56(b), where only two spectral values corresponding to the frequencies of 770 Hz and 1,209 Hz are above the threshold, and are encoded as logic 1. According to the key information in Figure 8.55, the final logic operation decodes the key as 7.

Program 8.20. DTMF detection using the Goertzel algoritm.

```
close all;clear all;
% DTMF tone generator
N=205;
fs=8000;t=[0:1:N-1]/fs;           % Sampling rate and time vector
x=zeros(1,length(t));x(1)=1;      % Generate the imupse function
%generation of tones
y697=filter([0 sin(2*pi*697/fs)],[1 -2*cos(2*pi*697/fs) 1],x);
y770=filter([0 sin(2*pi*770/fs)],[1 -2*cos(2*pi*770/fs) 1],x);
y852=filter([0 sin(2*pi*852/fs)],[1 -2*cos(2*pi*852/fs) 1],x);
y941=filter([0 sin(2*pi*941/fs)],[1 -2*cos(2*pi*941/fs) 1],x);
y1209=filter([0 sin(2*pi*1209/fs)],[1 -2*cos(2*pi*1209/fs) 1],x);
y1336=filter([0 sin(2*pi*1336/fs)],[1 -2*cos(2*pi*1336/fs) 1],x);
y1477=filter([0 sin(2*pi*1477/fs)],[1 -2*cos(2*pi*1477/fs) 1],x);
key=input('input of the following keys: 1,2,3,4,5,6,7,8,9,*,0,#=>','s');
yDTMF=[];
```

8.11 Generation and Detection of Dual-Tone Multifrequency Tones Using the Goertzel Algorithm

```
if key=='1'yDTMF=y697+y1209;end
if key=='2'yDTMF=y697+y1336;end
if key=='3'yDTMF=y697+y1477;end
if key=='4'yDTMF=y770+y1209;end
if key=='5'yDTMF=y770+y1336;end
if key=='6'yDTMF=y770+y1477;end
if key=='7'yDTMF=y852+y1209;end
if key=='8'yDTMF=y852+y1336;end
if key=='9'yDTMF=y852+y1477;end
if key=='*'yDTMF=y941+y1209;end
if key=='0'yDTMF=y941+y1336;end
if key=='#'yDTMF=y941+y1477;end
if size(yDTMF)==0 disp('Invalid input key');return;end
yDTMF=[yDTMF 0];          % DTMF signal appended with a zero
% DTMF detector (use Goertzel algorithm)
a697=[1 −2*cos(2*pi*18/N) 1];
a770=[1 −2*cos(2*pi*20/N) 1];
a852=[1 −2*cos(2*pi*22/N) 1];
a941=[1 −2*cos(2*pi*24/N) 1];
a1209=[1 −2*cos(2*pi*31/N) 1];
a1336=[1 −2*cos(2*pi*34/N) 1];
a1477=[1 −2*cos(2*pi*38/N) 1];
% Filter bank frequency responses
[w1, f]=freqz(1,a697,512,fs);
[w2, f]=freqz(1,a770,512,fs);
[w3, f]=freqz(1,a852,512,fs);
[w4, f]=freqz(1,a941,512,fs);
[w5, f]=freqz(1,a1209,512,fs);
[w6, f]=freqz(1,a1336,512,fs);
[w7, f]=freqz(1,a1477,512,fs);
subplot(2,1,1);plot(f,abs(w1),f,abs(w2),f,abs(w3), ...
f,abs(w4),f,abs(w5),f,abs(w6),f,abs(w7));grid
xlabel('Frequency (Hz)');ylabel('(a) Filter bank freq. responses');
% filter bank bandpass filtering
y697=filter(1,a697,yDTMF);
y770=filter(1,a770,yDTMF);
y852=filter(1,a852,yDTMF);
y941=filter(1,a941,yDTMF);
y1209=filter(1,a1209,yDTMF);
y1336=filter(1,a1336,yDTMF);
y1477=filter(1,a1477,yDTMF);
% Determine the absolute magnitudes of DFT coefficents
m(1)=sqrt(y697(206)^2+y697(205)^2- ...
  2*cos(2*pi*18/205)*y697(206)*y697(205));
m(2)=sqrt(y770(206)^2+y770(205)^2- ...
  2*cos(2*pi*20/205)*y770(206)*y770(205));
m(3)=sqrt(y852(206)^2+y852(205)^2- ...
  2*cos(2*pi*22/205)*y852(206)*y852(205));
m(4)=sqrt(y941(206)^2+y941(205)^2- ...
  2*cos(2*pi*24/205)*y941(206)*y941(205));
```

```
m(5)=sqrt(y1209(206)^2+y1209(205)^2- ...
   2*cos(2*pi*31/205)*y1209(206)*y1209(205));
m(6)=sqrt(y1336(206)^2+y1336(205)^2- ...
   2*cos(2*pi*34/205)*y1336(206)*y1336(205));
m(7)=sqrt(y1477(206)^2+y1477(205)^2- ...
   2*cos(2*pi*38/205)*y1477(206)*y1477(205));
% Convert the magnitudes of DFT coefficients to the single-side spectrum
m=2*m/205;
% Determine the threshold
th=sum(m)/4;
% Plot the DTMF spectrum with the threshold
f=[697 770 852 941 1209 1336 1477];
f1=[0 fs/2];
th=[ th th];
subplot(2,1,2);stem(f,m);grid;hold;plot(f1,th);
xlabel('Frequency (Hz)');ylabel(' (b) Spectral values');
m=round(m);          % Round to the binary pattern
if m== [1 0 0 0 1 0 0] disp('Detected Key 1');end
if m== [1 0 0 0 0 1 0] disp('Detected Key 2');end
if m== [1 0 0 0 0 0 1] disp('Detected Key 3');end
if m== [0 1 0 0 1 0 0] disp('Detected Key 4');end
if m== [0 1 0 0 0 1 0] disp('Detected Key 5');end
if m== [0 1 0 0 0 0 1] disp('Detected Key 6');end
if m== [0 0 1 0 1 0 0] disp('Detected Key 7');end
if m== [0 0 1 0 0 1 0] disp('Detected Key 8');end
if m== [0 0 1 0 0 0 1] disp('Detected Key 9');end
if m== [0 0 0 1 1 0 0] disp('Detected Key *');end
if m== [0 0 0 1 0 1 0] disp('Detected Key 0');end
if m== [0 0 0 1 0 0 1] disp('Detected Key #');end
```

The principle can easily be extended to transmit the ASCII (American Standard Code for Information Interchange) code or other types of code using the parallel Goertzel filter bank. If the calculated spectral value is larger than the threshold value, then the logic operation outputs logic 1; otherwise, it outputs logic 0. Finally, the logic operation at the last stage decodes the key information based on the 7-bit binary pattern.

8.12 Summary of Infinite Impulse Response (IIR) Design Procedures and Selection of the IIR Filter Design Methods in Practice

In this section, we first summarize the design procedures of the BLT design, impulse invariant design, and pole-zero placement design methods, and then discuss the selection of the particular filter for typical applications.

8.12 Summary of IIR Design Procedures and Selection of the IIR Filter Design Methods in Practice

The BLT design method:

1. Given the digital filter frequency specifications, prewarp each digital frequency edge to the analog frequency edge using Equations (8.18) and (8.19).

2. Determine the prototype filter order using Equation (8.29) for the Butterworth filter or Equation (8.35b) for the Chebyshev filter, and perform lowpass prototype transformation using the lowpass prototype in Table 8.3 (Butterworth function) or Tables 8.4 and 8.5 (Chebyshev functions) using Equations (8.20) to (8.23).

3. Apply the BLT to the analog filter using Equation (8.24) and output the transfer function.

4. Verify the frequency responses, and output the difference equation.

The impulse invariant design method:

1. Given the lowpass or bandpass filter frequency specifications, perform analog filter design. For the highpass or bandstop filter design, quit this method and use the BLT.

    a. Determine the prototype filter order using Equation (8.29) for the Butterworth filter or Equation (8.35b) for the Chebyshev filter.

    b. Perform lowpass prototype transformation using the lowpass prototype in Table 8.3 (Butterworth function) or Tables 8.4 and 8.5 (Chebyshev functions) using Equations (8.20) to (8.23).

    c. Skip step 1 if the analog filter transfer function is given to begin with.

2. Determine the impulse response by applying the partial fraction expansion technique to the analog transfer function and inverse Laplace transform using Equation (8.37).

3. Sample the analog impulse response using Equation (8.38) and apply the z-transform to the digital impulse function to obtain the digital filter transfer function.

4. Verify the frequency responses, and output the difference equation. If the frequency specifications are not met, quit the design method and use the BLT.

The pole-zero placement design method:

1. Given the filter cutoff frequency specifications, determine the pole-zero locations using the corresponding equations:

    a. Second-order bandpass filter: Equations (8.41) and (8.42)

b. Second-order notch filter: Equations (8.45) and (8.46)

c. First-order lowpass filter: Equation (8.49) or (8.50)

d. First-order highpass filter: Equation (8.53) or (8.54).

2. Apply the corresponding equation and scale factor to obtain the digital filter transfer function:

   a. Second-order bandpass filter: Equations (8.43) and (8.44)

   b. Second-order notch filter: Equations (8.47) and (8.48)

   c. First-order lowpass filter: Equations (8.51) and (8.52)

   d. First-order highpass filter: Equations (8.55) and (8.56).

3. Verify the frequency responses, and output the difference equation. If the frequency specifications are not met, quit the design method and use the BLT.

Table 8.13 compares the design parameters of the three design methods.

Performance comparisons using the three methods are given in Figure 8.57, where the bandpass filter is designed using the following specifications:

Passband ripple = 3 dB
Center frequency = 400 Hz
Bandwidth = 200 Hz
Sampling rate = 2,000 Hz
Butterworth IIR filter = second-order

As we expected, the BLT method satisfies the design requirement, and the pole-zero placement method has little performance degradation because $r = 1 - (f_0/f_s)\pi = 0.6858 < 0.9$, and this effect will also cause the center frequency to be shifted. For the bandpass filter designed using the impulse invariant method, the gain at the center frequency is scaled to 1 for a frequency response shape comparison. The performance of the impulse invariant method is satisfied in passband. However, it has significant performance degradation in stopband when compared with the other two methods. This is due to aliasing when sampling the analog impulse response in time domain.

Improvement in using the pole-zero placement and impulse invariant methods can be achieved by using a very high sampling rate. Example 8.29 describes the possible selection of the design method by a DSP engineer to solve a real-world problem.

8.12 Summary of IIR Design Procedures and Selection of the IIR Filter Design Methods in Practice

TABLE 8.13 Comparisons of three IIR design methods.

	BLT	Design Method Impulse Invariant	Pole-Zero Placement
Filter type	Lowpass, highpass, bandpass, bandstop	Appropriate for lowpass and bandpass	2nd-order for bandpass and bandstop; 1st-order for lowpass and highpass
Linear phase	No	No	No
Ripple and stopband specifications	Used for determining the filter order	Used for determining the filter order	Not required; 3 dB on passband offered.
Special requirement	None	Very high sampling relative to the cutoff frequency (LPF) or to upper cutoff frequency for (BPF)	Narrow band for BPF or notch filter; lower cutoff frequency or higher cutoff frequency for LPF or HPF.
Algorithm complexity	High: Frequency prewarping, analog filter design, BLT	Moderate: Analog filter design determining digital impulse response. Apply z-transform	Simple design equations
Minimum design tool	Calculator, algebra	Calculator, algebra	Calculator

BLT = bilinear transformation; LPF = lowpass filter; BPF = bandpass filter; HPF = highpass filter.

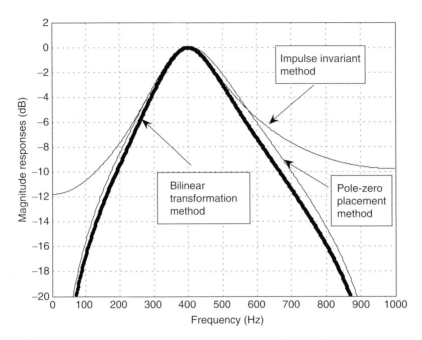

FIGURE 8.57 Performance comparisons for the BLT, pole-zero placement, and impulse invariant methods.

Example 8.29.

1. Determine an appropriate IIR filter design method for each of the following DSP applications. As described in a previous section, we apply a notch filter to remove 60 Hz interference and cascade a bandpass filter to remove noise in an ECG signal for heart rate detection. The following specifications are required:

    Notch filter:
    Harmonic to be removed = 60 Hz
    3 dB bandwidth for the notch filter = 4 Hz
    Bandpass filter:
    Passband frequency range = 0.25 to 40 Hz
    Passband ripple = 0.5 dB
    Sampling rate = 600 Hz.

The pole-zero placement method is the best choice, since the notch filter to be designed has a very narrow 3 dB bandwidth of 4 Hz. This simple design gives a quick solution. Since the bandpass filter requires a passband ripple of 0.5 dB from 0.25 to 40 Hz, the BLT can also be an appropriate choice. Even though the impulse invariant method could work for this case, since the sampling rate of 600 Hz is much larger than 40 Hz, aliasing cannot be prevented completely. Hence, the BLT is a preferred design method for the bandpass filter.

8.13 Summary

1. The BLT method is able to transform the transfer function of an analog filter to the transfer function of the corresponding digital filter in general.

2. The BLT maps the left half of an s-plane to the inside unit circle of the z-plane. Stability of mapping is guaranteed.

3. The BLT causes analog frequency warping. The analog frequency range from 0 Hz to infinity is warped to a digital frequency range from 0 Hz to the folding frequency.

4. Given the digital frequency specifications, analog filter frequency specifications must be developed using the frequency warping equation before designing the corresponding analog filter and applying the BLT.

5. An analog filter transfer function can be obtained by lowpass prototype, which can be selected from the Butterworth and Chebyshev functions.

6. The higher-order IIR filter can be designed using a cascade form.

7. The impulse invariant design method maps the analog impulse response to the digital equivalent impulse response. The method works for the lowpass and bandpass filter design with a very high sampling rate. It is not appropriate for the highpass and bandstop filter design.

8. The pole-zero placement method can be applied for a simple IIR filter design such as the second-order bandpass and bandstop filters with narrow band specifications, first-order lowpass and highpass filters with cutoff frequencies close to either DC or the folding frequency.

9. Quantizing IIR filter coefficients explores the fact that the quantization of the filter coefficients has more effect on the magnitude frequency response than on the phase frequency response. It may cause the quantized IIR filter to be unstable.

10. A simple audio equalizer uses bandpass IIR filter banks to create sound effects.

11. The 60-Hz interference eliminator is designed to enhance biomedical ECG signals for heart rate detection. It can also be adapted for audio humming noise elimination.

12. A single tone or a DTMF tone can be generated using the IIR filter with the impulse sequence as the filter input.

13. The Goertzel algorithm is applied for DTMF tone detection. This is an important application in the telecommunications industry.

14. The procedures for the BLT, impulse invariant, and pole-zero placement design methods were summarized, and their design feasibilities were compared, including the filter type, linear phase, ripple and stopband specifications, special requirements, algorithm complexity, design tool(s).

8.14 Problems

8.1. Given an analog filter with the transfer function

$$H(s) = \frac{1000}{s + 1000},$$

convert it to the digital filter transfer function and difference equation using the BLT if the DSP system has a sampling period of $T = 0.001$ second.

8.2. The lowpass filter with a cutoff frequency of 1 rad/sec is given as

$$H_P(s) = \frac{1}{s+1}.$$

a. Use $H_p(s)$ and the BLT to obtain a corresponding IIR digital lowpass filter with a cutoff frequency of 30 Hz, assuming a sampling rate of 200 Hz.

b. Use MATLAB to plot the magnitude and phase frequency responses of $H(z)$.

8.3. The normalized lowpass filter with a cutoff frequency of 1 rad/sec is given as

$$H_P(s) = \frac{1}{s+1}.$$

a. Use $H_p(s)$ and the BLT to obtain a corresponding IIR digital highpass filter with a cutoff frequency of 30 Hz, assuming a sampling rate of 200 Hz.

b. Use MATLAB to plot the magnitude and phase frequency responses of $H(z)$.

8.4. Consider the normalized lowpass filter with a cutoff frequency of 1 rad/sec:

$$H_P(s) = \frac{1}{s+1}.$$

a. Use $H_p(s)$ and the BLT to design a corresponding IIR digital notch (bandstop) filter with a lower cutoff frequency of 20 Hz, an upper cutoff frequency of 40 Hz, and a sampling rate of 120 Hz.

b. Use MATLAB to plot the magnitude and phase frequency responses of $H(z)$.

8.5. Consider the following normalized lowpass filter with a cutoff frequency of 1 rad/sec:

$$H_P(s) = \frac{1}{s+1}.$$

a. Use $H_p(s)$ and the BLT to design a corresponding IIR digital bandpass filter with a lower cutoff frequency of 15 Hz, an upper cutoff frequency of 25 Hz, and a sampling rate of 120 Hz.

b. Use MATLAB to plot the magnitude and phase frequency responses of $H(z)$.

8.6. Design a first-order digital lowpass Butterworth filter with a cutoff frequency of 1.5 kHz and a passband ripple of 3 dB at a sampling frequency of 8,000 Hz.

a. Determine the transfer function and difference equation.

b. Use MATLAB to plot the magnitude and phase frequency responses.

8.7. Design a second-order digital lowpass Butterworth filter with a cutoff frequency of 1.5 kHz and a passband ripple of 3 dB at a sampling frequency of 8,000 Hz.

a. Determine the transfer function and difference equation.

b. Use MATLAB to plot the magnitude and phase frequency responses.

8.8. Design a third-order digital highpass Butterworth filter with a cutoff frequency of 2 kHz and a passband ripple of 3 dB at a sampling frequency of 8,000 Hz.

a. Determine the transfer function and difference equation.

b. Use MATLAB to plot the magnitude and phase frequency responses.

8.9. Design a second-order digital bandpass Butterworth filter with a lower cutoff frequency of 1.9 kHz, an upper cutoff frequency of 2.1 kHz, and a passband ripple of 3 dB at a sampling frequency of 8,000 Hz.

a. Determine the transfer function and difference equation.

b. Use MATLAB to plot the magnitude and phase frequency responses.

8.10. Design a second-order digital bandstop Butterworth filter with a center frequency of 1.8 kHz, a bandwidth of 200 Hz, and a passband ripple of 3 dB at a sampling frequency of 8,000 Hz.

a. Determine the transfer function and difference equation.

b. Use MATLAB to plot the magnitude and phase frequency responses.

8.11. Design a first-order digital lowpass Chebyshev filter with a cutoff frequency of 1.5 kHz and 1 dB ripple on passband at a sampling frequency of 8,000 Hz.

a. Determine the transfer function and difference equation.

b. Use MATLAB to plot the magnitude and phase frequency responses.

8.12. Design a second-order digital lowpass Chebyshev filter with a cutoff frequency of 1.5 kHz and 0.5 dB ripple on passband at a sampling frequency of 8,000 Hz. Use MATLAB to plot the magnitude and phase frequency responses.

8.13. Design a third-order digital highpass Chebyshev filter with a cutoff frequency of 2 kHz and 1 dB ripple on the passband at a sampling frequency of 8,000 Hz.

a. Determine the transfer function and difference equation.

b. Use MATLAB to plot the magnitude and phase frequency responses.

8.14. Design a second-order digital bandpass Chebyshev filter with the following specifications:

Center frequency of 1.5 kHz
Bandwidth of 200 Hz
0.5 dB ripple on passband
Sampling frequency of 8,000 Hz.

 a. Determine the transfer function and difference equation.

 b. Use MATLAB to plot the magnitude and phase frequency responses.

8.15. Design a second-order bandstop digital Chebyshev filter with the following specifications:

Center frequency of 2.5 kHz
Bandwidth of 200 Hz
1 dB ripple on stopband
Sampling frequency of 8,000 Hz.

 a. Determine the transfer function and difference equation.

 b. Use MATLAB to plot the magnitude and phase frequency responses.

8.16. Design a fourth-order digital lowpass Butterworth filter with a cutoff frequency of 2 kHz and a passband ripple of 3 dB at a sampling frequency of 8,000 Hz.

 a. Determine the transfer function and difference equation.

 b. Use MATLAB to plot the magnitude and phase frequency responses.

8.17. Design a fourth-order digital lowpass Chebyshev filter with a cutoff frequency of 1.5 kHz and a 0.5 dB ripple at a sampling frequency of 8,000 Hz.

 a. Determine the transfer function and difference equation.

 b. Use MATLAB to plot the magnitude and phase frequency responses.

8.18. Design a fourth-order digital bandpass Chebyshev filter with a center frequency of 1.5 kHz, a bandwidth of 200 Hz, and a 0.5 dB ripple at a sampling frequency of 8,000 Hz.

 a. Determine the transfer function and difference equation.

 b. Use MATLAB to plot the magnitude and phase frequency responses.

8.19. Consider the following Laplace transfer function:

$$H(s) = \frac{10}{s+10}.$$

a. Determine $H(z)$ and the difference equation using the impulse invariant method if the sampling rate $f_s = 10$ Hz.

b. Use MATLAB to plot the magnitude frequency response $|H(f)|$ and the phase frequency response $\varphi(f)$ with respect to $H(s)$ for the frequency range from 0 to $f_s/2$ Hz.

c. Use MATLAB to plot the magnitude frequency response $|H(e^{j\Omega})| = |H(e^{j2\pi fT})|$ and the phase frequency response $\varphi(f)$ with respect to $H(z)$ for the frequency range from 0 to $f_s/2$ Hz.

8.20. Consider the following Laplace transfer function:

$$H(s) = \frac{1}{s^2 + 3s + 2}.$$

a. Determine $H(z)$ and the difference equation using the impulse invariant method if the sampling rate $f_s = 10$ Hz.

b. Use MATLAB to plot the magnitude frequency response $|H(f)|$ and the phase frequency response $\varphi(f)$ with respect to $H(s)$ for the frequency range from 0 to $f_s/2$ Hz.

c. Use MATLAB to plot the magnitude frequency response $|H(e^{j\Omega})| = |H(e^{j2\pi fT})|$ and the phase frequency response $\varphi(f)$ with respect to $H(z)$ for the frequency range from 0 to $f_s/2$ Hz.

8.21. Consider the following Laplace transfer function:

$$H(s) = \frac{s}{s^2 + 4s + 5}.$$

a. Determine $H(z)$ and the difference equation using the impulse invariant method if the sampling rate $f_s = 10$ Hz.

b. Use MATLAB to plot the magnitude frequency response $|H(f)|$ and the phase frequency response $\varphi(f)$ with respect to $H(s)$ for the frequency range from 0 to $f_s/2$ Hz.

c. Use MATLAB to plot the magnitude frequency response $|H(e^{j\Omega})| = |H(e^{j2\pi fT})|$ and the phase frequency response $\varphi(f)$ with respect to $H(z)$ for the frequency range from 0 to $f_s/2$ Hz.

8.22. A second-order bandpass filter is required to satisfy the following specifications:

Sampling rate = 8,000 Hz
3 dB bandwidth: $BW = 100$ Hz
Narrow passband centered at $f_0 = 2,000$ Hz
Zero gain at 0 Hz and 4,000 Hz.

Find the transfer function and difference equation by the pole-zero placement method.

8.23. A second-order notch filter is required to satisfy the following specifications:

Sampling rate = 8,000 Hz
3 dB bandwidth: $BW = 200$ Hz
Narrow passband centered at $f_0 = 1,000$ Hz.

Find the transfer function and difference equation by the pole-zero placement method.

8.24. A first-order lowpass filter is required to satisfy the following specifications:

Sampling rate = 8,000 Hz
3 dB cutoff frequency: $f_c = 200$ Hz
Zero gain at 4,000 Hz.

Find the transfer function and difference equation using the pole-zero placement method.

8.25. A first-order lowpass filter is required to satisfy the following specifications:

Sampling rate = 8,000 Hz
3 dB cutoff frequency: $f_c = 3,800$ Hz
Zero gain at 4,000 Hz.

Find the transfer function and difference equation by the pole-zero placement method.

8.26. A first-order highpass filter is required to satisfy the following specifications:

Sampling rate = 8,000 Hz
3 dB cutoff frequency: $f_c = 3,850$ Hz
Zero gain at 0 Hz.

Find the transfer function and difference equation by the pole-zero placement method.

8.27. A first-order highpass filter is required to satisfy the following specifications:

Sampling rate = 8,000 Hz
3 dB cutoff frequency: $f_c = 100$ Hz
Zero gain at 0 Hz.

Find the transfer function and difference equation by the pole-zero placement method.

8.28. Given a filter transfer function,

$$H(z) = \frac{0.3430z^2 + 0.6859z + 0.3430}{z^2 + 0.7075z + 0.7313},$$

a. realize the digital filter using direct form I and using direct form II;

b. determine the difference equations for each implementation.

8.29. Given a fourth-order filter transfer function,

$$H(z) = \frac{0.3430z^2 + 0.6859z + 0.3430}{z^2 + 0.7075z + 0.7313} \times \frac{0.4371z^2 + 0.8742z + 0.4371}{z^2 - 0.1316z + 0.1733},$$

a. realize the digital filter using the cascade (series) form via second-order sections using the direct form II;

b. determine the difference equations for implementation.

Use MATLAB to solve problems 8.30 to 8.36.

8.30. A speech sampled at 8,000 Hz is corrupted by a sine wave of 360 Hz. Design a notch filter to remove the noise with the following specifications:

Chebyshev notch filter
Center frequency: 360 Hz
Bandwidth: 60 Hz
Passband and ripple: 0.5 dB
Stopband attenuation: 5 dB at 355 Hz and 365 Hz, respectively.

Determine the transfer function and difference equation.

8.31. In Problem 8.30, if the speech is corrupted by a sine wave of 360 Hz and its third harmonic, cascading two notch filters can be applied to remove noise signals. The possible specifications are given as:

Chebyshev notch filter 1
Center frequency: 360 Hz

Bandwidth: 60 Hz
Passband and ripple: 0.5 dB
Stopband attenuation: 5 dB at 355 Hz and 365 Hz, respectively.
Chebyshev notch filter 2
Center frequency: 1,080 Hz
Bandwidth: 60 Hz
Passband and ripple: 0.5 dB
Stopband attenuation: 5 dB at 1,075 Hz 1,085 Hz, respectively.

Determine the transfer function and difference equation for each filter (Fig. 8.58).

8.32. In a speech recording system with a sampling frequency of 10,000 Hz, the speech is corrupted by random noise. To remove the random noise while preserving speech information, the following specifications are given:

Speech frequency range: 0–3,000 kHz
Stopband range: 4,000–5,000 Hz
Passband ripple: 3 dB
Stopband attenuation: 25 dB
Butterworth IIR filter.

Determine the filter order and transfer function.

8.33. In Problem 8.32, if we use a Chebyshev IIR filter with the following specifications:

Speech frequency range: 0–3,000 Hz
Stopband range: 4,000–5,000 Hz
Passband ripple: 1 dB
Stopband attenuation: 35 dB
Chebyshev IIR filter,

determine the filter order and transfer function.

8.34. Given a speech equalizer to compensate midrange frequency loss of hearing (Fig. 8.59) and the following specifications:

Sampling rate: 8,000 Hz
Second-order bandpass IIR filter

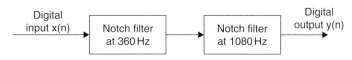

FIGURE 8.58. Cascaded notch filter in Problem 8.31.

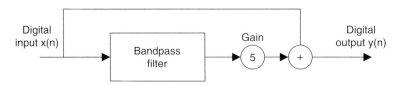

FIGURE 8.59. Speech equalizer in Problem 8.34.

Frequency range to be emphasized: 1,500–2,000 Hz
Passband ripple: 3 dB
Pole-zero placement design method,

determine the transfer function.

8.35. In Problem 8.34, if we use an IIR filter with the following specifications:

Sampling rate: 8,000 Hz
Butterworth IIR filter
Frequency range to be emphasized: 1,500–2,000 Hz
Lower stop band: 0–1,000 Hz
Upper stop band: 2,500–4,000 Hz
Passband ripple: 3 dB
Stopband attenuation: 20 dB,

determine the filter order and filter transfer function.

8.36. A digital crossover can be designed as shown in Figure 8.60. Given audio specifications as:

Sampling rate: 44,100 Hz
Crossover frequency: 1,000 Hz
Highpass filter: third-order Butterworth type at a cutoff frequency of 1,000 Hz
Lowpass filter: third-order Butterworth type at a cutoff frequency of 1,000 Hz,

use the MATLAB BLT design method to determine:

a. the transfer functions and difference equations for the highpass and lowpass filters

b. frequency responses for the highpass filter and the lowpass filter

c. combined frequency response for both filters.

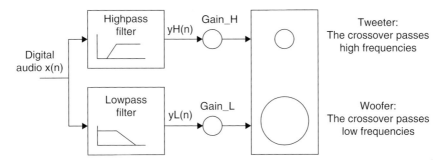

FIGURE 8.60. Two-band digital crossover system in Problem 8.36.

8.37. Given a DSP system with a sampling rate set up to be 8,000 Hz, develop an 800 Hz single-tone generator using a digital IIR filter by completing the following steps:

  a. Determine the digital IIR filter transfer function.

  b. Determine the DSP equation (difference equation).

  c. Write a MATLAB program using the MATLAB function filter() to generate and plot the 800-Hz tone for a duration of 0.01 sec.

8.38. Given a DSP system with a sampling rate set up to be 8,000 Hz, develop a DTMF tone generator for key "5" using digital IIR filters by completing the following steps:

  a. Determine the digital IIR filter transfer functions.

  b. Determine the DSP equations (difference equation).

  c. Write a MATLAB program using the MATLAB function filter() to generate and plot the DTMF tone for key 5 for 205 samples.

8.39. Given $x(0) = 1$, $x(1) = 1$, $x(2) = 0$, $x(3) = -1$, use the Goertzel algorithm to compute the following DFT coefficients and their amplitude spectra:

  a. $X(0)$

  b. $|X(0)|^2$

  c. A_0 (single sided)

  d. $X(1)$

  e. $|X(1)|^2$

  f. A_1 (single sided)

8.40. Given a DSP system with a sampling rate set up to be 8,000 Hz and a data size of 205 ($N = 205$), seven Goertzel IIR filters are implemented for DTMF tone detection. For the frequencies corresponding to key 5, determine:

a. the modified Goertzel filter transfer functions

b. the filter DSP equations for $v_k(n)$

c. the DSP equations for the squared magnitudes

$$|X(k)|^2 = |y_k(205)|^2$$

d. Using the data generated in Problem 8.38 (c), write a program using the MATLAB function filter() and Goertzel algorithm to detect the spectral values of the DTMF tone for key 5.

8.41. Given an input data sequence:

$x(n) = 1.2 \cdot \sin(2\pi(1000)n/10000)) - 1.5 \cdot \cos(2\pi(4000)n/10000)$

assuming a sampling frequency of 10 kHz, implement the designed IIR filter in Problem 8.33 to filter 500 data points of $x(n)$ with the following specified method, and plot the 500 samples of the input and output data.

a. Direct-form I implementation

b. Direct-form II implementation

References

Carr, J. J., and Brown, J. M. (2001). *Introduction to Biomedical Equipment Technology*, 4th ed. Upper Saddle River, NJ: Prentice Hall.

Ifeachor, E. C., and Jervis, B. W. (2002). *Digital Signal Processing: A Practical Approach*, 2nd ed. Upper Saddle River, NJ: Prentice Hall.

Lynn, P. A., and Fuerst, W. (1999). *Introductory Digital Signal Processing with Computer Applications*, 2nd ed. Chichester and New York: John Wiley & Sons.

Porat, B. (1997). *A Course in Digital Signal Processing*. New York: John Wiley & Sons.

Proakis, J. G., and Manolakis, D. G. (1996). *Digital Signal Processing: Principles, Algorithms, and Applications*, 3rd ed. Upper Saddle River, NJ: Prentice Hall.

Webster, J. G. (1998). *Medical Instrumentation: Application and Design*, 3rd ed. New York: John Wiley & Sons, Inc.

9

Hardware and Software for Digital Signal Processors

Objectives:

This chapter introduces basics of digital signal processors such as processor architectures and hardware units, investigates fixed-point and floating-point formats, and illustrates the implementation of digital filters in real time.

9.1 Digital Signal Processor Architecture

Unlike microprocessors and microcontrollers, digital signal (DS) processors have special features that require operations such as fast Fourier transform (FFT), filtering, convolution and correlation, and real-time sample-based and block-based processing. Therefore, DS processors use a different dedicated hardware architecture.

We first compare the architecture of the general microprocessor with that of the DS processor. The design of general microprocessors and microcontrollers is based on the *Von Neumann architecture,* which was developed from a research paper written by John von Neumann and others in 1946. Von Neumann suggested that computer instructions, as we shall discuss, be numerical codes instead of special wiring. Figure 9.1 shows the Von Neumann architecture.

As shown in Figure 9.1, a Von Neumann processor contains a single, shared memory for programs and data, a single bus for memory access, an arithmetic unit, and a program control unit. The processor proceeds in a serial fashion in terms of fetching and execution cycles. This means that the central processing unit (CPU) fetches an instruction from memory and decodes it to figure out

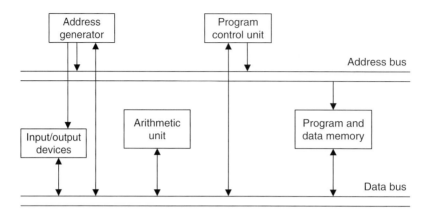

FIGURE 9.1 General microprocessor based on Von Neumann architecture.

what operation to do, then executes the instruction. The instruction (in machine code) has two parts: the *opcode* and the *operand.* The opcode specifies what the operation is, that is, tells the CPU what to do. The operand informs the CPU what data to operate on. These instructions will modify memory, or input and output (I/O). After an instruction is completed, the cycles will resume for the next instruction. One an instruction or piece of data can be retrieved at a time. Since the processor proceeds in a serial fashion, it causes most units to stay in a wait state.

As noted, the Von Neumann architecture operates the cycles of fetching and execution by fetching an instruction from memory, decoding it via the program control unit, and finally executing the instruction. When execution requires data movement—that is, data to be read from or written to memory—the next instruction will be fetched after the current instruction is completed. The Von Neumann–based processor has this bottleneck mainly due to the use of a single, shared memory for both program instructions and data. Increasing the speed of the bus, memory, and computational units can improve speed, but not significantly.

To accelerate the execution speed of digital signal processing, DS processors are designed based on the *Harvard architecture,* which originated from the Mark 1 relay-based computers built by IBM in 1944 at Harvard University. This computer stored its instructions on punched tape and data using relay latches. Figure 9.2 shows today's Harvard architecture. As depicted, the DS processor has two separate memory spaces. One is dedicated to the program code, while the other is employed for data. Hence, to accommodate two memory spaces, two corresponding address buses and two data buses are used. In this way, the program memory and data memory have their own

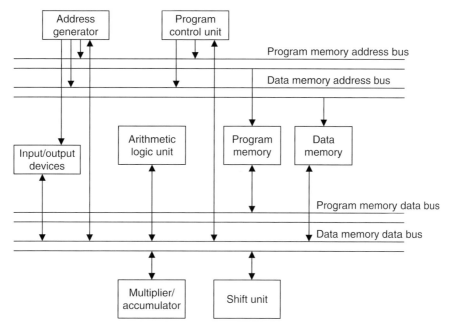

FIGURE 9.2 Digital signal processors based on the Harvard architecture.

connections to the program memory bus and data memory bus, respectively. This means that the Harvard processor can fetch the program instruction and data in parallel at the same time, the former via the program memory bus and the latter via the data memory bus. There is an additional unit called a *multiplier and accumulator* (MAC), which is the dedicated hardware used for the digital filtering operation. The last additional unit, the shift unit, is used for the scaling operation for fixed-point implementation when the processor performs digital filtering.

Let us compare the executions of the two architectures. The Von Neumann architecture generally has the execution cycles described in Figure 9.3. The fetch cycle obtains the opcode from the memory, and the control unit will decode the instruction to determine the operation. Next is the execute cycle. Based on the decoded information, execution will modify the content of the register or the memory. Once this is completed, the process will fetch the next instruction and continue. The processor operates one instruction at a time in a serial fashion.

To improve the speed of the processor operation, the Harvard architecture takes advantage of a common DS processor, in which one register holds the filter coefficient while the other register holds the data to be processed, as depicted in Figure 9.4.

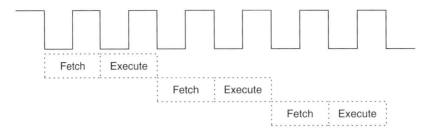

FIGURE 9.3 Execution cycle based on the Von Neumann architecture.

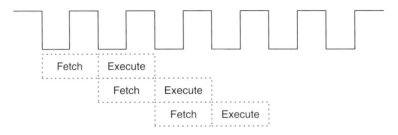

FIGURE 9.4 Execution cycle based on the Harvard architecture.

As shown in Figure 9.4, the execute and fetch cycles are overlapped. We call this the *pipelining* operation. The DS processor performs one execution cycle while also fetching the next instruction to be executed. Hence, the processing speed is dramatically increased.

The Harvard architecture is preferred for all DS processors due to the requirements of most DSP algorithms, such as filtering, convolution, and FFT, which need repetitive arithmetic operations, including multiplications, additions, memory access, and heavy data flow through the CPU.

For other applications, such as those dependent on simple microcontrollers with less of a timing requirement, the Von Neumann architecture may be a better choice, since it offers much less silica area and is thus less expansive.

9.2 Digital Signal Processor Hardware Units

In this section, we will briefly discuss special DS processor hardware units.

9.2.1 Multiplier and Accumulator

As compared with the general microprocessors based on the Von Neumann architecture, the DS processor uses the MAC, a special hardware unit for

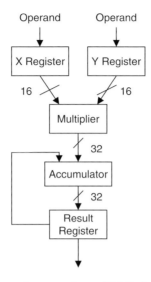

FIGURE 9.5   The multiplier and accumulator (MAC) dedicated to DSP.

enhancing the speed of digital filtering. This is dedicated hardware, and the corresponding instruction is generally referred to as MAC operation. The basic structure of the MAC is shown in Figure 9.5.

As shown in Figure 9.5, in a typical hardware MAC, the multiplier has a pair of input registers, each holding the 16-bit input to the multiplier. The result of the multiplication is accumulated in a 32-bit accumulator unit. The result register holds the double precision data from the accumulator.

9.2.2 Shifters

In digital filtering, to prevent overflow, a scaling operation is required. A simple scaling-down operation shifts data to the right, while a scaling-up operation shifts data to the left. Shifting data to the right is the same as dividing the data by 2 and truncating the fraction part; shifting data to the left is equivalent to multiplying the data by 2. As an example, for a 3-bit data word $011_2 = 3_{10}$, shifting 011 to the right gives $001_2 = 1$, that is, $3/2 = 1.5$, and truncating 1.5 results in 1. Shifting the same number to the left, we have $110_2 = 6_{10}$, that is, $3 \times 2 = 6$. The DS processor often shifts data by several bits for each data word. To speed up such operation, the special hardware shift unit is designed to accommodate the scaling operation, as depicted in Figure 9.2.

9.2.3 Address Generators

The DS processor generates the addresses for each datum on the data buffer to be processed. A special hardware unit for circular buffering is used (see the address generator in Figure 9.2). Figure 9.6 describes the basic mechanism of circular buffering for a buffer having eight data samples.

In circular buffering, a pointer is used and always points to the newest data sample, as shown in the figure. After the next sample is obtained from analog-to-digital conversion (ADC), the data will be placed at the location of $x(n-7)$, and the oldest sample is pushed out. Thus, the location for $x(n-7)$ becomes the location for the current sample. The original location for $x(n)$ becomes a location for the past sample of $x(n-1)$. The process continues according to the mechanism just described. For each new data sample, only one location on the circular buffer needs to be updated.

The circular buffer acts like a first-in/first-out (FIFO) buffer, but each datum on the buffer does not have to be moved. Figure 9.7 gives a simple illustration of the 2-bit circular buffer. In the figure, there is data flow to the ADC (*a, b, c, d, e, f, g,* ...) and a circular buffer initially containing *a, b, c,* and *d*. The pointer specifies the current data of *d*, and the equivalent FIFO buffer is shown on the right side with a current data of *d* at the top of the memory. When *e* comes in, as shown in the middle drawing in Figure 9.7, the circular buffer will change the pointer to the next position and update old *a* with a new datum *e*. It costs the pointer only one movement to update one datum in one step. However, on the right side, the FIFO has to move each of the other data down to let in the new

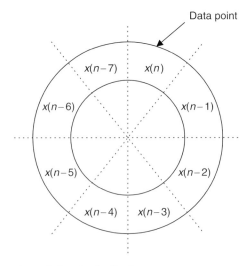

FIGURE 9.6 Illustration of circular buffering.

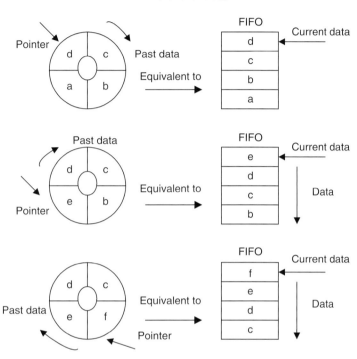

FIGURE 9.7 Circular buffer and equivalent FIFO.

datum *e* at the top. For this FIFO, it takes four data movements. In the bottom drawing in Figure 9.7, the incoming datum *f* for both the circular buffer and the FIFO buffer continues to confirm our observations.

Like finite impulse response (FIR) filtering, the data buffer size can reach several hundreds. Hence, using the circular buffer will significantly enhance the processing speed.

9.3 Digital Signal Processors and Manufacturers

DS processors are classified for general DSP and special DSP. The general-DSP processor is designed and optimized for applications such as digital filtering, correlation, convolution, and FFT. In addition to these applications, the special-DSP processor has features that are optimized for unique applications such as audio processing, compression, echo cancellation, and adaptive filtering. Here, we will focus on the general-DSP processor.

The major manufacturers in the DSP industry are Texas Instruments (TI), Analog Devices, and Motorola. TI and Analog Devices offer both fixed-point DSP families and floating-point DSP families, while Motorola offers fixed-point DSP families. We will concentrate on TI families, review their architectures, and study real-time implementation using the fixed- and floating-point formats.

9.4 Fixed-Point and Floating-Point Formats

In order to process real-world data, we need to select an appropriate DS processor, as well as a DSP algorithm or algorithms for a certain application. Whether a DS processor uses a fixed- or floating-point method depends on how the processor's CPU performs arithmetic. A fixed-point DS processor represents data in *2's complement integer format* and manipulates data using integer arithmetic, while a floating-point processor represents numbers using a mantissa (fractional part) and an exponent in addition to the integer format and operates data using floating-point arithmetic (discussed in a later section).

Since the fixed-point DS processor operates using the integer format, which represents only a very narrow dynamic range of the integer number, a problem such as overflow of data manipulation may occur. Hence, we need to spend much more coding effort to deal with such a problem. As we shall see, we may use floating-point DS processors, which offer a wider dynamic range of data, so that coding becomes much easier. However, the floating-point DS processor contains more hardware units to handle the integer arithmetic and the floating-point arithmetic, hence is more expensive and slower than fixed-point processors in terms of instruction cycles. It is usually a choice for prototyping or proof-of-concept development.

When it is time to make the DSP an application-specific integrated circuit (ASIC), a chip designed for a particular application, a dedicated hand-coded fixed-point implementation can be the best choice in terms of performance and small silica area.

The formats used by DSP implementation can be classified as fixed or floating point.

9.4.1 Fixed-Point Format

We begin with 2's complement representation. Considering a 3-bit 2's complement, we can represent all the decimal numbers shown in Table 9.1.

9.4 Fixed-Point and Floating-Point Formats

TABLE 9.1 A 3-bit 2's complement number representation.

Decimal Number	2's Complement
3	011
2	010
1	001
0	000
−1	111
−2	110
−3	101
−4	100

Let us review the 2's complement number system using Table 9.1. Converting a decimal number to its 2's complement requires the following steps:

1. Convert the magnitude in the decimal to its binary number using the required number of bits.

2. If the decimal number is positive, its binary number is its 2's complement representation; if the decimal number is negative, perform the 2's complement operation, where we negate the binary number by changing the logic 1's to logic 0's and logic 0's to logic 1's and then add a logic 1 to the data. For example, a decimal number of 3 is converted to its 3-bit 2's complement as 011; however, for converting a decimal number of −3, we first get a 3-bit binary number for the magnitude in the decimal, that is, 011. Next, negating the binary number 011 yields the binary number 100. Finally, adding a binary logic 1 achieves the 3-bit 2's complement representation of −3, that is, $100 + 1 = 101$, as shown in Table 9.1.

As we see, a 3-bit 2's complement number system has a dynamic range from −4 to 3, which is very narrow. Since the basic DSP operations include multiplications and additions, results of operation can cause overflow problems. Let us examine the multiplications in Example 9.1.

Example 9.1.

Given

1. $2 \times (-1)$

2. $2 \times (-3)$,

   a. Operate each using its 2's complement.

a. 1.
$$\begin{array}{r} 010 \\ \times\ 001 \\ \hline 010 \\ 000 \\ +\ 000 \\ \hline 00010 \end{array}$$

and 2's complement of $00010 = 11110$. Removing two extended sign bits gives 110.

The answer is 110 (-2), which is within the system.

2.
$$\begin{array}{r} 010 \\ \times\ 011 \\ \hline 010 \\ 010 \\ +\ 000 \\ \hline 00110 \end{array}$$

and 2's complement of $00110 = 11010$. Removing two extended sign bits achieves 010.

Since the binary number 010 is 2, which is not (-6) as we expect, overflow occurs; that is, the result of the multiplication (-6) is out of our dynamic range $(-4$ to $3)$.

Let us design a system treating all the decimal values as fractional numbers, so that we obtain the fractional binary 2's complement system shown in Table 9.2.

To become familiar with the fractional binary 2's complement system, let us convert a positive fraction number $\frac{3}{4}$ and a negative fraction number $-\frac{1}{4}$ in decimals to their 2's complements. Since

$$\frac{3}{4} = 0 \times 2^0 + 1 \times 2^{-1} + 1 \times 2^{-2},$$

its 2's complement is 011. Note that we did not mark the binary point for clarity. Again, since

$$\frac{1}{4} = 0 \times 2^0 + 0 \times 2^{-1} + 1 \times 2^{-2},$$

9.4 Fixed-Point and Floating-Point Formats

TABLE 9.2 A 3-bit 2's complement system using fractional representation.

Decimal Number	Decimal Fraction	2's Complement
3	3/4	0.11
2	2/4	0.10
1	1/4	0.01
0	0	0.00
−1	−1/4	1.11
−2	−2/4	1.10
−3	−3/4	1.01
−4	−4/4 = −1	1.00

its positive-number 2's complement is 001. For the negative number, applying the 2's complement to the binary number 001 leads to $110 + 1 = 111$, as we see in Table 9.2.

Now let us focus on the fractional binary 2's complement system. The data are normalized to the fractional range from -1 to $1 - 2^{-2} = \frac{3}{4}$. When we carry out multiplications with two fractions, the result should be a fraction, so that multiplication overflow can be prevented. Let us verify the multiplication $(010) \times (101)$, which is the overflow case in Example 9.1:

$$\begin{array}{r} 0.1\;0 \\ \times\;0.1\;1 \\ \hline 0\;1\;0 \\ 0\;1\;0 \\ +\;0\;0\;0 \\ \hline 0.0\;1\;1\;0 \end{array}$$

2's complement of $0.0110 = 1.1010$.
The answer in decimal form should be

$$1.1010 = (-1) \times (0.0110)_2 = -\left(0 \times (2)^{-1} + 1 \times (2)^{-2} + 1 \times (2)^{-3} + 0 \times (2)^{-4}\right)$$

$$= -\frac{3}{8}.$$

This number is correct, as we can verify from Table 9.2, that is, $\left(\frac{2}{4} \times \left(-\frac{3}{4}\right)\right) = -\frac{3}{8}$.

If we truncate the last two least-significant bits to keep the 3-bit binary number, we have an approximated answer as

$$1.10 = (-1) \times (0.10)_2 = -\left(1 \times (2)^{-1} + 0 \times (2)^{-2}\right) = -\frac{1}{2}.$$

The truncation error occurs. The error should be bounded by $2^{-2} = \frac{1}{4}$. We can verify that

$$|-1/2 - (-3/8)| = 1/8 < 1/4.$$

To use such a scheme, we can avoid the overflow due to multiplications but cannot prevent the additional overflow. In the following addition example,

$$\begin{array}{r} 0.11 \\ + 0.01 \\ \hline 1.00 \end{array}$$

where the result 1.00 is a negative number.

Adding two positive fractional numbers yields a negative number. Hence, overflow occurs. We see that this signed fractional number scheme partially solves the overflow in multiplications. Such fractional number format is called the signed Q-2 format, where there are 2 magnitude bits plus one sign bit. The additional overflow will be tackled using a scaling method discussed in a later section.

Q-format number representation is the most common one used in fixed-point DSP implementation. It is defined in Figure 9.8.

As indicated in Figure 9.8, Q-15 means that the data are in a sign magnitude form in which there are 15 bits for magnitude and one bit for sign. Note that after the sign bit, the dot shown in Figure 9.8 implies the binary point. The number is normalized to the fractional range from -1 to 1. The range is divided into 2^{16} intervals, each with a size of 2^{-15}. The most negative number is -1, while the most positive number is $1 - 2^{-15}$. Any result from multiplication is within the fractional range of -1 to 1. Let us study the following examples to become familiar with Q-format number representation.

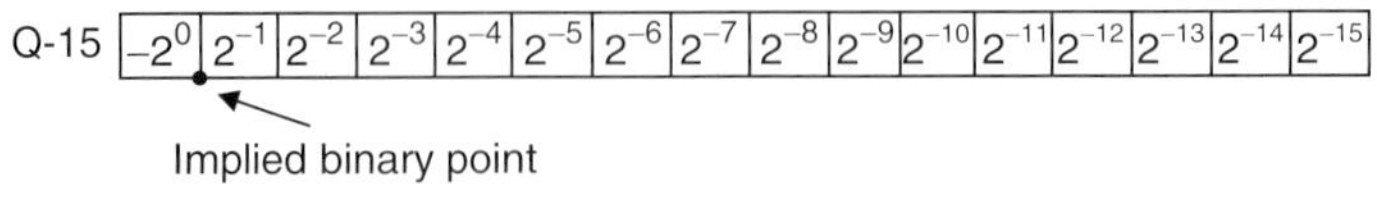

FIGURE 9.8 Q-15 (fixed-point) format.

Example 9.2.

a. Find the signed Q-15 representation for the decimal number 0.560123.

Solution:

a. The conversion process is illustrated using Table 9.3. For a positive fractional number, we multiply the number by 2 if the product is larger than 1, carry bit 1 as a most-significant bit (MSB), and copy the fractional part to the next line for the next multiplication by 2; if the product is less than 1, we carry bit 0 to MSB. The procedure continues to collect all 15 magnitude bits.

We yield the Q-15 format representation as

0.100011110110010.

Since we use only 16 bits to represent the number, we may lose accuracy after conversion. Like quantization, the truncation error is introduced. However, this error should be less than the interval size, in this case, $2^{-15} = 0.000030517$. We shall verify this in Example 9.5. An alternative way of conversion is to convert a fraction, let's say $\frac{3}{4}$, to Q-2 format, multiply it by 2^2, and then convert the truncated integer to its binary, that is,

$$(3/4) \times 2^2 = 3 = 011_2.$$

TABLE 9.3 Conversion process of Q-15 representation.

Number	Product	Carry
0.560123 × 2	1.120246	1 (MSB)
0.120246 × 2	0.240492	0
0.240492 × 2	0.480984	0
0.480984 × 2	0.961968	0
0.961968 × 2	1.923936	1
0.923936 × 2	1.847872	1
0.847872 × 2	1.695744	1
0.695744 × 2	1.391488	1
0.391488 × 2	0.782976	0
0.782976 × 2	1.565952	1
0.565952 × 2	1.131904	1
0.131904 × 2	0.263808	0
0.263808 × 2	0.527616	0
0.527616 × 2	1.055232	1
0.055232 × 2	0.110464	0 (LSB)

MSB, most-significant bit; LSB, least-significant bit.

In this way, it follows that

$$(0.560123) \times 2^{15} = 18354.$$

Converting 18354 to its binary representation will achieve the same answer. The next example illustrates the signed Q-15 representation for a negative number.

Example 9.3.

a. Find the signed Q-15 representation for the decimal number -0.160123.

Solution:

a. Converting the Q-15 format for the corresponding positive number with the same magnitude using the procedure described in Example 9.2, we have

$$0.160123 = 0.001010001111110.$$

Then, after applying 2's complement, the Q-15 format becomes

$$-0.160123 = 1.110101110000010.$$

Alternative way: Since $(-0.160123) \times 2^{15} = -5246.9$, converting the truncated number -5246 to its 16-bit 2's complement yields 1110101110000010.

Example 9.4.

a. Convert the Q-15 signed number 1.110101110000010 to the decimal number.

Solution:

a. Since the number is negative, applying the 2's complement yields

$$0.001010001111110.$$

Then the decimal number is

$$-(2^{-3} + 2^{-5} + 2^{-9} + 2^{-10} + 2^{-11} + 2^{-12} + 2^{-13} + 2^{-14}) = -0.160095.$$

Example 9.5.

a. Convert the Q-15 signed number 0.100011110110010 to the decimal number.

Solution:

a. The decimal number is

$$2^{-1} + 2^{-5} + 2^{-6} + 2^{-7} + 2^{-8} + 2^{-10} + 2^{-11} + 2^{-14} = 0.560120.$$

As we know, the truncation error in Example 9.2 is less than $2^{-15} = 0.000030517$. We verify that the truncation error is bounded by

$$|0.560120 - 0.560123| = 0.000003 < 0.000030517.$$

Note that the larger the number of bits used, the smaller the round-off error that may accompany it.

Examples 9.6 and 9.7 are devoted to illustrating data manipulations in the Q-15 format.

Example 9.6.

a. Add the two numbers in Examples 9.4 and 9.5 in Q-15 format.

Solution:

a. Binary addition is carried out as follows:

$$\begin{array}{r} 1.\ 1\ 1\ 0\ 1\ 0\ 1\ 1\ 1\ 0\ 0\ 0\ 0\ 0\ 1\ 0 \\ +\ 0.\ 1\ 0\ 0\ 0\ 1\ 1\ 1\ 1\ 0\ 1\ 1\ 0\ 0\ 1\ 0 \\ \hline 1\ 0.\ 0\ 1\ 1\ 0\ 0\ 1\ 1\ 0\ 0\ 1\ 1\ 0\ 1\ 0\ 0 \end{array}.$$

Then the result is

$$0.\ 0\ 1\ 1\ 0\ 0\ 1\ 1\ 0\ 0\ 1\ 1\ 0\ 1\ 0\ 0.$$

This number in the decimal form can be found to be

$$2^{-2} + 2^{-3} + 2^{-6} + 2^{-7} + 2^{-10} + 2^{-11} + 2^{-13} = 0.400024.$$

Example 9.7.

This is a simple illustration of fixed-point multiplication.

a. Determine the fixed-point multiplication of 0.25 and 0.5 in Q-3 fixed-point 2's complement format.

Solution:

a. Since $0.25 = 0.010$ and $0.5 = 0.100$, we carry out binary multiplication as follows:

$$
\begin{array}{r}
0.010 \\
\times\ 0.100 \\
\hline
0\ 000 \\
00\ 00 \\
001\ 0 \\
+\ 0\ 000 \\
\hline
0.001\ 000
\end{array}
$$

Truncating the least-significant bits to convert the result to Q-3 format, we have

$$0.010 \times 0.100 = 0.001.$$

Note that $0.001 = 2^{-3} = 0.125$. We can also verify that $0.25 \times 0.5 = 0.125$.

As a result, the Q-format number representation is a better choice than the 2's complement integer representation. But we need to be concerned with the following problems.

1. Converting a decimal number to its Q-N format, where N denotes the number of magnitude bits, we may lose accuracy due to the truncation error, which is bounded by the size of the interval, that is, 2^{-N}.

2. Addition and subtraction may cause overflow, where adding two positive numbers leads to a negative number, or adding two negative numbers yields a positive number; similarly, subtracting a positive number from a negative number gives a positive number, while subtracting a negative number from a positive number results in a negative number.

3. Multiplying two numbers in Q-15 format will lead to a Q-30 format, which has 31 bits in total. As in Example 9.7, the multiplication of Q-3 yields a Q-6 format, that is, 6 magnitude bits and a sign bit. In practice, it is common for a DS processor to hold the multiplication result using a double word size such as MAC operation, as shown in Figure 9.9 for multiplying two numbers in Q-15 format. In Q-30 format, there is one sign-extended bit. We may get rid of it by shifting left by one bit to obtain Q-31 format and maintaining the Q-31 format for each MAC operation.

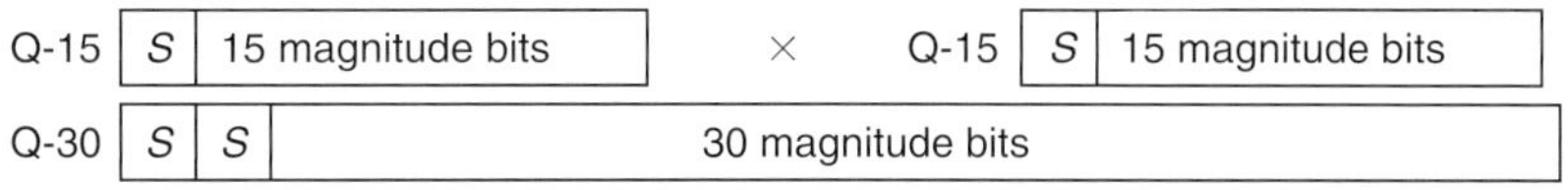

9.4 Fixed-Point and Floating-Point Formats

```
Q-15 | S | 15 magnitude bits          ×    Q-15 | S | 15 magnitude bits
Q-30 | S | S |             30 magnitude bits
```

FIGURE 9.9   Sign bit extended Q-30 format.

Sometimes, the number in Q-31 format needs to be converted to Q-15; for example, the 32-bit data in the accumulator needs to be sent for 16-bit digital-to-analog conversion (DAC), where the upper most-significant 16 bits in the Q-31 format must be used to maintain accuracy. We can shift the number in Q-30 to the right by 15 bits or shift the Q-31 number to the right by 16 bits. The useful result is stored in the lower 16-bit memory location. Note that after truncation, the maximum error is bounded by the interval size of 2^{-15}, which satisfies most applications. In using the Q-format in the fixed-point DS processor, it is costive to maintain the accuracy of data manipulation.

4. Underflow can happen when the result of multiplication is too small to be represented in the Q-format. As an example, in the Q-2 system shown in Table 9.2, multiplying 0.01×0.01 leads to 0.0001. To keep the result in Q-2, we truncate the last two bits of 0.0001 to achieve 0.00, which is zero. Hence, underflow occurs.

9.4.2 Floating-Point Format

To increase the dynamic range of number representation, a floating-point format, which is similar to scientific notation, is used. The general format for floating-point number representation is given by

$$x = M \cdot 2^E, \tag{9.1}$$

where M is the mantissa, or fractional part, in Q format, and E is the exponent. The mantissa and exponent are signed numbers. If we assign 12 bits for the mantissa and 4 bits for the exponent, the format looks like Figure 9.10.

Since the 12-bit mantissa has limits between -1 and $+1$, the dynamic range is controlled by the number of bits assigned to the exponent. The bigger the number of bits assigned to the exponent, the larger the dynamic range. The number of bits for the mantissa defines the interval in the normalized range; as shown in Figure 9.10, the interval size is 2^{-11} in the normalized range, which is smaller than the Q-15. However, when more mantissa bits are used, the smaller interval size will be achieved. Using the format in Figure 9.10, we can determine the most negative and most positive numbers as:

FIGURE 9.10 **Floating-point format.**

Most negative number $= (1.00000000000)_2 \cdot 2^{0111_2} = (-1) \times 2^7 = -128.0$

Most positive number $= (0.11111111111)_2 \cdot 2^{0111_2} = (1 - 2^{-11}) \times 2^7 = 127.9375$.

The smallest positive number is given by

Smallest positive number $= (0.00000000001)_2 \cdot 2^{1000_2} = (2^{-11}) \times 2^{-8} = 2^{-19}$.

As we can see, the exponent acts like a scale factor to increase the dynamic range of the number representation. We study the floating-point format in the following example.

Example 9.8.

a. Convert each of the following decimal numbers to the floating-point number using the format specified in Figure 9.10.

1. 0.1601230
2. -20.430527

Solution:

a. 1. We first scale the number 0.1601230 to $0.160123/2^{-2} = 0.640492$ with an exponent of -2 (other choices could be 0 or -1) to get $0.160123 = 0.640492 \times 2^{-2}$. Using 2's complement, we have $-2 = 1110$. Now we convert the value 0.640492 using Q-11 format to get 010100011111. Cascading the exponent bits and the mantissa bits yields

$$1110010100011111.$$

2. Since $-20.430527/2^5 = -0.638454$, we can convert it into the fractional part and exponent part as $-20.430527 = -0.638454 \times 2^5$.

Note that this conversion is not particularly unique; the forms $-20.430527 = -0.319227 \times 2^6$ and $-20.430527 = -0.1596135 \times 2^7 \ldots$ are still valid choices. Let us keep what we have now. Therefore, the exponent bits should be 0101. Converting the number 0.638454 using Q-11 format gives:

$$010100011011.$$

9.4 Fixed-Point and Floating-Point Formats

Using 2's complement, we obtain the representation for the decimal number -0.638454 as

$$101011100101.$$

Cascading the exponent bits and mantissa bits, we achieve

$$0101101011100101.$$

The floating arithmetic is more complicated. We must obey the rules for manipulating two floating-point numbers. Rules for arithmetic addition are given as:

$$x_1 = M_1 2^{E_1}$$
$$x_2 = M_2 2^{E_2}.$$

The floating-point sum is performed as follows:

$$x_1 + x_2 = \begin{cases} \left(M_1 + M_2 \times 2^{-(E_1 - E_2)}\right) \times 2^{E_1}, & \text{if } E_1 \geq E_2 \\ \left(M_1 \times 2^{-(E_2 - E_1)} + M_2\right) \times 2^{E_2} & \text{if } E_1 < E_2 \end{cases}$$

As a multiplication rule, given two properly normalized floating-point numbers:

$$x_1 = M_1 2^{E_1}$$
$$x_2 = M_2 2^{E_2},$$

where $0.5 \leq |M_1| < 1$ and $0.5 \leq |M_2| < 1$. Then multiplication can be performed as follows:

$$x_1 \times x_2 = (M_1 \times M_2) \times 2^{E_1 + E_2} = M \times 2^E.$$

That is, the mantissas are multiplied while the exponents are added:

$$M = M_1 \times M_2$$
$$E = E_1 + E_2.$$

Examples 9.9 and 9.10 serve to illustrate manipulators.

Example 9.9.

a. Add two floating-point numbers achieved in Example 9.8:

$$1110\ 010100011111 = 0.640136718 \times 2^{-2}$$
$$0101\ 101011100101 = -0.638183593 \times 2^5.$$

Solution:

a. Before addition, we change the first number to have the same exponent as the second number, that is,

$$0101\ 000000001010 = 0.005001068 \times 2^5.$$

Then we add two mantissa numbers:

$$0.000000001010$$
$$+\ 1.010111100101$$
$$\overline{1.010111101111}$$

and we get the floating number as

0101 101011101111.

We can verify the result by the following:

$$0101\ 101011101111 = -\left(2^{-1} + 2^{-3} + 2^{-7} + 2^{-11}\right) \times 2^5$$
$$= -0.633300781 \times 2^5 = -20.265625.$$

Example 9.10.

a. Multiply two floating-point numbers achieved in Example 9.8:

$$1110\ 010100011111 = 0.640136718 \times 2^{-2}$$
$$0101\ 101011100101 = -0.638183593 \times 2^5.$$

Solution:

a. From the results in Example 9.8, we have the bit patterns for these two numbers as

$$E_1 = 1110, E_2 = 0101, M_1 = 010100011111, M_2 = 101011100101.$$

Adding two exponents in 2's complement form leads to

$$E = E_1 + E_2 = 1110 + 0101 = 0011,$$

which is $+3$, as we expected, since in decimal domain $(-2) + 5 = 3$.

As previously shown in the multiplication rule, when multiplying two mantissas, we need to apply their corresponding positive values. If the sign for the final value is negative, then we convert it to its 2's complement form. In our example, $M_1 = 010100011111$ is a positive mantissa. However, $M_2 = 101011100101$ is a negative mantissa, since the MSB is 1. To perform multiplication, we use 2's complement to convert M_2 to its positive value, 010100011011, and note that the multiplication result is negative. We multiply two positive mantissas and truncate the result to 12 bits to give

$$010100011111 \times 010100011011 = 001101000100.$$

Now we need to add a negative sign to the multiplication result with 2's complement operation. Taking 2's complement, we have

$$M = 110010111100.$$

Hence, the product is achieved by cascading the 4-bit exponent and 12-bit mantissa as:

$$0011\ 110010111100.$$

converting this number back to the decimal number, we verify the result to be $0.408203125 \times 2^3 = -3.265625$.

Next, we examine overflow and underflow in the floating-point number system.

Overflow

During operation, overflow will occur when a number is too large to be represented in the floating-point number system. Adding two mantissa numbers may lead to a number larger than 1 or less than -1; and multiplying two numbers causes the addition of their two exponents, so that the sum of the two exponents could overflow. Consider the following overflow cases.

Case 1. Add the following two floating-point numbers:

$$0111\ 011000000000 + 0111\ 010000000000.$$

Note that two exponents are the same and they are the biggest positive number in 4-bit 2's complement representation. We add two positive mantissa numbers as

$$\begin{array}{r} 0.1\ 1\ 0\ 0\ 0\ 0\ 0\ 0\ 0\ 0\ 0 \\ +\ \ 0.1\ 0\ 0\ 0\ 0\ 0\ 0\ 0\ 0\ 0\ 0 \\ \hline 1.0\ 1\ 0\ 0\ 0\ 0\ 0\ 0\ 0\ 0\ 0. \end{array}$$

The result for adding mantissa numbers is negative. Hence, the overflow occurs.

Case 2. Multiply the following two numbers:

$$0111\ 011000000000 \times 0111\ 011000000000.$$

Adding two positive exponents gives

$$0111 + 0111 = 1000 \text{ (negative; the overflow occurs)}.$$

Multiplying two mantissa numbers gives:

$$0.11000000000 \times 0.11000000000 = 0.10010000000 \text{ (OK!)}.$$

Underflow

As we discussed before, underflow will occur when a number is too small to be represented in the number system. Let us divide the following two floating-point numbers:

$$1001\ 001000000000 \div 0111\ 010000000000.$$

First, subtracting two exponents leads to

$$1001\ (\text{negative}) - 0111\ (\text{positive}) = 1001 + 1001$$
$$= 0010\ (\text{positive; the underflow occurs}).$$

Then, dividing two mantissa numbers, it follows that

$$0.01000000000 \div 0.10000000000 = 0.10000000000\ (\text{OK!}).$$

However, in this case, the expected resulting exponent is -14 in decimal, which is too small to be presented in the 4-bit 2's complement system. Hence the underflow occurs.

Understanding basic principles of the floating-point formats, we can next examine two floating-point formats of the Institute of Electrical and Electronics Engineers (IEEE).

9.4.3 IEEE Floating-Point Formats

Single Precision Format

IEEE floating-point formats are widely used in many modern DS processors. There are two types of IEEE floating-point formats (IEEE 754 standard). One is the IEEE single precision format, and the other is the IEEE double precision format. The single precision format is described in Figure 9.11.

The format of IEEE single precision floating-point standard representation requires 23 fraction bits F, 8 exponent bits E, and 1 sign bit S, with a total of 32 bits for each word. F is the mantissa in 2's complement positive binary fraction represented from bit 0 to bit 22. The mantissa is within the normalized range limits between $+1$ and $+2$. The sign bit S is employed to indicate the sign of the number, where when $S = 1$ the number is negative, and when $S = 0$ the number is positive. The exponent E is in excess 127 form. The value of 127 is the offset from the 8-bit exponent range from 0 to 255, so that E-127 will have a range from -127 to $+128$. The formula shown in Figure 9.11 can be applied to convert the IEEE 754 standard (single precision) to the decimal number. The following simple examples also illustrate this conversion:

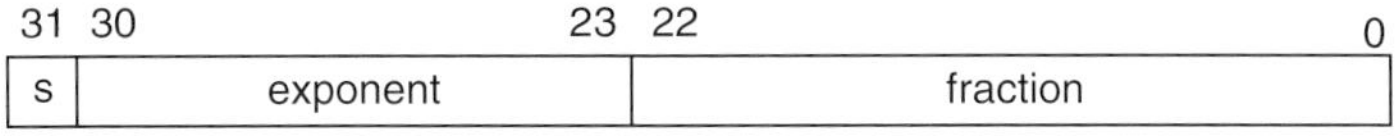

```
 31 30              23 22                        0
┌──┬─────────────────┬──────────────────────────┐
│ s│    exponent     │         fraction         │
└──┴─────────────────┴──────────────────────────┘
```
$$x = (-1)^s \times (1.F) \times 2^{E-127}$$

FIGURE 9.11 **IEEE single precision floating-point format.**

$0\ 10000000\ 00000000000000000000000 = (-1)^0 \times (1.0_2) \times 2^{128-127} = 2.0$

$0\ 10000001\ 10100000000000000000000 = (-1)^0 \times (1.101_2) \times 2^{129-127} = 6.5!$

$1\ 10000001\ 10100000000000000000000 = (-1)^1 \times (1.101_2) \times 2^{129-127} = -6.5.$

Let us look at Example 9.11 for more explanation.

Example 9.11.

a. Convert the following number in the IEEE single precision format to the decimal format:

$$110000000.010\ldots0000.$$

Solution:

a. From the bit pattern in Figure 9.11, we can identify the sign bit, exponent, and fractional as:

$$s = 1, E = 2^7 = 128$$
$$1.F = 1.01_2 = (2)^0 + (2)^{-2} = 1.25.$$

Then, applying the conversion formula leads to

$$x = (-1)^1(1.25) \times 2^{128-127} = -1.25 \times 2^1 = -2.5.$$

In conclusion, the value x represented by the word can be determined based on the following rules, including all the exceptional cases:

- If $E = 255$ and F is nonzero, then $x = $ *NaN* ("Not a number").
- If $E = 255$, F is zero, and S is 1, then $x = -$Infinity.
- If $E = 255$, F is zero, and S is 0, then $x = +$Infinity.
- If $0 < E < 255$, then $x = (-1)^s \times (1.F) \times 2^{E-127}$, where $1.F$ represents the binary number created by prefixing F with an implicit leading 1 and a binary point.

- If $E = 0$ and F is nonzero, then $x = (-1)^s \times (0.F) \times 2^{-126}$. This is an "unnormalized" value.
- If $E = 0$, F is zero, and S is 1, then $x = -0$.
- If $E = 0$, F is zero, and S is 0, then $x = 0$.

Typical and exceptional examples are shown as follows:

$$0\ 00000000\ 00000000000000000000000 = 0$$
$$1\ 00000000\ 00000000000000000000000 = -0$$
$$0\ 11111111\ 00000000000000000000000 = \text{Infinity}$$
$$1\ 11111111\ 00000000000000000000000 = -\text{Infinity}$$
$$0\ 11111111\ 00000100000000000000000 = \text{NaN}$$
$$1\ 11111111\ 00100010001001010101010 = \text{NaN}$$
$$0\ 00000001\ 00000000000000000000000 = (-1)^0 \times (1.0_2) \times 2^{1-127} = 2^{-126}$$
$$0\ 00000000\ 10000000000000000000000 = (-1)^0 \times (0.1_2) \times 2^{0-126} = 2^{-127}$$
$$0\ 00000000\ 00000000000000000000001 =$$
$$(-1)^0 \times (0.00000000000000000000001_2) \times 2^{0-126} = 2^{-149} \text{(smallest positive value)}$$

Double Precision Format

The IEEE double precision format is described in Figure 9.12.

The IEEE double precision floating-point standard representation requires a 64-bit word, which may be numbered from 0 to 63, left to right. The first bit is the sign bit S, the next eleven bits are the exponent bits E, and the final 52 bits are the fraction bits F. The IEEE floating-point format in double precision significantly increases the dynamic range of number representation, since there are eleven exponent bits; the double precision format also reduces the interval size in the mantissa normalized range of $+1$ to $+2$, since there are 52 mantissa bits as compared with the single precision case of 23 bits. Applying the conversion formula shown in Figure 9.12 is similar to the single precision case.

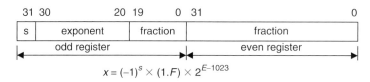

FIGURE 9.12 **IEEE double precision floating-point format.**

Example 9.12.

a. Convert the following number in IEEE double precision format to the decimal format:

$$001000\ldots0.110\ldots0000$$

Solution:

b. Using the bit pattern in Figure 9.12, we have

$$s = 0, E = 2^9 = 512 \text{ and}$$
$$1.F = 1.11_2 = (2)^0 + (2)^{-1} + (2)^{-2} = 1.75.$$

Then, applying the double precision formula yields

$$x = (-1)^0(1.75) \times 2^{512-1023} = 1.75 \times 2^{-511} = 2.6104 \times 10^{-154}.$$

For purposes of completeness, rules for determining the value x represented by the double precision word are listed as follows:

- If $E = 2047$ and F is nonzero, then $x = NaN$ ("Not a number")
- If $E = 2047$, F is zero, and S is 1, then $x = -\text{Infinity}$
- If $E = 2047$, F is zero, and S is 0, then $x = +\text{Infinity}$
- If $0 < E < 2047$, then $x = (-1)^s \times (1.F) \times 2^{E-1023}$, where $1.F$ is intended to represent the binary number created by prefixing F with an implicit leading 1 and a binary point
- If $E = 0$ and F is nonzero, then $x = (-1)^s \times (0.F) \times 2^{-1022}$. This is an "unnormalized" value
- If $E = 0$, F is zero, and S is 1, then $x = -0$
- If $E = 0$, F is zero, and S is 0, then $x = 0$

9.4.5 Fixed-Point Digital Signal Processors

Analog Device, Texas Instruments, and Motorola all manufacture fixed-point DS processors. Analog Devices offers a fixed-point DSP family such as ADSP21xx. Texas Instruments provides various generations of fixed-point DSP processors based on historical development, architectural features, and computational performance. Some of the most common ones are TMS320C1x (first generation), TMS320C2x, TMS320C5x, and TMS320C62x. Motorola manufactures varieties of fixed-point processors, such as the DSP5600x family. The new families of fixed-point DS processors are expected to continue to grow.

Since they share some basic common features such as program memory and data memory with associated address buses, arithmetic logic units (ALUs), program control units, MACs, shift units, and address generators, here we focus on an overview of the TMS320C54x processor. The typical TMS320C54x fixed-point DSP architecture appears in Figure 9.13.

The fixed-point TMS320C54x families supporting 16-bit data have on-chip program memory and data memory in various sizes and configurations. They include data RAM (random access memory) and program ROM (read-only memory) used for program code, instruction, and data. Four data buses and four address buses are accommodated to work with the data memories and program memories. The program memory address bus and program memory data bus are responsible for fetching the program instruction. As shown in Figure 9.13, the C and D data memory address buses and the C and D data

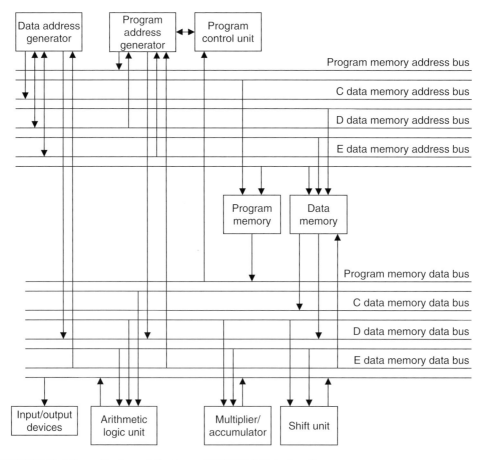

FIGURE 9.13 Basic architecture of TMS320C54x family.

memory data buses deal with fetching data from the data memory, while the E data memory address bus and the E data memory data bus are dedicated to moving data into data memory. In addition, the E memory data bus can access the I/O devices.

Computational units consist of an ALU, an MAC, and a shift unit. For TMS320C54x families, the ALU can fetch data from the C, D, and program memory data buses and access the E memory data bus. It has two independent 40-bit accumulators, which are able to operate 40-bit addition. The multiplier, which can fetch data from C and D memory data buses and write data via the E memory data bus, is capable of operating 17-bit $\times$ 17-bit multiplications. The 40-bit shifter has the same capability of bus access as the MAC, allowing all possible shifts for scaling and fractional arithmetic such as we have discussed for the Q-format.

The program control unit fetches instructions via the program memory data bus. Again, in order to speed up memory access, there are two address generators available: one responsible for program addresses and one for data addresses.

Advanced Harvard architecture is employed, where several instructions operate at the same time for a given single instruction cycle. Processing performance offers 40 MIPS (million instruction sets per second). To further explore this subject, the reader is referred to Dahnoun (2000), Embree (1995), Ifeachor and Jervis (2002), and Van der Vegte (2002), as well as the website for Texas Instruments (www.ti.com).

9.4.6 Floating-Point Processors

Floating-point DS processors perform DSP operations using floating-point arithmetic, as we discussed before. The advantages of using the floating-point processor include getting rid of finite word length effects such as overflows, round-off errors, truncation errors, and coefficient quantization errors. Hence, in terms of coding, we do not need to do scaling input samples to avoid overflow, shift the accumulator result to fit the DAC word size, scale the filter coefficients, or apply Q-format arithmetic. The floating-point DS processor with high-performance speed and calculation precision facilitates a friendly environment to develop and implement DSP algorithms.

Analog Devices provides floating-point DSP families such as ADSP210xx and TigerSHARC. Texas Instruments offers a wide range of floating-point DSP families, in which the TMS320C3x is the first generation, followed by the TMS320C4x and TMS320C67x families. Since the first generation of a floating-point DS processor is less complicated than later generations but still has the common basic features, we overview the first-generation architecture first.

Figure 9.14 shows the typical architecture of Texas Instruments' TMS320C3x families. We discuss some key features briefly. Further detail can

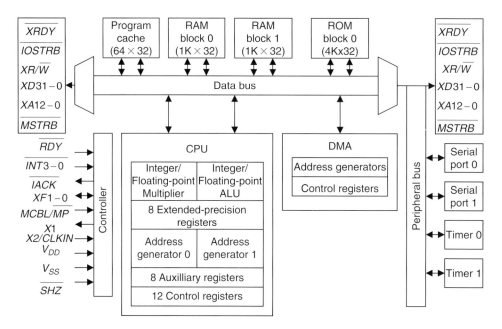

FIGURE 9.14 The typical TMS320C3x floating-point DS processor.

be found in the TMS320C3x User's Guide (Texas Instruments, 1991), the TMS320C6x CPU and Instruction Set Reference Guide (Texas Instruments, 1998), and other studies (Dahnoun, 2000; Embree, 1995; Ifeachor and Jervis, 2002; Kehtarnavaz and Simsek, 2000; Sorensen and Chen, 1997; van der Vegte, 2002). The TMS320C3x family consists of 32-bit single chip floating-point processors that support both integer and floating-point operations.

The processor has a large memory space and is equipped with dual-access on-chip memories. A program cache is employed to enhance the execution of commonly used codes. Similar to the fixed-point processor, it uses the Harvard architecture, in which there are separate buses used for program and data so that instructions can be fetched at the same time that data are being accessed. There also exist memory buses and data buses for direct-memory access (DMA) for concurrent I/O and CPU operations, and peripheral access such as serial ports, I/O ports, memory expansion, and an external clock.

The C3x CPU contains the floating-point/integer multiplier; an ALU, which is capable of operating both integer and floating-point arithmetic; a 32-bit barrel shifter; internal buses; a CPU register file; and dedicated auxiliary register arithmetic units (ARAUs). The multiplier operates single-cycle multiplications on 24-bit integers and on 32-bit floating-point values. Using parallel instructions to perform a multiplication, an ALU will cost a single cycle, which means that a multiplication and an addition are equally fast. The ARAUs support addressing modes, in which some of them are specific to DSP such as circular buffering and bit-reversal addressing (digital filtering and FFT operations). The CPU register file offers 28 registers, which can be operated on by the multiplier and ALU. The special functions of the registers include eight extended 40-bit precision registers for maintaining accuracy of the floating-point results. Eight auxiliary registers can be used for addressing and for integer arithmetic. These registers provide internal temporary storage of internal variables instead of external memory storage, to allow performance of arithmetic between registers. In this way, program efficiency is greatly increased.

The prominent feature of C3x is its floating-point capability, allowing operation of numbers with a very large dynamic range. It offers implementation of the DSP algorithm without worrying about problems such as overflows and coefficient quantization. Three floating-point formats are supported. A short 16-bit floating-point format has 4 exponent bits, 1 sign bit, and 11 mantissa bits. A 32-bit single precision format has 8 exponent bits, 1 sign bit, and 23 fraction bits. A 40-bit extended precision format contains 8 exponent bits, 1 sign bit, and 31 fraction bits. Although the formats are slightly different from the IEEE 754 standard, conversions are available between these formats.

The TMS320C30 offers high-speed performance with 60-nanosecond single-cycle instruction execution time, which is equivalent to 16.7 MIPS. For speech-quality applications with an 8 kHz sampling rate, it can handle over 2,000 single-cycle instructions between two samples (125 microseconds). With instruction enhancement such as pipelines executing each instruction in a single cycle (four cycles required from fetch to execution by the instruction itself) and a multiple interrupt structure, this high-speed processor validates implementation of real-time applications in floating-point arithmetic.

9.5 Finite Impulse Response and Infinite Impulse Response Filter Implementations in Fixed-Point Systems

With knowledge of the IEEE formats and of filter realization structures such as the direct form I, direct form II, and parallel and cascade forms (Chapter 6), we can study digital filter implementation in the fixed-point processor. In the fixed-point system, where only integer arithmetic is used, we prefer input data, filter coefficients, and processed output data to be in the Q-format. In this way, we avoid overflow due to multiplications and can prevent overflow due to addition by scaling input data. When the filter coefficients are out of the Q-format range, coefficient scaling must be taken into account to maintain the Q-format. We develop FIR filter implementation in Q-format first, and then infinite impulse response (IIR) filter implementation next. In addition, we assume that with a given input range in Q-format, the filter output is always in Q-format even if the filter passband gain is larger than 1.

First, to avoid overflow for an adder, we can scale the input down by a scale factor S, which can be safely determined by the equation

$$S = I_{max} \cdot \sum_{k=0}^{\infty} |h(k)| = I_{max} \cdot (|h(0)| + |h(1)| + |h(2)| + \cdots), \quad (9.2)$$

where $h(k)$ is the impulse response of the adder output and I_{max} the maximum amplitude of the input in Q-format. Note that this is not an optimal factor in terms of reduced signal-to-noise ratio. However, it shall prevent the overflow. Equation (9.2) means that the adder output can actually be expressed as a convolution output:

$$\text{adder output} = h(0)x(n) + h(1)x(n-1) + h(2)x(n-2) + \cdots.$$

Assuming the worst condition, that is, that all the inputs $x(n)$ reach a maximum value of I_{max} and all the impulse coefficients are positive, the sum of the adder gives the most conservative scale factor, as shown in Equation (9.2). Hence, scaling down of the input by a factor of S will guarantee that the output of the adder is in Q-format.

When some of the FIR coefficients are larger than 1, which is beyond the range of Q-format representation, coefficient scaling is required. The idea is that scaling down the coefficients will make them less than 1, and later the filtered output will be scaled up by the same amount before it is sent to DAC. Figure 9.15 describes the implementation.

In the figure, the scale factor B makes the coefficients b_k/B convertible to the Q-format. The scale factors of S and B are usually chosen to be a power of 2, so the simple shift operation can be used in the coding process. Let us implement an FIR filter containing filter coefficients larger than 1 in the fixed-point implementation.

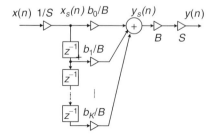

FIGURE 9.15 Direct-form I implementation of the FIR filter.

Example 9.13.

Given the FIR filter

$$y(n) = 0.9x(n) + 3x(n-1) - 0.9x(n-2),$$

with a passband gain of 4, and assuming that the input range occupies only 1/4 of the full range for a particular application,

a. Develop the DSP implementation equations in the Q-15 fixed-point system.

Solution:

a. The adder may cause overflow if the input data exist for $\frac{1}{4}$ of a full dynamic range. The scale factor is determined using the impulse response, which consists of the FIR filter coefficients, as discussed in Chapter 3.

$$S = \frac{1}{4}(|h(0)| + |h(1)| + |h(2)|) = \frac{1}{4}(0.9 + 3 + 0.9) = 1.2.$$

Overflow may occur. Hence, we select $S = 2$ (a power of 2). We choose $B = 4$ to scale all the coefficients to be less than 1, so the Q-15 format can be used. According to Figure 9.15, the developed difference equations are given by

$$x_s(n) = \frac{x(n)}{2}$$
$$y_s(n) = 0.225x_s(n) + 0.75x_s(n-1) - 0.225x_s(n-2)$$
$$y(n) = 8y_s(n)$$

Next, the direct-form I implementation of the IIR filter is illustrated in Figure 9.16.

As shown in Figure 9.16, the purpose of a scale factor C is to scale down the original filter coefficients to the Q-format. The factor C is usually chosen to be a power of 2 for using a simple shift operation in DSP.

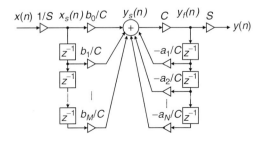

FIGURE 9.16 Direct-form I implementation of the IIR filter.

Example 9.14.

The following IIR filter,

$$y(n) = 2x(n) + 0.5y(n-1),$$

uses the direct form I, and for a particular application, the maximum input is $I_{max} = 0.010\ldots0_2 = 0.25$.

a. Develop the DSP implementation equations in the Q-15 fixed-point system.

Solution:

a. This is an IIR filter whose transfer function is

$$H(z) = \frac{2}{1 - 0.5z^{-1}} = \frac{2z}{z - 0.5}.$$

Applying the inverse z-transform, we have the impulse response

$$h(n) = 2 \times (0.5)^n u(n).$$

To prevent overflow in the adder, we can compute the S factor with the help of the Maclaurin series or approximate Equation (9.2) numerically. We get

$$S = 0.25 \times \left(2(0.5)^0 + 2(0.5)^1 + 2(0.5)^2 + \cdots\right) = \frac{0.25 \times 2 \times 1}{1 - 0.5} = 1.$$

MATLAB function **impz()** can also be applied to find the impulse response and the S factor:

```
≫ h = impz(2,[1 −0.5]); % Find the impulse response
≫ sf = 0.25* sum(abs(h)) % Determine the sum of absolute values of h(k)
  sf = 1
```

Hence, we do not need to perform input scaling. However, we need to scale down all the coefficients to use the Q-15 format. A factor of $C = 4$ is selected. From Figure 9.16, we get the difference equations as

$$x_s(n) = x(n)$$
$$y_s(n) = 0.5x_s(n) + 0.125y_f(n-1)$$
$$y_f(n) = 4y_s(n)$$
$$y(n) = y_f(n).$$

We can develop these equations directly. First, we divide the original difference equation by a factor of 4 to scale down all the coefficients to be less than 1, that is,

$$\frac{1}{4}y_f(n) = \frac{1}{4} \times 2x_s(n) + \frac{1}{4} \times 0.5 y_f(n-1),$$

and define a scaled output

$$y_s(n) = \frac{1}{4} y_f(n).$$

Finally, substituting $y_s(n)$ to the left side of the scaled equation and rescaling up the filter output as $y_f(n) = 4y_s(n)$ we have the same results we got before.

The fixed-point implementation for the direct form II is more complicated. The developed direct-form II implementation of the IIR filter is illustrated in Figure 9.17.

As shown in Figure 9.17, two scale factors A and B are designated to scale denominator coefficients and numerator coefficients to their Q-format representations, respectively. Here S is a special factor to scale down the input sample so that the numerical overflow in the first sum in Figure 9.17 can be prevented. The difference equations are given in Chapter 6 and listed here:

$$w(n) = x(n) - a_1 w(n-1) - a_2 w(n-2) - \cdots - a_M w(n-M)$$
$$y(n) = b_0 w(n) + b_1 w(n-1) + \cdots + b_M w(n-M).$$

The first equation is scaled down by the factor A to ensure that all the denominator coefficients are less than 1, that is,

$$w_s(n) = \frac{1}{A} w(n) = \frac{1}{A} x(n) - \frac{1}{A} a_1 w(n-1) - \frac{1}{A} a_2 w(n-2) - \cdots - \frac{1}{A} a_M w(n-M)$$
$$w(n) = A \times w_s(n).$$

Similarly, scaling the second equation yields

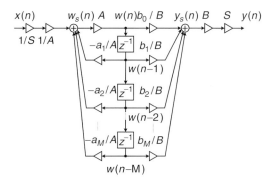

FIGURE 9.17   Direct-form II implementation of the IIR filter.

$$y_s(n) = \frac{1}{B}y(n) = \frac{1}{B}b_0 w(n) + \frac{1}{B}b_1 w(n-1) + \cdots + \frac{1}{B}b_M w(n-M)$$

and
$$y(n) = B \times y_s(n)$$

To avoid the first adder overflow (first equation), the scale factor S can be safely determined by Equation (9.3):

$$S = I_{\max}(|h(0)| + |h(1)| + |h(2)| + \cdots), \qquad (9.3)$$

where $h(k)$ is the impulse response of the filter whose transfer function is the reciprocal of the denominator polynomial, where the poles can cause a larger value to the first sum:

$$h(n) = Z^{-1}\left(\frac{1}{1 + a_1 z^{-1} + \cdots + a z^{-M}}\right). \qquad (9.4)$$

All the scale factors A, B, and S are usually chosen to be a power of 2, respectively, so that the shift operations can be used in the coding process. Example 9.15 serves for illustration.

Example 9.15.

Given the following IIR filter:
$$y(n) = 0.75x(n) + 1.49x(n-1) + 0.75x(n-2) - 1.52y(n-1) - 0.64y(n-2),$$
with a passband gain of 1 and a full range of input,

a. Use the direct-form II implementation to develop the DSP implementation equations in the Q-15 fixed-point system.

Solution:

a. The difference equations without scaling in the direct-form II implementation are given by
$$w(n) = x(n) - 1.52w(n-1) - 0.64w(n-2)$$
$$y(n) = 0.75w(n) + 1.49w(n-1) + 0.75w(n-2).$$

To prevent overflow in the first adder, we have the reciprocal of the denominator polynomial as

$$A(z) = \frac{1}{1 + 1.52z^{-1} + 0.64z^{-2}}.$$

Using MATLAB function leads to

```
>> h = impz(1,[1 1.52 0.64]);
>> sf = sum(abs(h))
sf = 10.4093.
```

We choose the S factor as $S = 16$ and we choose $A = 2$ to scale down the denominator coefficients by half. Since the second adder output after scaling is

$$y_s(n) = \frac{0.75}{B} w(n) + \frac{1.49}{B} w(n-1) + \frac{0.75}{B} w(n-2),$$

we have to ensure that each coefficient is less than 1, as well as the sum of the absolute values

$$\frac{0.75}{B} + \frac{1.49}{B} + \frac{0.75}{B} < 1$$

to avoid second adder overflow. Hence $B = 4$ is selected. We develop the DSP equations as

$$x_s(n) = x(n)/16$$
$$w_s(n) = 0.5x_s(n) - 0.76w(n-1) - 0.32w(n-2)$$
$$w(n) = 2w_s(n)$$
$$y_s(n) = 0.1875w(n) + 0.3725w(n-1) + 0.1875w(n-2)$$
$$y(n) = (B \times S)y_s(n) = 64y_s(n)$$

The implementation for cascading the second-order section filters can be found in Ifeachor and Jervis (2002).

A practical example will be presented in the next section. Note that if a floating-point DS processor is used, all the scaling concerns should be ignored, since the floating-point format offers a large dynamic range, so that overflow hardly ever happens.

9.6 Digital Signal Processing Programming Examples

In this section, we first review the TMS320C67x DSK (DSP Starter Kit), which offers floating-point and fixed-point arithmetic. We will then investigate real-time implementation of digital filters.

9.6.1 Overview of TMS320C67x DSK

In this section, a Texas Instruments TMS320C67x DSK is chosen for demonstration in Figure 9.18. This DSK board (Kehtarnavaz and Simsek, 2001; Texas Instruments, 1998) consists of the TMS320C67x chip, SDRAM (synchronous dynamic random access memory) and ROM for storing program code and data, and an ADC535 chip performing 16-bit ADC and DAC operations. The gain of

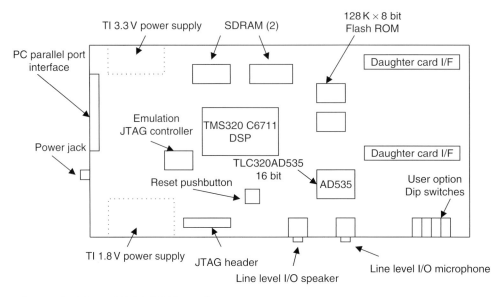

FIGURE 9.18 C6711 DSK board.

the ADC channel is programmable to provide microphone or other line inputs, such as from the function generator or other sensors. The DAC channel is also programmable to deliver the power gain to drive a speaker or other devices. The ADC535 chip sets a fixed sampling rate of 8 kHz for speech applications. The on-board daughter card connections facilitate the external units for advanced applications. For example, a daughter card designed using PCM3001/3 offers a variable high sampling rate, such as 44.1 kHz, and its own programmable ADC and DAC for CD-quality audio applications. The parallel port is used for connection between the DSK board and the host computer, where the user program is developed, compiled, and downloaded to the DSK for real-time applications using the user-friendly software called the Code Composer Studio, which we shall discuss later.

The TMS320C67x operates at a high clock rate of 300 MHz. Combining with high speed and multiple units operating at the same time has pushed its performance up to 2,400 MIPS at 300 MHz. Using this number, the C67x can handle 0.3 MIPS between two speech samples at a sampling rate of 8 kHz and can handle over 54,000 instructions between two audio samples with a sampling rate of 44.1 kHz. Hence, the C67x offers great flexibility for real-time applications with a high-level C language.

Figure 9.19 shows a C67x architecture overview, while Figure 9.20 displays a more detailed block diagram. C67x contains three main parts, which are the CPU, the memories, and the peripherals. As shown in Figure 9.19, these three main parts are joined by an external memory interface (EMIF) interconnected

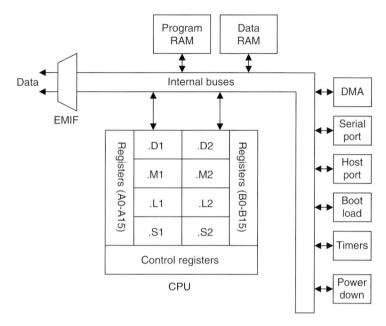

FIGURE 9.19 Block diagram of TMS320C67x floating-point DSP.

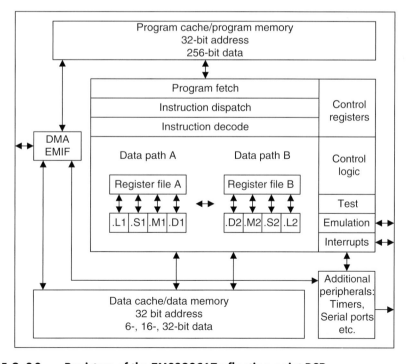

FIGURE 9.20 Registers of the TMS320C67x floating-point DSP.

by internal buses to facilitate interface with common memory devices; DMA; a serial port; and a host port interface (HPI).

Since this section is devoted to showing DSP coding examples, C67x key features and references are briefly listed here:

1. Architecture: The system uses Texas Instruments Veloci® architecture, which is an enhancement of the VLIW (very long instruction word architecture) (Dahnoun, 2000; Ifeachor and Jervis, 2002; Kehtarnavaz and Simsek, 2000).

2. CPU: As shown in Figure 9.20, the CPU has eight functional units divided into two sides A and B, each consisting of units .D, .M, .L, and .S. For each side, an .M unit is used for multiplication operations, an .L unit is used for logical and arithmetic operations, and a .D unit is used for loading/storing and arithmetic operations. Each side of the C67x CPU has sixteen 32-bit registers that the CPU must go through for interface. More detail can be found in Appendix D (Texas Instruments, 1991), as well as in Kehtarnavaz and Simsek (2000) and Texas Instruments (1998).

3. Memory and internal buses: Memory space is divided into internal program memory, internal data memory, and internal peripheral and external memory space. The internal buses include a 32-bit program address bus, a 256-bit program data bus to carry out eight 32-bit instructions (VLIW), two 32-bit data address buses, two 64-bit load data buses, two 64-bit store data buses, two 32-bit DMA buses, and two 32-bit DMA address buses responsible for reading and writing. There also exist a 22-bit address bus and a 32-bit data bus for accessing off-chip or external memory.

4. Peripherals:

   a. EMIF, which provides the required timing for accessing external memory

   b. DMA, which moves data from one memory location to another without interfering with the CPU operations

   c. Multichannel buffered serial port (McBSP) with a high-speed multichannel serial communication link

   d. HPI, which lets a host access internal memory

   e. Boot loader for loading code from off-chip memory or the HPI to internal memory

   f. Timers (two 32-bit counters)

   g. Power-down units for saving power for periods when the CPU is inactive.

The software tool for the C67x is the Code Composer Studio (CCS) provided by TI. It allows the user to build and debug programs from a user-friendly graphical user interface (GUI) and extends the capabilities of code development tools to include real-time analysis. Installation, tutorial, coding, and debugging can be found in the CCS Getting Started Guide (Texas Instruments, 2001) and in Kehtarnavaz and Simsek (2000).

9.6.2 Concept of Real-Time Processing

We illustrate real-time implementation in Figure 9.21, where the sampling rate is 8,000 samples per second; that is, the sampling period $T = 1/f_s = 125$ microseconds, which is the time between two samples.

As shown in Figure 9.21, the required timing includes an input sample clock and an output sample clock. The input sample clock maintains the accuracy of sampling time for each ADC operation, while the output sample clock keeps the accuracy of time instant for each DAC operation. Time between the input sample clock n and the output sample clock n consists of the ADC operation, algorithm processing, and the wait for the next ADC operation. The numbers of instructions for ADC and DSP algorithms must be estimated and verified to ensure that all instructions have been completed before the DAC begins. Similarly, the number of instructions for DAC must be verified so that DAC instructions will be finished between the output sample clock n and the next input sample clock $n + 1$. Timing usually is set up using the DSP interrupts (we will not pursue the interrupt setup here).

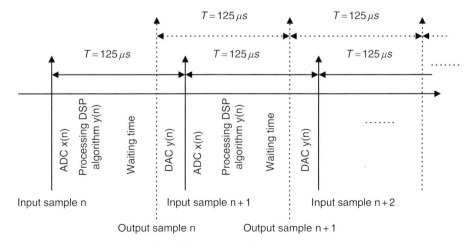

FIGURE 9.21 Concept of real-time processing.

Next, we focus on the implementation of the DSP algorithm in the floating-point system for simplicity.

9.6.3 Linear Buffering

During DSP such as digital filtering, past inputs and past outputs are required to be buffered and updated for processing the next input sample. Let us first study the FIR filter implementation.

Finite Impulse Response Filtering

Consider implementation for the following 3-tap FIR filter:

$$y(n) = 0.5x(n) + 0.2x(n-1) + 0.5x(n-2).$$

The buffer requirements are shown in Figure 9.22. The coefficient buffer b[3] contains 3 FIR coefficients, and the coefficient buffer is fixed during the process. The input buffer x[3], which holds the current and past inputs, is required to be updated. The FIFO update is adopted here with the segment of codes shown in Figure 9.22. For each input sample, we update the input buffer using FIFO, which begins at the end of the data buffer; the oldest sampled is kicked out first from the buffer and updated with the value from the upper location. When the FIFO completes, the first memory location x[0] will be free to be used to store the current input sample. The segment of code in Figure 9.22 explains implementation.

Note that in the code segment, x[0] holds the current input sample $x(n)$, while b[0] is the corresponding coefficient; x[1] and x[2] hold the past input samples $x(n-1)$ and $x(n-2)$, respectively; similarly, b[1] and b[2] are the corresponding coefficients.

Again, note that using the array and loop structures in the code segment is for simplicity in notations and assumes that the reader is not familiar with the C pointers in C-language. This concern for simplicity has to do mainly with the DSP algorithm. More coding efficiency can be achieved using the C pointers and circular buffer. The DSP-oriented coding implementation can be found in Kehtarnavaz and Simsek (2000).

Infinite Impulse Response Filtering

Similarly, we can implement an IIR filter. It requires an input buffer, which holds the current and past inputs; an output buffer, which holds the past outputs; a numerator coefficient buffer; and a denominator coefficient buffer. Considering the following IIR filter for implementation,

$$y(n) = 0.5x(n) + 0.7x(n-1) - 0.5x(n-2) - 0.4y(n-1) + 0.6y(n-2),$$

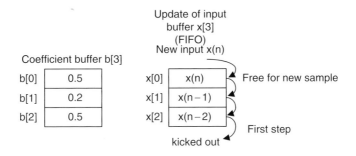

```
volatile int sample;
float x[3]={0.0, 0.0, 0.0};
float b[3]={0.5, 0.2, 0.5};
float y[1]={0.0};
interrupt void AtoD()
{
int i;
  sample = mcbsp0_read(); /*ADC*/
  for(i = 2; i > 0; i--)/* Update the input buffer x[3] */
  {
  x[i] = x[i − 1];
  }
  x[0]=(float) sample;
  y[0]=0;
  for(i = 0; i < 3; i++)
  {
  y[0] = y[0] + b[i]*x[i];
  }
  sample = (int) y[0]; /* The processed sample will be sent to DAC */
}
```

FIGURE 9.22 Example of FIR filtering with linear buffer update.

we accommodate the numerator coefficient buffer b[3], the denominator coefficient buffer a[3], the input buffer x[3], and the output buffer y[3] shown in Figure 9.23. The buffer updates for input x[3] and output y[3] are FIFO. The implementation is illustrated in the segment of code listed in Figure 9.23.

Again, note that in the code segment, x[0] holds the current input sample, while y[0] holds the current processed output, which will be sent to the DAC unit for conversion. The coefficient a[0] is never modified in the code. We keep that for a purpose of notation simplicity and consistency during the programming process.

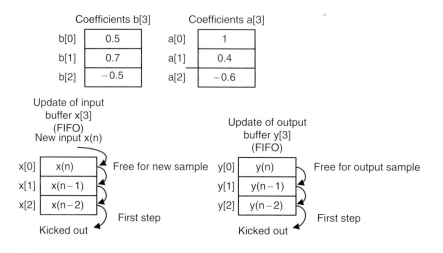

```
volatile int sample;
float b[3]={0.5, 0.7, −0.5};
float a[3]={1, 0.4, −0.6};
float x[3]={0.0, 0.0, 0.0};
float y[3]={0.0, 0.0, 0.0};
interrupt void AtoD()
{
  int i;
  sample=mcbsp0_read(); /* ADC */
  for(i = 2; i > 0; i--)  /* Update the input buffer */
  {
  x[i] = x[i − 1];
  }
  x[0]= (float) sample;
  for (i = 2; i > 0; i--) /* Update the output buffer */
  {
  y[i]=y[i − 1];
  }
  y[0] = b[0]*x[0] + b[1]*x[1] + b[2]*x[2]-a[1]*y[1]-a[2]*y[2];
  sample= (int) y[0]; /* the processed sample will be sent to DAC */
}
```

FIGURE 9.23 Example of IIR filtering using linear buffer update.

Digital Oscillation with Infinite Impulse Response Filtering

The principle for generating digital oscillation is described in Chapter 8, where the input to the digital filter is the impulse sequence, and the transfer function is obtained by applying the z-transform of the digital sinusoidal function.

Applications can be found in dual-tone multifrequency (DTMF) tone generation, digital carrier generation for communications, and so on. Hence, we can modify the implementation of IIR filtering for tone generation with the input generated internally instead of by using the ADC channel.

Let us generate an 800 Hz tone with a digital amplitude of 5,000. According to the section in Chapter 8 ("Applications: Generation and Detection of DTMF Tones Using the Goertzel Algorithm"), the transfer function, difference equation, and impulse input sequence are found to be, respectively,

$$H(z) = \frac{0.587785z^{-1}}{1 - 1.618034z^{-1} + z^{-2}}$$

$$y(n) = 0.587785x(n-1) + 1.618034y(n-1) - y(n-2)$$

$$x(n) = 5000\delta(n).$$

We define the numerator coefficient buffer b[2], the denominator coefficient buffer a[3], the input buffer x[2], and the output buffer y[3], shown in Figure 9.24, which also shows the modified implementation for the tone generation.

Initially, we set x[0] = 5000. Then it will be updated with x[0] = 0 for each current processed output sample y[0].

9.6.4 Sample C Programs

Floating-Point Implementation Example

Real-time DSP implementation using the floating-point processor is easy to program. The overflow problem hardly ever occurs. Therefore, we do not need to consider scaling factors, as described in the last section. The code segment shown in Figure 9.25 demonstrates the simplicity of coding the floating-point IIR filter using the direct-form I structure.

Fixed-Point Implementation Example

Where execution time is critical, fixed-point implementation is preferred in a floating-point processor. We implement the following IIR filter with a unit passband gain in direct form II:

$$H(z) = \frac{0.0201 - 0.0402z^{-2} + 0.0201z^{-4}}{1 - 2.1192z^{-1} + 2.6952z^{-2} - 1.6924z^{-3} + 0.6414z^{-4}}$$

$$w(n) = x(n) + 2.1192w(n-1) - 2.6952w(n-2) + 1.6924w(n-3) - 0.6414w(n-4)$$

$$y(n) = 0.0201w(n) - 0.0402w(n-2) + 0.0201w(n-4).$$

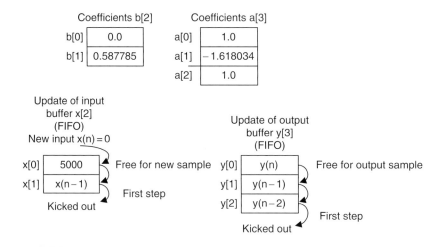

```
volatile int sample;
float b[2]= {0.0,0.587785};
float a[3]= {1,-1.618034,1};
float x[2]= {5000,0.0};  /*Set up the input as an impulse function */
float y[3]= {0.0,0.0,0.0};
interrupt void AtoD()
{
  int i;
  sample = mcbsp0_read(); /*ADC */
  y[0] = b[0]*x[0] + b[1]*x[1]-a[1]*y[1]-a[2]*y[2];
  sample= (int) y[0]; /*The processed sample will be sent to DAC */
  for(i = 1; i > 0; i--)/*Update the input buffer */
  {
  x[i] = x[i-1];
  }
  x[0] = 0;
  for(i = 2 > 0;i--)/* Update the output buffer */
  {
  y[i] = y[i - 1];
  }
}
```

FIGURE 9.24 Example of IIR filtering using linear buffer update and the impulse sequence input.

```c
volatile int sample;
float a[5] = {1.00, -2.1192, 2.6952, -1.6924, 0.6414};
float b[5] = {0.0201, 0.00, -0.0402, 0.00, 0.0201};
float x[5] = {0.0, 0.0, 0.0, 0.0, 0.0};
float y[5] = {0.0, 0.0, 0.0, 0.0, 0.0};
/*******************************************************************/
/* AtoD() Interrupt Service Routine (ISR)-> interrupt 12 defined in IST
of vectors.asm (read McBSP) */
/*******************************************************************/
interrupt void AtoD()
{
int i;
float temp, sum;
sample = mcbsp0_read(); /* ADC */
//Insert DSP Algorithm Here ()
temp = (float) sample;
for(i = 4; i > 0; i--)
{
  x[i] = x[i-1];
}
x[0] = temp;
for(i = 4; i > 0; i--)
{
y[i] = y[i - 1];
}
sum=b[0]*x[0]+b[1]*x[1]+b[2]*x[2]+b[3]*x[3]+b[4]*x[4]-a[1]*y[1]-a[2]*
y[2]-a[3]*y[3]-a[4]*y[4];

y[0] = sum;
sample = sum;
}
/*******************************************************************/
/* DtoA() Interrupt Service Routine (ISR)-> interrupt 11 defined in IST
of vectors.asm (write to McBSP)*/
/*******************************************************************/
interrupt void DtoA()
{
sample = sample & 0xfffe; /* set LSB to 0 for primary communication*/
mcbsp0_write(sample); /*DAC */
}
```

FIGURE 9.25   Sample C code for IIR filtering (float-point implementation).

Using MATLAB to calculate the scale factor S, it follows that

$$\gg h = \text{impz}([1],[1\ -2.1192\ \ 2.6952\ -1.6924\ \ 0.6414]);$$
$$\gg \text{sf} = \text{sum(abs(h))}$$
$$\text{sf} = 28.2196$$

Hence we choose $S = 32$. To scale the filter coefficients in the Q-15 format, we use the factors $A = 4$ and $B = 1$. Then the developed DSP equations are

$$x_s(n) = x(n)/32$$
$$w_s(n) = 0.25x_s(n) + 0.5298w_s(n-1) - 0.6738w_s(n-2) + 0.4231w_s(n-3)$$
$$+ 0.4231w_s(n-3) + 0.16035w_s(n-4)$$
$$w(n) = 4w_s(n)$$
$$y_s(n) = 0.0201w(n) - 0.0402w(n-2) + 0.0201w(n-4)$$
$$y(n) = 32y_s(n).$$

Using the method described in Section 9.5, we can convert filter coefficients into the Q-15 format; each coefficient is listed in Table 9.4.

The list of codes for the fixed-point implementation is displayed in Figure 9.26, and some coding notations are given in Figure 9.27.

Note that this chapter has provided only basic concepts and an introduction to real-time DSP implementation. The coding detail and real-time DSP applications will be treated in a separate DSP course, which deals with real-time implementations.

TABLE 9.4 Filter coefficients in Q-15 format.

IIR Filter	Filter Coefficients	Q-15 Format (Hex)
$-a_1$	0.5298	0x43D0
$-a_2$	−0.6738	0xA9C1
$-a_3$	0.4230	0x3628
$-a_4$	−0.16035	0xEB7A
b_0	0.0201	0×0293
b_1	0.0000	0×0000
b_2	−0.0402	$0 \times \text{FADB}$
b_3	0.0000	0×000
b_4	0.0201	0×0293

```c
volatile int sample;
/* float a[5]= {1.00, -2.1192, 2.6952, -1.6924, 0.6414};
   float b[5]= {0.0201, 0.00, -0.0402, 0.00, 0.0201};*/
short a[5]={0x2000, 0x43D0, 0xA9C1, 0x3628, 0xEB7A};/* coefficients in Q-15 format*/
short b[5] = {0x0293, 0x0000, 0xFADB, 0x0000, 0x0293};
int w[5]={0, 0, 0, 0, 0};
interrupt void AtoD()
{ int i, sum=0;
sample=mcbsp0_read(); /* ADC */
//Insert DSP Algorithm Here ()
sample = (sample << 16);/* Move to high 16 bits */
sample = (sample >> 5);/* Scaled down by 32 to avoid overflow */
for (i = 4; i > 0; i--)
{
w[i] = w[i - 1];
}
sum = (sample >> 2); /* Scaled down by 4 to use Q - 15*/
for (i = 1; i < 5; i++)
{
sum += (_mpyhl(w[i],a[i])) << 1;
}
sum = (sum << 2);/* scaled up by 4 */
w[0] =sum;
sum =0;
for (i = 0; i < 5; i++)
{
sum += (_mpyhl(w[i],b[i])) << 1;
}
sum = (sum << 5); /* Scaled up by 32 to get y(n) */
sample = (sum >> 16); /* Move to low 16 bits */
}

/*****************************************************************************/
/* DtoA() Interrupt Service Routine (ISR)-> interrupt 11 defined in IST of
vectors.asm (write to McBSP) */
/*****************************************************************************/
interrupt void DtoA()
{
sample = sample & 0xfffe; /* set LSB to 0 for primary communication*/
mcbsp0_write(sample); /* DAC */
}
```

FIGURE 9.26 Sample C code for IIR filtering (fixed-point implementation).

> short coefficient;   declaration of 16 bit signed integer
>
> int sample, result;   declaration of 32 bit signed integer
>
> MPYHL   assembly instruction (signed multiply high low 16 MSB × 16 LSB)
>
> > result = (_mpyhl(sample, coefficient)) <<1;
>
> sample must be shifted left by 16 bits to be stored in the high 16 MSB.
>
> coefficient is the 16 bit data to be stored in the low 16 LSB.
>
> result is shifted left by one bit to get rid of the extended sign bit, and high 16 MSB's are designated for the processed data.
>
> Final result will be shifted down to right by 16 bits before DAC conversion.
>
> > sample = (result>>16);

FIGURE 9.27   Some coding notations for the Q-15 fixed-point implementation.

9.7 Summary

1. The Von Neumann architecture consists of a single, shared memory for programs and data, a single bus for memory access, an arithmetic unit, and a program control unit. The Von Neumann processor operates fetching and execution cycles in series.

2. The Harvard architecture has two separate memory spaces dedicated to program code and to data, respectively, two corresponding address buses, and two data buses for accessing two memory spaces. The Harvard processor offers fetching and execution cycles in parallel.

3. The DSP special hardware units include an MAC dedicated to DSP filtering operations, a shifter unit for scaling, and address generators for circular buffering.

4. The fixed-point DS processor uses integer arithmetic. The data format Q-15 for the fixed-point system is preferred to avoid overflows.

5. The floating-point processor uses floating-point arithmetic. The standard floating-point formats include the IEEE single precision and double precision formats.

6. The architectures and features of fixed-point processors and floating-point processors were briefly reviewed.

7. Implementing digital filters in the fixed-point DSP system requires scaling filter coefficients so that the filters are in Q-15 format, and input scaling for the adder so that overflow during the MAC operations can be avoided.

8. The floating-point processor is easy to code using floating-point arithmetic and develops the prototype quickly. However, it is not efficient in terms of the number of instructions it has to complete compared with the fixed-point processor.

9. The fixed-point processor using fixed-point arithmetic takes much effort to code. But it offers the least number of instructions for the CPU to execute.

9.8 Problems

9.1. Find the signed Q-15 representation for the decimal number 0.2560123.

9.2. Find the signed Q-15 representation for the decimal number -0.2160123.

9.3. Convert the Q-15 signed number = 1.010101110100010 to the decimal.

9.4. Convert the Q-15 signed number = 0.001000111101110 to the decimal number.

9.5. Add the following two Q-15 numbers:

1. 101010111000001 + 0. 010001111011010

9.6. Convert each of the following decimal numbers to the floating-point number using the format specified in Figure 9.10.

   a. 0.1101235

   b. -10.430527

9.7. Add the following floating-point numbers whose formats are defined in Figure 9.10, and determine the sum in decimal format.

   1101 011100011011 + 0100 101111100101.

9.8. Convert the following number in IEEE single precision format to the decimal format:

   110100000.010...0000.

9.9. Convert the following number in IEEE double precision format to the decimal format:

   011000...0.1010...000

9.10. Given the following FIR filter:

$$y(n) = -0.36x(n) + 1.6x(n-1) + 0.36x(n-2),$$

with a passband gain of 2 and the input being half of range, develop the DSP implementation equations in the Q-15 fixed-point system.

9.11. Given the following IIR filter:

$$y(n) = 1.35x(n) + 0.3y(n-1),$$

with a passband gain of 2 and the input being half of range, use the direct-form I method to develop the DSP implementation equations in the Q-15 fixed-point system.

9.12. Given the following IIR filter:

$$y(n) = 0.72x(n) + 1.42x(n-2) + 0.72x(n-2) \\ - 1.35y(n-1) - 0.5y(n-2),$$

with a passband gain of 1 and a full range of input, use the direct-form II method to develop the DSP implementation equations in the Q-15 fixed-point system.

References

Dahnoun, N. (2000). *Digital Signal Processing Implementation Using the TMS320C6000TM DSP Platform.* Englewood Cliffs, NJ: Prentice Hall.

Embree, P. M. (1995). *C Algorithms for Real-Time DSP.* Upper Saddle River, NJ: Prentice Hall.

Ifeachor, E. C., and Jervis, B. W. (2002). *Digital Signal Processing: A Practical Approach*, 2nd ed. Upper Saddle River, NJ: Prentice Hall.

Kehtarnavaz, N., and Simsek, B. (2000). *C6X-Based Digital Signal Processing.* Upper Saddle River, NJ: Prentice Hall.

Sorensen, H. V., and Chen, J. P. (1997). *A Digital Signal Processing Laboratory Using TMS320C30.* Upper Saddle River, NJ: Prentice Hall.

Texas Instruments. (1991). *TMS320C3x User's Guide.* Dallas, TX: Author.

———. (1998). *TMS320C6x CPU and Instruction Set Reference Guide*, Literature ID# SPRU 189C. Dallas, TX: Author.

———. (2001). *Code Composer Studio: Getting Started Guide.* Dallas, TX: Author.

van der Vegte, J. (2002). *Fundamentals of Digital Signal Processing.* Upper Saddle River, NJ: Prentice Hall.

Wikipedia. (2007). *Harvard Mark I.* Retrieved March 14, 2007, from http://en.wikipedia.org/wiki/Harvard_Mark_I

10

Adaptive Filters and Applications

Objectives:

This chapter introduces principles of adaptive filters and adaptive least mean square algorithm and illustrates how to apply the adaptive filters to solve real-world application problems such as adaptive noise cancellation, system modeling, adaptive line enhancement, and telephone echo cancellation.

10.1 Introduction to Least Mean Square Adaptive Finite Impulse Response Filters

An *adaptive filter* is a digital filter that has self-adjusting characteristics. It is capable of adjusting its filter coefficients automatically to adapt the input signal via an adaptive algorithm. Adaptive filters play an important role in modern digital signal processing (DSP) products in areas such as telephone echo cancellation, noise cancellation, equalization of communications channels, biomedical signal enhancement, active noise control, and adaptive control systems. Adaptive filters work generally for adaptation of signal-changing environments, spectral overlap between noise and signal, and unknown, or time-varying, noise. For example, when interference noise is strong and its spectrum overlaps that of the desired signal, the conventional approach will fail to preserve the desired signal spectrum while removing the interference using a traditional filter, such as a notch filter with the fixed filter coefficients, as shown in Figure 10.1.

However, an adaptive filter will do the job. Note that adaptive filtering, with its applications, has existed for more than two decades in the research community and is still active there. This chapter can only introduce some fundamentals of the

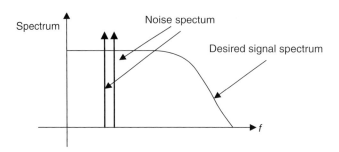

FIGURE 10.1 Spectrum illustration for using adaptive filters.

subject, that is, adaptive finite impulse response (FIR) filters with a simple and popular least mean square (LMS) algorithm. Further exploration into adaptive infinite impulse response (IIR) filters, adaptive lattice filters, their associated algorithms and applications, and so on can be found in comprehensive texts by Haykin (1991), Stearns (2003), and Widrow and Stearns (1985).

To get the concept of adaptive filtering, we will first look at an illustrative example of the simplest noise canceler to see how it works before diving into detail. The block diagram for such a noise canceler is shown in Figure 10.2.

As shown in Figure 10.2, first, the DSP system consists of two analog-to-digital conversion (ADC) channels. The first microphone with ADC is used to capture the desired speech $s(n)$. However, due to a noisy environment, the signal is contaminated and the ADC channel produces a signal with the noise; that is, $d(n) = s(n) + n(n)$. The second microphone is placed where only noise is picked up and the second ADC channel captures noise $x(n)$, which is fed to the adaptive filter.

Note that the corrupting noise $n(n)$ in the first channel is uncorrelated to the desired signal $s(n)$, so that separation between them is possible. The noise signal $x(n)$ from the second channel is correlated to the corrupting noise $n(n)$ in the first

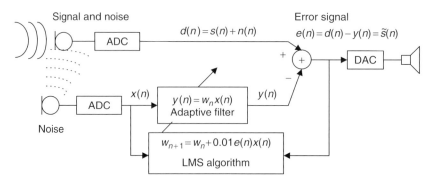

FIGURE 10.2 Simplest noise canceler using a one-tap adaptive filter.

channel, since both come from the same noise source. Similarly, the noise signal $x(n)$ is not correlated to the desired speech signal $s(n)$.

We assume that the corrupting noise in the first channel is a linear filtered version of the second-channel noise, since it has a different physical path from the second-channel noise, and the noise source is time varying, so that we can estimate the corrupting noise $n(n)$ using an adaptive filter. The adaptive filter contains a digital filter with adjustable coefficient(s) and the LMS algorithm to modify the value(s) of the coefficient(s) for filtering each sample. The adaptive filter then produces an estimate of noise $y(n)$, which will be subtracted from the corrupted signal $d(n) = s(n) + n(n)$. When the noise estimate $y(n)$ equals or approximates the noise $n(n)$ in the corrupted signal, that is, $y(n) \approx n(n)$, the error signal $e(n) = s(n) + n(n) - y(n) \approx \tilde{s}(n)$ will approximate the clean speech signal $s(n)$. Hence, the noise is canceled.

In our illustrative numerical example, the adaptive filter is set to be a one-tap FIR filter to simplify numerical algebra. The filter adjustable coefficient w_n is adjusted based on the LMS algorithm (discussed later in detail) in the following:

$$w_{n+1} = w_n + 0.01 \cdot e(n) \cdot x(n),$$

where w_n is the coefficient used currently, while w_{n+1} is the coefficient obtained from the LMS algorithm and will be used for the next coming input sample. The value of 0.01 controls the speed of the coefficient change. To illustrate the concept of the adaptive filter in Figure 10.2, the LMS algorithm has the initial coefficient set to be $w_0 = 0.3$ and leads to:

$$y(n) = w_n x(n)$$
$$e(n) = d(n) - y(n)$$
$$w_{n+1} = w_n + 0.01 e(n) x(n).$$

The corrupted signal is generated by adding noise to a sine wave. The corrupted signal and noise reference are shown in Figure 10.3, and their first 16 values are listed in Table 10.1.

Let us perform adaptive filtering for several samples using the values for the corrupted signal and reference noise in Table 10.1. We see that

$n = 0$, $y(0) = w_0 x(0) = 0.3 \times (-0.5893) = -0.1768$
$e(0) = d(0) - y(0) = -0.2947 - (-0.1768) = -0.1179 = \tilde{s}(0)$
$w_1 = w_0 + 0.01 e(0) x(0) = 0.3 + 0.01 \times (-0.1179) \times (-0.5893) = 0.3007$
$n = 1$, $y(1) = w_1 x(1) = 0.3007 \times 0.5893 = 0.1772$
$e(1) = d(1) - y(1) = 1.0017 - 0.1772 = 0.8245 = \tilde{s}(1)$
$w_2 = w_1 + 0.01 e(1) x(1) = 0.3007 + 0.01 \times 0.8245 \times 0.5893 = 0.3056$
$n = 2$, $y(2) = w_2 x(2) = 0.3056 \times 3.1654 = 0.9673$

TABLE 10.1 Adaptive filtering results for the simplest noise canceler example.

n	$d(n)$	$x(n)$	$\tilde{s}(n) = e(n)$	Original $s(n)$	w_{n+1}
0	−0.2947	−0.5893	−0.1179	0	0.3000
1	1.0017	0.5893	0.8245	0.7071	0.3007
2	2.5827	3.1654	1.6155	1.0000	0.3056
3	−1.6019	−4.6179	0.0453	0.7071	0.3567
4	0.5622	1.1244	0.1635	0.0000	0.3546
5	0.4456	2.3054	−0.3761	−0.7071	0.3564
6	−4.2674	−6.5348	−1.9948	−1.0000	0.3478
7	−0.8418	−0.2694	−0.7130	−0.7071	0.4781
8	−0.3862	−0.7724	−0.0154	−0.0000	0.4800
9	1.2274	1.0406	0.7278	0.7071	0.4802
10	0.6021	−0.7958	0.9902	1.0000	0.4877
11	1.1647	0.9152	0.7255	0.7071	0.4799
12	0.9630	1.9260	0.0260	0.0000	0.4865
13	−1.5065	−1.5988	−0.7279	−0.7071	0.4870
14	−0.1329	1.7342	−0.9976	−1.0000	0.4986
15	0.8146	3.0434	−0.6503	−0.7071	0.4813

$$e(2) = d(2) - y(2) = 2.5827 - 0.9673 = 1.6155 = \tilde{s}(2)$$
$$w_3 = w_2 + 0.01e(2)x(2) = 0.3056 + 0.01 \times 1.6155 \times 3.1654 = 0.3567$$
$$n = 3, \cdots$$

For comparison, results of the first 16 processed output samples, original samples, and filter coefficient values are also included in Table 10.1. Figure 10.3 also shows the original signal samples, reference noise samples corrupted signal samples, enhanced signal samples, and filter coefficient values for each incoming sample, respectively.

As shown in Figure 10.3, after 7 adaptations, the adaptive filter learns noise characteristics and cancels the noise in the corrupted signal. The adaptive coefficient is close to the optimal value of 0.5. The processed output is close to the original signal. The first 16 processed values for the corrupted signal, reference noise, clean signal, original signal, and adaptive filter coefficient used at each step are listed in Table 10.1.

Clearly, the enhanced signal samples look much like the sinusoidal input samples. Now our simplest one-tap adaptive filter works for this particular case. In general, the FIR filter with multiple-taps is used and has the following format:

$$y(n) = \sum_{i=0}^{N-1} w_n(i)x(n-i) \qquad (10.1)$$
$$= w_n(0)x(n) + w_n(1)x(n-1) + \cdots + w_n(N-1)x(n-N+1).$$

The LMS algorithm for the adaptive FIR filter will be developed next.

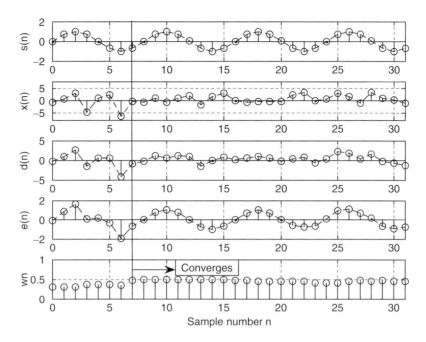

FIGURE 10.3 Original signal, reference noise, corrupted signal, enhanced signal, and adaptive coefficient in the noise cancellation.

10.2 Basic Wiener Filter Theory and Least Mean Square Algorithm

Many adaptive algorithms can be viewed as approximations of the discrete Wiener filter shown in Figure 10.4.

The Wiener filter adjusts its weight(s) to produce filter output $y(n)$, which would be as close as the noise $n(n)$ contained in the corrupted signal $d(n)$. Hence, at the subtracted output, the noise is canceled, and the output $e(n)$ contains clean signal.

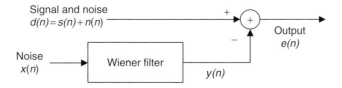

FIGURE 10.4 Wiener filter for noise cancellation.

Consider a single-weight case of $y(n) = wx(n)$, and note that the error signal $e(n)$ is given by

$$e(n) = d(n) - wx(n). \quad (10.2)$$

Now let us solve the best weight w^*. Taking the square of the output error leads to

$$e^2(n) = (d(n) - wx(n))^2 = d^2(n) - 2d(n)wx(n) + w^2 x^2(n). \quad (10.3)$$

Taking the statistical expectation of Equation (10.3), we have

$$E(e^2(n)) = E(d^2(n)) - 2wE(d(n)x(n)) + w^2 E(x^2(n)). \quad (10.4)$$

Using the notations in statistics, we define

$J = E(e^2(n)) = $ MSE (mean squared error)
$\sigma^2 = E(d^2(n)) = $ power of corrupted signal
$P = E(d(n)x(n)) = $ cross-correlation between $d(n)$ and $x(n)$
$R = E(x^2(n)) = $ autocorrelation

We can view the statistical expectation as an average of the N signal terms, each being a product of two individual signal samples:

$$E(e^2(n)) = \frac{e^2(0) + e^2(1) + \cdots + e^2(N-1)}{N}$$

or

$$E(d(n)x(n)) = \frac{d(0)x(0) + d(1)x(1) + \cdots + d(N-1)x(N-1)}{N}$$

for a sufficiently large sample number of N. We can write Equation (10.4) as

$$J = \sigma^2 - 2wP + w^2 R. \quad (10.5)$$

Since σ^2, P, and R are constants, J is a quadratic function of w that may be plotted in Figure 10.5.

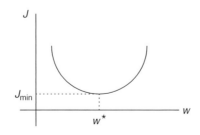

FIGURE 10.5 **Mean square error quadratic function.**

The best weight (optimal) w^* is at the location where the minimum MSE J_{min} is achieved. To obtain w^*, taking a derivative of J and setting it to zero leads to

$$\frac{dJ}{dw} = -2P + 2wR = 0. \tag{10.6}$$

Solving Equation 10.6, we get the best weight solution as

$$w^* = R^{-1}P. \tag{10.7}$$

Example 10.1.

Given a quadratic MSE function for the Wiener filter:

$$J = 40 - 20w + 10w^2,$$

a. Find the optimal solution for w^* to achieve the minimum MSE J_{min} and determine J_{min}.

Solution:

a. Taking a derivative of the MSE function and setting it to zero, we have

$$\frac{dJ}{dw} = -20 + 10 \times 2w = 0.$$

Solving the equation leads to

$$w^* = 1.$$

Finally, substituting $w^* = 1$ into the MSE function, we get the minimum J_{min} as

$$J_{min} = J|_{w=w^*} = 40 - 20w + 10w^2|_{w=1} = 40 - 20 \times 1 + 10 \times 1^2 = 30.$$

Notice that a few points need to be clarified for Equation (10.7):

1. Optimal coefficient(s) can be different for every block of data, since the corrupted signal and reference signal are unknown. The autocorrelation and cross-correlation may vary.

2. If a larger number of coefficients (weights) are used, the inverse matrix of R^{-1} may require a larger number of computations and may come to be ill-conditioned. This will make real-time implementation impossible.

3. The optimal solution is based on the statistics, assuming that the size of the data block, N, is sufficiently long. This will cause a long processing delay that will make real-time implementation impossible.

As we pointed out, solving the Wiener solution Equation (10.7) requires a lot of computations, including matrix inversion for a general multiple-taps FIR filter. The well-known textbook by Widrow and Stearns (1985) describes a powerful LMS algorithm by using the steepest descent algorithm to minimize the MSE sample by sample and locate the filter coefficient(s). We first study the steepest descent algorithm as illustrated in Equation (10.8):

$$w_{n+1} = w_n - \mu \frac{dJ}{dw} \qquad (10.8)$$

where $\mu =$ constant controlling speed of convergence.

The illustration of the steepest descent algorithm for solving the optimal coefficient(s) is described in Figure 10.6.

As shown in the first plot in Figure 10.6, if $\frac{dJ}{dw} < 0$, notice that $-\mu \frac{dJ}{dw} > 0$. The new coefficient w_{n+1} will be increased to approach the optimal value w^* by Equation (10.8). On the other hand, if $\frac{dJ}{dw} > 0$, as shown in the second plot in Figure 10.6, we see that $-\mu \frac{dJ}{dw} < 0$. The new coefficient w_{n+1} will be decreased to approach the optimal value w^*. When $\frac{dJ}{dw} = 0$, the best coefficient w_{n+1} is reached.

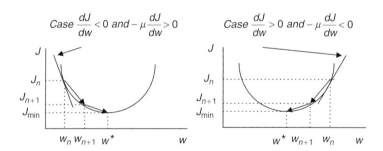

FIGURE 10.6 Illustration of the steepest descent algorithm.

Example 10.2.

Given a quadratic MSE function for the Wiener filter:

$$J = 40 - 20w + 10w^2,$$

a. Use the steepest descent method with an initial guess as $w_0 = 0$ and $\mu = 0.04$ to find the optimal solution for w^* and determine J_{min} by iterating three times.

Solution:

a. Taking the derivative of the MSE function, we have

$$\frac{dJ}{dw} = -20 + 10 \times 2w_n.$$

10.2 Basic Wiener Filter Theory and Least Mean Square Algorithm

When $n = 0$, we calculate

$$\mu \frac{dJ}{dw} = 0.04 \times (-20 + 10 \times 2w_0)|_{w_0=0} = -0.8.$$

Applying the steepest descent algorithm, it follows that

$$w_1 = w_0 - \mu \frac{dJ}{dw} = 0 - (-0.8) = 0.8.$$

Similarly for $n = 1$, we get

$$\mu \frac{dJ}{dw} = 0.04 \times (-20 + 10 \times 2w_1)|_{w_1=0.8} = -0.16$$

$$w_2 = w_1 - \mu \frac{dJ}{dw} = 0.8 - (-0.16) = 0.96,$$

and for $n = 2$, it follows that

$$\mu \frac{dJ}{dw} = 0.04 \times (-20 + 10 \times 2w_2)|_{w_2=0.96} = -0.032$$

$$w_3 = w_2 - \mu \frac{dJ}{dw} = 0.96 - (-0.032) = 0.992.$$

Finally, substituting $w^* \approx w_3 = 0.992$ into the MSE function, we get the minimum $J_{\min}$ as

$$J_{\min} \approx 40 - 20w + 10w^2|_{w=0.992} = 40 - 20 \times 0.992 + 10 \times 0.992^2 = 30.0006.$$

As we can see, after three iterations, the filter coefficient and minimum MSE values are very close to the theoretical values obtained in Example 10.1.

Application of the steepest descent algorithm still needs an estimation of the derivative of the MSE function that could include statistical calculation of a block of data. To change the algorithm to do sample-based processing, an LMS algorithm must be used. To develop the LMS algorithm in terms of sample-based processing, we take the statistical expectation out of J and then take the derivative to obtain an approximate of $\frac{dJ}{dw}$, that is,

$$J = e^2(n) = (d(n) - wx(n))^2 \tag{10.9}$$

$$\frac{dJ}{dw} = 2(d(n) - wx(n)) \frac{d(d(n) - wx(n))}{dw} = -2e(n)x(n). \tag{10.10}$$

Substituting $\frac{dJ}{dw}$ into the steepest descent algorithm in Equation (10.8), we achieve the LMS algorithm for updating a single-weight case as

$$w_{n+1} = w_n + 2\mu e(n)x(n), \tag{10.11}$$

where μ is the convergence parameter controlling speed of convergence. For example, let us choose $2\mu = 0.01$. In general, with an adaptive FIR filter of length N, we extend the single-tap LMS algorithm without going through derivation, as shown in the following equations:

$$y(n) = w_n(0)x(n) + w_n(1)x(n-1) + \cdots + w_n(N-1)x(n-N+1) \qquad (10.12)$$

for $i = 0, \ldots, N-1$

$$w_{n+1}(i) = w_n(i) + 2\mu e(n)x(n-i). \qquad (10.13)$$

The convergence factor is chosen to be

$$0 < \mu < \frac{1}{NP_x}, \qquad (10.14)$$

where P_x is the input signal power. In practice, if the ADC has 16-bit data, the maximum signal amplitude should be $A = 2^{15}$. Then the maximum input power must be less than

$$P_x < \left(2^{15}\right)^2 = 2^{30}.$$

Hence, we may make a selection of the convergence parameter as

$$\mu = \frac{1}{N \times 2^{30}} \approx \frac{9.3 \times 10^{-10}}{N}. \qquad (10.15)$$

We further neglect time index for $w_n(i)$ and use the notation $w(i) = w_n(i)$, since only the currently updated coefficients are needed for the next sample adaptation. We conclude the implementation of the LMS algorithm by the following steps:

1. Initialize $w(0), w(1), \ldots, w(N-1)$ to arbitrary values.
2. Read $d(n)$, $x(n)$, and perform digital filtering:

$$y(n) = w(0)x(n) + w(1)x(n-1) + \cdots + w(N-1)x(n-N+1).$$

3. Compute the output error:

$$e(n) = d(n) - y(n).$$

4. Update each filter coefficient using the LMS algorithm:

for $i = 0, \ldots, N-1$

$$w(i) = w(i) + 2\mu e(n)x(n-i).$$

We will apply the adaptive filter to solve real-world problems in the next section.

10.3 Applications: Noise Cancellation, System Modeling, and Line Enhancement

We now examine several applications of the LMS algorithm, such as noise cancellation, system modeling, and line enhancement via application examples. First, we begin with the noise cancellation problem to illustrate operations of the LMS adaptive FIR filter.

10.3.1 Noise Cancellation

The concept of noise cancellation was introduced in the previous section. Figure 10.7 shows the main idea.

The DSP system consists of two ADC channels. The first microphone with ADC captures the noisy speech, $d(n) = s(n) + n(n)$, which contains the clean speech $s(n)$ and noise $n(n)$ due to a noisy environment, while the second microphone with ADC resides where it picks up only the correlated noise and feeds the noise reference $x(n)$ to the adaptive filter. The adaptive filter uses the LMS algorithm to adjust its coefficients to produce the best estimate of noise $y(n) \approx n(n)$, which will be subtracted from the corrupted signal $d(n) = s(n) + (n)$. The output of the error signal $e(n) = s(n) + n(n) - y(n) \approx \tilde{s}(n)$ is expected to be a best estimate of the clean speech signal. Through digital-to-analog conversion (DAC), the cleaned digital speech becomes analog voltage, which drives the speaker.

We first study the noise cancellation problem using a simple two-tap adaptive filter via Example 10.3 and assumed data. The purpose of doing so is to become familiar with the setup and operations of the adaptive filter and LMS algorithm. The simulation for real adaptive noise cancellation follows.

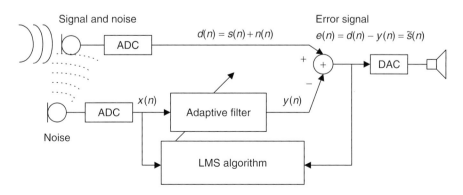

FIGURE 10.7 Simplest noise canceler using a one-tap adaptive filter.

Example 10.3.

Given the DSP system for the noise cancellation application using an adaptive filter with two coefficients shown in Figure 10.8,

a. Set up the LMS algorithm for the adaptive filter.
b. Perform adaptive filtering to obtain outputs $e(n) = n = 0, 1, 2$ given the following inputs and outputs:

$$x(0) = 1, \ x(1) = 1 x(2) = -1, \ d(0) = 2, \ d(1) = 1, \ d(2) = -2$$

and initial weights:

$$w(0) = w(1) = 0,$$

convergence factor is set to be $\mu = 0.1$.

Solution:

a. The adaptive LMS algorithm is set up as:

Initialization: $w(0) = 0, \ w(1) = 0$
Digital filtering: $y(n) = w(0)x(n) + w(1)x(n-1)$
Computing the output error = output: $e(n) = d(n) - y(n)$.
Updating each weight for the next coming sample:

$$w(i) = w(i) + 2\mu e(n) x(n-i), \text{ for } i = 0, 1$$

or

$$w(0) = w(0) + 2\mu e(n) x(n)$$
$$w(1) = w(1) + 2\mu e(n) x(n-1).$$

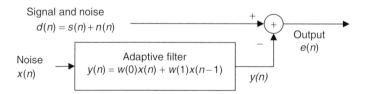

FIGURE 10.8 Noise cancellation in Example 10.3.

b. We can see the adaptive filtering operations as follows:
  a. For $n = 0$
  Digital filtering:
  $$y(0) = w(0)x(0) + w(1)x(-1) = 0 \times 1 + 0 \times 0 = 0$$
  Computing the output:
  $$e(0) = d(0) - y(0) = 2 - 0 = 2$$
  Updating coefficients:
  $$w(0) = w(0) + 2 \times 0.1 \times e(0)x(0) = 0 + 2 \times 0.1 \times 2 \times 1 = 0.4$$
  $$w(1) = w(1) + 2 \times 0.1 \times e(0)x(-1) = 0 + 2 \times 0.1 \times 2 \times 0 = 0.0$$
  For $n = 1$
  Digital filtering:
  $$y(1) = w(0)x(1) + w(1)x(0) = 0.4 \times 1 + 0 \times 1 = 0.4$$
  Computing the output:
  $$e(1) = d(1) - y(1) = 1 - 0.4 = 0.6$$
  Updating coefficients:
  $$w(0) = w(0) + 2 \times 0.1 \times e(1)x(1) = 0.4 + 2 \times 0.1 \times 0.6 \times 1 = 0.52$$
  $$w(1) = w(1) + 2 \times 0.1 \times e(1)x(0) = 0 + 2 \times 0.1 \times 0.6 \times 1 = 0.12$$
  For $n = 2$
  Digital filtering:
  $$y(2) = w(0)x(2) + w(1)x(1) = 0.52 \times (-1) + 0.12 \times 1 = -0.4$$
  Computing the output:
  $$e(2) = d(2) - y(2) = -2 - (-0.4) = -1.6$$
  Updating coefficients:
  $$w(0) = w(0) + 2 \times 0.1 \times e(2)x(2) = 0.52 + 2 \times 0.1 \times (-1.6) \times (-1) = 0.84$$
  $$w(1) = w(1) + 2 \times 0.1 \times e(2)x(1) = 0.12 + 2 \times 0.1 \times (-1.6) \times 1 = -0.2.$$

Hence, the adaptive filter outputs for the first three samples are listed as
$$e(0) = 2,\ e(1) = 0.6,\ e(2) = -1.6.$$

Next we examine the MSE function assuming the following statistical data:

$$\sigma^2 = E[d^2(n)] = 4,\ E[x^2(n)] = E[x^2(n-1)] = 1,\ E[x(n)x(n-1)] = 0$$
$$E[d(n)x(n)] = 1,\ \text{ and } E[d(n)x(n-1)] = -1$$

for the two-tap adaptive filter $y(n) = w(0)x(n) + w(1)x(n-1)$. We follow Equations (10.2) to (10.5) to achieve the minimum MSE function in two dimensions as

$$J = 4 + w^2(0) + w^2(1) - 2w(0) + 2w(1).$$

Figure 10.9 shows the MSE function versus the weights, where the optimal weights and the minimum MSE are $w^*(0) = 1$, $w^*(1) = -1$, and $J_{\min} = 2$. If the adaptive filter continues to process the data, it will converge to the optimal weights, which locate the minimum MSE. The plot also indicates that the function is quadratic and that there exists only one minimum of the MSE surface.

Next, a simulation example is given to illustrate this idea and its results. The noise cancellation system is assumed to have the following specifications:

- Sample rate = 8,000 Hz
- Original speech data: wen.dat

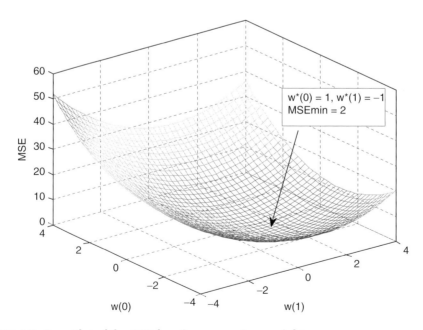

FIGURE 10.9 Plot of the MSE function versus two weights.

- Speech corrupted by Gaussian noise with a power of 1 delayed by 5 samples from the noise reference
- Noise reference containing Gaussian noise with a power of 1
- Adaptive FIR filter used to remove the noise
- Number of FIR filter taps $= 21$
- Convergence factor for the LMS algorithm chosen to be $0.01(<1/21)$.

The speech waveforms and spectral plots for the original, corrupted, and reference noise and for the cleaned speech are plotted in Figures 10.10a and 10.10b. From the figures, it is observed that the enhanced speech waveform and spectrum are very close to the original ones. The LMS algorithm converges after approximately 400 iterations. The method is a very effective approach for noise canceling. MATLAB implementation is detailed in Program 10.1.

Program 10.1. MATLAB program for adaptive noise cancellation.

```
close all; clear all
load wen.dat % Given by the instructor
fs=8000;            % Sampling rate
t=0:1:length(wen)-1;        % Create the index array
t=t/fs;        % Convert indices to time instants
x=randn(1,length(wen));        % Generate the random noise
n=filter([ 0 0 0 0 0 0.5 ],1,x);        % Generate the corruption noise
d=wen+n;        % Generate the signal plus noise
mu=0.01;        % Initialize the step size
w=zeros(1,21);        % Initialize the adaptive filter coefficients
y=zeros(1,length(t));        % Initialize the adaptive filter output array
e=y;        % Initialize the output array
% Adaptive filtering using the LMS algorithm
for m=22:1:length(t)-1
sum=0;
for i=1:1:21
sum = sum + w(i)*x(m − i);
end
y(m)=sum;
e(m)=d(m)-y(m);
for i=1:1:21
w(i) = w(i) + 2*mu*e(m)*x(m − i);
end
end
```

(Continued)

```
% Calculate the single-sided amplitude spectrum for the original signal
WEN = 2*abs(fft(wen))/length(wen);WEN(1)=WEN(1)/2;
% Calculate the single-sided amplitude spectrum for the corrupted signal
D = 2*abs(fft(d))/length(d);D(1)=D(1)/2;
f=[0:1:length(wen)/2]*fs/length(wen);
% Calculate the single-sided spectrum for the noise-canceled signal
E = 2*abs(fft(e))/length(e);E(1)=E(1)/2;
% Plot signals and spectra
subplot(4,1,1), plot(wen);grid;ylabel('Orig. speech');
subplot(4,1,2),plot(d);grid;ylabel('Corrupt. speech')
subplot(4,1,3),plot(x);grid;ylabel('Ref. noise');
subplot(4,1,4),plot(e);grid;ylabel('Clean speech');
xlabel('Number of samples');
figure
subplot(3,1,1),plot(f,WEN(1:length(f)));grid
ylabel('Orig. spectrum')
subplot(3,1,2),plot(f,D(1:length(f)));grid;ylabel('Corrupt.spectrum')
subplot(3,1,3),plot(f,E(1:length(f)));grid
ylabel('Clean spectrum');xlabel('Frequency (Hz)');
```

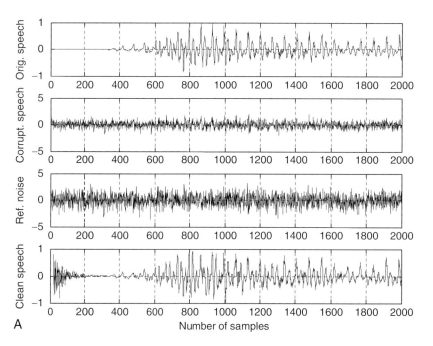

FIGURE 10.10A Waveforms for original speech, corrupted speech, reference noise, and clean speech.

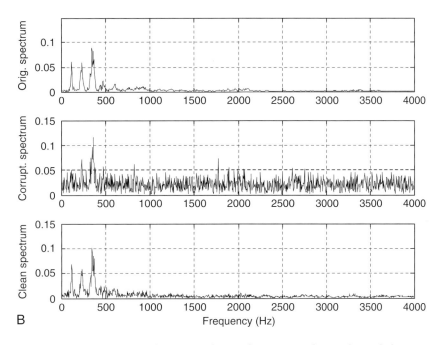

FIGURE 10.10B Spectrum for original speech, corrupted speech, and clean speech.

Other interference cancellations include that of 60 Hz interference in electrocardiography (ECG) (Chapter 8) and echo cancellation in long-distance telephone circuits, which will be described in a later section.

10.3.2 System Modeling

Another application of the adaptive filter is system modeling. The adaptive filter can keep tracking the behavior of an unknown system by using the unknown system's input and output, as depicted in Figure 10.11.

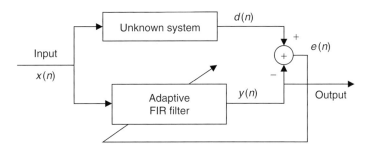

FIGURE 10.11 Adaptive filter for system modeling.

As shown in the figure, $y(n)$ is going to be as close as the unknown system's output. Since both the unknown system and the adaptive filter use the same input, the transfer function of the adaptive filter will approximate that of the unknown system.

Example 10.4.

Given the system modeling described and using the single-weight adaptive filter $y(n) = wx(n)$ to perform the system modeling task,

a. Set up the LMS algorithm to implement the adaptive filter, assuming that the initial $w = 0$ and $\mu = 0.5$.

b. Perform adaptive filtering to obtain $y(0)$, $y(1)$, $y(2)$, and $y(3)$ given
$$d(0) = 1,\ d(1) = 2,\ d(2) = -2,\ d(3) = 2,$$
$$x(0) = 0.5,\ x(1) = 1,\ x(2) = -1,\ x(3) = 1.$$

Solution:

a. Adaptive filtering equations are set up as
$$w = 0 \text{ and } 2\mu = 2 \times 0.5 = 1$$
$$y(n) = wx(n)$$
$$e(n) = d(n) - y(n)$$
$$w = w + e(n)x(n)$$

b. Adaptive filtering:
$$n = 0,\ y(0) = wx(0) = 0 \times 0.5 = 0$$
$$e(0) = d(0) - y(0) = 1 - 0 = 1$$
$$w = w + e(0)x(0) = 0 + 1 \times 0.5 = 0.5$$
$$n = 1,\ y(1) = wx(1) = 0.5 \times 1 = 0.5$$
$$e(1) = d(1) - y(1) = 2 - 0.5 = 1.5$$
$$w = w + e(1)x(1) = 0.5 + 1.5 \times 1 = 2.0$$
$$n = 2,\ y(2) = wx(2) = 2 \times (-1) = -2$$
$$e(2) = d(2) - y(2) = -2 - (-2) = 0$$
$$w = w + e(2)x(2) = 2 + 0 \times (-1) = 2$$
$$n = 3,\ y(3) = wx(3) = 2 \times 1 = 2$$
$$e(3) = d(3) - y(3) = 2 - 2 = 0$$
$$w = w + e(3)x(3) = 2 + 0 \times 1 = 2$$

For this particular case, the system is actually a digital amplifier with a gain of 2.

Next, we assume that the unknown system is a fourth-order bandpass IIR filter whose 3 dB lower and upper cutoff frequencies are 1,400 Hz and 1,600 Hz operating at 8,000 Hz. We use an input consisting of tones of 500, 1,500, and 2,500 Hz. The unknown system's frequency responses are shown in Figure 10.12.

The input waveform $x(n)$ with three tones is shown as the first plot in Figure 10.13. We can predict that the output of the unknown system will contain a 1,500 Hz tone only, since the other two tones are rejected by the unknown system. Now, let us look at adaptive filter results. We use an adaptive FIR filter with the number of taps being 21, and a convergence factor set to be 0.01. In time domain, the output waveforms of the unknown system $d(n)$ and adaptive filter output $y(n)$ are almost identical after 70 samples when the LMS algorithm converges. The error signal $e(n)$ is also plotted to show that the adaptive filter keeps tracking the unknown system's output with no difference after the first 50 samples.

Figure 10.14 depicts the frequency domain comparisons. The first plot displays the frequency components of the input signal, which clearly shows 500, 1,500, and 2,500 Hz. The second plot shows the unknown system's output spectrum, which contains only a 1,500 Hz tone, while the third plot displays the spectrum of the adaptive filter output. As we can see, in frequency domain, the

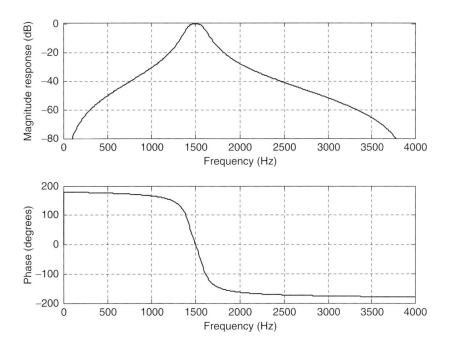

FIGURE 10.12 The unknown system's frequency responses.

482 10 ADAPTIVE FILTERS AND APPLICATIONS

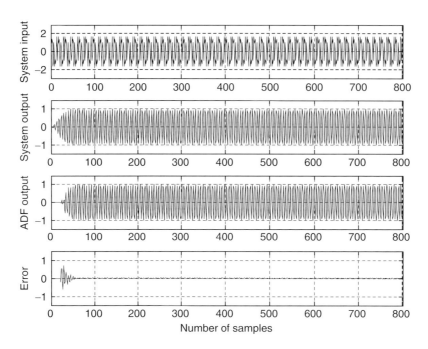

FIGURE 10.13 The waveforms for the unknown system's output, adaptive filter output, and error output.

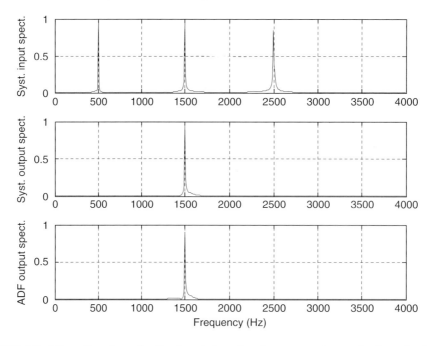

FIGURE 10.14 Spectrum for the input signal, unknown system output, and adaptive filter output.

adaptive filter tracks the characteristics of the unknown system. The MATLAB implementation is given in Program 10.2.

Program 10.2. MATLAB program for adaptive system identification.

```
close all; clear all
%Design unknown system
fs = 8000; T = 1/fs;         % Sampling rate and sampling period
% Bandpass filter design
%for the assumed unknown system using the bilinear transformation
%(BLT) method (see Chapter 8)
wd1 = 1400*2*pi;wd2 = 1600*2*pi;
wa1 = (2/T)*tan(wd1*T/2);wa2 = (2/T)*tan(wd2*T/2);
BW=wa2-wa1;
w0 = sqrt(wa2*wa1);
[B,A]=lp2bp([1],[1 1.4141 1],w0,BW);
[b,a]=bilinear(B,A,fs);
freqz1(b,a,512,fs); axis([0 fs/2 -80 1]);      % Frequency response plots
figure
t=0:T:0.1;          % Generate the time vector
x = cos(2*pi*500*t) + sin(2*pi*1500*t) + cos(2*pi*2500*t + pi/4);
d=filter(b,a,x);           % Produce the unknown system output
mu= 0.01;           % Convergence factor
w=zeros(1,21); y=zeros(1,length(t)); % Initialize the coefficients and output
e=y;% initialize the error vector
% Perform adaptive filtering using the LMS algorithm
for m=22:1:length(t)-1
sum=0;
for i=1:1:21
sum = sum + w(i)*x(m - i);
end
y(m)=sum;
e(m) = d(m)-y(m);
for i=1:1:21
w(i) = w(i) + 2*mu*e(m)*x(m-i);
end
end
% Calculate the single-sided amplitude spectrum for the input
X = 2*abs(fft(x))/length(x);X(1)=X(1)/2;
% Calculate the single-sided amplitude spectrum for the unknown system output
D = 2*abs(fft(d))/length(d);D(1)=D(1)/2;
% Calculate the single-sided amplitude spectrum for the adaptive filter output
Y = 2*abs(fft(y))/length(y);Y(1)=Y(1)/2;
% Map the frequency index to its frequency in Hz
f=[0:1:length(x)/2]*fs/length(x);
```

(*Continued*)

```
% Plot signals and spectra
subplot(4,1,1), plot(x);grid;axis([0 length(x) -3 3]);
ylabel('System input');
subplot(4,1,2), plot(d);grid;axis([0 length(x) -1.5 1.5]);
ylabel('System output');
subplot(4,1,3),plot(y);grid;axis([0 length(y) -1.5 1.5]);
ylabel('ADF output')
subplot(4,1,4),plot(e);grid;axis([0 length(e) -1.5 1.5]);
ylabel('Error'); xlabel('Number of samples')
figure
subplot(3,1,1),plot(f,X(1:length(f)));grid;ylabel('Syst. input spect.')
subplot(3,1,2),plot(f,D(1:length(f)));grid;ylabel('Syst. output spect.')
subplot(3,1,3),plot(f,Y(1:length(f)));grid
ylabel('ADF output spect.');xlabel('Frequency (Hz)');
```

10.3.3 Line Enhancement Using Linear Prediction

We study adaptive filtering via another application example of line enhancement. If a signal frequency content is very narrow compared with the bandwidth and changes with time, then the signal can efficiently be enhanced by the adaptive filter, which is line enhancement. Figure 10.15 shows line enhancement using the adaptive filter where the LMS algorithm is used. As illustrated in the figure, the signal $d(n)$ is the corrupted sine wave signal by the white Gaussian noise $n(n)$. The enhanced line consists of the delay element to delay the corrupted signal by Δ samples to produce an input to the adaptive filter. The adaptive filter is actually a linear predictor of the desired narrow band signal. A two-tap adaptive FIR filter can predict one sinusoid (proof is beyond the scope of this text). The value of Δ is usually determined by experiments or experience in practice to achieve the best enhanced signal.

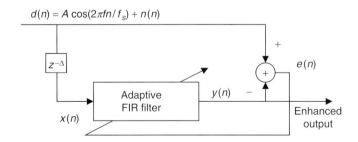

FIGURE 10.15   Line enhancement using an adaptive filter.

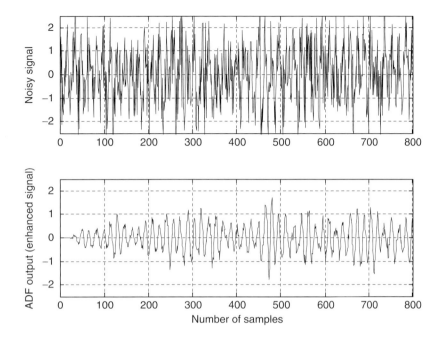

FIGURE 10.16 Noisy signal and enhanced signal.

Our simulation example has the following specifications:

- Sampling rate = 8,000 Hz
- Corrupted signal = 500 Hz tone with unit amplitude added with white Gaussian noise
- Adaptive filter = FIR type, 21 taps
- Convergence factor = 0.001
- Delay value $\Delta = 7$
- LMS algorithm applied

Figure 10.16 shows time domain results. The first plot is the noisy signal, while the second plot clearly demonstrates the enhanced signal. Figure 10.17 describes the frequency domain point of view. The spectrum of the noisy signal is shown in the top plot, where we can see that white noise is populated over the entire bandwidth. The bottom plot is the enhanced signal spectrum. Since the method is adaptive, it is especially effective when the enhanced signal frequency is changing with time. Program 10.3 lists the MATLAB program for this simulation.

Program 10.3. MATLAB program for adaptive line enhancement.

```
close all; clear all
fs = 8000; T = 1/fs;          % Sampling rate and sampling period
t = 0:T:0.1;                  % 1 second time instants
n =randn(1,length(t));        % Generate the Gaussian random noise
d = cos (2*pi*500*t) + n;     % Generate the 500-Hz tone plus noise
x = filter([0 0 0 0 0 0 0 1], 1, d);       %Delay filter
mu = 0.001;           % Initialize the step size for the LMS algorithms
w = zeros(1, 21);            % Initialize the adaptive filter coefficients
y = zeros(1,length(t));       % Initialize the adaptive filter output
e = y;            % Initialize the error vector
%Perform adaptive filtering using the LMS algorithm
for m = 22:1:length(t)-1
sum = 0;
for i = 1:1:21
sum = sum + w(i)*x(m − i);
end
y(m) = sum;
e(m) = d(m)-y(m);
for i = 1:1:21
w(i) = w(i) + 2*mu*e(m)*x(m-i);
end
end
% Calculate the single-sided amplitude spectrum for the corrupted signal
D = 2*abs(fft(d))/length(d);D(1) = D(1)/2;
% Calculate the single-sided amplitude spectrum for the enhanced signal
Y = 2*abs(fft(y))/length(y);Y(1) = Y(1)/2;
% Map the frequency index to its frequency in Hz
f=[0:1:length(x)/2]*8000/length(x);
% Plot the signals and spectra
subplot(2,1,1),plot(d);grid;axis([0 length(x) -2.5 2.5]);ylabel('Noisy signal');
subplot(2,1,2),plot(y);grid;axis([0 length(y) −2.5 2.5]);
ylabel('ADF output (enhanced signal)');xlabel('Number of samples')
figure
subplot(2,1,1),plot(f,D(1:length(f)));grid;axis([0 fs/2 0 1.5]);
ylabel('Noisy signal spectrum')
subplot(2,1,2),plot(f,Y(1:length(f)));grid;axis([0 fs/2 0 1.5]);
ylabel('ADF output spectrum');xlabel('Frequency (Hz)');
```

10.4 Other Application Examples

This section continues to explore other adaptive filter applications briefly, without showing computer simulations. The topics include periodic interference cancellation, ECG interference cancellation, and echo cancellation in long-distance

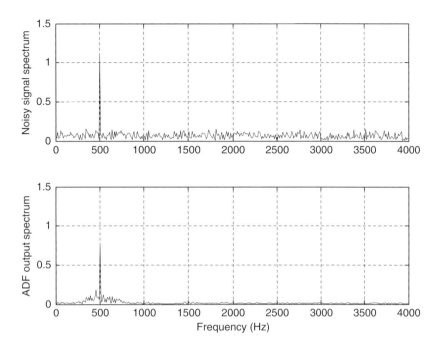

FIGURE 10.17 Spectrum plots for the noisy signal and enhanced signal.

telephone circuits. Detailed information can also be explored in Haykin (1991), Ifeachor and Jervis (2002), Stearns (2003), and Widrow and Stearns (1985).

10.4.1 Canceling Periodic Interferences Using Linear Prediction

An audio signal may be corrupted by periodic interference and no noise reference is available. Such examples include the playback of speech or music with the interference of tape hum, turntable rumble, or vehicle engine or power line interference. We can use the modified line enhancement structure as shown in Figure 10.18.

The adaptive filter uses the delayed version of the corrupted signal $x(n)$ to predict the periodic interference. The number of delayed samples is selected by the experiment of the adaptive filter performance. Note that a two-tap adaptive FIR filter can predict a one sinusoid, as noted earlier. After convergence, the adaptive filter would predict the interference as

$$y(n) = \sum_{i=0}^{N-1} w(i)x(n-i) \approx A\cos(2\pi fn/f_s). \qquad (10.16)$$

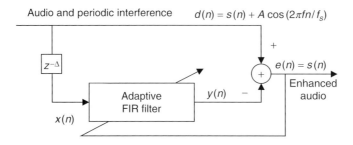

FIGURE 10.18   Canceling periodic interference using the adaptive filter.

Therefore, the error signal contains only the desired audio signal

$$e(n) \approx s(n). \tag{10.17}$$

10.4.2 Electrocardiography Interference Cancellation

As we discussed in Chapters 1 and 8, in recording of electrocardiograms, there often exists unwanted 60-Hz interference, along with its harmonics, in the recorded data. This interference comes from the power line, including effects from magnetic induction, displacement currents in leads or in the body of the patient, and equipment interconnections and imperfections.

Figure 10.19 illustrates the application of adaptive noise canceling in ECG. The primary input is taken from the ECG preamplifier, while a 60-Hz reference

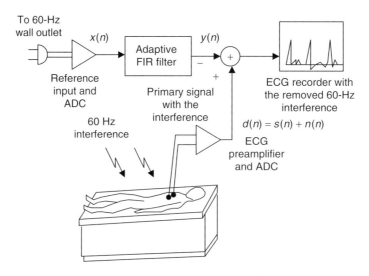

FIGURE 10.19   Illustration of canceling 60-Hz interference in ECG.

input is taken from a wall outlet with proper attenuation. After proper signal conditioning, the digital interference $x(n)$ is acquired by the digital signal (DS) processor. The digital adaptive filter uses this reference input signal to produce an estimate, which approximates the 60-Hz interference $n(n)$ sensed from the ECG amplifier:

$$y(n) \approx n(n). \tag{10.18}$$

Here, an FIR adaptive filter with N taps and the LMS algorithm can be used for this application:

$$y(n) = w(0)x(n) + w(1)x(n-1) + \cdots + w(N-1)x(n-N+1). \tag{10.19}$$

Then after convergence of the adaptive filter, the estimated interference is subtracted from the primary signal of the ECG preamplifier to produce the output signal $e(n)$, in which the 60-Hz interference is canceled:

$$e(n) = d(n) - y(n) = s(n) + n(n) - x(n) \approx s(n). \tag{10.20}$$

With enhanced ECG recording, doctors in clinics can give more accurate diagnoses for patients.

10.4.3 Echo Cancellation in Long-Distance Telephone Circuits

Long-distance telephone transmission often suffers from impedance mismatches. This occurs primarily at the hybrid circuit interface. Balancing electric networks within the hybrid can never perfectly match the hybrid to the subscriber loop due to temperature variations, degradation of the transmission line, and so on. As a result, a small portion of the received signal is leaked for transmission. For example, in Figure 10.20a, if speaker B talks, the speech indicated as $x_B(n)$ will pass the transmission line to reach user A, and a portion of $x_B(n)$ at site A is leaked and transmitted back to the user B, forcing caller B to hear his or her own

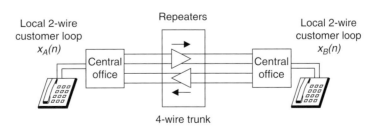

FIGURE 10.20A   **Simplified long-distance circuit.**

voice. This is known as an echo for speaker B. A similar echo illustration can be conducted for speaker A. When the telephone call is made over a long distance (more than 1,000 miles, such as with geostationary satellites), the echo can be delayed by as much as 540 ms. The echo impairment can be annoying to the customer and increases with the distance.

To circumvent the problem of echo in long-distance communications, an adaptive filter is applied at each end of the communication system, as shown in Figure 10.20b. Let us examine the adaptive filter installed at the speaker A site. The incoming signal is $x_B(n)$ from speaker B, while the outgoing signal contains the speech from speaker A and a portion of leakage from the hybrid circuit $d_A(n) = x_A(n) + \bar{x}_B(n)$. If the leakage $\bar{x}_B(n)$ returns back to speaker B, it becomes an annoying echo. To prevent the echo, the adaptive filter at the speaker A site uses the incoming signal from speaker B as an input and makes its output approximate to the leaked speaker B signal by adjusting its filter coefficients; that is,

$$y_A(n) = \sum_{i=0}^{N-1} w(i)x_B(n-i) \approx \bar{x}_B(n). \qquad (10.21)$$

As shown in Figure 10.20(b), the estimated echo $y_A(n) \approx \bar{x}_B(n)$ is subtracted from the outgoing signal, thus producing the signal that contains only speech A; that is, $e_A(n) \approx x_A(n)$. As a result, the echo of speaker B is removed. We can illustrate similar operations for the adaptive filter used at the speaker B site. In practice, the FIR adaptive filter with several hundred coefficients or more is

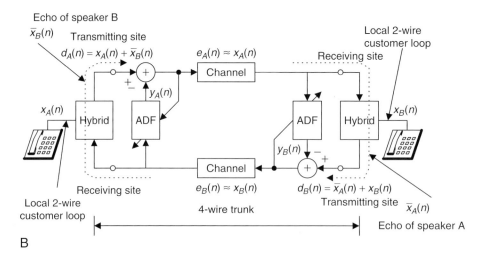

FIGURE 10.20B   **Adaptive echo cancelers.**

commonly used to effectively cancel the echo. If nonlinearities are concerned in the echo path, a corresponding nonlinear adaptive canceler can be used to improve the performance of the echo cancellation.

Other forms of adaptive filters and other applications are beyond the scope of this book. The reader is referred to the references for further development.

10.5 Summary

1. Adaptive filters can be applied to signal-changing environments, spectral overlap between noise and signal, and unknown, or time-varying, noises.

2. Wiener filter theory provides optimal weight solutions based on statistics. It involves collection of a large block of data, calculation of an autocorrelation matrix and a cross-correlation matrix, and inversion of a large size of the autocorrelation matrix.

3. The steepest descent algorithm can find the optimal weight solution using an iterative method, so a large matrix inversion is not needed. But it still requires calculating an autocorrelation and cross-correlation matrix.

4. The LMS is a sample-based algorithm, which does not need collection of data or computation of statistics and does not involve matrix inversion.

5. The convergence factor for the LMS algorithm is bounded by the reciprocal of the product of the number of filter coefficients and input signal power.

6. The LMS adaptive FIR filter can be effectively applied for noise cancellation, system modeling, and line enhancement.

7. Further exploration includes other applications such as cancellation of periodic interference, biomedical ECG signal enhancement, and adaptive telephone echo cancellation.

10.6 Problems

10.1. Given a quadratic MSE function for the Wiener filter:

$$J = 50 - 40w + 10w^2,$$

find the optimal solution for w^* to achieve the minimum MSE J_{min} and determine J_{min}.

10.2. Given a quadratic MSE function for the Wiener filter:
$$J = 15 + 20w + 10w^2,$$
find the optimal solution for w^* to achieve the minimum MSE J_{min} and determine J_{min}.

10.3. Given a quadratic MSE function for the Wiener filter:
$$J = 50 - 40w + 10w^2,$$
use the steepest descent method with an initial guess as $w_0 = 0$ and the convergence factor $\mu = 0.04$ to find the optimal solution for w^* and determine J_{min} by iterating three times.

10.4. Given a quadratic MSE function for the Wiener filter:
$$J = 15 + 20w + 10w^2,$$
use the steepest descent method with an initial guess as $w_0 = 0$ and the convergence factor $\mu = 0.04$ to find the optimal solution for w^* and determine J_{min} by iterating three times.

10.5. Given the following adaptive filter used for noise cancellation application (Fig. 10.21), in which $d(0) = 3$, $d(1) = -2$, $d(2) = 1$, $x(0) = 3$, $x(1) = -1$, $x(2) = 2$, and an adaptive filter with two taps: $y(n) = w(0)x(n) + w(1)x(n-1)$ with initial values $w(0) = 0$, $w(1) = 1$, and $\mu = 0.1$,

a. determine the LMS algorithm equations
$y(n) =$
$e(n) =$
$w(0) =$
$w(1) =$

b. perform adaptive filtering for each of $n = 0, 1, 2$.

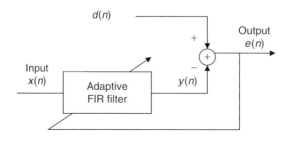

FIGURE 10.21.   Noise cancellation in Problem 10.5.

10.6. Given a DSP system with a sampling rate set up to be 8,000 samples per second, implement an adaptive filter with 5 taps for system modeling.

As shown in Figure 10.22, assume that the unknown system transfer function is

$$H(z) = \frac{0.25 + 0.25z^{-1}}{1 - 0.5z^{-1}}.$$

Determine the DSP equations using the LMS algorithm

$y(n) =$

$e(n) =$

$w(i) =$

for $i = 0, 1, 2, 3, 4$; that is, write the equations for all adaptive coefficients:

$w(0) =$

$w(1) =$

$w(2) =$

$w(3) =$

$w(4) =$

10.7. Given a DSP system for noise cancellation application with a sampling rate set up to be 8,000 Hz, as shown in Figure 10.23, the desired signal of a 1,000 Hz tone is generated internally via a tone generator; and the generated tone is corrupted by the noise captured from a microphone. An adaptive FIR filter with 25 taps is applied to reduce the noise in the corrupted tone.

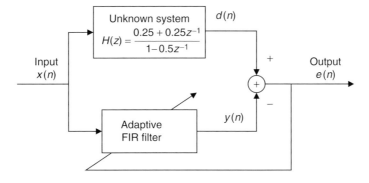

FIGURE 10.22. System modeling in Program 10.6.

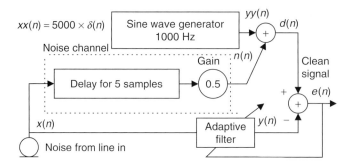

FIGURE 10.23.   Noise cancellation in Problem 10.7.

   a. Determine the DSP equation for the channel noise $n(n)$.

   b. Determine the DSP equation for signal tone $yy(n)$.

   c. Determine the DSP equation for the corrupted tone $d(n)$.

   d. Set up the LMS algorithm for the adaptive FIR filter.

10.8. An audio playback application is described in Figure 10.24.

   Due to the interference environment, the audio is corrupted by 15 different periodic interferences. The DSP engineer uses an FIR adaptive filter to remove such interferences, as shown in Figure 10.24. What is the minimum number of filter coefficients?

10.9. In a noisy ECG acquisition environment, the DSP engineer uses an adaptive FIR filter with 20 coefficients to remove 60 Hz interferences. The system is set up as shown in Figure 10.19, where the corrupted ECG and enhanced ECG signals are represented as $d(n)$ and $e(n)$, respectively; $x(n)$ is the captured reference signal from the 60-Hz interference; and $y(n)$ is the adaptive filter output. Determine all difference equations to implement the adaptive filter.

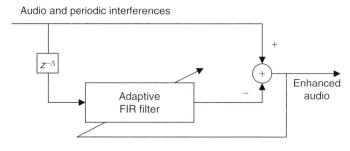

FIGURE 10.24.   Interference cancellation in Problem 10.8.

10.10. Given an application of the echo cancellation shown in Figure 10.20b,

  a. explain the concepts and benefits of using the echo canceler;

  b. explain the operations of the adaptive filter at the speaker B site;

  c. determine all difference equations to implement the adaptive filter at the speaker A site.

MATLAB Problems

10.11. Write a MATLAB program for minimizing the two-weight MSE (mean squared error) function

$$J = 100 + 100w_1^2 + 4w_2^2 - 100w_1 + 8w_2 + 10w_1w_2$$

by applying the steepest descent algorithm for 500 iterations. The derivatives are derived as

$$\frac{dJ}{dw_1} = 200w_1 - 100 + 10w_2 \text{ and } \frac{dJ}{dw_2} = 8w_2 - 8 + 10w_1$$

and the initial weights are assumed as $w_1(0) = 0$, $w_2(0) = 0$, $\mu = 0.001$. Plot $w_1(k)$, $w_2(k)$, and $J(k)$ versus the number of iterations, respectively. Summarize your results.

10.12. In Problem 10.6, the unknown system is assumed as a fourth-order Butterworth bandpass filter with a lower cut-off frequency of 700 Hz and an upper cut-off frequency of 900 Hz. Design the bandpass filter by the bilinear transformation method for simulating the unknown system with a sampling rate of 8,000 Hz.

  a. Generate the input signal for 0.1 second using a sum of three sinusoids having 100 Hz, 800 Hz, and 1500 Hz with a sampling rate of 8,000 Hz.

  b. Use the generated input as the unknown system input to produce the system output.

The adaptive FIR filter is then applied to model the designed bandpass filter. The following parameters are assumed:

Adaptive FIR filter
Number of taps: 15 coefficients
Algorithm: LMS algorithm
Convergence factor: 0.01

c. Implement the adaptive FIR filter, plot the system input, system output, adaptive filter output, and error signal, respectively.

d. Plot the input spectrum, system output spectrum, and adaptive filter output spectrum, respectively.

10.13. Use the following MATLAB code to generate the reference noise and the signal of 300 Hz corrupted by the noise with a sampling rate of 8000 Hz.

```
fs=8000; T=1/fs;        % Sampling rate and sampling period
t=0:T:1;                % Create time instants
x=randn(1,length(t));   % Generate the reference noise
n=filter([0 0 0 0 0 0 0 0 0 0.8],1,x); % Generate the corrupting noise
d=sin(2*pi*300*t)+n;    % Generate the signal plus noise
```

a. Implement an adaptive FIR filter to remove the noise. The adaptive filter specifications are as follows:

Sample rate = 8,000 Hz
Signal corrupted by Gaussian noise delayed by 9 samples from the reference noise as shown in above MATLAB code.
Reference noise: Gaussian noise with a power of 1
Number of FIR filter tap:16
Convergence factor for the LMS algorithm: 0.01

b. Plot the corrupted signal, reference noise, and enhanced signal, respectively.

c. Compare the spectral plots for the corrupted signal, and enhanced signal, respectively.

References

Haykin, S. (1991). *Adaptive Filter Theory,* 2nd ed. Englewood Cliffs, NJ: Prentice Hall.

Ifeachor, E. C., and Jervis, B. W. (2002). *Digital Signal Processing: A Practical Approach,* 2nd ed. Upper Saddle River, NJ: Prentice Hall.

Stearns, S. D. (2003). *Digital Signal Processing with Examples in MATLAB.* Boca Raton, FL: CRC Press LLC.

Widrow, B., and Stearns, S. (1985). *Adaptive Signal Processing.* Upper Saddle River, NJ: Prentice Hall.

11

Waveform Quantization and Compression

Objectives:

This chapter studies speech quantization and compression techniques such as signal companding, differential pulse code modulation, and adaptive differential pulse code modulation. The chapter continues to explore the discrete-cosine transform (DCT) and modified DCT and shows how to apply the developed concepts to understand the MP3 audio format. The chapter also introduces industry standards that are widely used in the digital signal processing field.

11.1 Linear Midtread Quantization

As we discussed in Chapter 2, in the digital signal processing (DSP) system, the first step is to sample and quantize the continuous signal. Quantization is the process of rounding off the sampled signal voltage to the predetermined levels that will be encoded by analog-to-digital conversion (ADC). We have described the quantization process in Chapter 2, in which we studied unipolar and bipolar linear quantizers in detail. In this section, we focus on a linear midtread quantizer, which is used in digital communications (Roddy and Coolen, 1997; Tomasi, 2004), and its use to quantize speech waveform. The linear midtread quantizer is similar to the bipolar linear quantizer discussed in Chapter 2 except that the midtread quantizer offers the same decoded magnitude range for both positive and negative voltages.

Let us look at a midtread quantizer. The characteristics and binary codes for a 3-bit midtread quantizer are depicted in Figure 11.1, where the code is in a sign magnitude format. Positive voltage is coded using a sign bit of logic 1, while negative voltage is coded by a sign bit of logic 0; the next two bits are the magnitude bits. The key feature of the linear midtread quantizer is noted as follows: When $0 \le x < \Delta/2$, the binary code of 100 is produced; when $-\Delta/2 \le x < 0$, the binary code of 000 is generated, where Δ is the quantization step size. However, the quantized values for both codes of 100 and 000 are the same and equal to $x_q = 0$. We can also see details in Table 11.1. For the 3-bit midtread quantizer, we expect 7 quantized values instead of 8; that is, there are $2^n - 1$ quantization levels for the n-bit midtread quantizer. Notice that quantization signal range is $(2^n - 1)\Delta$ and the magnitudes of the quantized values are symmetric, as shown in Table 11.1. We apply the midtread quantizer particularly for speech waveform coding.

The following example serves to illustrate coding principles of the 3-bit midtread quantizer.

Example 11.1.

For the 3-bit midtread quantizer described in Figure 11.1 and the analog signal with a range from -5 to 5 volts,

a. Determine the quantization step size.

b. Determine the binary codes, recovered voltages, and quantization errors when the input is -3.6 volts and 0.5 volt, respectively.

Solution:

a. The quantization step size is calculated as

$$\Delta = \frac{5 - (-5)}{2^3 - 1} = 1.43 \text{ volts.}$$

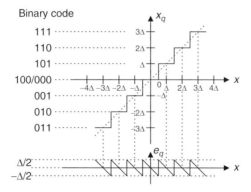

FIGURE 11.1 Characteristics of a 3-bit midtread quantizer.

TABLE 11.1 Quantization table for the 3-bit midtread quantizer.

Binary Code	Quantization Level x_q (V)	Input Signal Subrange (V)
0 1 1	-3Δ	$-3.5\Delta \leq x < -2.5\Delta$
0 1 0	-2Δ	$-2.5\Delta \leq x < -1.5\Delta$
0 0 1	$-\Delta$	$-1.5\Delta \leq x < -0.5\Delta$
0 0 0	0	$-0.5 \leq x < 0$
1 0 0	0	$0 \leq x < 0.5\Delta$
1 0 1	Δ	$0.5\Delta \leq x < 1.5\Delta$
1 1 0	2Δ	$1.5\Delta \leq x < 2.5\Delta$
1 1 1	3Δ	$2.5\Delta \leq x < 3.5\Delta$

(Note that: step size = $\Delta = x_{max} - x_{min}/2^3 - 1$; x_{max} = maximum voltage; and $x_{min} = -x_{max}$; and coding format: a. Sign bit: 1 = plus; 0 = minus; b. 2 magnitude bits)

b. For $x = -3.6$ volts, we have $x = \frac{-3.6}{1.43}\Delta = -2.52\Delta$. From quantization characteristics, it follows that the binary code = 011 and the recovered voltage is $x_q = -3\Delta = -4.29$ volts. Thus the quantization error is computed as

$$e_q = x_q - x = -4.28 - (-3.6) = -0.69 \text{ volt.}$$

For $x = 0.5 = \frac{0.5}{1.43}\Delta = 0.35\Delta$, we get a binary code = 100. Based on Figure 11.1 or Table 11.1, the recovered voltage and quantization error are found to be

$$x_q = 0 \text{ and } e_q = 0 - 0.5 = -0.5 \text{ volt.}$$

As discussed in Chapter 2, the linear midtread quantizer introduces the quantization noise, as shown in Figure 11.1; and the signal-to-noise power ratio (SNR) is given by

$$SNR = 10.79 + 20 \cdot \log_{10}\left(\frac{x_{rms}}{\Delta}\right) \text{ dB,} \qquad (11.1)$$

where x_{rms} designates the root mean squared value of the speech data to be quantized. The practical equation for estimating the SNR for the speech data sequence $x(n)$ of N data points is written as:

$$SNR = \left(\frac{\sum_{n=0}^{N-1} x^2(n)}{\sum_{n=0}^{N-1}\left(x_q(n) - x(n)\right)^2}\right) \qquad (11.2)$$

$$SNR \, dB = 10 \cdot \log_{10}(SNR) \text{ dB.} \qquad (11.3)$$

Notice that $x(n)$ and $x_q(n)$ are the speech data to be quantized and the quantized speech data, respectively. Equation (11.2) gives the absolute SNR, and Equation (11.3) produces the SNR in terms of decibel (dB). Quantization error is the difference between the quantized speech data (or quantized voltage level) and speech data (or analog voltage), that is $x_q(n) - x(n)$. Also note that from Equation (11.1), increasing 1 bit to the linear quantizer would improve SNR by approximately 6 dB. Let us examine performance of the 5-bit linear midtread quantizer.

In the following simulation, we use a 5-bit midtread quantizer to quantize the speech data. The original speech, quantized speech, and quantized error after quantization are plotted in Figure 11.2. Since the program calculates $x_{rms}/x_{max} = 0.203$, we yield x_{rms} as $x_{rms} = 0.203 \times x_{max} = 0.0203 \times 5 = 1.015$ and $\Delta = 10/(2^5 - 1) = 0.3226$. Applying Equation (11.1) gives SNR = 21.02 dB. The SNR using Equations (11.2) and (11.3) is approximately 21.6 dB.

The first plot in Figure 11.2 is the original speech, and the second plot shows the quantized speech. Quantization error is displayed in the third plot, where the error amplitude interval is uniformly distributed between -0.1613 and 0.1613, indicating the bounds of the quantized error ($\Delta/2$). The details of the MATLAB implementation are given in Program 11.1.

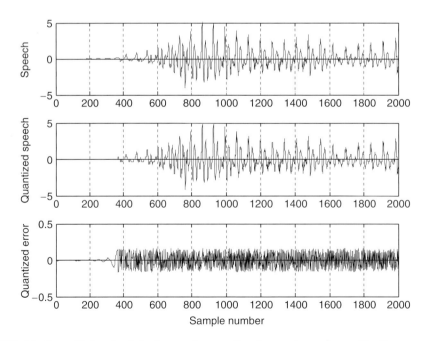

FIGURE 11.2   Plots of original speech, quantized speech, and quantization error.

To improve the SNR, the number of bits must be increased. However, increasing the number of encoding bits will cost an expansive ADC device, larger storage media for storing the speech data, and a larger bandwidth for transmitting the digital data. To gain a more efficient quantization approach, we will study the μ-law companding in the next section.

11.2 μ-Law Companding

In this section, we will study analog μ-law companding, which takes analog input signal; and digital μ-law companding, which deals with linear pulse code modulation (PCM) codes.

11.2.1 Analog μ-Law Companding

To save the number of bits to encode each speech datum, μ-law companding, called log-PCM coding, is applied. μ-Law Companding (Roddy and Coolen, 1997; Tomasi, 2004) was first used in the United States and Japan in the telephone industry (G.711 standard). μ-Law companding is a compression process. It explores the principle that the higher amplitudes of analog signals are compressed before ADC while expanded after digital-to-analog conversion (DAC). As studied in the linear quantizer, the quantization error is uniformly distributed. This means that the maximum quantization error stays the same no matter how much bigger or smaller the speech samples are. μ-Law companding can be employed to make the quantization error smaller when the sample amplitude is smaller and to make the quantization error bigger when the sample amplitude is bigger, using the same number of bits per sample. It is described in Figure 11.3.

As shown in Figure 11.3, x is the original speech sample, which is the input to the compressor, while y is the output from the μ-law compressor; then the output y is uniformly quantized. Assuming that the quantized sample y_q is encoded and sent to the μ-law expander, the expander will perform the reverse process to obtain the quantized speech sample x_q. The compression and decompression processes cause the maximum quantization error

FIGURE 11.3 Block diagram for μ-law compressor and μ-law expander.

$|x_q - x|_{\max}$ to be small for the smaller sample amplitudes and large for the larger sample amplitudes.

The equation for the μ-law compressor is given by

$$y = sign(x) \frac{\ln\left(1 + \mu \frac{|x|}{|x|_{\max}}\right)}{\ln(1 + \mu)}, \quad (11.4)$$

where $|x|_{\max}$ is the maximum amplitude of the inputs, while μ is the positive parameter to control the degree of the compression. $\mu = 0$ corresponds to no compression, while $\mu = 255$ is adopted in the industry. The compression curve with $\mu = 255$ is plotted in Figure 11.4. And note that the sign function sign(x) in Equation (11.4) is defined as

$$sign(x) = \begin{cases} 1 & x \geq 0 \\ -1 & x < 0. \end{cases} \quad (11.5)$$

Solving Equation (11.4) by substituting the quantized value, that is, $y = y_q$ we achieve the expander equation as

$$x_q = |x|_{\max} sign(y_q) \frac{(1+\mu)^{|y_q|} - 1}{\mu}. \quad (11.6)$$

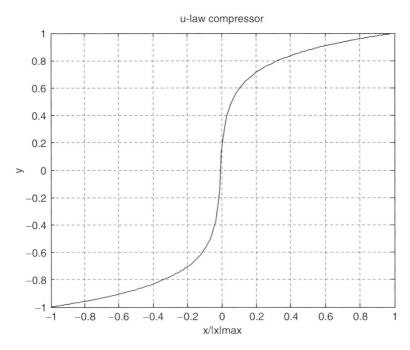

FIGURE 11.4 Characteristics for the μ-law compander.

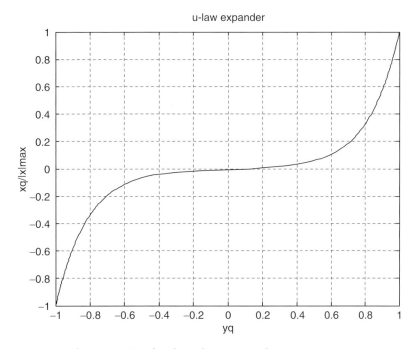

FIGURE 11.5 Characteristics for the μ-law expander.

For the case of $\mu = 255$, the expander curve is plotted in Figure 11.5.

Let's look at Example 11.2 for μ-law compression.

Example 11.2.

For the μ-law compression and expansion process shown in Figure 11.3, with $\mu = 255$, the 3-bit midtread quantizer described in Figure 11.1, and an analog signal ranging from -5 to 5 volts,

    a. Determine the binary codes, recovered voltages, and quantization errors when the input for each is -3.6 volts and 0.5 volt.

Solution:

    a. For μ-law compression and $x = -3.6$ volts, we can determine the quantization input as

$$y = sign(-3.6) \frac{\ln\left(1 + 255 \frac{|-3.6|}{|5|_{max}}\right)}{\ln(1 + 255)} = -0.94.$$

As shown in Figure 11.4, the range of y is 2, thus the quantization step size is calculated as

$$\Delta = \frac{2}{2^3 - 1} = 0.286 \text{ and } y = \frac{-0.94}{0.286}\Delta = -3.28\Delta.$$

From quantization characteristics, it follows that the binary code $= 011$ and the recovered signal is $y_q = -3\Delta = -0.858$.

Applying the μ-law expander leads to

$$x_q = |5|_{\max}\text{sign}(-0.858)\frac{(1+255)^{|-0.858|}-1}{255} = -2.264.$$

Thus the quantization error is computed as

$$e_q = x_q - x = -2.264 - (-3.6) = 1.336 \text{ volts}.$$

Similarly, for $x = 0.5$, we get

$$y = \text{sign}(0.5)\frac{\ln\left(1 + 255 \frac{|0.5|}{|5|_{\max}}\right)}{\ln(1+255)} = 0.591.$$

In terms of the quantization step, we get

$$y = \frac{0.519}{0.286}\Delta = 2.1\Delta \text{ and binary code} = 110.$$

Based on Figure 11.1, the recovered signal is

$$y_q = 2\Delta = 0.572$$

and the expander gives

$$x_q = |5|_{\max}\text{sign}(0.572)\frac{(1+255)^{|0.572|}-1}{255} = 0.448 \text{ volt}.$$

Finally, the quantization error is given by

$$e_q = 0.448 - 0.5 = -0.052 \text{ volt}.$$

As we can see, with 3 bits per sample, the strong signal is encoded with bigger quantization error, while the weak signal is quantized with smaller quantization error.

In the following simulation, we apply a 5-bit μ-law compander with $\mu = 255$ in order to quantize and encode the speech data used in the last section. Figure 11.6 is a block diagram of compression and decompression.

Figure 11.7 shows the original speech data, the quantized speech data using μ-law compression, and the quantization error for comparisons. The quantized speech wave is very close to the original speech wave. From the plots in Figure 11.7, we can observe that the amplitude of the quantization error changes

11.2 μ-Law Companding

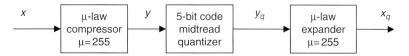

FIGURE 11.6 The 5-bit midtread uniform quantizer with $\mu = 255$ used for simulation.

according to the amplitude of the speech being quantized. The bigger quantization error is introduced when the amplitude of speech data is larger; on the other hand, the smaller quantization error is produced when the amplitude of speech data is smaller.

Compared with the quantized speech using the linear quantizer shown in Figure 11.2, the decompressed signal using the μ-law compander looks and sounds much better, since the quantized signal can better keep tracking the original larger amplitude signal and original small amplitude signal as well. MATLAB implementation is shown in Program 11.2.

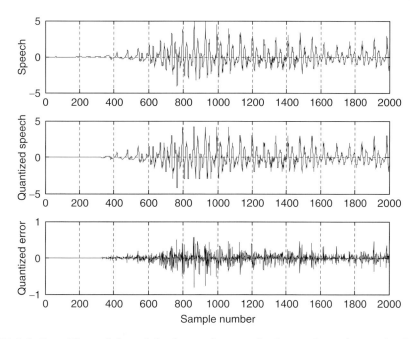

FIGURE 11.7 Plots of the original speech, quantized speech, and quantization error with the μ-law compressor and expander.

11.2.2 Digital μ-Law Companding

In many multimedia applications, the analog signal is first sampled and then it is digitized into a linear PCM code with a larger number of bits per sample. Digital μ-law companding further compresses the linear PCM code using the compressed PCM code with a smaller number of bits per sample without losing sound quality. The block diagram of a digital μ-law compressor and expander is shown in Figure 11.8.

The typical digital μ-law companding system compresses a 12-bit linear PCM code to an 8-bit compressed code. This companding characteristics is depicted in Figure 11.9, where it closely resembles an analog compression curve with $\mu = 255$ by approximating the curve using a set of 8 straight-line segments. The slope of each successive segment is exactly one-half that of the previous segment. Figure 11.9 shows the 12-bit to 8-bit digital companding curve for the positive portion only. There are 16 segments, accounting for both positive and negative portions.

However, like the midtread quantizer discussed in the first section of this chapter, only 13 segments are used, since segments $+0, -0, +1$, and -1 form a straight line with a constant slope and are considered as one segment. As shown in Figure 11.9, when the relative input is very small, such as in segment 0 or segment 1, there is no compression, while when the relative input is getting bigger such that it is in segment 3 or segment 4, the compression occurs with the compression ratios of 2:1 and 4:1, respectively. The format of the 12-bit linear PCM code is in sign-magnitude form with the most significant bit (MSB) as the sign bit (1 = positive value and 0 = negative value) plus 11 magnitude bits. The compressed 8-bit code has a format shown in Table 11.2, where it consists of a sign bit, a 3-bit segment identifier, and a 4-bit quantization interval within the specified segment. Encoding and decoding procedures are very simple, as illustrated in Tables 11.3 and 11.4, respectively.

As shown in those two tables, the prefix "S" is used to indicate the sign bit, which could be either 1 or 0; A, B, C, and D are transmitted bits; and the bit position with an "X" is the truncated bit during compression and hence would be lost during decompression. For the 8-bit compressed PCM code in Table 11.3, the 3 bits between "S" and "ABCD" indicate the segment number that is obtained by subtracting the number of consecutive zeros (less than or equal to 7) after the "S" bit in the original 12-bit PCM code from 7. Similarly, to recover

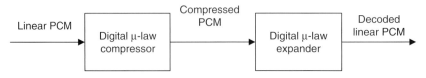

FIGURE 11.8 The block diagram for μ-law compressor and expander.

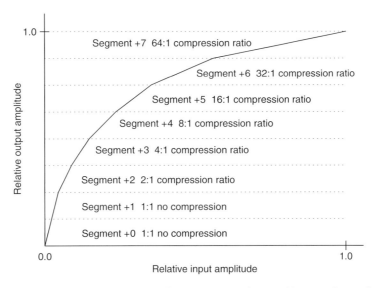

FIGURE 11.9 $\mu = 255$ **compression characteristics (for positive portion only).**

the 12-bit linear code in Table 11.4, the number of consecutive zeros after the "S" bit can be determined by subtracting the segment number in the 8-bit compressed code from 7. We illustrate the encoding and decoding processes in Examples 11.3 and 11.4.

Example 11.3.

a. In a digital companding system, encode each of the following 12-bit linear PCM codes into 8-bit compressed PCM codes.

   1. 1 0 0 0 0 0 0 0 0 1 0 1
   2. 0 0 0 0 1 1 1 0 1 0 1 0

Solution:

a. 1. Based on Table 11.3, we identify the 12-bit PCM code as: S = 1, A = 0, B = 1, C = 0, and D = 1, which is in segment 0. From the fourth column in Table 11.3, we get the 8-bit compressed code as 10000101.

TABLE 11.2 The format of 8-bit compressed PCM code.

Sign Bit	3-Bit	4-Bit
1 = + 0 = −	Segment identifier 000 to 111	Quantization interval A B C D 0000 to 111

11 WAVEFORM QUANTIZATION AND COMPRESSION

TABLE 11.3 $\mu = 255$ encoding table.

Segment	12-Bit Linear Code	12-Bit Amplitude Range in Decimal	8-Bit Compressed Code
0	S0000000ABCD	0 to 15	S000ABCD
1	S0000001ABCD	16 to 31	S001ABCD
2	S000001ABCDX	32 to 63	S010ABCD
3	S00001ABCDXX	64 to 127	S011ABCD
4	S0001ABCDXXX	128 to 255	S100ABCD
5	S001ABCDXXXX	256 to 511	S101ABCD
6	S01ABCDXXXXX	512 to 1023	S110ABCD
7	S1ABCDXXXXXX	1023 to 2047	S111ABCD

2. For the second 12-bit PCM code, we note that S = 0, A = 1, B = 1, C = 0, D = 1, and XXX = 010, and the code belongs to segment 4. Thus, from the fourth column in Table 11.3, we have

01001101.

Example 11.4.

a. In a digital companding system, decode each of the following 8-bit compressed PCM codes into 12-bit linear PCM code.

1. 1 0 0 0 0 1 0 1
2. 0 1 0 0 1 1 0 1

Solution:

a. 1. Using the decoding Table 11.4, we notice that S = 1, A = 0, B = 1, C = 0, and D = 1, and the code is in segment 0. Decoding leads to

100000000101,

TABLE 11.4 $\mu = 255$ decoding table.

8-Bit Compressed Code	8-Bit Amplitude Range in Decimal	Segment	12-Bit Linear Code
S000ABCD	0 to 15	0	S0000000ABCD
S001ABCD	16 to 31	1	S0000001ABCD
S010ABCD	32 to 47	2	S000001ABCD1
S011ABCD	48 to 63	3	S00001ABCD10
S100ABCD	64 to 79	4	S0001ABCD100
S101ABCD	80 to 95	5	S001ABCD1000
S110ABCD	96 to 111	6	S01ABCD10000
S111ABCD	112 to 127	7	S1ABCD100000

which is identical to the 12-bit PCM code in (1) in Example 11.3. We expect this result, since there is no compression for segment 0 and segment 1.

2. Applying Table 11.4, it follows that S = 0, A = 1, B = 1, C = 0, and D = 1, and the code resides in segment 4. Decoding achieves

000011101100.

As expected, this code is an approximation of the code in (2) in Example 11.3. Since segment 4 has compression, the last 3 bits in the original 12-bit linear code, that is, XXX = 010 = 2 in decimal, are discarded during transmission or storage. When we recover these three bits, the best guess should be the middle value, XXX = 100 = 4 in decimal for the 3-bit coding range from 0 to 7.

Now we apply the $\mu = 255$ compander to compress the 12-bit speech data as shown in Figure 11.10(a). The 8-bit compressed code is plotted in Figure 11.10(b). Plots (c) and (d) in the figure show the 12-bit speech after decoding and quantization error, respectively. We can see that the quantization error follows the amplitude of speech data relatively. The decoded speech sounds no different when compared with the original speech. Programs 11.8 to 11.10 show the detail of the MATLAB implementation.

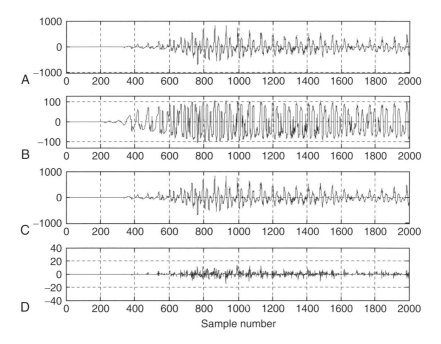

FIGURE 11.10 The $\mu = 255$ compressor and expander: (a) 12-bit speech data; (b) 8-bit compressed data; (c) 12-bit decoded speech; (d) quantization error.

11.3 Examples of Differential Pulse Code Modulation (DPCM), Delta Modulation, and Adaptive DPCM G.721

Data compression can be further achieved using *differential pulse code modulation* (DPCM). The general idea is to use past recovered values as the basis to predict the current input data and to then encode the difference between the current input and the predicted input. Since the difference has a significantly reduced signal dynamic range, it can be encoded with fewer bits per sample. Therefore, we obtain data compression. First, we study the principles of the DPCM concept that will help us understand adaptive DPCM in the next subsection.

11.3.1 Examples of Differential Pulse Code Modulation and Delta Modulation

Figure 11.11 shows a schematic diagram for the DPCM encoder and decoder. We denote the original signal $x(n)$; the predicted signal $\tilde{x}(n)$; the quantized, or recovered signal $\hat{x}(n)$; the difference signal to be quantized $d(n)$; and the quantized difference signal $d_q(n)$. The quantizer can be chosen as a uniform quantizer, a midtread quantizer (e.g., see Table 11.5), or others available. The encoding block produces binary bit stream in the DPCM encoding. The predictor uses the past predicted signal and quantized difference signal to predict the current input

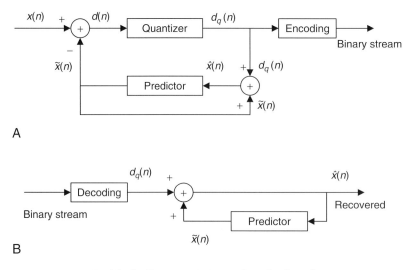

FIGURE 11.11   DPCM block diagram: (a) encoder; (b) decoder.

TABLE 11.5 Quantization table for the 3-bit quantizer in Example 11.5.

Binary Code	Quantization Value $d_q(n)$	Subrange in $d(n)$
0 1 1	−11	$-15 \le d(n) < -7$
0 1 0	−5	$-7 \le d(n) < -3$
0 0 1	−2	$-3 \le d(n) < -1$
0 0 0	0	$-1 \le d(n) < 0$
1 0 0	0	$0 \le d(n) \le 1$
1 0 1	2	$1 < d(n) \le 3$
1 1 0	5	$3 < d(n) \le 7$
1 1 1	11	$7 < d(n) \le 15$

value $x(n)$ as close as possible. The digital filter or adaptive filter can serve as the predictor. On the other hand, the decoder recovers the quantized difference signal, which can be added to the predictor output signal to produce the quantized and recovered signal, as shown in Figure 11.11(b).

In Example 11.5, we examine a simple DPCM coding system via the process of encoding and decoding numerical actual data.

Example 11.5.

A DPCM system has the following specifications:
Encoder scheme: $\tilde{x}(n) = \hat{x}(n-1)$; predictor

$$d(n) = x(n) - \tilde{x}(n)$$
$$d_q(n) = Q[d(n)] = \text{quantizer in Table 11.5}$$
$$\hat{x}(n) = \tilde{x}(n) + d_q(n)$$

Decoder scheme: $\tilde{x}(n) = \hat{x}(n-1)$; predictor

$$d_q(n) = \text{quantizer in Table 11.5}$$
$$\hat{x}(n) = \tilde{x}(n) + d_q(n)$$

5-bit input data: $x(0) = 6$, $x(1) = 8$, $x(2) = 13$.

a. Perform DPCM encoding to produce the binary code for each input datum.

b. Perform DPCM decoding to recover the data using the binary code in (1).

Solution:

a. Let us perform encoding according to the encoding scheme.
For $n = 0$, we have

$$\tilde{x}(0) = \hat{x}(-1) = 0$$
$$d(0) = x(0) - \tilde{x}(0) = 6 - 0 = 6$$
$$d_q(0) = Q[d(0)] = 5$$
$$\hat{x}(0) = \tilde{x}(0) + d_q(0) = 0 + 5 = 5$$
Binary code = 110.

For $n = 1$, it follows that
$$\tilde{x}(1) = \hat{x}(0) = 5$$
$$d(1) = x(1) - \tilde{x}(1) = 8 - 5 = 3$$
$$d_q(1) = Q[d(1)] = 2$$
$$\hat{x}(1) = \tilde{x}(1) + d_q(1) = 5 + 2 = 7$$
Binary code = 101.

For $n = 2$, results are
$$\tilde{x}(2) = \hat{x}(1) = 7$$
$$d(2) = x(2) - \tilde{x}(2) = 13 - 7 = 6$$
$$d_q(2) = Q[d(2)] = 5$$
$$\hat{x}(2) = \tilde{x}(2) + d_q(2) = 7 + 5 = 12$$
Binary code = 110.

b. We conduct the decoding scheme as follows.
For $n = 0$, we get:

Binary code = 110
$d_q(0) = 5$; from Table 11.5
$\tilde{x}(0) = \hat{x}(-1) = 0$
$\hat{x}(0) = \tilde{x}(0) + d_q(0) = 0 + 5 = 5$ (recovered).

For $n = 1$, decoding shows:

Binary code = 101
$d_q(1) = 2$; from Table 11.5
$\tilde{x}(1) = \hat{x}(0) = 5$
$\hat{x}(1) = \tilde{x}(1) + d_q(1) = 5 + 2 = 7$ (recovered).

For $n = 2$, we have:

Binary code = 110
$d_q(2) = 5$; from Table 11.5

$$\tilde{x}(2) = \hat{x}(1) = 7$$
$$\hat{x}(2) = \tilde{x}(2) + d_q(2) = 7 + 5 = 12 \text{ (recovered)}.$$

From this example, we can verify that the 5-bit code is compressed to the 3-bit code. However, we can see that each recovered data has a quantization error. Hence, the DPCM is a lossy data compression scheme.

DPCM for which a single bit is used in the quantization table becomes *delta modulation* (DM). The quantization table contains two quantized values, A and $-A$, quantization step size. Delta modulation quantizes the difference of the current input sample and the previous input sample using a 1-bit code word. To conclude the idea, we list the equations for encoding and decoding as follows:

Encoder scheme: $\tilde{x}(n) = \hat{x}(n-1)$; predictor

$$d(n) = x(n) - \tilde{x}(n)$$
$$d_q(n) = \begin{cases} +A & d_q(n) \geq 0, \text{ output bit: } 1 \\ -A & d_q(n) < 0, \text{ output bit: } 0 \end{cases}$$
$$\hat{x}(n) = \tilde{x}(n) + d_q(n)$$

Decoder scheme: $\tilde{x}(n) = \hat{x}(n-1)$; predictor

$$d_q(n) = \begin{cases} +A & \text{input bit: } 1 \\ -A & \text{input bit: } 0 \end{cases}$$
$$\hat{x}(n) = \tilde{x}(n) + d_q(n).$$

Note that the predictor has a sample delay.

Example 11.6.

For a DM system with 5-bit input data

$$x(0) = 6, \ x(1) = 8, \ x(2) = 13$$

and the quantized constant as $A = 7$,

a. Perform the DM encoding to produce the binary code for each input datum.

b. Perform the DM decoding to recover the data using the binary code in (a).

Solution:

a. Applying encoding accordingly, we have:
   For $n = 0$,

$$\tilde{x}(0) = \hat{x}(-1) = 0, \; d(0) = x(0) - \tilde{x}(0) = 6 - 0 = 6$$
$$d_q(0) = 7, \; \hat{x}(0) = \tilde{x}(0) + d_q(0) = 0 + 7 = 7$$
Binary code = 1.

For $n = 1$,
$$\tilde{x}(1) = \hat{x}(0) = 7, \; d(1) = x(1) - \tilde{x}(1) = 8 - 7 = 1$$
$$d_q(1) = 7, \; \hat{x}(1) = \tilde{x}(1) + d_q(1) = 7 + 7 = 14$$
Binary code = 1.

For $n = 2$,
$$\tilde{x}(2) = \hat{x}(1) = 14, \; d(2) = x(2) - \tilde{x}(2) = 13 - 14 = -1$$
$$d_q(2) = -7, \; \hat{x}(2) = \tilde{x}(2) + d_q(2) = 14 - 7 = 7$$
Binary code = 0.

b. Applying the decoding scheme leads to:
For $n = 0$,

Binary code = 1
$$d_q(0) = 7, \; \tilde{x}(0) = \hat{x}(-1) = 0$$
$$\hat{x}(0) = \tilde{x}(0) + d_q(0) = 0 + 7 = 7 \text{ (recovered)}.$$

For $n = 1$,

Binary code = 1
$$d_q(1) = 7, \; \tilde{x}(1) = \hat{x}(0) = 7$$
$$\hat{x}(1) = \tilde{x}(1) + d_q(1) = 7 + 7 = 14 \text{ (recovered)}.$$

For $n = 2$,

Binary code = 0
$$d_q(2) = -7, \; \tilde{x}(2) = \hat{x}(1) = 14$$
$$\hat{x}(2) = \tilde{x}(2) + d_q(2) = 14 - 7 = 7 \text{ (recovered)}.$$

We can see that the coding causes a larger quantization error for each recovered sample. In practice, this can be solved using a very high sampling rate (much larger than the Nyquist rate) and making the quantization step size A adaptive. The quantization step size increases by a factor when the slope magnitude of the input sample curve becomes bigger, that is, the condition in which the encoder produces continuous logic 1's or generates continuous logic 0's in the coded bit stream. Similarly, the quantization step decreases by a factor when the encoder generates logic 1 and logic 0 alternately. Hence, the resultant DM is called *adaptive DM*. In practice, the DM chip replaces the predictor, feedback path, and summer (see Fig. 11.11) with an integrator for both the encoder and the decoder. Detailed information can be found in Li and Drew (2004), Roddy and Coolen (1997), and Tomasi (2004).

11.3.2 Adaptive Differential Pulse Code Modulation G.721

In this subsection, an efficient compression technique for speech waveform is described, that is, *adaptive DPCM* (ADPCM), per recommendation G.721 of the CCITT (the Comité Consultatif International Téléphonique et Télégraphique). General discussion can be found in Li and Drew (2004), Roddy and Coolen (1997), and Tomasi (2004). The simplified block diagrams of the ADPCM encoder and decoder are shown in Figures 11.12a and 11.12b.

As shown in Figure 11.12a for the ADPCM encoder, first a difference signal $d(n)$ is obtained, by subtracting an estimate of the input signal $\tilde{x}(n)$ from the input signal $x(n)$. An adaptive 16-level quantizer is used to assign four binary digits $I(n)$ to the value of the difference signal for transmission to the decoder. An inverse quantizer produces a quantized difference signal $d_q(n)$ from the same four binary digits $I(n)$. The adaptive quantizer and inverse quantizer operate based on the quantization table and the scale factor obtained from the quantizer adaptation to keep tracking the energy change of the difference signal to be quantized. The input signal estimate is then added to this quantized difference signal to produce the reconstructed version of the input $\hat{x}(n)$. Both the reconstructed signal and the quantized difference signal are operated on by an adaptive predictor, which

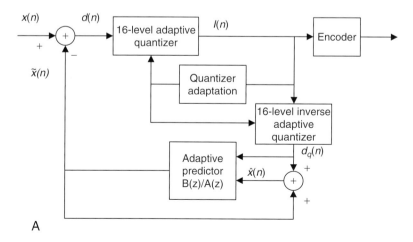

FIGURE 11.12A The ADPCM encoder.

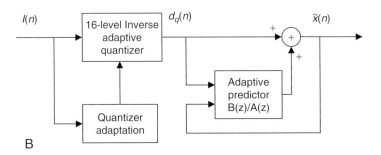

FIGURE 11.12B The ADPCM decoder.

generates the estimate of the next input signal, thereby completing the feedback loop.

The decoder shown in Figure 11.12b includes a structure identical to the feedback part of the encoder as depicted in Figure 11.12a. It first converts the received 4-bit data $I(n)$ to the quantized difference signal $d_q(n)$ using the adaptive quantizer. Then, at the second stage, the adaptive predictor uses the recovered quantized difference signal $d_q(n)$ and recovered current output $\tilde{x}(n)$ to generate the next output. Notice that the adaptive predictors of both the encoder and the decoder change correspondingly based on the signal to be quantized. The details of the adaptive predictor will be discussed.

Now, let us examine the ADPCM encoder principles. As shown in Figure 11.12a, the difference signal is computed as

$$d(n) = x(n) - \tilde{x}(n). \tag{11.7}$$

A 16-level non-uniform adaptive quantizer is used to quantize the difference signal $d(n)$. Before quantization, $d(n)$ is converted to a base-2 logarithmic representation and scaled by $y(n)$, which is computed by the scale factor algorithm. Four binary codes $I(n)$ are used to specify the quantized signal level representing $d_q(n)$, and the quantized difference $d_q(n)$ is also fed to the inverse adaptive quantizer. Table 11.6 shows the quantizer normalized input and output characteristics.

The scaling factor for the quantizer and the inverse quantizer $y(n)$ is computed according to the 4-bit quantizer output $I(n)$ and the adaptation speed control parameter $a_l(n)$, the fast (unlocked) scale factor $y_u(n)$, the slow (locked) scale factor $y_l(n)$, and the discrete function $W(I)$, defined in Table 11.7:

$$y_u(n) = (1 - 2^{-5})y(n) + 2^{-5}W(I(n)), \tag{11.8}$$

where $1.06 \leq y_u(n) \leq 10.00$.

TABLE 11.6 Quantizer normalized input and output characteristics.

Normalized Quantizer Input Range: $\log_2 |d(n)| - y(n)$	Magnitude: $|I(n)|$	Normalized Quantizer Output: $\log_2 |d_q(n)| - y(n)$
$[3.12, +\infty)$	7	3.32
$[2.72, 3.12)$	6	2.91
$[2.34, 2.72)$	5	2.52
$[1.91, 2.34)$	4	2.13
$[1.38, 1.91)$	3	1.66
$[0.62, 1.38)$	2	1.05
$[-0.98, 0.62)$	1	0.031
$(-\infty, -0.98)$	0	$-\infty$

TABLE 11.7 Discrete function $W(I)$.

$|I(n)|$	7	6	5	4	3	2	1	0
$W(I)$	70.13	22.19	12.38	7.00	4.0	2.56	1.13	-0.75

The slow scale factor $y_l(n)$ is derived from the fast scale factor $y_u(n)$ using a lowpass filter as follows:

$$y_l(n) = (1 - 2^{-6})y_l(n-1) + 2^{-6}y_u(n). \qquad (11.9)$$

The fast and slow scale factors are then combined to compute the scale factor

$$y(n) = a_l(n)y_u(n-1) + (1 - a_l(n))y_l(n-1), \ 0 \le a_l(n) \le 1. \qquad (11.10)$$

Next the controlling parameter $0 \le a_l(n) \le 1$ tends toward unity for speech signals and toward zero for voice band data signals and tones. It is updated based on the following parameters: $d_{ms}(n)$, which is the relatively short term average of $F(I(n))$; $d_{ml}(n)$, which is the relatively long term average of $F(I(n))$; and the variable $a_p(n)$, where $F(I(n))$ is defined as in Table 11.8.

Hence, we have

$$d_{ms}(n) = (1 - 2^{-5})d_{ms}(n-1) + 2^{-5}F(I(n)) \qquad (11.11)$$

and

$$d_{ml}(n) = (1 - 2^{-7})d_{ml}(n-1) + 2^{-7}F(I(n)), \qquad (11.12)$$

TABLE 11.8 Discrete function $F(I(n))$.

$|I(n)|$	7	6	5	4	3	2	1	0
$F(I(n))$	7	3	1	1	1	0	0	0

while the variable $a_p(n)$ is given by

$$a_p(n) = \begin{cases} (1-2^{-4})a_p(n-1) + 2^{-3} & \text{if } |d_{ms}(n) - d_{ml}(n)| \geq 2^{-3}d_{ml}(n) \\ (1-2^{-4})a_p(n-1) + 2^{-3} & \text{if } y(n) < 3 \\ (1-2^{-4})a_p(n) + 2^{-3} & \text{if } t_d(n) = 1 \\ 1 & \text{if } t_r(n) = 1 \\ (1-2^{-4})a_p(n) & \text{otherwise} \end{cases} \quad (11.13)$$

a_p approaches the value of 2 when the difference between $d_{ms}(n)$ and d_{ml} is large and the value of 0 when the difference is small. Also, $a_p(n)$ approaches 2 for an idle channel (indicated by $y(n) < 3$) or partial band signals (indicated by $t_d(n) = 1$). Finally, $a_p(n)$ is set to 1 when the partial band signal transition is detected ($t_r(n) = 1$).

a_l used in Equation (11.10) is defined as

$$\text{Finally, } a_l(n) = \begin{cases} 1 & a_p(n-1) > 1 \\ a_p(n-1) & a_p(n-1) \leq 1. \end{cases} \quad (11.14)$$

The partial band signal (tone) $t_d(n)$ and the partial band signal transition $t_r(n)$ in Equation (11.13) will be discussed later.

The predictor is to compute the signal estimate $\tilde{x}(n)$ from the quantized difference signal $d_q(n)$. The predictor z-transfer function, which is effectively suitable for a variety of input signals, is given by

$$\frac{B(z)}{A(z)} = \frac{b_0 + b_1 z^{-1} + b_2 z^{-2} + b_3 z^{-3} + b_4 z^{-4} + b_5 z^{-5}}{1 - a_1 z^{-1} - a_2 z^{-2}}. \quad (11.15)$$

It consists of a sixth-order portion that models the zeros and a second-order portion that models poles of the input signals. The input signal estimate is expressed in terms of the processed signal $\hat{x}(n)$ and the signal $x_z(n)$ processed by the finite impulse response (FIR) filter as follows:

$$\tilde{x}(n) = a_1(n)\hat{x}(n-1) + a_2(n)\hat{x}(n-2) + x_z(n), \quad (11.16)$$

where

$$\hat{x}(n-i) = \tilde{x}(n-i) + d_q(n-i) \quad (11.17)$$

$$x_z(n) = \sum_{i=0}^{5} b_i(n) d_q(n-i). \quad (11.18)$$

Both sets of predictor coefficients are updated using a simplified gradient algorithm:

$$a_1(n) = (1 - 2^{-8})a_1(n-1) + 3 \cdot 2^{-8} \text{signn}(p(n))\text{sign}(p(n-1)) \quad (11.19)$$

$$a_2(n) = (1 - 2^{-7})a_2(n-1) + 2^{-7}\{signn(p(n))sign(p(n-2)) \\ - f(a_1(n-1))signn(p(n))sign(p(n-1))\}, \quad (11.20)$$

where $p(n) = d_q(n) + x_z(n)$ and

$$f(a_1(n)) = \begin{cases} 4a_1(n) & |a_1(n)| \leq 2^{-1} \\ 2sign(a_1(n)) & |a_1(n)| > 2^{-1}, \end{cases} \quad (11.21)$$

Note that the function $sign(x)$ is defined in Equation(11.5) and $signn(x) = 1$ when $x > 0$; $signn(x) = 0$ when $x = 0$; $signn(x) = -1$ when $x < 0$ with stability constraints as

$$|a_2(n)| \leq 0.75 \text{ and } |a_1(n)| \leq 1 - 2^{-4} - a_2(n) \quad (11.22)$$

$$a_1(n) = a_2(n) = 0 \text{ if } t_r(n) = 1. \quad (11.23)$$

Also, the equations for updating the coefficients for the zero-order portion are given by

$$b_i(n) = (1 - 2^{-8})b_i(n-1) + 2^{-7}signn(d_q(n))sign(d_q(n-i)) \\ \text{for } i = 0, 1, 2, \ldots, 5 \quad (11.24)$$

with the following constraints:

$$b_0(n) = b_1(n) = b_2(n) = b_3(n) = b_4(n) = b_5(n) = 0 \text{ if } t_r(n) = 1. \quad (11.25)$$

$$t_d(n) = \begin{cases} 1 & a_2(n) < -0.71875 \\ 0 & \text{otherwise} \end{cases} \quad (11.26)$$

$$t_r(n) = \begin{cases} 1 & a_2(n) < -0.71875 \text{ and } |d_q(n)| > 24 \cdot 2^{y_l}. \\ 0 & \text{otherwise} \end{cases} \quad (11.27)$$

$t_d(n)$ is the indicator of detecting a partial band signal (tone). If a tone is detected ($t_d(n) = 1$); Equation (11.13) is invoked to drive the quantizer into the fast mode of adaptation. $t_r(n)$ is the indicator for a transition from a partial band signal. If it is detected ($t_r(n) = 1$), setting the predictor coefficients to be zero as shown in Equations (11.23) and (11.25) will force the quantizer into the fast mode of adaptation.

Simulation Example

To illustrate the performance, we apply the ADPCM encoder to the speech data used in Section 11.1 and then operate the ADPCM decoder to recover the speech signal. As described, the ADPCM uses 4 bits to encode each speech sample. The MATLAB implementations for the encoder and decoder are listed

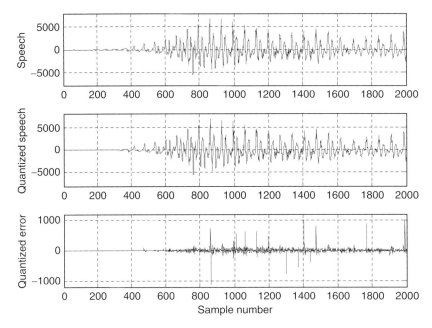

FIGURE 11.13 Original speech, quantized speech, and quantization error using ADPCM.

in Programs 11.11 to 11.13. Figure 11.13 plots the original speech samples, decoded speech samples, and quantization errors. From the figure, we see that the decoded speech data are very close to the original speech data; the quantization error is very small as compared with the speech sample, and its amplitude follows the change in amplitude of the speech data. In practice, we cannot tell the difference between the original speech and the decoded speech by listening to them. However, ADPCM encodes each speech sample using 4 bits per sample, while the original data are presented using 16 bits, thus the compression ratio (CR) of 4:1 is achieved.

In practical applications, data compression can reduce the storage media and bit rate for efficient digital transmission. To measure performance of data compression, we use

- the data CR, which is the ratio of the original data file size to the compressed data file size, or the ratio of the original code size in bits to the compressed code size in bits for the fixed length coding, and
- the bit rate, which is in terms of bits per second (bps) and can be calculated by:

$$bitrate = m \times f_s \text{ (bps)}, \tag{11.28}$$

where m = number of bits per sample (bits) and f_s = sampling rate (samples per second). Now let us look at an application example.

Example 11.7.

Speech is sampled at 8 kHz and each sample is encoded by 12 bits per sample. Using the following encoding methods:

1. noncompression
2. standard μ-law compression
3. standard ADPCM encoding (4 bits per sample),

   a. Determine the CR and the bit rate for each of the encoders and decoders.

   b. Determine the number of channels that the phone company can carry if a telephone system can transport a digital voice channel over a digital link having a capacity of 1.536 Mbps.

Solution:

a.1. For noncompression:

$$CR = 1:1$$

$$\text{Bit rate} = 12 \frac{bits}{sample} \times 8000 \frac{samples}{second} = 96 \text{(kbps)}.$$

b.1. Number of channels = $\frac{1.536}{96} \frac{MBPS}{KBPS} = 16$.

a.2. For the standard μ-law compression, each sample is encoded using 8 bits per sample. Hence, we have

$$CR = \frac{12}{8} \frac{bits/sample}{bits/sample} = 1.5:1$$

$$\text{Bit rate} = 8 \frac{bits}{sample} \times 8000 \frac{samples}{second} = 64 \text{ (kbps)}.$$

b.2. Number of channels = $\frac{1.536}{64} \frac{MBPS}{KBPS} = 24$.

a.3. For standard ADPCM with 4 bits per sample, it follows that

$$CR = \frac{12}{4} \frac{bits/sample}{bits/sample} = 3:1$$

$$\text{Bit rate} = 4 \frac{bits}{sample} \times 8000 \frac{samples}{second} = 32 \text{ (kbps)}.$$

b.3. Number of channels = $\frac{1.536}{32} \frac{MBPS}{KBPS} = 48$.

11.4 Discrete Cosine Transform, Modified Discrete Cosine Transform, and Transform Coding in MPEG Audio

This section introduces *discrete cosine transform* (DCT) and explains how to apply it in transform coding. The section will also show how to remove the block effects in transform coding using a modified DCT (MDCT). Finally, we will examine how the MDCT coding is used in MPEG (Motion Picture Experts Group) audio format, which is used as part of MPEG audio, such as in MP3 (MPEG-1 layer 3).

11.4.1 Discrete Cosine Transform

Given N data samples, we define the one-dimensional (1D) DCT pair given:
Forward transform:

$$X_{DCT}(k) = \sqrt{\frac{2}{N}} C(k) \sum_{n=0}^{N-1} x(n) \cos\left[\frac{(2n+1)k\pi}{2N}\right], \quad k = 0, 1, \ldots, N-1 \quad (11.29)$$

Inverse transform:

$$x(n) = \sqrt{\frac{2}{N}} \sum_{k=0}^{N-1} C(k) X_{DCT}(k) \cos\left[\frac{(2n+1)k\pi}{2N}\right], \quad n = 0, 1, \ldots, N-1 \quad (11.30)$$

$$C(k) = \begin{cases} \frac{\sqrt{2}}{2} & k = 0 \\ 1 & \text{otherwise,} \end{cases} \quad (11.31)$$

where $x(n)$ is the input data sample and $X_{DCT}(k)$ is the DCT coefficient. The DCT transforms the time domain signal to frequency domain coefficients. However, unlike the discrete Fourier transform (DFT), there is no complex number operations for both forward and inverse transforms. Both forward and inverse transforms use the same scale factor:

$$\sqrt{\frac{2}{N}} C(k).$$

In terms of transform coding, the DCT decomposes a block of data into the direct-current (DC) coefficient corresponding to the average of the data samples and the alternating-current (AC) coefficients, each corresponding to the frequency component (fluctuation). The terms "DC" and "AC" come from basic electrical engineering. In transform coding, we can quantize the DCT coefficients and encode them into binary information. The inverse DCT can transform the DCT coefficients back to the input data. Let us proceed to Examples 11.8 and 11.9.

Example 11.8.

Assuming that the following input data each can be encoded by 5 bits, including a sign bit:

$$x(0) = 10, \ x(1) = 8, \ x(2) = 10, \ \text{and} \ x(3) = 12,$$

a. Determine the DCT coefficients.
b. Use the MATLAB function **dct()** to verify all the DCT coefficients.

Solution:

a. Using Equation (11.29) leads to

$$X_{DCT}(k) = \sqrt{\frac{1}{2}C(k)} \left[x(0)\cos\left(\frac{\pi k}{8}\right) + x(1)\cos\left(\frac{3\pi k}{8}\right) + x(2)\cos\left(\frac{5\pi k}{8}\right) \right.$$
$$\left. + x(3)\cos\left(\frac{7\pi k}{8}\right) \right].$$

When $k = 0$, we see that the DC component is calculated as

$$X_{DCT}(0) = \sqrt{\frac{1}{2}C(0)} \left[x(0)\cos\left(\frac{\pi \times 0}{8}\right) + x(1)\cos\left(\frac{3\pi \times 0}{8}\right) + x(2)\cos\left(\frac{5\pi \times 0}{8}\right) \right.$$
$$\left. + x(3)\cos\left(\frac{7\pi \times 0}{8}\right) \right].$$

$$= \sqrt{\frac{1}{2}} \times \sqrt{\frac{2}{2}}[x(0) + x(1) + x(2) + x(3)] = \frac{1}{2}(10 + 8 + 10 + 12) = 20.$$

We clearly see that the first DCT coefficient is a scaled average value.

For $k = 1$,

$$X_{DCT}(1) = \sqrt{\frac{1}{2}C(1)} \left[x(0)\cos\left(\frac{\pi \times 1}{8}\right) + x(1)\cos\left(\frac{3\pi \times 1}{8}\right) + x(2)\cos\left(\frac{5\pi \times 1}{8}\right) \right.$$
$$\left. + x(3)\cos\left(\frac{7\pi \times 1}{8}\right) \right].$$

$$= \sqrt{\frac{1}{2}} \times 1 \left[10 \times \cos\left(\frac{\pi}{8}\right) + 8 \times \cos\left(\frac{3\pi}{8}\right) \right.$$
$$\left. + 10 \times \cos\left(\frac{5\pi}{8}\right) + 12 \times \cos\left(\frac{7\pi}{8}\right) \right] = -1.8478$$

Similarly, we have the rest as

$$X_{DCT}(2) = 2 \text{ and } X_{DCT}(3) = 0.7654.$$

b. Using the MATLAB 1D-DCT function **dct()**, we can verify that

≫ dct([10 8 10 12])

ans = 20.0000 −1.8478 2.0000 0.7654.

Example 11.9.

Assuming the following DCT coefficients:

$$X_{DCT}(0) = 20, \ X_{DCT}(1) = -1.8478, \ X_{DCT}(0) = 2, \text{ and } X_{DCT}(0) = 0.7654,$$

a. Determine $x(0)$.

b. Use the MATLAB function **idct()** to verify all the recovered data samples.

Solution:

a. Applying Equations (11.30) and (11.31), we have

$$x(0) = \sqrt{\frac{1}{2}} \left[C(0)X_{DCT}(0)\cos\left(\frac{\pi}{8}\right) + C(1)X_{DCT}(1)\cos\left(\frac{3\pi}{8}\right) \right.$$
$$\left. + C(2)X_{DCT}(2)\cos\left(\frac{5\pi}{8}\right) + C(3)X_{DCT}(3)\cos\left(\frac{7\pi}{8}\right) \right]$$

$$= \sqrt{\frac{1}{2}} \left[\frac{\sqrt{2}}{2} \times 20 \times \cos\left(\frac{\pi}{8}\right) + 1 \times (-1.8478) \times \cos\left(\frac{3\pi}{8}\right) \right.$$
$$\left. + 1 \times 2 \times \cos\left(\frac{5\pi}{8}\right) + 1 \times 0.7654 \times \cos\left(\frac{7\pi}{8}\right) \right] = 10.$$

b. With the MATLAB 1D inverse DCT function **idct()**, we obtain

≫ idct([20 −1.8478 2 0.7654])

ans = 10.0000 8.0000 10.0000 12.0000.

We verify that the input data samples are as the ones in Example 11.8.

In Example 11.9, we can get an exact recovery of the input data from the DCT coefficients, since infinite precision of each DCT coefficient is preserved. However, in transform coding, each DCT coefficient is quantized using the number of bits per sample assigned by a bit allocation scheme. Usually the DC coefficient requires a larger number of bits to encode, since it carries more energy of the signal, while each AC coefficient requires a smaller number of bits to encode. Hence, the quantized DCT coefficients approximate the DCT coefficients in infinite precision, and the recovered input data with the quantized DCT coefficients will certainly have quantization errors.

Example 11.10.

Assuming the following DCT coefficients in infinite precision:

$$X_{DCT}(0) = 20, \ X_{DCT}(1) = -1.8478, \ X_{DCT}(2) = 2, \text{ and } X_{DCT}(3) = 0.7654,$$

we had exact recovered data as: 10, 8, 10, 12, verified in Example 11.9. If a bit allocation scheme quantizes the DCT coefficients using a scale factor of 4 in the following form:

$$X_{DCT}(0) = 4 \times 5 = 20, \ X_{DCT}(1) = 4 \times (-0) = 0, \ X_{DCT}(2) = 4 \times 1$$
$$= 4, \text{ and } X_{DCT}(3) = 4 \times 0 = 0,$$

We can code the scale factor of 4 by 3 bits (magnitude bits only), the scaled DC coefficient of 5 by 4 bits (including a sign bit), and scaled AC coefficients of 0, 1, 0 using 2 bits each. 13 bits in total are required.

a. Use the MATLAB function **idct()** to recover the input data samples.

Solution:

a. Using the MATLAB function **idct()** and the quantized DCT coefficients, it follows that

$$\gg \text{idct}([20 \ 0 \ 4 \ 0])$$
$$\text{ans} = 12 \quad 8 \quad 8 \quad 12.$$

As we see, the original sample requires 5 bits (4 magnitude bits and 1 sign bit) to encode each of 10, 8, 10, and 12 with a total of 20 bits. Hence, 7 bits are saved for coding this data block using the DCT. We expect many more bits to be saved in practice, in which a longer frame of the correlated data samples is used. However, quantization errors are introduced.

For comprehensive coverage of the topics on DCT, see Li and Drew (2004), Nelson (1992), Sayood (2000), and Stearns (2003).

11.4.2 Modified Discrete Cosine Transform

In the previous section, we have seen how a 1D-DCT is adopted for coding a block of data. When we apply the 1D-DCT to audio coding, we first divide the audio samples into blocks and then transform each block of data with DCT.

The DCT coefficients for each block are quantized according to the bit allocation scheme. However, when we decode DCT blocks back, we encounter edge artifacts at boundaries of the recovered DCT blocks, since the DCT coding is block based. This effect of edge artifacts produces periodic noise and is annoying in the decoded audio. To solve for such a problem, the windowed MDCT has been developed (described in Pan, 1995; Princen and Bradley, 1986). The principles are illustrated in Figure 11.14. As we shall see, the windowed MDCT is used in MPEG-1 MP3 audio coding.

We describe and discuss only the main steps for coding data blocks using the windowed MDCT (W-MDCT) based on Figure 11.14.

Encoding stage:

1. Divide data samples into blocks that each have N (must be an even number) samples, and further divide each block into two subblocks, each with $N/2$ samples for the data overlap purpose.

2. Apply the window function for the overlapped blocks. As shown in Figure 11.14, if one block contains the subblocks A and B, the next one

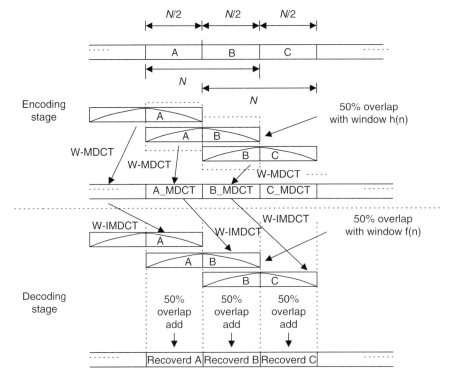

FIGURE 11.14 Modified discrete cosine transform (MDCT).

would consist of subblocks B and C. The subblock B is the overlapped block. This procedure continues. A window function $h(n)$ is applied to each N sample block to reduce the possible edge effects. Next, the W-MDCT is applied. The W-MDCT is given by:

$$X_{MDCT}(k) = 2\sum_{n=0}^{N-1} x(n)h(n)\cos\left[\frac{2\pi}{N}(n+0.5+N/4)(k+0.5)\right]$$
$$\text{for } k = 0, 1, \ldots, N/2 - 1. \quad (11.32)$$

Note that we need to compute and encode only half of the MDCT coefficients (since the other half can be reconstructed based on the first half of the MDCT coefficients).

3. Quantize and encode the MDCT coefficients.

Decoding stage:

1. Receive the $N/2$ MDCT coefficients, and use Equation (11.33) to recover the second half of the coefficients:

$$X_{MDCT}(k) = (-1)^{\frac{N}{2}+1} X_{MDCT}(N-1-k),$$
$$\text{for } k = N/2, N/2+1, \ldots, N-1. \quad (11.33)$$

2. Apply the windowed inverse MDCT (W-IMDCT) to each N MDCT coefficient block using Equation (11.34) and then apply a decoding window function $f(n)$ to reduce the artifacts at the block edges:

$$x(n) = \frac{1}{N}f(n)\sum_{k=0}^{N-1} X_{MDCT}(k)\cos\left[\frac{2\pi}{N}(n+0.5+N/4)(k+0.5)\right]$$
$$\text{for } n = 0, 1, \ldots, N-1. \quad (11.34)$$

Note that the recovered sequence contains the overlap portion. As shown in Figure 11.14, if a decoded block has the decoded subblocks A and B, the next one would have subblocks B and C, where the subblock B is an overlapped block. The procedure continues.

3. Reconstruct the subblock B using the overlap and add operation, as shown in Figure 11.14, where two subblocks labeled B are overlapped and added to generate the recovered subblock B. Note that the first subblock B comes from the recovered N sample block containing A and B, while the second subblock B belongs to the next recovered N sample block with B and C.

In order to obtain the perfect reconstruction, that is, all aliasing introduced by the MDCT being fully canceled, the following two conditions must be met for selecting the window functions, in which one is used for encoding while the other is used for decoding (Princen and Bradley, 1986):

$$f\left(n+\frac{N}{2}\right)h\left(n+\frac{N}{2}\right)+f(n)h(n)=1 \qquad (11.35)$$

$$f\left(n+\frac{N}{2}\right)h(N-n-1)-f(n)h\left(\frac{N}{2}-n-1\right)=0. \qquad (11.36)$$

Here, we choose the following simple function for the W-MDCT given by

$$f(n)=h(n)=\sin\left(\frac{\pi}{N}(n+0.5)\right). \qquad (11.37)$$

Equation (11.37) must satisfy the conditions described in Equations (11.35) and (11.36). This will be left for an exercise in the Problems section at the end of this chapter. The MATLAB functions **wmdct()** and **wimdct()** are written for use and listed in the MATLAB Programs section at the end of the chapter. Now, let us examine the W-MDCT in Example 11.11.

Example 11.11.

Given the data 1, 2, −3, 4, 5, −6, 4, 5 . . . ,

    a. Determine the W-MDCT coefficients for the first three blocks using a block size of 4.

    b. Determine the first two overlapped subblocks, and compare the results with the original data sequence using the W-MDCT coefficients in (1).

Solution:

    a. We divided the first two data blocks using the overlapping of two samples:

        First data block: 1 2 −3 4
        Second data block: −3 4 5 −6
        Third data block: 5 −6 4 5

    We apply the W-MDCT to get

```
≫ wmdct([1 2 −3 4])
ans = 1.1716  3.6569
≫ wmdct([−3 4 5 −6])
ans = −8.0000  7.1716
≫ wmdct([5 −6 4 5])
ans = −4.6569  −18.0711.
```

b. We show the results from W-IMDCT as:

>> x1=wimdct([1.1716 3.6569])
x1 = −0.5607 1.3536 −1.1465 − 0.4749
>> x2=wimdct([−8.0000 7.1716])
x2 = −1.8536 4.4749 2.1464 0.8891
>> x3=wimdct([−4.6569 − 18.0711])
x3 = 2.8536 − 6.8891 5.1820 2.1465.

Applying the overlap and add operation, we have

>> [x1 0 0 0 0]+ [0 0 x2 0 0]+ [0 0 0 0 x3]
ans = −0.5607 1.3536 −3.0000 4.0000 5.0000 −6.0000 5.1820 2.1465.

The recovered first two subblocks have values of −3, 4, 5, −6, which are consistent with the input data.

Figure 11.15 shows coding of speech data we.dat using the DCT transform and W-MDCT transform. To be able to see the block edge artifacts, the following parameters are used for both DCT and W-MDCT transform coding:

Speech data: 16 bits per sample, 8,000 samples per second
Block size: 16 samples
Scale factor: 2-bit nonlinear quantizer
Coefficients: 3-bit linear quantizer

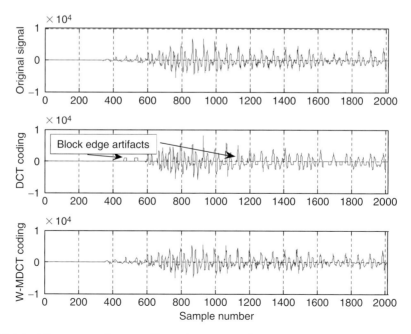

FIGURE 11.15   Waveform coding using DCT and W-MDCT.

Note that we assume a lossless scheme will further compress the quantized scale factors and coefficients. This stage does not affect the simulation results.

We use a 2-bit nonlinear quantizer with four levels to select the scale factor so that the block artifacts can be clearly displayed in Figure 11.15. We also apply a 3-bit linear quantizer to the scaled coefficients for both DCT and W-MDCT coding. As shown in Figure 11.15, the W-MDCT demonstrates significant improvement in smoothing out the block edge artifacts. The MATLAB simulation list is given in Programs 14 to 16, where Program 11.6 is the main program.

11.4.3 Transform Coding in MPEG Audio

With the DCT and MDCT concepts developed, we now explore the MPEG audio data format, where the DCT plays a key role. MPEG was established in 1988 to develop a standard for delivery of digital video and audio. Since MPEG audio compression contains so many topics, we focus here on examining its data format briefly, using the basic concepts developed in this book. Readers can further explore this subject by reading Pan's (1985) tutorial on MPEG audio compression, as well as Li and Drew (2004).

Figure 11.16 shows the MPEG audio frame. First, the input PCM samples—with a possible selection of sampling rates of 32 kHz, 44.1 kHz, and 48 kHz—are divided into 32 frequency subbands. All the subbands have equal bandwidths. The sum of their bandwidths covers up to the folding frequency, that is, $f_s/2$, which is the Nyquist limit in the DSP system. The subband filters are designed to minimize aliasing in frequency domain. Each subband filter outputs one sample for every 32 input PCM samples continuously, and forms a data segment for

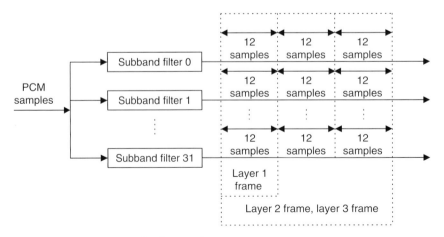

FIGURE 11.16 MPEG audio frame size.

every 12 output samples. The purpose of the filter banks is to separate the data into different frequency bands so that the psycho-acoustic model of the human auditory system (Yost, 1994) can be applied to activate the bit allocation scheme for a particular subband. The data frames are formed before quantization.

There are three types of data frames, as shown in Figure 11.16. Layer 1 contains 32 data segments, each coming from one subband with 12 samples, so the total frame has 384 data samples. As we see, layer 2 and layer 3 have the same size data frame, consisting of 96 data segments, where each filter outputs 3 data segments of 12 samples. Hence, layer 2 and layer 3 each have 1,152 data samples.

Next, let us examine briefly the content of each data frame, as shown in Figure 11.17. Layer 1 contains 384 audio samples from 32 subbands, each having 12 samples. It begins with a header followed by a cyclic redundancy check (CRC) code. The numbers within parentheses indicate the possible number of bits to encode each field. The bit allocation informs the decoder of the number of bits used for each encoded sample in the specific band. Bit allocation can also be set to zero number of bits for a particular subband if analysis of the psycho-acoustic model finds that the data in the band can be discarded without affecting audio quality. In this way, the encoder can achieve more data compression. Each scale factor is encoded with 6 bits. The decoder will multiply the scale factor by the decoded quantizer output to get the quantized subband value. Using the scale factor can make use of the full range of the quantizer. The field "ancillary data" is reserved for "extra" information.

The layer 2 encoder takes 1,152 samples per frame, with each subband channel having 3 data segments of 12 samples. These 3 data segments may have a bit allocation and up to three scale factors. Using one scale factor for 3 data segments would be called for when values of the scale factors per subband are sufficiently close and the encoder applies temporal noise masking (a type of

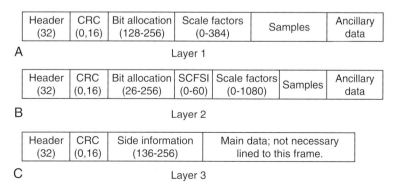

FIGURE 11.17 **MPEG audio frame formats.**

noise masking by the human auditory system) to hide any distortion. In Figure 11.17, the field "SCFSI" (scale-factor selection information) contains the information to inform the decoder. A different scale factor is used for each subband channel when avoidance of audio distortion is required. The bit allocation can also provide a possible single compact code word to represent three consecutive quantized values.

The layer 3 frame contains side information and main data that come from Huffman encoding (lossless coding having an exact recovery) of the W-MDCT coefficients to gain improvement over layer 1 and layer 2.

Figure 11.18 shows the MPEG-1 layers 1 and 2 encoder and the layer 3 encoder. For the MPEG-1 layer 1 and layer 2, the encoder examines the audio input samples using a 1,024-point fast Fourier transform (FFT). The psycho-acoustic model is analyzed based on the FFT coefficients. This includes possible frequency masking (hiding noise in frequency domain) and noise temporal masking (hiding noise in time domain). The result of the analysis of the psycho-acoustic model instructs the bit allocation scheme.

The major difference in layer 3, called MP3 (the most popular format in the multimedia industry), is that it adopts the MDCT. First, the encoder can gain further data compression by transforming the data segments using DCT from each subband channel and then quantizing the DCT coefficients, which, again, are losslessly compressed using Huffman coding. As shown in Examples 11.8 to 11.11, since the DCT is block-based processing, it produces block edge effects, where the beginning samples and ending samples show discontinuity and cause

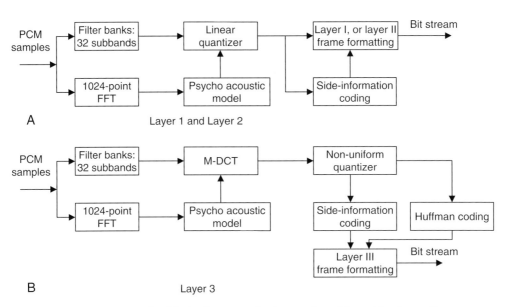

FIGURE 11.18 Encoder block diagrams for layers 1 and 2 and for layer 3.

audible periodic noise. This periodic edge noise can be alleviated, as discussed in the previous section, by using the W-MDCT, in which there is 50% overlap between successive transform windows.

There are two sizes of windows. One has 36 samples and the other 12 samples used in MPEG-1 layer 3 (MP3) audio. The larger block length offers better frequency resolution for low-frequency tonelike signals, hence it is used for the lowest two subbands. For the rest of the subbands, the shorter block is used, since it allows better time resolution for noiselike transient signals. Other improvements of MP3 over layers 1 and 2 include use of the scale-factor band, where the W-MDCT coefficients are regrouped from the original 32 uniformly divided subbands into 25 actual critical bands based on the human auditory system. Then the corresponding scale factors are assigned, and a nonlinear quantizer is used.

Finally, Huffman coding is applied to the quantizer outputs to obtain more compression. Particularly in CD-quality audio, MP3 (MPEG-1 layer 3) can achieve CRs varying from 12:1 to 8:1, corresponding to the bit rates from 128 kbps to 192 kbps. Besides the use of DCT in MP3, MPEG-2 audio coding methods such as AC-2, AC-3, ATRAC, and PAC/MPAC also use W-MDCT coding. Readers can further explore these subjects in Brandenburg (1997) and Li and Drew (2004).

11.5 Summary

1. The linear midtread quantizer used in PCM coding has an odd number of quantization levels, that is, $2^n - 1$. It accommodates the same decoded magnitude range for quantizing the positive and negative voltages.

2. The analog and digital μ-law compressions improve coding efficiency. The 8-bit μ-law compression of speech is equivalent to 12-bit linear PCM coding, with no difference in the sound quality. These methods are widely used in the telecommunications industry and multimedia system applications.

3. DPCM encodes the difference between the input sample and the predicted sample using a predictor to achieve coding efficiency.

4. DM coding is essentially a 1-bit DPCM.

5. ADPCM is similar to DPCM except that the predictor transfer function has six zeros and two poles and is an adaptive filter. ADPCM is superior to 8-bit μ-law compression, since it provides the same sound quality with only 4 bits per code.

6. Data compression performance is measured in terms of the data compression ratio and the bit rate.
7. The DCT decomposes a block of data to the DC coefficient (average) and AC coefficients (fluctuation) so that different numbers of bits are assigned to encode DC coefficients and AC coefficients to achieve data compression.
8. W-MDCT alleviates the block effects introduced by the DCT.
9. The MPEG-1 audio formats such as MP3 (MPEG-1, layer 3) include W-MDCT, filter banks, a psycho-acoustic model, bit allocation, a nonlinear quantizer, and Huffman lossless coding.

11.6 MATLAB Programs

Program 11.1. MATLAB program for the linear midtread quantizer

```
clear all;close all
disp('load speech: We');
load we.dat;            % Provided by your instructor
sig = we;
lg=length(sig);         % Length of the speech data
t=[0:1:lg-1];           % Time index
sig = 5*sig/max(abs(sig));        % Normalize signal to the range between -5 to 5
Emax = max(abs(sig));
Erms = sqrt( sum(sig .* sig) / length(sig))
k=Erms/Emax
disp('20*log10(k)=>');
k = 20* log10 (k)
bits = input('input number of bits =>');
lg = length(sig);
% Encoding
for x=1:lg
[indx(x) qy] = mtrdenc(bits, 5, sig(x));
end
disp('Finished and transmitted');
% Decoding
for x=1:lg
qsig(x) = mtrddec(bits, 5, indx(x));
end
disp('decoding finished');
qerr = sig-qsig;        % Calculate quantization errors
subplot(3,1,1);plot(t, sig);grid
ylabel('Speech');axis([0 lg -5 5]);
```

```
subplot(3,1,2);plot(t, qsig);grid
ylabel('Quantized speech');axis([0 2000 -5 5]);
subplot(3,1,3);plot(qerr);grid
axis([0 lg -0.5 0.5]);
ylabel('Qunatized error');xlabel('Sample number');
disp('signal to noise ratio due to quantization noise')
snr(sig,qsig);        % Calculate signal to noise ratio due to quantization
```

Program 11.2. MATLAB program for μ-law encoding and decoding

```
close all; clear all
disp('load speech file');
load we.dat;           % Provided by your instructor
lg=length(we);         % Length of the speech data
we = 5*we/max(abs(we));        % Normalize the speech data
we_nor=we/max(abs(we));        % Normalization
t=[0:1:lg-1];          % Time indices
disp('mu-law companding')
mu=input('input mu =>');
for x=1:lg
ymu(x) = mu-law(we_nor(x),1,mu);
end
disp('finished mu-law companding');
disp('start to quantization')
bits = input('input bits=>');
% Midtread quantization and encoding
for x=1:lg
[indx(x) qy] = mtrdenc(bits, 1, ymu(x));
end
disp('finished and transmitted');
%
% Midtread decoding
for x=1:lg
qymu(x) = mtrddec(bits, 1, indx(x));
end
disp('expander');
for x=1:lg
dymu(x) = muexpand(qymu(x),1,mu)*5;
end
disp('finished')
qerr = dymu-we;        % Quantization error
subplot(3,1,1);plot(we);grid
ylabel('Speech');axis([0 lg -5 5]);
subplot(3,1,2);plot(dymu);grid
ylabel('recovered speech');axis([0 lg -5 5]);
```

(Continued)

```
subplot(3,1,3);plot(qerr);grid
ylabel('Quantized error');xlabel('Sample number');
axis([0 lg -1 1]);
snr(we,dymu);          % Calculate signal to noise ratio due to quantization
```

Program 11.3. MATLAB function for μ-law companding

```
function qvalue = mu law(vin, vmax, mu)
% This function performs mulaw companding
% Usage:
% function qvalue = mu law(vin, vmax, mu)
% vin = input value
% vmax = input value
% mu = parameter for controlling the degree of compression
% qvalue = output value from the mu-law compander
% as the mu-law expander
%
  vin = vin/vmax;           % Normalization
% mu-law companding formula
  qvalue = vmax*sign(vin)*log(1+mu*abs(vin))/log(1+mu);
```

Program 11.4. MATLAB program for μ-law expanding

```
function rvalue = muexpand(y,vmax, mu)
% This function performs mu-law expanding
% usage:
% function rvalue = muexpand(y,vmax, mu)
% y = input signal
% vmax = maximum amplitude
% mu = parameter for controlling the degree of compression
% rvalue = output value from the mu-law expander
%
y=y/vmax;            % Normalization
% mu-law expanding
rvalue = sign(y)*(vmax/mu)*((1+mu)^abs(y) − 1);
```

Program 11.5. MATLAB function for midtread quantizer encoding

```
function [indx, pq ] = mtrdenc(NoBits,Xmax,value)
% function pq = mtrdenc(NoBits, Xmax, value)
% This routine is created for simulation of the midtread uniform quantizer.
%
% NoBits: number of bits used in quantization.
% Xmax: overload value.
% value: input to be quantized.
% pq: output of the quantized value
% indx: integer index
%
% Note: the midtread method is used in this quantizer.
%
if NoBits == 0
pq = 0;
indx=0;
else
delta = 2*abs(Xmax)/(2^NoBits − 1);
Xrmax = delta*(2^NoBits/2 − 1);
if abs(value) >= Xrmax
tmp = Xrmax;
else
tmp = abs(value);
end
indx=round(tmp/delta);
pq = indx*delta;
if value < 0
pq = −pq;
indx = −indx;
end
end
```

Program 11.6. MATLAB function for midtread quantizer decoding

```
function pq = mtrddec(NoBits,Xmax,indx)
% function pq = mtrddec(NoBits, Xmax, value)
% This routine is the dequantizer.
%
% NoBits: number of bits used in quantization.
% Xmax: overload value.
% pq: output of the quantized value
```

(*Continued*)

```
% indx: integer index
%
% Note: the midtread method is used in this quantizer.
%
delta = 2*abs(Xmax)/(2^NoBits − 1);
pq = indx*delta;
```

Program 11.7. MATLAB function for calculation of signal to quantization noise ratio (SNR)

```
function snr = calcsnr(speech, qspeech)
% function snr = calcsnr(speech, qspeech)
% This routine is created for calculation of SNR
%
% speech: original speech waveform.
% qspeech: quantized speech.
% snr: output SNR in dB.
%
qerr = speech-qspeech;
snr = 10*log10(sum(speech.*speech)/sum(qerr.*qerr))
```

Program 11.8. Main program for digital μ-law encoding and decoding

```
load we12b.dat
for i=1:1:length(we12b)
  code8b(i)=dmuenc(12, we12b(i));        % Encoding
  qwe12b(i)=dmudec(code8b(i));           % Decoding
end
subplot(4,1,1),plot(we12b);grid
ylabel('a');axis([0 length(we12b) −1024 1024]);
subplot(4,1,2),plot(code8b);grid
ylabel('b');axis([0 length(we12b) −128 128]);
subplot(4,1,3),plot(qwe12b);grid
ylabel('c');axis([0 length(we12b) −1024 1024]);
subplot(4,1,4),plot(qwe12b-we12b);grid
ylabel('d');xlabel('Sample number');axis([0 length(we12b) −40 40]);
```

Program 11.9. The digital μ-law compressor

```
function [cmp_code] = dmuenc(NoBits, value)
% This routine is created for simulation of 8-bit mu-law compression.
% function [cmp_code] = dmuenc(NoBits, value)
% NoBits = number of bits for the data
% value = input value
% cmp_code = output code
%
scale = NoBits-12;
value = value*2^(-scale);            % Scale to 12 bit
if (abs(value) >=0) & (abs(value)<16)
  cmp_code=value;
end
if (abs(value) >=16) & (abs(value)<32)
cmp_code =sgn(value)*(16+fix(abs(value)-16));
end
if (abs(value) >=32) & (abs(value)<64)
  cmp_code =sgn(value)*(32+fix((abs(value)-32)/2));
end
if (abs(value) >=64) & (abs(value)<128)
  cmp_code =sgn(value)*(48+fix((abs(value)-64)/4));
end
if (abs(value) >=128) & (abs(value)<256)
  cmp_code =sgn(value)*(64+fix((abs(value)-128)/8));
end
if (abs(value) >=256) & (abs(value)<512)
  cmp_code =sgn(value)*(80+fix((abs(value)-256)/16));
end
if (abs(value) >=512) & (abs(value)<1024)
  cmp_code=sgn(value)*(96+fix((abs(value)-512)/32));
end
if (abs(value) >=1024) & (abs(value)<2048)
  cmp_code=sgn(value)*(112+fix((abs(value)-1024)/64));
end
```

Program 11.10. The digital μ-law expander

```
function [value] = dmudec(cmp_code)
% This routine is created for simulation of 8-bit mu-law decoding.
% Usage:
% function [value] = dmudec(cmp_code)
```

(*Continued*)

```
% cmp_code = input mu-law encoded code
% value = recovered output value
%
if (abs(cmp_code) >=0) & (abs(cmp_code) <16)
  value =cmp_code;
end
if (abs(cmp_code) >=16) & (abs(cmp_code) <32)
  value=sgn(cmp_code)*(16+ (abs(cmp_code) − 16));
end
if (abs(cmp_code) >=32) & (abs(cmp_code) <48)
  value =sgn(cmp_code)*(32+ (abs(cmp_code) − 32)*2 + 1);
end
if (abs(cmp_code) >=48) & (abs(cmp_code) <64)
  value=sgn(cmp_code)*(64+ (abs(cmp_code) − 48)*4 + 2);
end
if (abs(cmp_code) >=64) & (abs(cmp_code) <80)
  value=sgn(cmp_code)*(128+ (abs(cmp_code) − 64)*8 + 4);
end
if (abs(cmp_code) >= 80) & (abs(cmp_code) < 96)
  value=sgn(cmp_code)*(256+ (abs(cmp_code) − 80)*16 + 8);
end
if (abs(cmp_code) >=96) & (abs(cmp_code)<112)
  value=sgn(cmp_code)*(512+(abs(cmp_code)-96)*32 + 16);
end
if (abs(cmp_code) >=112) & (abs(cmp_code)<128)
  value=sgn(cmp_code)*(1024+(abs(cmp_code)-112)*64 + 32);
end
```

Program 11.11. Main program for ADPCM coding

```
% This program is written for offline simulation
clear all; close all
load we.dat % Provided by the instructor
speech=we;
desig= speech;
lg=length(desig);            % Length of speech data
enc = adpcmenc(desig);       % ADPCM encoding
%ADPCM finished
dec = adpcmdec(enc);         % ADPCM decoding
snrvalue = snr(desig,dec) % Calculate signal to noise ratio due to quantization
subplot(3,1,1);plot(desig);grid;
ylabel('Speech');axis([0 lg −8000 8000]);
subplot(3,1,2);plot(dec);grid;
ylabel('Quantized speech');axis([0 lg −8000 8000]);
```

```
subplot(3,1,3);plot(desig-dec);grid
ylabel('Quantized error');xlabel('Sample number');
axis([0 lg -1200 1200]);
```

Program 11.12. MATLAB function for ADPCM encoding

```
function iiout = adpcmenc(input)
% This function performs ADPCM encoding
% function iiout = adpcmenc(input)
% Usage:
% input = input value
% iiout = output index
%
% Quantization tables
fitable = [0 0 0 1 1 1 1 3 7];
witable = [-0.75 1.13 2.56 4.00 7.00 12.38 22.19 70.13 ];
qtable = [-0.98 0.62 1.38 1.91 2.34 2.72 3.12];
invqtable = [0.031 1.05 1.66 2.13 2.52 2.91 3.32];
lgth = length(input);
sr = zeros(1,2);pk = zeros(1,2);
a = zeros(1,2);b = zeros(1,6);
dq = zeros(1,6);ii= zeros(1,lgth);
y = 0; ap = 0;al = 0;yu = 0;yl = 0;dms = 0;dml = 0;tr = 0;td = 0;
for k = 1:lgth
sl = input(k);
%
sez = b(1)*dq(1);
for i = 2:6
sez = sez + b(i)*dq(i);
end
se = a(1)*sr(1)+a(2)*sr(2)+ sez;
d = sl - se;
%
% Perform quantization
%
dqq = log10(abs(d))/log10(2.0)-y;
ik = 0;
for i = 1:7
if dqq > qtable(i)
  ik = i;
end
end
```

(Continued)

```
if d < 0
ik = -ik;
end
ii(k) = ik;
yu = (31.0/32.0)*y+ witable(abs(ik)+1)/32.0;
if yu > 10.0
yu = 10.0;
end
if yu < 1.06
yu = 1.06;
end
yl = (63.0/64.0)*yl+yu/64.0;
%
%Inverse quantization
%
if ik == 0
dqq = 2^(-y);
else
dqq = 2^(invqtable(abs(ik))+y);
end
if ik < 0
dqq = -dqq;
end
srr = se + dqq;
dqsez = srr+sez-se;
%
% Update state
%
pk1 = dqsez;
%
% Obtain adaptive predictor coefficients
%
  if tr == 1
    a = zeros(1,2);b = zeros(1,6);
    tr = 0;
    td = 0;              %Set for the time being
  else
% Update predictor poles
% Update a2 first
  a2p = (127.0/128.0)*a(2);
  if abs(a(1)) <= 0.5
    fa1 = 4.0*a(1);
  else
    fa1 = 2.0*sgn(a(1));
  end
  a2p = a2p+(sign(pk1)*sgn(pk(2))-fa1*sign(pk1)*sgn(pk(1)))/128.0;
  if abs(a2p) > 0.75
    a2p = 0.75*sgn(a2p);
  end
```

```matlab
    a(2) = a2p;
%
% Update a1
  a1p = (255.0/256.0)*a(1);
    a1p = a1p +3.0*sign(pk1)*sgn(pk(1))/256.0;
  if abs(a1p) > 15.0/16.0-a2p
    a1p = 15.0/16.0-a2p;
  end
a(1) = a1p;
%
% Update b coefficients
%
for i = 1:6
b(i) = (255.0/256.0)*b(i)+sign(dqq)*sgn(dq(i))/128.0;
end
if a2p < -0.7185
td = 1;
else
td = 0;
end
if a2p < -0.7185 & abs(dq(6)) > 24.0*2^(y1)
tr = 1;
else
tr = 0;
end
for i = 6:-1:2
dq(i) = dq(i-1);
end
dq(1) = dqq;pk(2) = pk(1);pk(1) = pk1;sr(2) = sr(1);sr(1) = srr;
%
% Adaptive speed control
%
dms = (31.0/32.0)*dms; dms = dms + fitable(abs(ik)+1)/32.0;
dml = (127.0/128.0)*dml; dml = dml + fitable(abs(ik)+1)/128.0;
if ap > 1.0
al = 1.0;
else
al = ap;
end
ap = (15.0/16.0)*ap;
if abs(dms-dml) >= dml/8.0
ap = ap + 1/8.0;
end
if y < 3
ap = ap + 1/8.0;
end
if td == 1
ap = ap + 1/8.0;
```

(*Continued*)

```
end
if tr == 1
ap = 1.0;
end
y = al*yu+ (1.0-al)*yl;
end
end
iiout = ii;
```

Program 11.13. MATLAB function for ADPCM decoding

```
function iiout = adpcmdec(ii)
% This function performs ADPCM decoding
% function iiout = adpcmdec(ii)
% Usage:
% ii = input ADPCM index
% iiout = decoded output value
%
% Quantization tables:
fitable = [0 0 0 1 1 1 1 3 7];
witable = [−0.75 1.13 2.56 4.00 7.00 12.38 22.19 70.13 ];
qtable = [ −0.98 0.62 1.38 1.91 2.34 2.72 3.12 ];
invqtable = [0.031 1.05 1.66 2.13 2.52 2.91 3.32 ];
lgth = length(ii);
sr = zeros(1,2);pk = zeros(1,2);
a = zeros(1,2);b = zeros(1,6);
dq = zeros(1,6);out= zeros(1,lgth);
y = 0;ap = 0;al = 0;yu = 0;yl = 0;dms = 0;dml = 0;tr = 0;td = 0;
for k = 1:lgth
%
sez = b(1)*dq(1);
for i = 2:6
sez = sez + b(i)*dq(i);
end
se = a(1)*sr(1) + a(2)*sr(2) + sez;
%
%Inverse quantization
%
ik = ii(k);
yu = (31.0/32.0)*y + witable(abs(ik)+1)/32.0;
if yu > 10.0
yu = 10.0;
end
if yu < 1.06
yu = 1.06;
end
```

```
yl = (63.0/64.0)*yl + yu/64.0;
if ik == 0
dqq = 2^(-y);
else
dqq = 2^(invqtable(abs(ik))+y);
end
if ik < 0
dqq = -dqq;
end
srr = se + dqq;
dqsez = srr+sez-se;
out(k) =srr;
%
% Update state
%
pk1 = dqsez;
%
% Obtain adaptive predictor coefficients
%
  if tr == 1
    a = zeros(1,2);
    b = zeros(1,6);
    tr = 0;
    td = 0;          %set for the time being
  else
% Update predictor poles
% Update a2 first;
  a2p = (127.0/128.0)*a(2);
  if abs(a(1)) <= 0.5
    fa1 = 4.0*a(1);
  else
    fa1 = 2.0*sgn(a(1));
  end
  a2p=a2p+(sign(pk1)*sgn(pk(2))-fa1*sign(pk1)*sgn(pk(1)))/128.0;
  if abs(a2p) > 0.75
    a2p = 0.75*sgn(a2p);
  end
    a(2) = a2p;
%
% Update a1
  a1p = (255.0/256.0)*a(1);
    a1p = a1p +3.0*sign(pk1)*sgn(pk(1))/256.0;
  if abs(a1p) > 15.0/16.0-a2p
    a1p = 15.0/16.0-a2p;
  end
  a(1) = a1p;
```

(Continued)

```
%
% Update b coefficients
%
for i=1: 6
b(i) = (255.0/256.0)*b(i)+sign(dqq)*sgn(dq(i))/128.0;
end
if a2p < -0.7185
td = 1;
else
td = 0;
end
if a2p < -0.7185 & abs(dq(6)) > 24.0*2^(yl)
tr = 1;
else
tr = 0;
end
for i=6:-1:2
dq(i) = dq(i-1);
end
dq(1) = dqq; pk(2) = pk(1); pk(1) = pk1; sr(2) = sr(1); sr(1) = srr;
%
% Adaptive speed control
%
dms = (31.0/32.0)*dms;
dms = dms + fitable(abs(ik)+1)/32.0;
dml = (127.0/128.0)*dml;
dml = dml + fitable(abs(ik)+1)/128.0;
if ap > 1.0
al = 1.0;
else
al = ap;
end
ap = (15.0/16.0)*ap;
if abs(dms-dml) >= dml/8.0
ap = ap + 1/8.0;
end
if y < 3
ap = ap + 1/8.0;
end
if td == 1
ap = ap + 1/8.0;
end
if tr == 1
ap = 1.0;
end
y = al*yu+ (1.0-al)*yl;
end
end
iiout = out;
```

Program 11.14. W-MDCT function

```
function [ tdac_coef ] = wmdct(ipsig)
%
% This function transforms the signal vector using the W-MDCT
% Usage:
% ipsig: input signal block of N samples (N=even number)
% tdac_coe: W-MDCT coefficients (N/2 coefficients)
%
N = length(ipsig);
NN =N;
for i = 1:NN
h(i) = sin((pi/NN)*(i − 1 + 0.5));
end
for k = 1:N/2
tdac_coef(k) = 0.0;
for n = 1:N
tdac_coef(k) = tdac_coef(k) +...
h(n)*ipsig(n)*cos((2*pi/N)*(k−1 + 0.5)*(n−1 + 0.5 + N/4));
end
end
tdac_coef= 2*tdac_coef;
```

Program 11.15. Inverse W-MDCT function

```
function [ opsig ] = wimdct(tdac_coef)
%
% This function transforms the W-MDCT coefficients back to the signal
% Usage:
% tdac_coeff: N/2 W-MDCT coefficients
% opsig: output signal block with N samples
%
N = length(tdac_coef);
tmp_coef = ((−1)^(N + 1))*tdac_coef(N:-1:1);
tdac_coef = [ tdac_coef tmp_coef];
N = length(tdac_coef);
NN =N;
for i = 1:NN
f(i) = sin((pi/NN)*(i − 1 + 0.5));
end
for n = 1:N
opsig(n) = 0.0;
for k = 1:N
```

(Continued)

```
opsig(n) = opsig(n) +...
tdac_coef(k)*cos((2*pi/N)*(k − 1 + 0.5)*(n − 1 + 0.5 + N/4));
end
opsig(n) = opsig(n)*f(n)/N;
end
```

Program 11.16. Waveform coding using DCT and W-MDCT

```
% Waveform coding using DCT and MDCT for a block size of 16 samples
% Main program
close all; clear all
load we.dat% Provided by the instructor
% Create scale factors
N=16            %Blocksize
scalef4bits=sqrt (2*N)*[1  2 4 8 16 32 64 128...
                        256 512 1024 2048 4096 8192 16384 32768];
scalef3bits=sqrt(2*N)*[256 512 1024 2048 4096 8192 16384 32768];
scalef2bits=sqrt(2*N)*[4096 8192 16384 32768];
scalef1bit=sqrt(2*N)*[16384 32768];
scalef=scalef2bits;
nbits =3;
% Ensure the block size to be 16 samples.
x=[we zeros(1,16-mod(length(we),16))];
Nblock=length(x)/16;
DCT_code=[]; scale_code=[];
% DCT transform coding
% Encoder
for i=1:Nblock
xblock_DCT=dct(x((i − 1)*16 + 1:i*16));
diff=(scalef-(max(abs(xblock_DCT))));
iscale(i)=min(find(diff==min(diff(find(diff>=0))))); %Find a scale factor
xblock_DCT=xblock_DCT/scalef(iscale(i)); % Scale the input vector
for j=1:16
[DCT_coeff(j) pp]=biquant(nbits,−1,1,xblock_DCT(j));%biquant(): Program 2.4
end
DCT_code=[DCT_code DCT_coeff ];
end
%Decoder
Nblock=length(DCT_code)/16;
xx=[];
for i=1:Nblock
DCT_coefR=DCT_code((i − 1)*16 + 1:i*16);
for j=1:16
xrblock_DCT(j)=biqtdec(nbits,−1,1,DCT_coefR(j)); %bigtdet(): Program 2.5
```

```
end
xrblock=idct(xrblock_DCT.*scalef(iscale(i)));
xx=[xx xrblock];
end
% Transform coding using the MDCT
xm=[zeros(1,8) we zeros(1,8-mod(length(we),8)), zeros(1,8)];
Nsubblock=length(x)/8;
MDCT_code=[];
% Encoder
for i=1:Nsubblock
xsubblock_DCT = wmdct(xm((i − 1)*8 + 1:(i + 1)*8));
diff=(scalef-max(abs(xsubblock_DCT)));
iscale(i)=min(find(diff==min(diff(find(diff>=0))))); %Find a scale factor
xsubblock_DCT=xsubblock_DCT/scalef(iscale(i)); %Scale the input vector
for j=1:8
[MDCT_coeff(j) pp]=biquant(nbits,−1,1,xsubblock_DCT(j));
end
MDCT_code=[MDCT_code MDCT_coeff];
end
%Decoder
% Recover the first subblock
Nsubblock=length(MDCT_code)/8;
xxm=[];
MDCT_coeffR=MDCT_code(1:8);
for j=1:8
xmrblock_DCT(j)=biqtdec(nbits,−1,1,MDCT_coeffR(j));
end
xmrblock=wimdct(xmrblock_DCT*scalef(iscale(1)));
xxr_pre=xmrblock(9:16) % recovered first block for overlap and add
for i=2:Nsubblock
MDCT_coeffR = MDCT_code((i − 1)*8 + 1:i*8);
for j=1:8
xmrblock_DCT(j)=biqtdec(nbits,−1,1,MDCT_coeffR(j));
end
xmrblock=wimdct(xmrblock_DCT*scalef(iscale(i)));
xxr_cur=xxr_pre+xmrblock(1:8);% overlap and add
xxm=[xxm xxr_cur];
xxr_pre=xmrblock(9:16);% set for the next overlap
end
subplot(3,1,1);plot(x);grid; axis([0 length(x) −10000 10000])
ylabel('Original signal');
subplot(3,1,2);plot(xx);grid;axis([0 length(xx) −10000 10000]);
ylabel('DCT coding')
subplot(3,1,3);plot(xxm);grid;axis([0 length(xxm) −10000 10000]);
ylabel('W-MDCT coding');
xlabel('Sample number');
```

Program 11.17. Sign function

```
function sgn = sgn(sgninp)
%
% sign function
% if sginp >=0 then sgn=1
% else sgn = -1
%
if sgninp >=0
opt = 1;
else
opt = -1;
end
sgn = opt;
```

11.7 Problems

11.1. For the 3-bit midtread quantizer described in Figure 11.1, and the analog signal range from −2.5 to 2.5 volts, determine

  a. the quantization step size

  b. the binary codes, recovered voltages, and quantization errors when each input is 1.6 volts and −0.2 volt.

11.2. For the 3-bit midtread quantizer described in Figure 11.1, and the analog signal range from −4 to 4 volts, determine

  a. the quantization step size

  b. the binary codes, recovered voltages, and quantization errors when each input is −2.6 volts and 0.1 volt.

11.3. For the μ-law compression and expanding process shown in Figure 11.3 with $\mu = 255$ and the 3-bit midtread quantizer described in Figure 11.1, with analog signal range from −2.5 to 2.5 volts, determine the binary codes, recovered voltages, and quantization errors when each input is 1.6 volts and −0.2 volt.

11.4. For the μ-law compression and expanding process shown in Figure 11.3 with $\mu = 255$ and the 3-bit midtread quantizer described in Figure 11.1, with the analog signal range from −4 to 4 volts, determine the binary codes, recovered voltages, and quantization errors when each input is −2.6 volts and 0.1 volt.

11.5. In a digital companding system, encode each of the following 12-bit linear PCM codes into the 8-bit compressed PCM code.

a. 0 0 0 0 0 0 0 1 0 1 0 1

b. 1 0 1 0 1 1 1 0 1 0 1 0

11.6. In a digital companding system, decode each of the following 8-bit compressed PCM codes into the 12-bit linear PCM code.

a. 0 0 0 0 0 1 1 1

b. 1 1 1 0 1 0 0 1

11.7. For a 3-bit DPCM encoding system with the following specifications (Fig. 11.19):

Encoder scheme: $\tilde{x}(n) = \hat{x}(n-1)$ (predictor)

$$d(n) = x(n) - \tilde{x}(n)$$
$$d_q(n) = Q[d(n)] = \text{quantizer in Table 11.9}$$
$$\hat{x}(n) = \tilde{x}(n) + d_q(n)$$

5-bit input data: $x(0) = -6$, $x(1) = -8$, $x(2) = -13$, perform DPCM encoding to produce the binary code for each input data.

11.8. For a 3-bit DPCM decoding system with the following specifications:

Decoding scheme: $\tilde{x}(n) = \hat{x}(n-1)$ (predictor)

$$d_q(n) = \text{quantizer in Table 11.9}$$
$$\hat{x}(n) = \tilde{x}(n) + d_q(n)$$

Received 3 binary codes: 110, 100, 101, perform DPCM decoding to recover each digital value using its binary code.

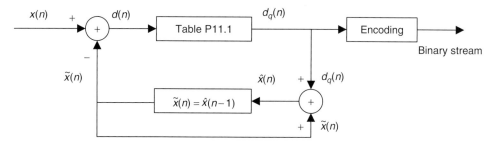

FIGURE 11.19 DPCM encoding in problem 11.7.

TABLE 11.9 Quantization table for the 3-bit quantizer in Problem 11.7.

Binary Code	Quantization Value $d_q(n)$	Subrange in $d(n)$
0 1 1	−11	$-15 \leq d(n) < -7$
0 1 0	−5	$-7 \leq d(n) < -3$
0 0 1	−2	$-3 \leq d(n) < -1$
0 0 0	0	$-1 \leq d(n) < 0$
1 0 0	0	$0 \leq d(n) \leq 1$
1 0 1	2	$1 < d(n) \leq 3$
1 1 0	5	$3 < d(n) \leq 7$
1 1 1	11	$7 < d(n) \leq 15$

11.9. Assuming that a speech waveform is sampled at 8 kHz and each sample is encoded by 16 bits, determine the compression ratio for each of the encoding methods.

   a. noncompression

   b. standard μ-law compression (8 bits per sample)

   c. standard ADPCM encoding (4 bits per sample).

11.10. Suppose that a speech waveform is sampled at 8 kHz into 16 bits per sample. Determine the bit rate for each of the following encoding methods:

   a. noncompression

   b. standard μ-law companding (8 bits per sample)

   c. standard ADPCM encoding (4 bits per sample).

11.11. Speech is sampled at 8 kHz and each sample is encoded by 16 bits. The telephone system can transport the digital voice channel over a digital link having a capacity of 1.536 Mbps. Determine the number of channels that the phone company can carry for each of the following encoding methods:

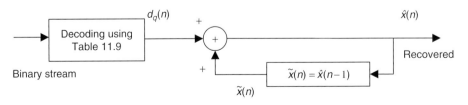

FIGURE 11.20. DPCM decoding in problem 11.8.

a. noncompression

b. standard μ-law companding (8 bits per sample)

c. standard ADPCM encoding (4 bits per sample).

11.12. Given the following input data:

$$x(0) = 25,\ x(1) = 30,\ x(2) = 28,\ \text{and}\ x(3) = 25,$$

determine the DCT coefficients.

11.13. Assuming the following DCT coefficients with infinite precision:

$$X_{DCT}(0) = 14,\ X_{DCT}(1) = 6,\ X_{DCT}(3) = -6,\ \text{and}\ X_{DCT}(0) = 8,$$

a. determine the input data using the MATLAB function **idct()**;

b. recover the input data samples using the MATLAB function **idct()** if a bit allocation scheme quantizes the DCT coefficients as follows: 2 magnitude bits plus 1 sign bit (3 bits) for the DC coefficient, 1 magnitude bit plus 1 sign bit (2 bits) for each AC coefficient and a scale factor of 8, that is,

$$X_{DCT}(0) = 8 \times 2 = 16,\ X_{DCT}(1) = 8 \times 1 = 8,\ X_{DCT}(2) = 8 \times (-1) = -8,\ \text{and}\ X_{DCT}(3) = 8 \times 1 = 8;$$

c. compute the quantized error in part b. of this question.

11.14. a. Verify the window functions

$$f(n) = h(n) = \sin\left(\frac{\pi}{N}(n + 0.5)\right)$$

used in MDCT that are satisfied with Equations (11.35) and (11.36).

b. Verify W-MDCT coefficients

$$X_{MDCT}(k) = (-1)^{\frac{N}{2}+1} X_{MDCT}(N - 1 - k) \quad \text{for } k = N/2,\ N/2 + 1,\ldots,\ N - 1.$$

11.15. Given data 1, 2, 3, 4, 5, 4, 3, 2, ...,

a. determine the W-MDCT coefficients for the first three blocks using a block size of 4;

b. determine the first two overlapped subblocks and compare the results with the original data sequence using the W-MDCT coefficients in part (a) of Problem 11.15.

Computer Problems with MATLAB

Use the MATLAB programs in the program section for Problems 11.16 to 11.18.

11.16. Given the data file "speech.dat" with 16 bits per sample and a sampling rate of 8 kHz,

  a. Use the PCM coding (midtrad quantizer) to perform compression and decompression and apply the MATLAB function sound() to evaluate the sound quality in terms of "excellent," "good," "intelligent," and "unacceptable" for the following bit rates:

  1. 4 bits/sample (32 Kbits per second)
  2. 6 bits/sample (48 Kbits per second)
  3. 8 bits/sample (64 Kbits per second)

  b. Use the μ-law PCM coding to perform compression and decompression and apply the MATLAB function sound() to evaluate the sound quality.

  1. 4 bits/sample (32 Kbits per second)
  2. 6 bits/sample (48 Kbits per second)
  3. 8 bits/sample (64 Kbits per second)

11.17. Given the data file "speech.dat" with 16 bits per sample, a sampling rate of 8 kHz, and ADCPM coding, perform compression, and decompression and apply the MATLAB function sound() to evaluate the sound quality.

11.18. Given the data file "speech.dat" with 16 bits per sample, a sampling rate of 8 kHz, and DCT and M-DCT coding in the program section, perform compression and decompression using the following specified parameters in Program 11.16 to compare the sound quality.

  a. nbits = 3, scalef = scalef2bits
  b. nbits = 3, scalef = scalef3bits
  c. nbits = 4, scalef = scalef2bits
  d. nbits = 4, scalef = scalef3bits

References

Brandenburg, K. (1997). Overview of MPEG audio: Current and future standards for low-bit-rate audio coding. *Journal of the Audio Engineering Society*, 45 (1/2).

Li, Z.-N., and Drew, M. S. (2004). *Fundamentals of Multimedia.* Upper Saddle River, NJ: Pearson/Prentice Hall.

Nelson, M. (1992). *The Data Compression Book.* Redwood City, CA: M&T Publishing.

Pan, D. (1995). A tutorial on MPEG/audio compression. *IEEE Multimedia*, 2: 60–74.

Princen, J., and Bradley, A. B. (1986). Analysis/synthesis filter bank design based on time domain aliasing cancellation. *IEEE Transactions on Acoustics, Speech, and Signal Processing*, ASSP 34 (5).

Roddy, D., and Coolen, J. (1997). *Electronic Communications*, 4th ed. Englewood Cliffs, NJ: Prentice Hall.

Sayood, K. (2000). *Introduction to Data Compression*, 2nd ed. San Francisco: Morgan Kaufmann Publishers.

Stearns, S. D. (2003). *Digital Signal Processing with Examples in MATLAB.* Boca Raton, FL: CRC Press LLC.

Tomasi, W. (2004). *Electronic Communications Systems: Fundamentals Through Advanced*, 5th ed. Upper Saddle River, NJ: Pearson/Prentice Hall.

Yost, W. A. (1994). *Fundamentals of Hearing: An Introduction*, 3rd ed. San Diego: Academic Press.

12

Multirate Digital Signal Processing, Oversampling of Analog-to-Digital Conversion, and Undersampling of Bandpass Signals

Objectives:

This chapter investigates basics of multirate digital signal processing, illustrates how to change a sampling rate for speech and audio signals, and describes the polyphase implementation for the decimation filter and interpolation filter. Next, the chapter introduces the advanced analog-to-digital conversion system with the oversampling technique and sigma-delta modulation. Finally, the chapter explores the principles of undersampling of bandpass signals.

12.1 Multirate Digital Signal Processing Basics

In many areas of digital signal processing (DSP) applications—such as communications, speech, and audio processing—rising or lowering of a sampling rate is required. The principle that deals with changing the sampling rate belongs essentially to *multirate signal processing* (Ifeachor and Jervis, 2002; Porat, 1997; Proakis and Manolakis, 1996; Sorensen and Chen, 1997). As an introduction, we will focus on sampling rate conversion; that is, sampling rate reduction or increase.

12.1.1 Sampling Rate Reduction by an Integer Factor

The process of reducing a sampling rate by an integer factor is referred to as *downsampling* of a data sequence. We also refer to downsampling as "decimation" (not taking one of ten). The term "decimation" used for the downsampling process has been accepted and used in many textbooks and fields. To downsample a data sequence $x(n)$ by an integer factor of M, we use the following notation:

$$y(m) = x(mM), \qquad (12.1)$$

where $y(m)$ is the downsampled sequence, obtained by taking a sample from the data sequence $x(n)$ for every M samples (discarding $M-1$ samples for every M samples). As an example, if the original sequence with a sampling period $T = 0.1$ second (sampling rate $= 10$ samples per sec) is given by

$$x(n): 8\ 7\ 4\ 8\ 9\ 6\ 4\ 2\ -2\ -5\ -7\ -7\ -6\ -4\ldots$$

and we downsample the data sequence by a factor of 3, we obtain the downsampled sequence as

$$y(m): 8\ \ 8\ \ 4\ \ -5\ \ -6\ldots,$$

with the resultant sampling period $T = 3 \times 0.1 = 0.3$ second (the sampling rate now is 3.33 samples per second). Although the example is straightforward, there is a requirement to avoid aliasing noise. We will illustrate this next.

From the Nyquist sampling theorem, it is known that aliasing can occur in the downsampled signal due to the reduced sampling rate. After downsampling by a factor of M, the new sampling period becomes MT, and therefore the new sampling frequency is

$$f_{sM} = \frac{1}{MT} = \frac{f_s}{M}, \qquad (12.2)$$

where f_s is the original sampling rate.

Hence, the folding frequency after downsampling becomes

$$f_{sM}/2 = \frac{f_s}{2M}. \qquad (12.3)$$

This tells us that after downsampling by a factor of M, the new folding frequency will be decreased M times. If the signal to be downsampled has frequency components larger than the new folding frequency, $f > f_s/(2M)$, aliasing noise will be introduced into the downsampled data.

To overcome this problem, it is required that the original signal $x(n)$ be processed by a lowpass filter $H(z)$ before downsampling, which should have a stop frequency edge at $f_s/(2M)$ (Hz). The corresponding normalized stop frequency edge is then converted to be

$$\Omega_{stop} = 2\pi \frac{f_s}{2M} T = \frac{\pi}{M} \text{ radians}. \tag{12.4}$$

In this way, before downsampling, we can guarantee that the maximum frequency of the filtered signal satisfies

$$f_{\max} < \frac{f_s}{2M}, \tag{12.5}$$

such that no aliasing noise is introduced after downsampling. A general block diagram of decimation is given in Figure 12.1, where the filtered output in terms of the z-transform can be written as

$$W(z) = H(z)X(z), \tag{12.6}$$

where $X(z)$ is the z-transform of the sequence to be decimated, $x(n)$, and $H(z)$ is the lowpass filter transfer function. After anti-aliasing filtering, the down-sampled signal $y(m)$ takes its value from the filter output as:

$$y(m) = w(mM). \tag{12.7}$$

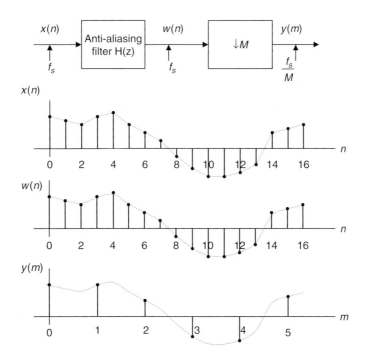

FIGURE 12.1 Block diagram of the downsampling process with $M = 3$.

12 MULTIRATE DIGITAL SIGNAL PROCESSING

The process of reducing the sampling rate by a factor of 3 is shown in Figure 12.1. The corresponding spectral plots for $x(n)$, $w(n)$, and $y(m)$ in general are shown in Figure 12.2.

To verify this principle, let us consider a signal $x(n)$ generated by the following:

$$x(n) = 5\sin\left(\frac{2\pi \times 1000n}{8000}\right) + \cos\left(\frac{2\pi \times 2500n}{8000}\right), \qquad (12.8)$$

with a sampling rate of $f_s = 8{,}000$ Hz, the spectrum of $x(n)$ is plotted in the first graph in Figure 12.3a, where we observe that the signal has components at frequencies of 1,000 and 2,500 Hz. Now we downsample $x(n)$ by a factor of 2,

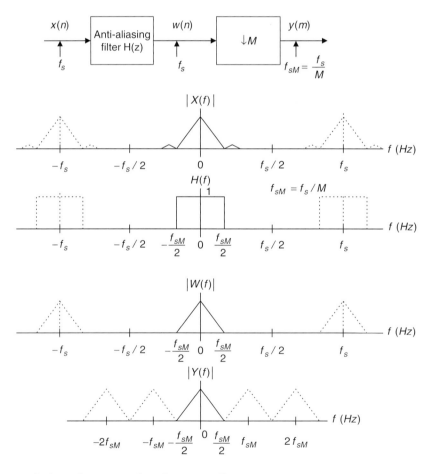

FIGURE 12.2 Spectrum after downsampling.

that is, $M = 2$. According to Equation (12.3), we know that the new folding frequency is $4000/2 = 2000$ Hz. Hence, without using the anti-aliasing lowpass filter, the spectrum would contain the aliasing frequency of $4\,\text{kHz} - 2.5\,\text{kHz} = 1.5\,\text{kHz}$ introduced by 2.5 kHz, plotted in the second graph in Figure 12.3a.

Now we apply a finite impulse response (FIR) lowpass filter designed with a filter length of $N = 27$ and a cutoff frequency of 1.5 kHz to remove the 2.5-kHz signal before downsampling to avoid aliasing. How to obtain such specifications will be discussed in a later example. The normalized cutoff frequency used for design is given by

$$\Omega_c = 2\pi \times 1500 \times (1/8000) = 0.375\pi.$$

Thus, the aliasing noise is avoided. The spectral plots are given in Figure 12.3b, where the first plot shows the spectrum of $w(n)$ after anti-aliasing filtering, while the second plot describes the spectrum of $y(m)$ after downsampling. Clearly, we prevent aliasing noise in the downsampled data by sacrificing the original 2.5-kHz signal. Program 12.1 gives the detail of MATLAB implementation.

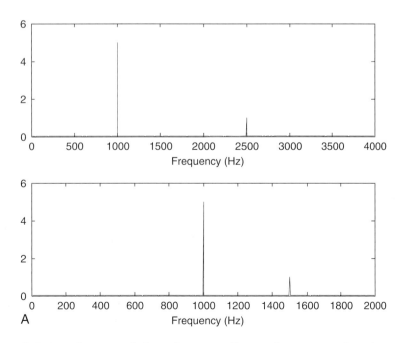

FIGURE 12.3A Spectrum before downsampling and spectrum after downsampling without using the anti-aliasing filter.

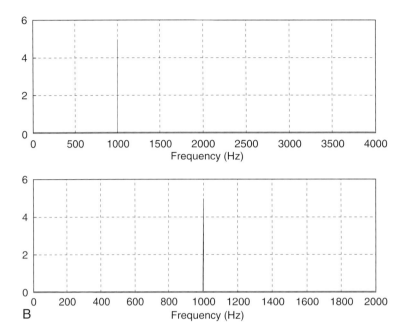

FIGURE 12.3B Spectrum before downsampling and spectrum after downsampling using the anti-aliasing filter.

Program 12.1. MATLAB program for decimation.

```
close all;clear all;
% Downsampling filter (see Chapter 7 for FIR filter design)
B=[0.00074961181416   0.00247663033476   0.00146938649416  -0.00440446121505...
  -0.00910635730662   0.00000000000000   0.02035676831506   0.02233710562885...
  -0.01712963672810  -0.06376620649567  -0.03590670035210   0.10660980550088...
  -0.29014909103794   0.37500000000000   0.29014909103794   0.10660980550088...
  -0.03590670035210  -0.06376620649567  -0.01712963672810   0.02233710562885...
  -0.02035676831506   0.00000000000000  -0.00910635730662  -0.00440446121505...
  -0.00146938649416   0.00247663033476   0.00074961181416];
% Generate the 2048 samples
fs=8000;         % Sampling rate
N=2048;          % Number of samples
M=2;             % Downsample factor
n=0:1:N-1;
x = 5*sin(n*pi/4)+cos(5*n*pi/8);
% Compute the single-sided amplitude spectrum
% AC component will be doubled, and DC component will be kept the same value
X = 2*abs(fft(x,N))/N;X(1)=X(1)/2;
```

```
% Map the frequency index up to the folding frequency in Hz
f=[0:1:N/2-1]*fs/N;
%Downsampling
y=x(1:M:N);
NM=length(y);         % Length of the downsampled data
% Compute the single-sided amplitude spectrum for the downsampled signal
Y = 2*abs(fft(y,NM))/length(y);Y(1)=Y(1)/2;
% Map the frequency index to the frequency in Hz
fsM=[0:1:NM/2-1]*(fs/M)/NM;
subplot(2,1,1);plot(f,X(1:1:N/2));grid;xlabel('Frequency (Hz)');
subplot(2,1,2);plot(fsM,Y(1:1:NM/2));grid;xlabel('Frequency (Hz)');
figure
w=filter(B,1,x);      % Anti-aliasing filtering
% Compute the single-sided amplitude spectrum for the filtered signal
W = 2*abs(fft(w,N))/N;W(1)=W(1)/2;
% Downsampling
y=w(1:M:N);
NM=length(y);
% Compute the single-sided amplitude spectrum for the downsampled signal
Y = 2*abs(fft(y,NM))/NM;Y(1)=Y(1)/2;
% plot spectra
subplot(2,1,1);plot(f,W(1:1:N/2));grid;xlabel('Frequency (Hz)');
subplot(2,1,2);plot(fsM,Y(1:1:NM/2));grid;xlabel('Frequency (Hz)');
```

Now we focus on how to design the anti-aliasing FIR filter, or decimation filter. We will discuss this topic via the following example.

Example 12.1.

Given a DSP downsampling system with the following specifications:

Sampling rate = 6,000 Hz
Input audio frequency range = 0–800 Hz
Passband ripple = 0.02 dB
Stopband attenuation = 50 dB
Downsample factor $M = 3$,

a. Determine the FIR filter length, cutoff frequency, and window type if the window method is used.

Solution:

a. Specifications are reorganized as:

Anti-aliasing filter operating at the sampling rate = 6000 Hz
Passband frequency range = 0–800 Hz

Stopband frequency range = 1–3 kHz

Passband ripple = 0.02 dB

Stopband attenuation = 50 dB

Filter type = FIR.

The block diagram and specifications are depicted in Figure 12.4. The Hamming window is selected, since it provides 0.019 dB ripple and 53 dB stopband attenuation. The normalized transition band is given by

$$\Delta f = \frac{f_{stop} - f_{pass}}{f_s} = \frac{1000 - 800}{6000} = 0.033.$$

The length of the filter and the cutoff frequency can be determined by

$$N = \frac{3.3}{\Delta f} = \frac{3.3}{0.033} = 100.$$

We choose the odd number, that is, $N = 101$, and

$$f_c = \frac{f_{pass} + f_{stop}}{2} = \frac{800 + 1000}{2} = 900 \text{ Hz}.$$

12.1.2 Sampling Rate Increase by an Integer Factor

Increasing a sampling rate is a process of upsampling by an integer factor of L. This process is described as follows:

$$y(m) = \begin{cases} x\left(\frac{m}{L}\right) & m = nL \\ 0 & otherwise \end{cases}, \quad (12.9)$$

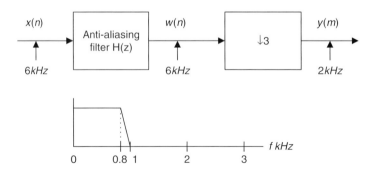

FIGURE 12.4 Filter specifications for Example 12.1.

where $n = 0, 1, 2, \ldots$, $x(n)$ is the sequence to be upsampled by a factor of L, and $y(m)$ is the upsampled sequence. As an example, suppose that the data sequence is given as follows:

$$x(n): 8 \quad 8 \quad 4 \quad -5 \quad -6 \ldots.$$

After upsampling the data sequence $x(n)$ by a factor of 3 (adding $L - 1$ zeros for each sample), we have the upsampled data sequence $w(m)$ as:

$$w(m): 8\ 0\ 0 \quad 8\ 0\ 0 \quad 4\ 0\ 0 \quad -5\ 0\ 0 \quad -6\ 0\ 0 \ldots.$$

The next step is to smooth the upsampled data sequence via an interpolation filter. The process is illustrated in Figure 12.5a.

Similar to the downsampling case, assuming that the data sequence has the current sampling period of T, the Nyquist frequency is given by $f_{max} = f_s/2$. After upsampling by a factor of L, the new sampling period becomes T/L, thus the new sampling frequency is changed to be

$$f_{sL} = Lf_s. \tag{12.10}$$

This indicates that after upsampling, the spectral replicas originally centered at $\pm f_s, \pm 2f_s, \ldots$ are included in the frequency range from 0 Hz to the new Nyquist limit $Lf_s/2$ Hz, as shown in Figure 12.5b. To remove those included spectral replicas, an interpolation filter with a stop frequency edge of $f_s/2$ in Hz must be attached, and the normalized stop frequency edge is given by

$$\Omega_{stop} = 2\pi \left(\frac{f_s}{2}\right) \times \left(\frac{T}{L}\right) = \frac{\pi}{L} \text{ radians}. \tag{12.11}$$

After filtering via the interpolation filter, we will achieve the desired spectrum for $y(n)$, as shown in Figure 12.5b. Note that since the interpolation is to remove the high-frequency images that are aliased by the upsampling operation, it is essentially an anti-aliasing lowpass filter.

To verify the upsampling principle, we generate the signal $x(n)$ with 1 kHz and 2.5 kHz as follows:

$$x(n) = 5 \sin\left(\frac{2\pi \times 1000n}{8000}\right) + \cos\left(\frac{2\pi \times 2500n}{8000}\right),$$

with a sampling rate of $f_s = 8{,}000$ Hz. The spectrum of $x(n)$ is plotted in Figure 12.6. Now we upsample $x(n)$ by a factor of 3, that is, $L = 3$. We know that the sampling rate is increased to be $3 \times 8000 = 24{,}000$ Hz. Hence, without using the interpolation filter, the spectrum would contain the image frequencies originally centered at the multiple frequencies of 8 kHz. The top plot in Figure 12.6 shows the spectrum for the sequence after upsampling and before applying the interpolation filter.

Now we apply an FIR lowpass filter designed with a length of 53, a cutoff frequency of 3,250 Hz, and a new sampling rate of 24,000 Hz as the interpolation filter, whose normalized frequency should be

$$\Omega_c = 2\pi \times 3250 \times \left(\frac{1}{24000}\right) = 0.2708\pi.$$

The bottom plot in Figure 12.6 shows the spectrum for $y(m)$ after applying the interpolation filter, where only the original signals with frequencies of 1 kHz and 2.5 kHz are presented. Program 12.2 shows the implementation detail in MATLAB.

Now let us study how to design the interpolation filter via Example 12.2.

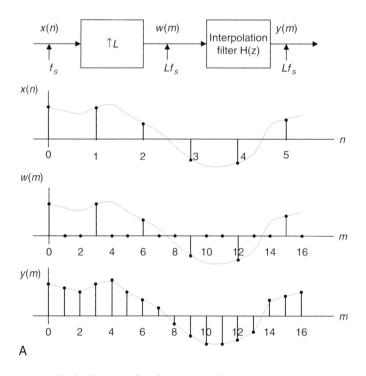

FIGURE 12.5A Block diagram for the upsampling process with $L = 3$.

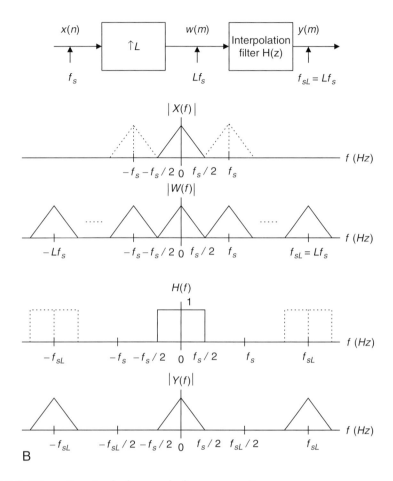

FIGURE 12.5B Spectra before and after upsampling.

Example 12.2.

Given a DSP upsampling system with the following specifications:

Sampling rate = 6,000 Hz
Input audio frequency range = 0–800 Hz
Passband ripple = 0.02 dB
Stopband attenuation = 50 dB
Upsample factor $L = 3$,

    a. Determine the FIR filter length, cutoff frequency, and window type if the window design method is used.

(Continued on page 569)

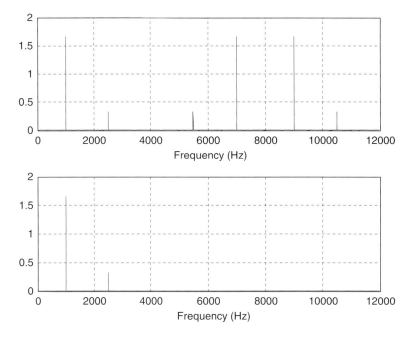

FIGURE 12.6 (*Top*) The spectrum after upsampling and before applying the interpolation filter; (*bottom*) the spectrum after applying the interpolation filter.

Program 12.2. MATLAB program for interpolation.

```
close all; clear all
%Upsampling filter (see Chapter 7 for FIR filter design)
B=[-0.00012783931504   0.00069976044649   0.00123831516738   0.00100277549136...
   -0.00025059018468  -0.00203448515158  -0.00300830295487  -0.00174101657599...
    0.00188598835011   0.00578414933758   0.00649330625041   0.00177982369523...
   -0.00670672686935  -0.01319379342716  -0.01116855281442   0.00123034314117...
    0.01775600060894   0.02614700427364   0.01594155162392  -0.01235169936557...
   -0.04334322148505  -0.05244745563466  -0.01951094855292   0.05718573279009...
    0.15568416401644   0.23851539047347   0.27083333333333   0.23851539047347...
    0.15568416401644   0.05718573279009  -0.01951094855292  -0.05244745563466...
   -0.04334322148505  -0.01235169936557   0.01594155162392   0.02614700427364...
    0.01775600060894   0.00123034314117  -0.01116855281442  -0.01319379342716...
   -0.00670672686935   0.00177982369523   0.00649330625041   0.00578414933758...
    0.00188598835011  -0.00174101657599  -0.00300830295487  -0.00203448515158...
   -0.00025059018468   0.00100277549136   0.00123831516738   0.00069976044649...
   -0.00012783931504];
% Generate the 2048 samples with fs = 8000 Hz
fs = 8000;          % Sampling rate
```

```
N = 2048;          % Number of samples
L = 3;             % Upsampling factor
n=0:1:N-1;
x = 5*sin(n*pi/4) + cos(5*n*pi/8);
% Upsampling by a factor of L
w=zeros(1,L*N);
for n=0:1:N-1
w(L*n + 1) = x(n + 1);
end
NL = length(w);    % Length of the upsampled data
W = 2*abs(fft(w,NL))/NL;W(1)=W(1)/2; %Compute the one-sided amplitude spectrum
f=[0:1:NL/2-1]*fs*L/NL; % Map the frequency index to the frequency (Hz)
%Interpolation
y=filter(B,1,w);   % Apply the interpolation filter
Y = 2*abs(fft(y,NL))/NL;Y(1)=Y(1)/2; %Compute the one-sided amplitude spectrum
fsL=[0:1:NL/2-1]*fs*L/NL; % Map the frequency index to the frequency (Hz)
subplot(2,1,1);plot(f,W(1:1:NL/2));grid;xlabel('Frequency (Hz)');
subplot(2,1,2);plot(fsL,Y(1:1:NL/2));grid;xlabel('Frequency (Hz)');
```

Solution:

a. The specifications are reorganized as follows:

Interpolation filter operating at the sampling rate = 18,000 Hz
Passband frequency range = 0–800 Hz
Stopband frequency range = 3–9 kHz
Passband ripple = 0.02 dB
Stopband attenuation = 50 dB
Filter type = FIR.

The block diagram and filter frequency specifications are given in Figure 12.7.

We choose the Hamming window for this application. The normalized transition band is

$$\Delta f = \frac{f_{stop} - f_{pass}}{f_{sL}} = \frac{3000 - 800}{18000} = 0.1222.$$

The length of the filter and the cutoff frequency can be determined by

$$N = \frac{3.3}{\Delta f} = \frac{3.3}{0.1222} = 27,$$

and the cutoff frequency is given by

$$f_c = \frac{f_{pass} + f_{stop}}{2} = \frac{3000 + 800}{2} = 1900 \text{ Hz}.$$

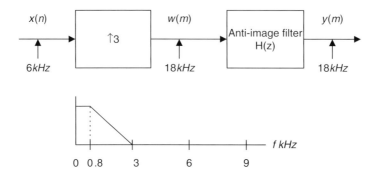

FIGURE 12.7 Filter frequency specifications for Example 12.2.

12.1.3 Changing Sampling Rate by a Non-Integer Factor L/M

With an understanding of the downsampling and upsampling processes, we now study the sampling rate conversion by a non-integer factor of L/M. This can be viewed as two sampling conversion processes. In step 1, we perform the upsampling process by a factor of integer L following application of an interpolation filter $H_1(z)$; in step 2, we continue filtering the output from the interpolation filter via an anti-aliasing filter $H_2(z)$, and finally operate downsampling. The entire process is illustrated in Figure 12.8.

Since the interpolation and anti-aliasing filters are in a cascaded form and operate at the same rate, we can select one of them. We choose the one with the lower stop frequency edge and choose the most demanding requirement for passband gain and stopband attenuation for the filter design. A lot of computational saving can be achieved by using one lowpass filter. We illustrate the procedure via the following simulation. Let us generate the signal $x(n)$ by:

$$x(n) = 5\sin\left(\frac{2\pi \times 1000n}{8000}\right) + \cos\left(\frac{2\pi \times 2500n}{8000}\right),$$

with a sampling rate of $f_s = 8,000$ Hz and frequencies of 1 kHz and 2.5 kHz. Now we resample $x(n)$ to 3,000 Hz by a non-integer factor of 0.375, that is,

$$\left(\frac{L}{M}\right) = 0.375 = \frac{3}{8}.$$

Upsampling is at a factor of $L = 3$ and the upsampled sequence is filtered by an FIR lowpass filter designed with the filter length $N = 53$ and a cutoff frequency of 3,250 Hz at the sampling rate of $3 \times 8000 = 24,000$ Hz. The spectrum for the upsampled sequence and the spectrum after application of the interpolation filter are plotted in Figure 12.9a.

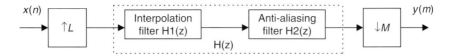

FIGURE 12.8 Block diagram for sampling rate conversion.

The sequence from step 1 can be filtered via another FIR lowpass filter designed with the filter length $N = 159$ and a cutoff frequency of 1,250 Hz, followed by downsampling by a factor of $M = 8$. The spectrum after the anti-aliasing filter and the spectrum for the final output $y(m)$ are plotted in Figure 12.9b. Note that the anti-aliasing filter removes the frequency component of 2.5 kHz to avoid aliasing. This is because after downsampling, the Nyquist limit is 1.5 kHz. As we discussed previously, we can select one filter for implementation. We choose the FIR lowpass with $N = 159$ and a cutoff frequency of 1,250 Hz because its bandwidth is smaller than that of the interpolation filter. The MATLAB implementation is listed in Program 12.3.

Program 12.3. MATLAB program for changing sampling rate with a non-integer factor.

```
close all; clear all;clc;
% Downsampling filter
Bdown=firwd(159,1,2*pi*1250/24000,0,4);
% Generate the 2048 samples with fs = 8000 Hz
fs = 8000;           % Original sampling rate
N = 2048;            % The number of samples
L = 3;        % Upsampling factor
M = 8;        % Downsampling factor
n = 0:1:N-1;         % Generate the time index
x = 5*sin(n*pi/4) +cos (5*n*pi/8);      % Generate the test signal
% upsampling by a factor of L
w1=zeros(1,L*N);
for n = 0:1:N-1
w1(L*n + 1) = x(n + 1);
end
NL= length(w1);          % Length of the upsampled data
W1 = 2*abs(fft(w1,NL))/NL;W1(1) = W1(1)/2; % Compute the one-sided amplitude spectrum
f = [0:1:NL/2-1]*fs*L/NL;         % Map frequency index to its frequency in Hz
subplot(3,1,1);plot(f,W1(1:1:NL/2));grid
xlabel('Frequency (Hz)');
w2 = filter(Bdown,1,w1); % Perform the combined anti-aliasing filter
```

(Continued)

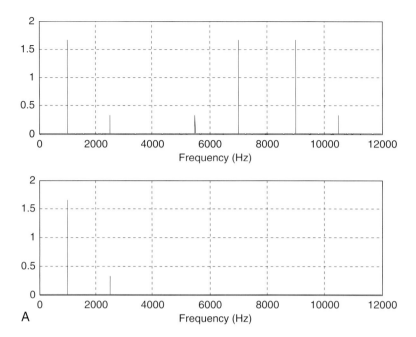

FIGURE 12.9A *(Top)* Spectrum after upsampling and *(bottom)* spectrum after interpolation filtering.

```
W2 = 2*abs(fft(w2,NL))/NL;W2(1) = W2(1)/2; % Compute the one-sided amplitude spectrum
y2=w2(1:M:NL);
NM=length(y2); % Length of the downsampled data
Y2 = 2*abs(fft(y2,NM))/NM;Y2(1) = Y2(1)/2;% Compute the one-sided amplitude spectrum
% Map frequency index to its frequency in Hz before downsampling
fbar = [0:1:NL/2-1]*24000/NL;
% Map frequency index to its frequency in Hz
fsM = [0:1:NM/2-1]*(fs*L/M)/NM;
subplot(3,1,2);plot(f,W2(1:1:NL/2));grid;xlabel('Frequency (Hz)');
subplot(3,1,3);plot(fsM,Y2(1:1:NM/2));grid;xlabel('Frequency (Hz)');
```

Therefore, three steps are required to accomplish the process:

1. Upsampling by a factor of $L = 3$

2. Filtering the upsampled sequence by an FIR lowpass filter designed with the filter length $N = 159$ and a cutoff frequency of 1,250 Hz at the sampling rate of $3 \times 8000 = 24,000$ Hz

3. Downsampling by a factor of $M = 8$.

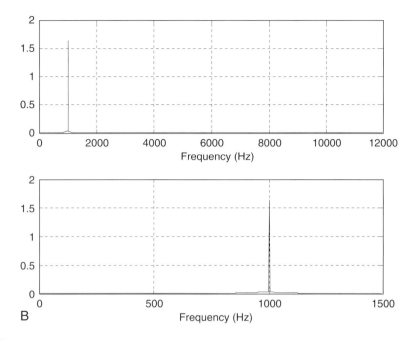

FIGURE 12.9B (*Top*) Spectrum after anti-aliasing filtering and (*bottom*) spectrum after downsampling.

Example 12.3.

Given a sampling conversion DSP system with the following specifications:

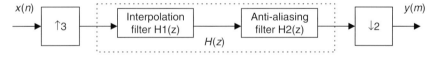

Audio input $x(n)$ is sampled at the rate of 6,000 Hz,

A Audio output $y(m)$ is operated at the rate of 9,000 Hz.

FIGURE 12.10A Sampling conversion in Example 12.3.

a. Determine the filter length and cutoff frequency for the combined anti-aliasing filter $H(z)$, and window types, respectively, if the window design method is used.

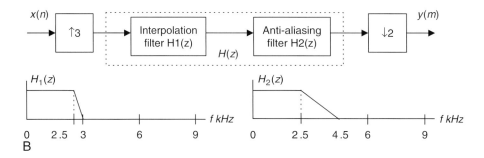

FIGURE 12.10B Filter frequency specifications for Example 12.3.

Solution:

a. The filter frequency specifications and corresponding block diagram are developed in Figure 12.10b.
   Specifications for the interpolation filter $H_1(z)$:

   Passband frequency range = 0–2500 Hz
   Passband ripples for $H_1(z) = 0.04$ dB
   Stopband frequency range = 3000–9000 Hz
   Stopband attenuation = 42 dB

   Specifications for the anti-aliasing filter $H_2(z)$:

   Passband frequency range = 0–2500 Hz
   Passband ripples for $H_2(z) = 0.02$ dB
   Stopband frequency range = 4500–9000 Hz
   Stopband attenuation = 46 dB

   Combined specifications $H(z)$:

   Passband frequency range = 0–2500 Hz
   Passband ripples for $H(z) = 0.02$ dB
   Stopband frequency range = 3000–9000 Hz
   Stopband attenuation = 46 dB.

   We use the FIR filter with the Hamming window. Since

   $$\Delta f = \frac{f_{stop} - f_{pass}}{f_{sL}} = \frac{3000 - 2500}{18000} = 0.0278,$$

   the length of the filter and the cutoff frequency can be determined by

$$N = \frac{3.3}{\Delta f} = \frac{3.3}{0.0278} = 118.8.$$

We choose $N = 119$, and

$$f_c = \frac{f_{pass} + f_{stop}}{2} = \frac{3000 + 2500}{2} = 2750 \text{ Hz}.$$

12.1.4 Application: CD Audio Player

In this application example, we will discuss principles of the upsampling and interpolation-filter processes used in the CD audio system to help the reconstruction filter design.

Each raw digital sample recorded on the CD audio system contains 16 bits and is sampled at the rate of 44.1 kHz. Figure 12.11 describes a portion of one channel of the CD player in terms of a simplified block diagram.

Let us consider the situation without upsampling and application of a digital interpolation filter. We know that the audio signal has a bandwidth of 22.05 kHz, that is, the Nyquist frequency, and digital-to-analog conversion (DAC) produces the sample-and-hold signals that contain the desired audio band and images thereof. To achieve the audio band signal, we need to apply a *reconstruction filter* (also called a smooth filter or anti-image filter) to remove all image frequencies beyond the Nyquist frequency of 22.05 kHz. Due to the requirement of the sharp transition band, a higher-order analog filter design becomes a requirement.

The design of the higher-order analog filter is complex and expensive to implement. In order to relieve such design constraints, as shown in Figure 12.11, we can add the upsampling process before DAC, followed by application of the digital interpolation filter (assume $L = 4$). Certainly, the interpolation filter design must satisfy the specifications studied in the previous section on increasing the sampling rate by an integer factor. Again, after digital interpolation, the audio band is kept the same, while the sampling frequency is increased fourfold ($L = 4$), that is, $44.1 \times 4 = 176.4$ kHz.

Since the audio band of 22.05 kHz is now relatively low compared with the new folding frequency ($176.4/2 = 88.2$ kHz), the use of a simple first-order or second-order analog anti-image filter may be sufficient. Let us look at the following simulation.

A test audio signal with a frequency of 16 kHz and a sampling rate of 44.1 kHz is generated using the formula

$$x(n) = \sin\left(\frac{2\pi \times 16000 n}{441000}\right).$$

If we use an upsampling factor of 4, then the bandwidth would increase to 88.2 kHz. Based on the audio frequency of 16 kHz, the original Nyquist

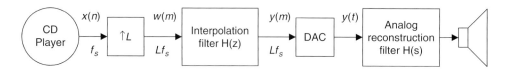

FIGURE 12.11 Sample rate conversion in the CD audio player system.

frequency of 22.05 kHz, and the new sampling rate of 176.4 kHz, we can determine the filter length as

$$\Delta f = \frac{22.05 - 16}{176.4} = 0.0343.$$

Using the Hamming window for FIR filter design leads to

$$N = \frac{3.3}{\Delta f} = 96.2.$$

We choose $N = 97$. The cutoff frequency therefore is

$$f_c = \frac{16 + 22.05}{2} = 19.025 \, \text{kHz}.$$

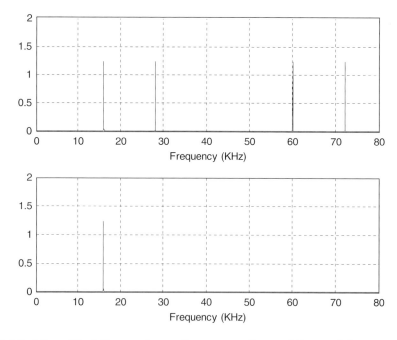

FIGURE 12.12 (*Top*) The spectrum after upsampling and (*bottom*) the spectrum after applying the interpolation filter.

The spectrum of the interpolated audio test signal is shown in Figure 12.12, where the top plot illustrates that after upsampling, the audio test signal has the frequency of 16 kHz, along with image frequencies coming from 44.1 kHz − 16 = 28.1 kHz, 44.1 + 16 = 60.1 kHz, 88.2 − 16 = 72.2 kHz, and so on. The bottom graph describes the spectrum after the interpolation filter. From lowpass FIR filtering, the interpolated audio signal with a frequency of 16 kHz is observed.

Let us examine the corresponding process in time domain, as shown in Figure 12.13. The upper left plot shows the original samples. The upper right plot describes the upsampled signals. The lower left plot shows the signals after the upsampling process and digital interpolation filtering. Finally, the lower right plot shows the sample-and-hold signals after DAC. Clearly, we can easily design a reconstruction filter to smooth the sample-and-hold signals and obtain the original audio test signal. The advantage of reducing hardware is illustrated. The MATLAB implementation can be seen in Program 12.4.

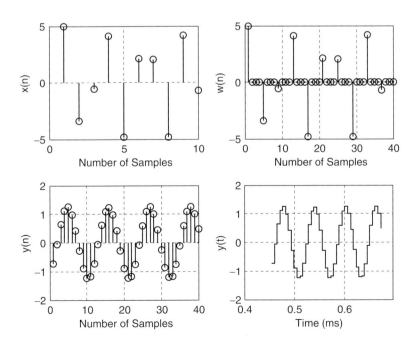

FIGURE 12.13 Plots of signals at each stage according to the block diagram in Figure 12.11.

Program 12.4. MATLAB program for CD player example.

```
close all; clear all; clc
% Generate the 2048 samples with fs = 44100 Hz
fs = 44100;          % Original sampling rate
T = 1/fs;            % Sampling period
N = 2048;            % Number of samples
L = 4;
fsL = fs*L;          % Upsampling rate
%Upsampling filter (see Chapter 7 for FIR filter design)
Bup = firwd(97,1,2*19025*pi/fsL,0,4);
n = 0:1:N-1;         % Generate the time indices
x = 5*sin(2*pi*16000*n*T);        % Generate the test signal
% Upsampling by a factor of L
w = zeros(1,L*N);
for n = 0:1:N-1
w(L*n + 1) = x(n + 1);
end
NL=length(w);        % Number of the upsampled data
W = 2*abs(fft(w,NL))/NL;W(1)=W(1)/2; % Compute the one-sided amplitude spectrum
f = [0:1:NL/2-1]*fs*L/NL; % Map the frequency index to its frequency in Hz
f=f/1000;            % Convert to kHz
%Interpolation
y=filter(Bup,1,w);             % Perform the interpolation filter
Y=2*abs(fft(y,NL))/NL;Y(1)=Y(1)/2; % Compute the one-sided amplitude spectrum
subplot(2,1,1);plot(f,W(1:1:NL/2));grid;
xlabel('Frequency (kHz)'); axis([0 f(length(f)) 0 2]);
subplot(2,1,2);plot(f,Y(1:1:NL/2));grid;
xlabel('Frequency (kHz)');axis([0 f(length(f)) 0 2]);
figure
subplot(2,2,1);stem(x(21:30));grid
xlabel('Number of Samples');ylabel('x(n)');
subplot(2,2,2);stem(w(81:120));grid
xlabel('Number of Samples'); ylabel('w(n)');
subplot(2,2,3);stem(y(81:120));grid
xlabel('Number of Samples'); ylabel('y(n)')
subplot(2,2,4);stairs([80:1:119]*1000*T,y(81:120));grid
xlabel('Time (ms)'); ylabel('y(t)')
```

12.1.5 Multistage Decimation

The multistage approach for downsampling rate conversion can be used to dramatically reduce the anti-aliasing filter length. Figure 12.14 describes a two-stage decimator.

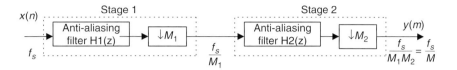

FIGURE 12.14 Multistage decimation.

As shown in Figure 12.14, a total decimation factor is $M = M_1 \times M_2$. Here, even though we develop a procedure for a two-stage case, a similar principle can be applied to general multistage cases.

Using the two-stage decimation in Figure 12.15, the final Nyquist limit is $f_s/2M$ after final downsampling. So our useful information bandwidth should stop at the frequency edge of $f_s/2M$. Next, we need to determine the stop frequency edge for the anti-aliasing lowpass filter at stage 1 before the first decimation process begins. This stop frequency edge is actually the lower frequency edge of the first image replica centered at the sampling frequency of f_s/M_1 after the stage 1 decimation. This lower frequency edge of the first image replica is then determined by

$$\frac{f_s}{M_1} - \frac{f_s}{2M},$$

After downsampling, we expect that some frequency components from $\frac{f_s}{M_1} - \frac{f_s}{2M}$ to $\frac{f_s}{M_1}$ to be folded over to the frequency band between $\frac{f_s}{2M}$ and $\frac{f_s}{2M_1}$. However, these aliased frequency components do not affect the final useful band between 0 Hz to $\frac{f_s}{2M}$ and will be removed by the anti-aliasing filter(s) in the future stage(s). As illustrated in Figure 12.15, any frequency components beyond the edge $\frac{f_s}{M_1} - \frac{f_s}{2M}$ can fold over into the final useful information band to create aliasing distortion. Therefore, we can use this frequency as the lower stop frequency edge of the anti-aliasing filter to prevent the aliasing distortion at the final stage. The upper stopband edge (Nyquist limit) for the anti-image filter at stage 1 is clearly $\frac{f_s}{2}$, since the filter operates at f_s samples per second. So the stopband frequency range at stage 1 is from $\frac{f_s}{M_1} - \frac{f_s}{2M}$ to $\frac{f_s}{2}$. The aliasing distortion, introduced into the frequency band from $\frac{f_s}{2M}$ to $\frac{f_s}{2M_1}$, will be filtered out after future decimation stage(s).

Similarly, for stage 2, the lower frequency edge of the first image developed after stage 2 downsampling is

$$\frac{f_s}{M_1 M_2} - \frac{f_s}{2M} = \frac{f_s}{2M}.$$

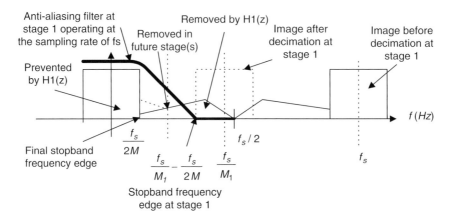

FIGURE 12.15 Stopband frequency edge for the anti-aliasing filter at stage 1 for two-stage decimation.

As is evident in our two-stage scheme, the stopband frequency range for the second anti-aliasing filter at stage 2 should be from $\frac{f_s}{2M}$ to $\frac{f_s}{2M_1}$.

We summarize specifications for the two-stage decimation as follows:

Filter requirements for stage 1:

- Passband frequency range $= 0$ to f_p
- Stopband frequency range $= \frac{f_s}{M_1} - \frac{f_s}{2M}$ to $\frac{f_s}{2}$
- Passband ripple $= \delta_p/2$, where δ_p is the combined absolute ripple on passband
- Stopband attenuation $= \delta_s$

Filter requirements for stage 2:

- Passband frequency range $= 0$ to f_p
- Stopband frequency range $= \frac{f_s}{M_1 \times M_2} - \frac{f_s}{2M}$ to $\frac{f_s}{2M_1}$
- Passband ripple $= \delta_p/2$, where δ_p is the combined absolute ripple on passband
- Stopband attenuation $= \delta_s$.

Example 12.4 illustrates the two-stage decimator design.

Example 12.4.

a. Determine the anti-aliasing FIR filter lengths and cutoff frequencies for the two-stage decimator with the following specifications and block diagram:

> Original sampling rate: $f_s = 240 \text{ kHz}$
> Audio frequency range: 0–3,400 Hz
> Passband ripple: $\delta_p = 0.05$ (absolute)
> Stopband attenuation: $\delta_s = 0.005$ (absolute)
> FIR filter design using the window method
> New sampling rate: $f_{sM} = 8 \text{ kHz}$

Solution:

a. $M = \frac{240 \, kHz}{8 \, kHz} = 30 = 10 \times 3$

We choose $M_1 = 10$ and $M_2 = 3$; there could be other choices. Figure 12.16b shows the block diagram and filter frequency specifications.

Filter specification for $H_1(z)$:

> Passband frequency range: 0–3,400 Hz
> Passband ripples: $0.05/2 = 0.025$ ($\delta_s \, dB = 20 log_{10}(1 + \delta_p) = 0.212 \, dB$)
> Stopband frequency range: 20,000–120,000 Hz
> Stopband attenuation: 0.005, $\delta_s \, dB = -20 \times \log_{10}(\delta_s) = 46 \, dB$
> Filter type: FIR, Hamming window.

Note that the lower stopband edge can be determined as

$$f_{stop} = \frac{f_s}{M_1} - \frac{f_s}{2 \times M} = \frac{240000}{10} - \frac{240000}{2 \times 30} = 20000 \, Hz$$

$$\Delta f = \frac{f_{stop} - f_{pass}}{f_s} = \frac{20000 - 3400}{240000} = 0.06917.$$

The length of the filter and the cutoff frequency can be determined by

$$N = \frac{3.3}{\Delta f} = 47.7.$$

We choose $N = 49$, and

$$f_c = \frac{f_{pass} + f_{stop}}{2} = \frac{20000 + 3400}{2} = 11700 \, Hz.$$

Filter specification for $H_2(z)$:

> Passband frequency range: 0–3,400 Hz
> Passband ripples: $0.05/2 = 0.025$ (0.212 dB)
> Stopband frequency range: 4,000–12,000 Hz

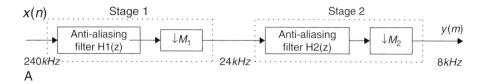

FIGURE 12.16A Multistage decimation in Example 12.4.

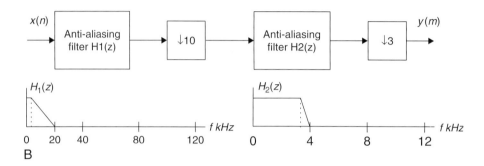

FIGURE 12.16B Filter frequency specifications for Example 12.4.

Stopband attenuation: 0.005, $\delta_s\ dB = 46\ dB$
Filter type: FIR, Hamming window

Note that

$$\Delta f = \frac{f_{stop} - f_{pass}}{f_{sM1}} = \frac{4000 - 3400}{24000} = 0.025.$$

The length of the filter and the cutoff frequency can be determined by

$$N = \frac{3.3}{\Delta f} = 132.$$

We choose $N = 133$, and

$$f_c = \frac{f_{pass} + f_{stop}}{2} = \frac{4000 + 3400}{2} = 3700\ Hz.$$

The reader can verify for the case by using only one stage with a decimation factor of $M = 30$. Using the Hamming window for the FIR filter, the resulting number of taps is 1,321, and the cutoff frequency is 3,700 Hz. Thus, such a filter requires a huge number of computations and causes a large delay during implementation compared with the two-stage case.

The multistage scheme is very helpful for sampling rate conversion between audio systems. For example, to convert the CD audio at the sampling rate of 44.1 kHz to the MP3 or digital audio type system (professional audio rate), in which the sampling rate of 48 kHz is used, the conversion factor $L/M = 48/44.1 = 160/147$ is required. Using the single-stage scheme may cause impractical FIR filter sizes for interpolation and downsampling. However, since $L/M = 160/147 = (4/3)(8/7)(5/7)$, we may design an efficient three-stage system, in which stages 1, 2, and 3 use the conversion factors $L/M = 8/7, L/M = 5/7$, and $L/M = 4/3$, respectively.

12.2 Polyphase Filter Structure and Implementation

Due to the nature of the decimation and interpolation processes, polyphase filter structures can be developed to efficiently implement the decimation and interpolation filters (using fewer numbers of multiplications and additions). As we will explain, these filters are all-pass filters with different phase shifts (Proakis and Manolakis, 1996), thus we call them *polyphase filters*.

Here, we skip their derivations and illustrate implementations of decimation and interpolation using simple examples. Consider the interpolation process shown in Figure 12.17, where $L = 2$. We assume that the FIR interpolation filter has four taps, shown as:

$$H(z) = h(0) + h(1)z^{-1} + h(2)z^{-2} + h(3)z^{-3}$$

and the filter output is

$$y(m) = h(0)w(m) + h(1)w(m-1) + h(2)w(m-2) + h(3)w(m-3).$$

For the purpose of comparison, the direct interpolation process shown in Figure 12.17 is summarized in Table 12.1, where $w(m)$ is the upsampled signal and $y(m)$ the interpolated output. Processing each input sample $x(n)$ requires applying the difference equation twice to obtain $y(0)$ and $y(1)$. Hence, for this example, we need eight multiplications and six additions.

The output results in Table 12.1 can be easily obtained by using the polyphase filters shown in Figure 12.18.

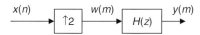

FIGURE 12.17 Upsampling by a factor of 2 and a four-tap interpolation filter.

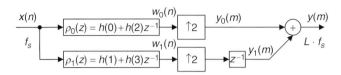

FIGURE 12.18 Polyphase filter implementation for the interpolation in Figure 12.17 (4 multiplications and 4 additions for processing each input sample $x(n)$).

In general, there are L polyphase filters. Having the designed interpolation filter $H(z)$ of the N taps, we can determine each bank of filter coefficients as follows:

$$\rho_k(n) = h(k + nL) \quad \text{for } k = 0, 1, \ldots, L-1 \text{ and } n = 0, 1, \ldots, \frac{N}{L} - 1. \quad (12.12)$$

For our example $L = 2$ and $N = 4$, we have $L - 1 = 1$, and $N/L - 1 = 1$, respectively. Hence, there are two filter banks, $\rho_0(z)$ and $\rho_1(z)$, each having a length of 2, as illustrated in Figure 12.18. When $k = 0$ and $n = 1$, the upper limit of time index required for $h(k + nL)$ is $k + nL = 0 + 1 \times 2 = 2$. When $k = 1$ and $n = 1$, the upper limit of the time index for $h(k + nL)$ is 3. Hence, the first filter $\rho_0(z)$ has the coefficients $h(0)$ and $h(2)$. Similarly, the second filter $\rho_1(z)$ has coefficients $h(1)$ and $h(3)$. In fact, the filter coefficients of $\rho_0(z)$ are a decimated version of $h(n)$ starting at $k = 0$, while the filter coefficients of $\rho_1(z)$ are a decimated version of $h(n)$ starting at $k = 1$, and so on.

As shown in Figure 12.18, we can reduce the computational complexity from eight multiplications and six additions down to four multiplications and four additions for processing each input sample $x(n)$. Generally, the computation can be reduced by a factor of L as compared with the direct process.

The commutative model for the polyphase interpolation filter is given in Figure 12.19.

TABLE 12.1 Results of the direct interpolation process in Figure 12.17 (8 multiplications and 6 additions for processing each input sample $x(n)$).

n	$x(n)$	m	$w(m)$	$y(m)$
$n = 0$	$x(0)$	$m = 0$	$w(0) = x(0)$	$y(0) = h(0)x(0)$
		$m = 1$	$w(1) = 0$	$y(1) = h(1)x(0)$
$n = 1$	$x(1)$	$m = 2$	$w(2) = x(1)$	$y(2) = h(0)x(1) + h(2)x(0)$
		$m = 3$	$w(3) = 0$	$y(3) = h(1)x(1) + h(3)x(0)$
$n = 2$	$x(2)$	$m = 4$	$w(4) = x(2)$	$y(4) = h(0)x(2) + h(2)x(1)$
		$m = 5$	$w(5) = 0$	$y(5) = h(1)x(2) + h(3)x(1)$
...	...	...	...	...

Example 12.5.

a. Verify $y(1)$ in Table 12.1 using the polyphase filter implementation in Figures 12.18 and 12.19, respectively.

Solution:

a. Applying the polyphase interpolation filter as shown in Figure 12.18 leads to

$$w_0(n) = h(0)x(n) + h(2)x(n-1)$$
$$w_1(n) = h(1)x(n) + h(3)x(n-1);$$

when $n = 0$,

$$w_0(0) = h(0)x(0)$$
$$w_1(0) = h(1)x(0).$$

After interpolation, we have

$$y_0(m): w_0(0) \quad 0 \quad \ldots$$

and

$$y_1(m): 0 \quad w_1(0) \quad 0 \quad \ldots$$

Note: there is a unit delay for the second filter bank. Hence

$$m = 0, \ y_0(0) = h(0)x(0), \ y_1(0) = 0$$
$$m = 1, \ y_0(1) = 0, \ y_1(1) = h(1)x(0).$$

Combining two channels, we finally get

$$m = 0, \ y(0) = y_0(0) + y_1(0) = h(0)x(0),$$
$$m = 1, \ y(1) = y_0(1) + y_1(1) = h(1)x(0).$$

Therefore, $y(1)$ matches that in the direct interpolation process given in Table 12.1.

Applying the polyphase interpolation filter using the commutative model in Figure 12.19, we have

$$y_0(n) = h(0)x(n) + h(2)x(n-1)$$
$$y_1(n) = h(1)x(n) + h(3)x(n-1);$$

when $n = 0$,

$$m = 0, \ y(0) = y_0(0) = h(0)x(0) + h(2)x(-1) = h(0)x(0)$$
$$m = 1, \ y(1) = y_1(0) = h(1)x(0) + h(3)x(-1) = h(1)x(0).$$

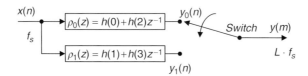

FIGURE 12.19 Commutative model for the polyphase interpolation filter.

Clearly, $y(1) = h(1)x(0)$ matches the $y(1)$ result in Table 12.1.

Next, we will explain the properties of polyphase filters (all-pass gain and possible different phases). Each polyphase filter $\rho_k(n)$ operating at the original sampling rate f_s (assuming 8 kHz) is a downsampled version of the interpolation filter $h(n)$ operating at the upsampling rate Lf_s (32 kHz, assuming an interpolation factor of $L = 4$). Considering that the designed interpolation FIR filter coefficients $h(n)$ are the impulse response sequence having a flat frequency spectrum up to a bandwidth of $f_s/2$ (assume a bandwidth of 4 kHz with a perfect flat frequency magnitude response, theoretically) at the sampling rate of Lf_s (32 kHz), we then downsample $h(n)$ to obtain polyphase filters by a factor of $L = 4$ and operate them at a sampling rate of f_s (8 kHz).

The Nyquist frequency after downsampling should be $(Lf_s/2)/L = f_s/2$ (4 kHz); at the same time, each downsampled sequence $\rho_k(n)$ operating at f_s (8 kHz) has a flat spectrum up to $f_s/2$ (4 kHz) due to the $f_s/2$ (4 kHz) band-limited sequence of $h(n)$ at the sampling rate of f_s (32 kHz). Hence, all of the polyphase filters are all-pass filters. Since each polyphase $\rho_k(n)$ filter has different coefficients, each may have a different phase. Therefore, these polyphase filters are the all-pass filters having possible different phases, theoretically.

Next, consider the following decimation process in Figure 12.20.

FIGURE 12.20 Decimation by a factor of 2 and a three-tap anti-aliasing filter.

Assuming a three-tap decimation filter, we have

$$H(z) = h(0) + h(1)z^{-1} + h(2)z^{-2}$$

$$w(n) = h(0)x(n) + h(1)x(n-1) + h(2)x(n-2).$$

12.2 Polyphase Filter Structure and Implementation

The direct decimation process is shown in Table 12.2 for the purpose of comparison. Obtaining each output $y(m)$ requires processing filter difference equations twice, resulting in six multiplications and four additions for this particular example.

The efficient way to implement a polyphase filter is given in Figure 12.21.

Similarly, there are M polyphase filters. With the designed decimation filter $H(z)$ of the N taps, we can obtain filter bank coefficients by

$$\rho_k(n) = h(k + nM)$$

$$\text{for } k = 0, 1, \ldots, M - 1 \text{ and } n = 0, 1, \ldots, \frac{N}{M} - 1. \tag{12.13}$$

For our example, we see that $M - 1 = 1$ and $N/M - 1 = 1$ (rounded up). Thus we have two filter banks. Since $k = 0$ and $n = 1$, $k + nM = 0 + 1 \times 2 = 2$. The time index upper limit required for $h(k + nM)$ is 2 for the first filter bank $\rho_0(z)$. Hence $\rho_0(z)$ has filter coefficients $h(0)$ and $h(2)$.

However, when $k = 1$ and $n = 1$, $k + nM = 1 + 1 \times 2 = 3$, the time index upper limit required for $h(k + nM)$ is 3 for the second filter bank, and the corresponding filter coefficients are required to be $h(1)$ and $h(3)$. Since our direct decimation filter $h(n)$ does not contain the coefficient $h(3)$, we set $h(3) = 0$ to get the second filter bank with one tap only, as shown in Figure 12.21. Also as shown in that figure, achieving each $y(m)$ needs three multiplications and one addition. In general, the number of multiplications can be reduced by a factor of M.

The commutative model for the polyphase decimator is shown in Figure 12.22.

TABLE 12.2 Results of direct decimation process in Figure 12.20 (6 multiplications and 4 additions for obtaining each output $y(m)$).

n	$x(n)$	$w(n)$	m	$y(m)$
$n = 0$	$x(0)$	$w(0) = h(0)x(0)$	$m = 0$	$y(0) = h(0)x(0)$
$n = 1$	$x(1)$	$w(1) = h(0)x(1) + h(1)x(0)$ discard		
$n = 2$	$x(2)$	$w(2) = h(0)x(2) + h(1)x(1) + h(2)x(0)$	$m = 1$	$y(1) = h(0)x(2) +$
$n = 3$	$x(3)$	$w(3) = h(0)x(3) + h(1)x(2) + h(2)x(1)$ discard		$h(1)x(1) +$
				$h(2)x(0)$
$n = 4$	$x(5)$	$w(4) = h(0)x(4) + h(1)x(3) + h(2)x(2)$	$m = 2$	$y(2) = h(0)x(4) +$
$n = 5$	$x(6)$	$w(5) = h(0)x(5) + h(1)x(4) + h(2)x(3)$ discard		$h(1)x(3) +$
				$h(2)x(2)$
...	...	...	...	...

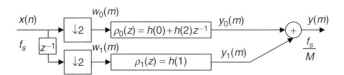

FIGURE 12.21 Polyphase filter implementation for the decimation in Figure 12.20 (3 multiplications and 1 addition for obtaining each output $y(m)$).

Example 12.6.

a. Verify $y(1)$ in Table 12.2 using the polyphase decimation filter implementation in Figure 12.21.

Solution:

a. Using Figure 12.21, we write the difference equations as

$$y_0(m) = h(0)w_0(m) + h(2)w_0(m-1)$$
$$y_1(m) = h(1)w_1(m).$$

Assuming $n = 0$, $n = 1$, $n = 2$, and $n = 3$, we have the inputs $x(0)$, $x(1)$, $x(2)$, and $x(3)$, and

$$w_0(m): \ x(0) \ x(2) \ \ldots.$$

Delaying $x(n)$ by one sample and decimating it by a factor of 2 leads to

$$w_1(m): \ \ 0 \ \ x(1) \ \ x(3) \ \ \ldots.$$

Hence, applying the filter banks yields the following:
$m = 0$, we have inputs for each filter as

$$w_0(0) = x(0) \text{ and } w_1(0) = 0$$

then

$$y_0(0) = h(0)w_0(0) + h(2)w_0(-1) = h(0)x(0)$$
$$y_1(0) = h(1)w_1(0) = h(1) \times 0 = 0.$$

Combining two channels, we obtain

$$y(1) = y_0(1) + y_1(1) = h(0)x(0) + 0 = h(0)x(0)$$

$m = 1$, we get inputs for each filter as

$$w_0(1) = x(2) \text{ and } w_1(1) = x(1),$$

then

$$y_0(1) = h(0)w_0(1) + h(2)w_0(0) = h(0)x(2) + h(2)x(0)$$
$$y_1(1) = h(1)w_1(1) = h(1)x(1).$$

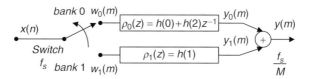

FIGURE 12.22 Commutative model for the polyphase decimation filter.

Combining two channels leads to

$$y(1) = y_0(1) + y_1(1) = h(0)x(2) + h(2)x(0) + h(1)x(1).$$

We note that $y(1)$ is the same as that shown in Table 12.2. Similar analysis can be done for the commutative model shown in Figure 12.22.

Note that wavelet transform and subband coding are also in the area of multirate signal processing. We do not pursue these subjects in this book. The reader can find useful fundamental information in Akansu and Haddad (1992), Stearns (2003), Van der Vegte (2002), and Vetterli and Kovacevic (1995).

12.3 Oversampling of Analog-to-Digital Conversion

Oversampling of the analog signal has become popular in DSP industry to improve resolution of analog-to-digital conversion (ADC). Oversampling uses a sampling rate, which is much higher than the Nyquist rate. We can define an oversampling ratio as

$$\frac{f_s}{2f_{\max}} \gg 1. \tag{12.14}$$

The benefits from an oversampling ADC include:

1. helping to design a simple analog anti-aliasing filter before ADC, and
2. reducing the ADC noise floor with possible noise shaping so that a low-resolution ADC can be used.

12.3.1 Oversampling and Analog-to-Digital Conversion Resolution

To begin with developing the relation between oversampling and ADC resolution, we first summarize the regular ADC and some useful definitions discussed in Chapter 2:

$$\text{Quantization noise power} = \sigma_q^2 = \frac{\Delta^2}{12} \qquad (12.15)$$

$$\text{Quantization step} = \Delta = \frac{A}{2^n} \qquad (12.16)$$

$A = $ full range of the analog signal to be digitized

$n = $ number of bits per sample (ADC resolution).

Substituting Equation (12.16) into Equation (12.15), we have:

$$\text{Quantization noise power} = \sigma_q^2 = \frac{A^2}{12} \times 2^{-2n}. \qquad (12.17)$$

The power spectral density of the quantization noise with an assumption of uniform probability distribution is shown in Figure 12.23. Note that this assumption is true for quantizing a uniformly distributed signal in a full range with a sufficiently long duration. It is not generally true in practice. See research papers by Lipshitz et al. (1992) and Maher (1992). However, using the assumption will guide us for some useful results for oversampling systems.

The quantization noise power is the area obtained from integrating the power spectral density function in the range of $-f_s/2$ to $f_s/2$. Now let us examine the oversampling ADC, where the sampling rate is much bigger than that of the regular ADC; that is $f_s \gg 2f_{\max}$. The scheme is shown in Figure 12.24.

As we can see, oversampling can reduce the level of noise power spectral density. After the decimation process with the decimation filter, only a portion of quantization noise power in the range from $-f_{\max}$ to $f_{\max}$ is kept in the DSP system. We call this an *in-band frequency range*.

In Figure 12.24, the shaded area, which is the quantization noise power, is given by

$$\text{quantization noise power} = \int_{-\infty}^{\infty} P(f)df = \frac{2f_{\max}}{f_s} \times \sigma_q^2$$

$$= \frac{2f_{\max}}{f_s} \cdot \frac{A^2}{12} \times 2^{-2m}. \qquad (12.18)$$

Assuming that the regular ADC shown in Figure 12.23 and the oversampling ADC shown in Figure 12.24 are equivalent, we set their quantization noise powers to be the same to obtain

12.3 Oversampling of Analog-to-Digital Conversion

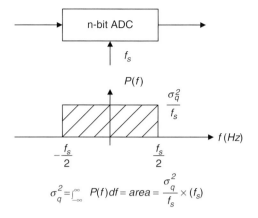

FIGURE 12.23 Regular ADC system.

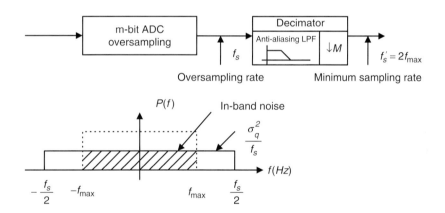

FIGURE 12.24 Oversampling ADC system.

$$\frac{A^2}{12} \cdot 2^{-2n} = \frac{2f_{\max}}{f_s} \cdot \frac{A^2}{12} \times 2^{-2m}. \qquad (12.19)$$

Equation (12.19) leads to two useful equations for applications:

$$n = m + 0.5 \times \log_2\left(\frac{f_s}{2f_{\max}}\right) \text{ and} \qquad (12.20)$$

$$f_s = 2f_{\max} \times 2^{2(n-m)}, \qquad (12.21)$$

where

f_s = sampling rate in the oversampling DSP system
$f_{\max}$ = maximum frequency of the analog signal

m = number of bits per sample in the oversampling DSP system
n = number of bits per sample in the regular DSP system using the minimum sampling rate

From Equation (12.20) and given the number of bits (m) used in the oversampling scheme, we can determine the number of bits per sample equivalent to the regular ADC. On the other hand, given the number of bits in the oversampling ADC, we can determine the required oversampling rate so that the oversampling ADC is equivalent to the regular ADC with the larger number of bits per sample (n). Let us look at the following examples.

Example 12.7.

Given an oversampling audio DSP system with maximum audio input frequency of 20 kHz and ADC resolution of 14 bits,

a. Determine the oversampling rate to improve the ADC to 16-bit resolution.

Solution:

a. Based on the specifications, we have

$$f_{max} = 20\,\text{kHz},\ m = 14\ \text{bits, and}\ n = 16\ \text{bits}.$$

Using Equation (12.21) leads to

$$f_s = 2f_{max} \times 2^{2(n-m)} = 2 \times 20 \times 2^{2(16-14)} = 640\,\text{kHz}.$$

Since $\frac{f_s}{2f_{max}} = 2^4$, we see that each doubling of the minimum sampling rate ($2f_{max} = 40\,\text{kHz}$) will increase the resolution by a half bit.

Example 12.8.

Given an oversampling audio DSP system with

Maximum audio input frequency = 4 kHz
ADC resolution = 8 bits
Sampling rate = 80 MHz,

a. Determine the equivalent ADC resolution.

Solution:

a. Since $f_{max} = 4\,\text{kHz}$, $f_s = 80\,\text{kHz}$, and $m = 8\,\text{bits}$, applying Equation (12.20) yields

$$n = m + 0.5 \times \log_2\left(\frac{f_s}{2f_{\max}}\right) = 8 + 0.5 \times \log_2\left(\frac{80000\,kHz}{2 \times 4\,kHz}\right) \approx 15 \text{ bits}.$$

12.3.2 Sigma-Delta Modulation Analog-to-Digital Conversion

To further improve ADC resolution, *sigma-delta modulation* (SDM) ADC is used. The principles of the first-order SDM are described in Figure 12.25.

First, the analog signal is sampled to obtain the discrete-time signal $x(n)$. This discrete-time signal is subtracted by the analog output from the m-bit DAC, converting the m bit oversampled digital signal $y(n)$. Then the difference is sent to the discrete-time analog integrator, which is implemented by the switched-capacitor technique, for example. The output from the discrete-time analog integrator is converted using an m-bit ADC to produce the oversampled digital signal. Finally, the decimation filter removes outband quantization noise. Further decimation process can change the oversampling rate back to the desired sampling rate for the output digital signal $w(m)$.

To examine the SDM, we need to develop a DSP model for the discrete-time analog filter described in Figure 12.26.

As shown in Figure 12.26, the input signal $c(n)$ designates the amplitude at time instant n, while the output $d(n)$ is the area under the curve at time instant n, which can be expressed as a sum of the area under the curve at time instant $n - 1$ and area increment:

$$d(n) = d(n - 1) + \text{area incremental}. \tag{12.22}$$

Using the extrapolation method, we have

$$d(n) = d(n - 1) + 1 \times c(n). \tag{12.23}$$

Applying the z-transform to Equation (12.23) leads to a transfer function of the discrete-time analog filter as

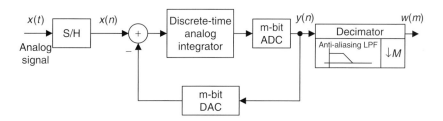

FIGURE 12.25 Block diagram of SDM ADC.

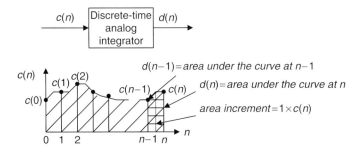

FIGURE 12.26 Illustration of discrete-time analog integrator.

$$H(z) = \frac{D(z)}{C(z)} = \frac{1}{1-z^{-1}}. \tag{12.24}$$

Again, considering that the *m*-bit quantization requires one sample delay, we get the DSP model for the first-order SDM depicted in Figure 12.27, where $y(n)$ is the oversampling data encoded by *m* bits each, and $e(n)$ represents quantization error.

The SDM DSP model represents a feedback control system. Applying the z-transform leads to

$$Y(z) = \frac{1}{1-z^{-1}}\left(X(z) - z^{-1}Y(z)\right) + E(z). \tag{12.25}$$

After simple algebra, we have

$$Y(z) = \underbrace{X(z)}_{\substack{\text{Original} \\ \text{digital signal} \\ \text{transform}}} + \underbrace{(1-z^{-1})}_{\substack{\text{Highpass} \\ \text{filter}}} \cdot \underbrace{E(z)}_{\substack{\text{Quantization} \\ \text{error} \\ \text{transform}}}. \tag{12.26}$$

In Equation (12.26), the indicated highpass filter pushes quantization noise to the high-frequency range, where later the quantization noise can be removed by the decimation filter. Thus we call this highpass filter $(1 - z^{-1})$ the *noise shaping filter,* illustrated in Figure 12.28.

Shaped-in-band noise power after use of the decimation filter can be estimated by the solid area under the curve. We have

$$\text{Shaped-in-band noise power} = \int_{-\Omega_{\max}}^{\Omega_{\max}} \frac{\sigma_q^2}{2\pi}\left|1 - e^{-j\Omega}\right|^2 d\Omega. \tag{12.27}$$

Using the Maclaurin series expansion and neglecting the higher-order term due to the small value of $\Omega_{\max}$, we yield

12.3 Oversampling of Analog-to-Digital Conversion

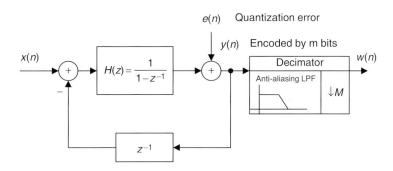

FIGURE 12.27 DSP model for first-order SDM ADC.

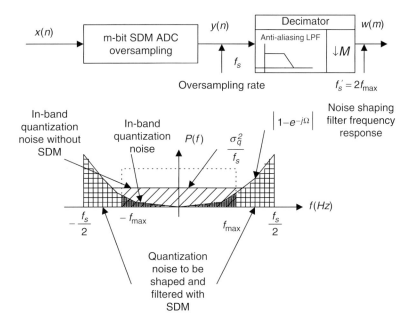

FIGURE 12.28 Noise shaping of quantization noise for SDM ADC.

$$1 - e^{-j\Omega} = 1 - \left(1 + \frac{(-j\Omega)}{1!} + \frac{(-j\Omega)^2}{2!} + \ldots\right) \approx j\Omega.$$

Applying this approximation to Equation (12.27) leads to

$$\text{Shaped-in-band noise power} \approx \int_{-\Omega_{\max}}^{\Omega_{\max}} \frac{\sigma_q^2}{2\pi} |j\Omega|^2 d\Omega = \frac{\sigma_q^2}{3\pi} \Omega_{\max}^3. \qquad (12.28)$$

After simple algebra, we have

$$\text{Shaped-in-band noise power} \approx \frac{\pi^2 \sigma_q^2}{3}\left(\frac{2f_{\max}}{f_s}\right)^3 \qquad (12.29)$$

$$= \frac{\pi^2}{3} \cdot \frac{A^2 2^{-2m}}{12}\left(\frac{2f_{\max}}{f_s}\right)^3.$$

If we let the shaped-in-band noise power equal the quantization noise power from the regular ADC using a minimum sampling rate, we have

$$\frac{\pi^2}{3} \cdot \frac{A^2 2^{-2m}}{12}\left(\frac{2f_{\max}}{f_s}\right)^3 = \frac{A^2}{12} \cdot 2^{-2n}. \qquad (12.30)$$

We modify Equation (12.30) into the following useful formats for applications:

$$n = m + 1.5 \times \log_2\left(\frac{f_s}{2f_{\max}}\right) - 0.86 \qquad (12.31)$$

$$\left(\frac{f_s}{2f_{\max}}\right)^3 = \frac{\pi^2}{3} \times 2^{2(n-m)}. \qquad (12.32)$$

Example 12.9.

Given the following DSP system specifications:

Oversampling rate system
First-order SDM with 2-bit ADC
Sampling rate $= 4\,\text{MHz}$
Maximum audio input frequency $= 4\,\text{kHz}$,

a. Determine the equivalent ADC resolution.

Solution:

a. Since $m = 2$ bits, and

$$\frac{f_s}{2f_{\max}} = \frac{4000\,kHz}{2 \times 4\,kHz} = 500.$$

we calculate

$$n = m + 1.5 \times \log_2\left(\frac{f_s}{2f_{\max}}\right) - 0.86$$
$$= 2 + 1.5 \times \log_2(500) - 0.86 \approx 15\,\text{bits}.$$

We can also extend the first-order SDM DSP model to the second-order SDM DSP model by cascading one section of the first-order discrete-time analog filter, as depicted in Figure 12.29.

Similarly to the first-order SDM DSP model, applying the z-transform leads to the following relationship:

$$Y(z) = \underbrace{X(z)}_{\substack{Original \\ digital\ signal \\ transform}} + \underbrace{(1-z^{-1})^2}_{\substack{Highpass \\ noise\ shaping \\ filter}} \times \underbrace{E(z)}_{\substack{Quantization \\ error \\ transform}}. \quad (12.33)$$

Notice that the noise shaping filter becomes a second-order highpass filter; hence, the more quantization noise is pushed to the high-frequency range, the better ADC resolution is expected to be. In a similar analysis to the first-order SDM, we get the following useful formulas:

$$n = m + 2.5 \times \log_2\left(\frac{f_s}{2f_{max}}\right) - 2.14 \quad (12.34)$$

$$\left(\frac{f_s}{2f_{max}}\right)^5 = \frac{\pi^4}{5} \times 2^{2(n-m)}. \quad (12.35)$$

In general, the Kth-order SDM DSP model and ADC resolution formulas are given as:

$$Y(z) = \underbrace{X(z)}_{\substack{Original \\ digital\ signal \\ transform}} + \underbrace{(1-z^{-1})^K}_{\substack{Highpass \\ noise\ shaping \\ filter}} \times \underbrace{E(z)}_{\substack{Quantization \\ error \\ transform}} \quad (12.36)$$

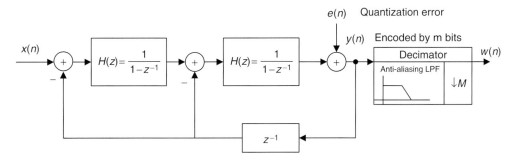

FIGURE 12.29   **DSP model for the second-order SDM ADC.**

$$n = m + 0.5 \cdot (2K+1) \times \log_2\left(\frac{f_s}{2f_{\max}}\right) - 0.5 \times \log_2\left(\frac{\pi^{2K}}{2K+1}\right) \quad (12.37)$$

$$\left(\frac{f_s}{2f_{\max}}\right)^{2K+1} = \frac{\pi^{2K}}{2K+1} \times 2^{2(n-m)}. \quad (12.38)$$

Example 12.10.

Given the oversampling rate DSP system with the following specifications:

Second-order SDM = 1-bit ADC
Sampling rate = 1 MHz
Maximum audio input frequency = 4 kHz,

a. Determine the effective ADC resolution.

Solution:

a. $n = 1 + 2.5 \times \log_2\left(\frac{1000\,kHz}{2 \times 4\,kHz}\right) - 2.14 \approx 16$ bits.

Next, we review the application of the oversampling ADC used in industry. Figure 12.30 illustrates a function diagram for the MAX1402 low-power, multi-channel oversampling sigma-delta analog-to-digital converter used in industry. It applies a sigma-delta modulator with a digital decimation filter to achieve 16-bit accuracy. The device offers three fully differential input channels, which can be independently programmed. It can also be configured as five pseudo-differential input channels. It comprises two chopper buffer amplifiers and a programmable gain amplifier, a DAC unit with predicted input subtracted from the analog input to acquire the differential signal, and a second-order switched-capacitor sigma-delta modulator.

The chip produces a 1-bit data stream, which will be filtered by the integrated digital filter to complete ADC. The digital filter's user-selectable decimation factor offers flexibility for conversion resolution to be reduced in exchange for a higher data rate, or vice versa. The integrated digital lowpass filter is first-order or third-order Sinc infinite impulse response. Such a filter offers notches corresponding to its output data rate and its frequency harmonics, so it can effectively reduce the developed image noises in the frequency domain. (The Sinc filter is beyond the scope of our discussion.) The MAX1402 can provide 16-bit accuracy at 480 samples per second and 12-bit accuracy at 4,800 samples per second. The chip finds wide application in sensors and instrumentation. Its detailed features can be found in the MAX1402 data sheet (Maxim Integrated Products, 2007).

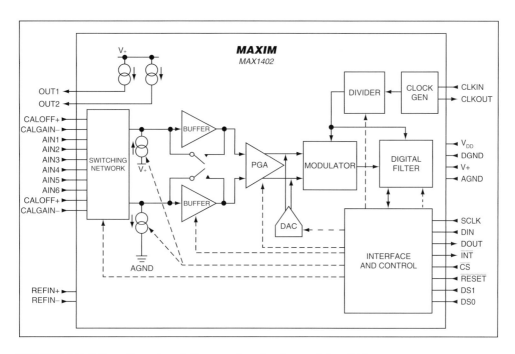

FIGURE 12.30 Functional diagram for the sigma-delta ADC.

12.4 Application Example: CD Player

Figure 12.31 illustrates a CD playback system, also described earlier in this chapter. A laser optically scans the tracks on a CD to produce a digital signal. The digital signal is then demodulated, and parity bits are used to detect bit errors due to manufacturing defects, dust, and so on and to correct them. The demodulated signal is again oversampled by a factor of 4 and hence the sampling rate is increased to 176.4 kHz for each channel. Each digital sample then passes through a 14-bit DAC, which produces the sample-and-hold voltage signals that pass the anti-image lowpass filter. The output from each analog filter is fed to its corresponding loudspeaker. Oversampling relaxes the design requirements of the analog anti-image lowpass filter, which is used to smooth out the voltage steps.

The earliest system used a third-order Bessel filter with a 3 dB passband at 30 kHz. Notice that the first-order sigma-delta modulation (first-order SDM) is added to the 14-bit DAC unit to further improve the 14-bit DAC to 16-bit DAC.

Let us examine the single-channel DSP portion shown in Figure 12.32.

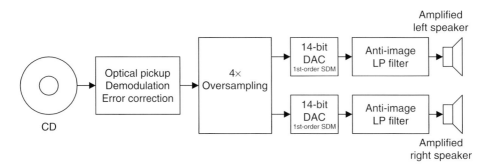

FIGURE 12.31 Simplified decoder of a CD recording system.

The spectral plots for the oversampled and interpolated signal $\bar{x}(n)$, the 14-bit SDM output $y(n)$, and the final analog output audio signal are given in Figure 12.33. As we can see in plot (a) in the figure, the quantization noise is uniformly distributed, and only in-band quantization noise (0 to 22.05 kHz) is expected. Again, 14 bits for each sample are kept after oversampling. Without using the first-order SDM, we expect the effective ADC resolution due to oversampling to be

$$n = 14 + 0.5 \times \log_2\left(\frac{176.4}{44.1}\right) = 15 \text{ bits},$$

which is fewer than 16 bits. To improve quality further, the first-order SDM is used. The in-band quantization noise is then shaped. The first-SDM pushes

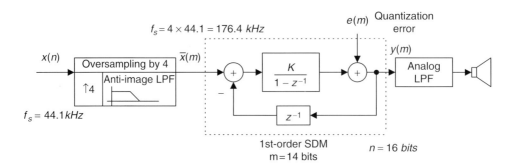

FIGURE 12.32 Illustration of oversampling and SDM ADC used in the decoder of a CD recording system.

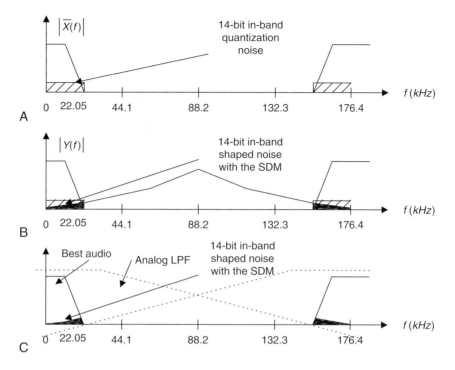

FIGURE 12.33 Spectral illustrations for oversampling and SDM ADC used in the decoder of a CD recording system.

quantization noise to the high-frequency range, as illustrated in plot (b) in Figure 12.33. The effective ADC resolution now becomes

$$n = 14 + 1.5 \times \log_2 \left(\frac{176.4}{44.1}\right) - 0.86 \approx 16 \text{ bits}.$$

Hence, 16-bit ADC audio quality is preserved. On the other hand, from plot (c) in Figure 12.33, the audio occupies a frequency range up to 22.05 kHz, while the DSP Nyquist limit is 88.2, so the low-order analog anti-image filter can satisfy the design requirement.

12.5 Undersampling of Bandpass Signals

As we discussed in Chapter 2, the sampling theorem requires that the sampling rate be twice as large as the highest frequency of the analog signal to be sampled. The sampling theorem ensures the complete reconstruction of the analog signal without aliasing distortion. In some applications, such as modulated signals in communications systems, the signal exists in only a small portion of the bandwidth. Figure 12.34 shows an amplitude-modulated (AM) signal in both time domain and frequency domain. Assuming that the message signal has a bandwidth of 4 kHz and a carrier frequency of 96 kHz, the upper frequency edge is therefore 100 kHz. Then the traditional sampling process requires that the sampling rate be larger than 200 kHz at a high cost. The baseband signal of 4 kHz requires a sampling rate of only 8 kHz.

If a certain condition is satisfied at the undersampling stage, we are able to make use of the aliasing signal to recover the message signal, since the aliasing signal contains the folded original message information (which we used to consider as distortion). The reader is referred to the undersampling technique discussed in Ifeachor and Jervis (2002) and Porat (1997). Let the message to be recovered have a bandwidth of B, the theoretical minimum sampling rate be $f_s = 2B$, and the carrier frequency of the modulated signal be f_c. We discuss the following cases.

Case 1.

If $f_c =$ even integer $\times B$ and $f_s = 2B$, the sampled spectrum with all the replicas will be as shown in Figure 12.35(a).

As an illustrative example in time domain for Case 1, suppose we have a bandpass signal with a carrier frequency of 20 Hz; that is,

$$x(t) = \cos(2\pi \times 20t)m(t), \qquad (12.39)$$

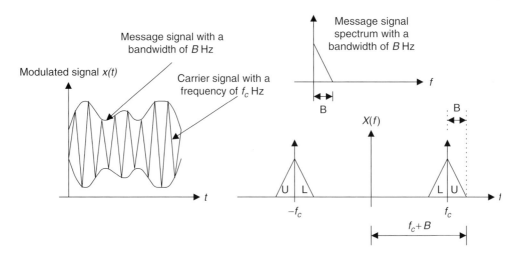

FIGURE 12.34 Message signal, modulated signal, and their spectra.

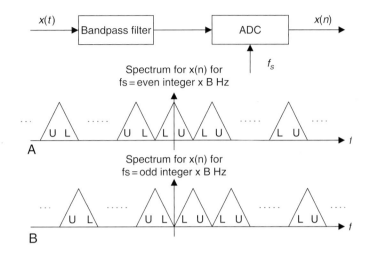

FIGURE 12.35 Spectrum of the undersampled signal.

where $m(t)$ is the message signal with a bandwidth of 2 Hz. Using a sampling rate of 4 Hz by substituting $t = nT$, where $T = 1/f_s$ into Equation (12.39), we get the sampled signal as

$$x(nT) = \cos(2\pi \times 20t)m(t)\big|_{t=nT} = \cos(2\pi \times 20n/4)m(nT). \quad (12.40)$$

Since $10n\pi = 5n(2\pi)$ is the multiple of 2π,

$$\cos(2\pi \times 20n/4) = \cos(10\pi n) = 1, \quad (12.41)$$

we obtained the undersampled signal as

$$x(nT) = \cos(2\pi \times 20n/4)m(nT) = m(nT), \quad (12.42)$$

which is a perfect digital message signal. Figure 12.36 shows the bandpass signal and its sampled signal when the message signal is 1 Hz, given as

$$m(t) = \cos(2\pi t). \quad (12.43)$$

Case 2.

If $f_c = $ odd integer $\times B$ and $f_s = 2B$, the sampled spectrum with all the replicas will be as shown in Figure 12.35(b), where the spectral portion L and U are reversed. Hence, frequency reversal will occur. Then a further digital modulation in which the signal is multiplied by the digital oscillator with a frequency of B Hz can be used to adjust the spectrum to be the same as that in Case 1.

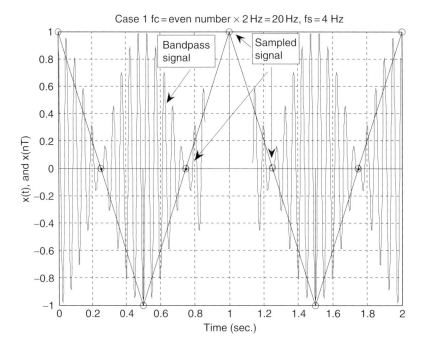

FIGURE 12.36   Plots of the bandpass signal and sampled signal for Case 1.

As another illustrative example for Case 2, let us sample the following bandpass signal with a carrier frequency of 22 Hz, given by

$$x(t) = \cos(2\pi \times 22t)m(t). \qquad (12.44)$$

Applying undersampling using a sampling rate of 4 Hz, it follows that

$$x(nT) = \cos(2\pi \times 22n/4)m(nT) = \cos(11n\pi)m(nT). \qquad (12.45)$$

Since $11n\pi$ can be either an odd or an even integer multiple of π, we have

$$\cos(11\pi n) = \begin{cases} -1 & n = \text{odd} \\ 1 & n = \text{even}. \end{cases} \qquad (12.46)$$

We see that Equation (12.46) causes the message samples to change sign alternately with a carrier frequency of 22 Hz, which is the odd integer multiple of the message bandwidth of 2 Hz. This in fact will reverse the baseband message spectrum. To correct the spectrum reversal, we multiply an oscillator with a frequency of $B = 2$ Hz by the bandpass signal, that is

$$x(t)\cos(2\pi \times 2t) = \cos(2\pi \times 22t)m(t)\cos(2\pi \times 2t). \qquad (12.47)$$

Then the undersampled signal is given by

$$x(nT)\cos(2\pi \times 2n/4) = \cos(2\pi \times 22n/4)m(nT)\cos(2\pi \times 2n/4)$$
$$= \cos(11n\pi)m(nT)\cos(n\pi) \quad (12.48)$$

Since $\cos(11\pi n)\cos(\pi n) = 1$, it follows that (12.49)

$$x(nT)\cos(2\pi \times 2n/4) = \cos(\pi \times 11n)m(nT)\cos(\pi \times n) = m(nT), \quad (12.50)$$

which is the recovered message signal. Figure 12.37 shows the sampled bandpass signals with the reversed message spectrum and the corrected message spectrum, respectively, for a message signal having a frequency of 0.5 Hz; that is,

$$m(t) = \cos(2\pi \times 0.5t). \quad (12.51)$$

Case 3.
If $f_c =$ non-integer $\times B$, we can extend the bandwidth B to $\bar{B}$ such that

$$f_c = \text{integer} \times \bar{B} \text{ and } f_s = 2\bar{B}. \quad (12.52)$$

Then we can apply Case 1 or Case 2. An illustration of Case 3 is included in the following example.

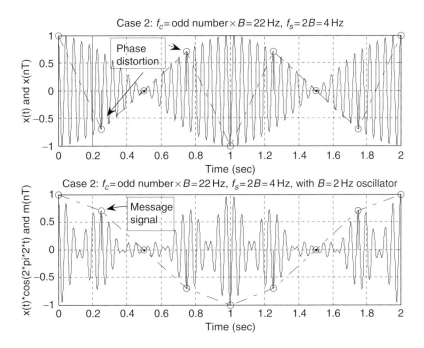

FIGURE 12.37 Plots of the bandpass signals and sampled signals for Case 2.

Example 12.11.

Given a bandpass signal with the spectrum and the carry frequency f_c shown in Figure 12.38(a), 12.38(b), and 12.38(c), respectively, and assuming the baseband bandwidth $B = 4\,\text{kHz}$,

a. Select the sampling rate and sketch the sampled spectrum ranging from 0 Hz to the carrier frequency for each of the following carrier frequencies:

1. $f_c = 16\,\text{kHz}$
2. $f_c = 12\,\text{kHz}$
3. $f_c = 18\,\text{kHz}$

Solution:

a. 1. Since $f_c/B = 4$ is an even number, which is Case 1, we select $f_s = 8\,\text{kHz}$ and sketch the sampled spectrum shown in Figure 12.38a.

2. Since $f_c/B = 3$ is an odd number, we select $f_s = 8\,\text{kHz}$ and sketch the sampled spectrum shown in Figure 12.38b.

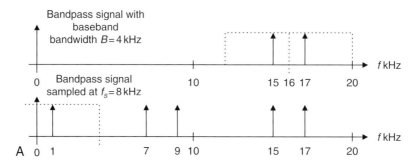

FIGURE 12.38A   Sampled signal spectrum for $f_c = 16\,\text{kHz}$.

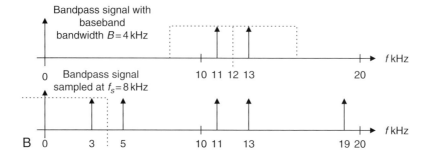

FIGURE 12.38B   Sampled signal spectrum for $f_c = 12\,\text{kHz}$.

12.5 Undersampling of Bandpass Signals

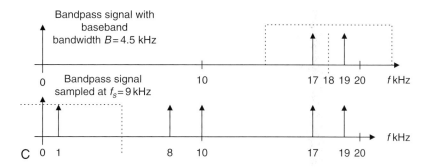

FIGURE 12.38C Sampled signal spectrum for $f_c = 18$ kHz.

3. Now, $f_c/B = 4.5$, which is a non-integer. We extend the bandwidth $\bar{B} = 4.5$ kHz, so $f_c/\bar{B} = 4$ and $f_s = 2\bar{B} = 9$ kHz. Then the sketched spectrum is shown in Figure 12.38c.

Simulation Example

An AM with a 1-kHz message signal is given as:

$$x(t) = [1 + 0.8 \times \sin(2\pi \times 1000t)] \cos(2\pi \times f_c t). \tag{12.53}$$

Assuming a message bandwidth of 4 kHz, determine the sampling rate, use MATLAB to sample the AM signal, and sketch the sampled spectrum up to the sampling frequency for each the following carrier frequencies:

1. $f_c = 96$ kHz
2. $f_c = 100$ kHz
3. $f_c = 99$ kHz

1. For this case, $f_c/B = 24$ is an even number. We select $f_s = 8$ kHz. Figure 12.39a describes the simulation, where the upper left plot is the AM signal, the upper right plot is the spectrum of the AM signal, the lower left plot is the undersampled signal, and the lower right plot is the spectrum of the undersampled signal displayed from 0 to 8 kHz.
2. $f_c/B = 25$ is an odd number, so we choose $f_s = 8$ kHz, and a further process is needed. We can multiply the undersampled signal by a digital oscillator with a frequency of $B = 4$ kHz to achieve the 1-kHz baseband signal. The plots of the AM signal spectrum, the undersampled signal spectrum, and the oscillator mixed signal and its spectrum are shown in Figure 12.39b.
3. For $f_c = 99$ kHz, $f_c/B = 24.75$. We extend the bandwidth to $\bar{B} = 4.125$ so that $f_c/\bar{B} = 24$. Hence, the undersampling rate is used as $f_s = 8.25$ kHz.

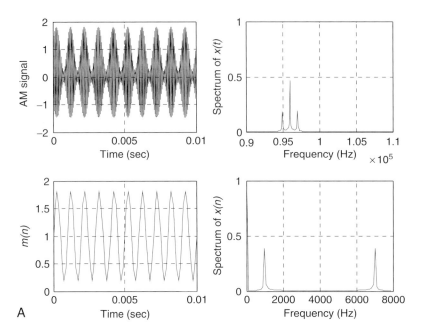

FIGURE 12.39A Sampled AM signal and spectrum for $f_c = 96$ kHz.

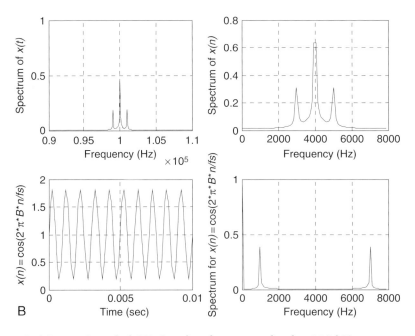

FIGURE 12.39B Sampled AM signal and spectrum for $f_c = 100$ kHz.

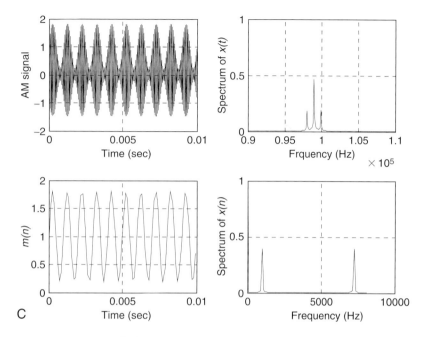

FIGURE 12.39C Sampled AM signal and spectrum for $f_c = 99$ kHz.

Figure 12.39c shows the plots for the AM signal, the AM signal spectrum, the undersampled signal based on the extended baseband width, and the sampled signal spectrum ranging from 0 to 8.25 kHz, respectively.

This example verifies principles of undersampling of bandpass signals.

12.6 Summary

1. Downsampling (decimation) by an integer factor of M means taking one sample from the data sequence $x(n)$ for every M sample and discarding the last $M - 1$ sample.

2. Upsampling (interpolation) by an integer factor of L means inserting $L - 1$ zeros for every sample in the data sequence $x(n)$.

3. Downsampling requires a decimation (anti-aliasing) filter to avoid frequency aliasing before downsampling.

4. Upsampling requires an interpolation (anti-image) filter to remove the images after interpolation.

5. Changing the sampling rate by a non-integer factor of L/M requires two stages: an interpolation stage and a downsampling stage.

6. Two-stage decimation can dramatically reduce the anti-aliasing filter length.

7. Polyphase implementations of the decimation filter and interpolation filter can reduce complexity of the filter operations, that is, fewer multiplications and additions.

8. Using oversampling can improve the regular ADC resolution. Sigma-delta modulation ADC can achieve even higher ADC resolution, using noise shaping effect for further reduction of quantization noise.

9. The audio CD player uses multirate signal processing and oversampling.

10. Undersampling can be used to sample the bandpass signal, and finds its application in communications.

12.7 Problems

12.1. For a single-stage decimator with the following specifications:

Original sampling rate $= 8\,\text{kHz}$
Decimation factor $M = 4$
Frequency of interest $= 0 - 800\,\text{Hz}$
Passband ripple $= 0.02\,\text{dB}$
Stopband attenuation $= 46\,\text{dB}$,

a. draw the block diagram for the decimator;

b. determine the window type, filter length, and cutoff frequency if the window method is used for the anti-aliasing FIR filter design.

12.2. For a single-stage interpolator with the following specifications:

Original sampling rate $= 8\,\text{kHz}$
Interpolation factor $L = 3$
Frequency of interest $= 0 - 3{,}400\,\text{Hz}$
Passband ripple $= 0.02\,\text{dB}$
Stopband attenuation $= 46\,\text{dB}$,

a. draw the block diagram for the interpolator;

b. determine the window type, filter length, and cutoff frequency if the window method is used for the anti-image FIR filter design.

12.3. For the sampling conversion from 6 kHz to 8 kHz with the following specifications:

> Original sampling rate = 6 kHz
> Interpolation factor $L = 4$
> Decimation factor $M = 3$
> Frequency of interest = 0–2,400 Hz
> Passband ripple = 0.02 dB
> Stopband attenuation = 46 dB,

a. draw the block diagram for the processor;

b. determine the window type, filter length, and cutoff frequency if the window method is used for the combined FIR filter $H(z)$.

12.4. For the design of a two-stage decimator with the following specifications:

> Original sampling rate = 320 kHz
> Frequency of interest = 0–3,400 Hz
> Passband ripple = 0.05 (absolute)
> Stopband attenuation = 0.005 (absolute)
> Final sampling rate = 8,000 Hz,

a. draw the decimation block diagram;

b. specify the sampling rate for each stage;

c. determine the window type, filter length, and cutoff frequency for the first stage if the window method is used for the anti-aliasing FIR filter design ($H_1(z)$);

d. determine the window type, filter length, and cutoff frequency for the second stage if the window method is used for the anti-aliasing FIR filter design ($H_2(z)$).

12.5.

a. Given an interpolator filter as

$$H(z) = 0.25 + 0.4z^{-1} + 0.5z^{-2},$$

draw the block diagram for interpolation polyphase filter implementation for the case of $L = 2$.

b. Given a decimation filter as

$$H(z) = 0.25 + 0.4z^{-1} + 0.5z^{-2} + 0.6z^{-3},$$

draw the block diagram for decimation polyphase filter implementation for the case of $M = 2$.

12.6. Using the commutative models for the polyphase interpolation and decimation filters,

a. draw the block diagram for interpolation polyphase filter implementation for the case of $L = 2$, and $H(z) = 0.25 + 0.4z^{-1} + 0.5z^{-2}$;

b. draw the block diagram for decimation polyphase filter implementation for the case of $M = 2$, and $H(z) = 0.25 + 0.4z^{-1} + 0.5z^{-2} + 0.6z^{-3}$.

12.7. Given the audio system with the following specifications:

Audio input frequency range: $0 - 15\,\text{kHz}$
ADC resolution = 16 bits
Current sampling rate = $30\,\text{kHz}$,

a. determine the oversampling rate if the 12-bit ADC chip is used to replace the audio system;

b. draw the block diagram.

12.8. Given the audio system with the following specifications:

Audio input frequency range: $0 - 15\,\text{kHz}$
ADC resolution = 6 bits
Oversampling rate = $45\,\text{MHz}$,

a. draw the block diagram;

b. determine the actual effective ADC resolution (number of bits per sample).

12.9. Given the following specifications of an oversampling DSP system:

Audio input frequency range: $0 - 4\,\text{kHz}$
First-order SDM with a sampling rate of $128\,\text{kHz}$
ADC resolution in SDM = 1 bit,

a. draw the block diagram using the DSP model;

b. determine the equivalent (effective) ADC resolution.

12.10. Given the following specifications of an oversampling DSP system:

Audio input frequency range: $0 - 20\,\text{kHz}$
Second-order SDM with a sampling rate of $160\,\text{kHz}$
ADC resolution in SDM = 10 bits,

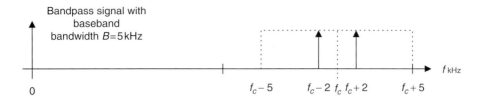

FIGURE 12.40 Spectrum of the bandpass signal in Problem 12.11.

a. draw the block diagram using the DSP model;

b. determine the equivalent (effective) ADC resolution.

12.11. Given a bandpass signal with its spectrum shown in Figure 12.40, and assuming the bandwidth $B = 5\,\text{kHz}$, select the sampling rate and sketch the sampled spectrum ranging from 0 Hz to the carrier frequency for each of the following carrier frequencies:

a. $f_c = 30$ kHz

b. $f_c = 25$ kHz

c. $f_c = 33$ kHz

MATLAB Problems

Use MATLAB to solve Problems 12.12 to 12.16

12.12. Generate a sinusoid with a 1000 Hz for 0.05 sec using a sampling rate of 8 kHz,

a. Design a decimator to change the sampling rate to 4 kHz with specifications below:

Signal frequency range: 0–1800 Hz

Hamming window required for FIR filter design

b. Write a MATLAB program to implement the downsampling scheme, and plot the original signal and the downsampled signal versus the sample number, respectively.

12.13. Generate a sinusoid with a 1000 Hz for 0.05 sec using a sampling rate of 8 kHz,

  a. Design an interpolator to change the sampling rate to 16 kHz with following specifications:

  Signal frequency range: 0 – 3600 Hz

  Hamming window required for FIR filter design

  b. Write a MATLAB program to implement the upsampling scheme, and plot the original signal and the downsampled signal versus the sample number, respectively.

12.14. Generate a sinusoid with a frequency of 500 Hz for 0.1 sec using a sampling rate of 8 kHz,

  a. design an interpolation and decimation processing algorithm to change the sampling rate to 22 kHz

  Signal frequency range: 0 – 3400 Hz

  Hamming window required for FIR filter design

  b. Write a MATLAB program to implement the scheme, and plot the original signal and the sampled signal at the rate of 22 kHz versus the sample number, respectively.

12.15. Repeat Problem 12.12 using the polyphase form for the decimator.

12.16. Repeat Problem 21.13 using the polyphase form for the interpolator.

References

Akansu, A. N., and Haddad, R. A. (1992). *Multiresolution Signal Decomposition: Transforms, Subbands, and Wavelets.* Boston: Academic Press.

Ifeachor, E. C., and Jervis, B. W. (2002). *Digital Signal Processing: A Practical Approach*, 2nd ed. Upper Saddle River, NJ: Prentice Hall.

Lipshitz, S. P., Wannamaker, R. A., and Vanderkooy, J. (1992). Quantization and dither: A theoretical survey. *Journal of the Audio Engineering Society*, 40 (5): 355–375.

Maher, R. C. (1992). On the nature of granulation noise in uniform quantization systems. *Journal of the Audio Engineering Society*, 40 (1/2): 12–20.

Maxim Integrated Products. (2007). *MAXIM +5V, 18-Bit, Low-Power, Multichannel, Oversampling (Sigma-Delta) ADC.* Retrieved March 3, 2007, from http://datasheets.maxim-ic.com/en/ds/MAX1402.pdf

Porat, B. (1997). *A Course in Digital Signal Processing.* New York: John Wiley & Sons.

Proakis, J. G., and Manolakis, D. G. (1996). *Digital Signal Processing: Principles, Algorithms, and Applications*, 3rd ed. Upper Saddle River, NJ: Prentice Hall.

Sorensen, H. V., and Chen, J. P. (1997). *A Digital Signal Processing Laboratory Using TMS320C30.* Upper Saddle River, NJ: Prentice Hall.

Stearns, S. D. (2003). *Digital Signal Processing with Examples in MATLAB.* Boca Raton, FL: CRC Press LLC.

van der Vegte, J. (2002). *Fundamentals of Digital Signal Processing.* Upper Saddle River, NJ: Prentice Hall.

Vetterli, M., and Kovacevic, J. (1995). *Wavelets and Subband Coding.* Englewood Cliffs, NJ: Prentice Hall.

13

Image Processing Basics

Objectives: In today's modern computers, media information such as audio, images, and video have come to be necessary for daily business operations and entertainment. In this chapter, we will study the digital image and its processing techniques. This chapter introduces the basics of image processing, including image enhancement using histogram equalization and filtering methods, and proceeds to study pseudo-color generation for object detection and recognition. Finally, the chapter investigates image compression techniques and basics of video signals.

13.1 Image Processing Notation and Data Formats

The digital image is picture information in digital form. The image can be filtered to remove noise and obtain enhancement. It can also be transformed to extract features for pattern recognition. The image can be compressed for storage and retrieval, as well as transmitted via a computer network or a communication system.

The digital image consists of pixels. The position of each pixel is specified in terms of an index for the number of columns and another for the number of rows. Figure 13.1 shows that the pixel $p(2, 8)$ has a level of 86 and is located in the second row, eighth column. We express it in notation as

$$p(2, 8) = 86. \tag{13.1}$$

The number of pixels in the presentation of a digital image is its *spatial resolution*, which relates to the image quality. The higher the spacial resolution,

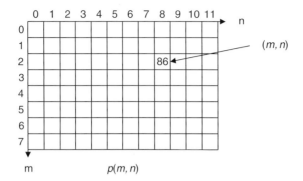

FIGURE 13.1 Image pixel notation.

the better quality the image has. The spacial resolution can be fairly high, for instance, as high as $1600 \times 1200\,(1{,}920{,}000$ pixels $= 1.92$ megapixels), or as low as $320 \times 200\,(64{,}000$ pixels $= 64$ kilopixels). In notation, the number to the left of the multiplication symbol represents the width, and that to the right of the symbol represents the height. Image quality also depends on the numbers of bits used in encoding each pixel level, which will be discussed in the next section.

13.1.1 8-Bit Gray Level Images

If a pixel is encoded on a gray scale from 0 to 255, where $0 =$ black and $255 =$ white, the numbers in between represent levels of gray forming a *grayscale image*. For a 640×480 8-bit image, 307.2 kilobytes are required for storage. Figure 13.2 shows a grayscale image format. As shown in the figure, the pixel indicated in the box has an 8-bit value of 25.

The image of a cruise ship with spatial resolution of 320×240 in an 8-bit grayscale level is shown in Figure 13.3.

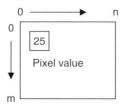

FIGURE 13.2 Grayscale image format.

FIGURE 13.3   Grayscale image (8-bit 320 × 240).

13.1.2   24-Bit Color Images

In a 24-bit color image representation, each pixel is recoded with red, green, and blue (RGB) components. With each component value encoded in 8 bits, resulting in 24 bits in total, we achieve a full color RGB image. With such an image, we can have $2^{24} = 16.777216 \times 10^6$ different colors. A 640×480 24-bit color image requires 921.6 kilobytes for storage. Figure 13.4 shows the format for the 24-bit color image where the indicated pixel has 8-bit RGB components.

Figure 13.5 shows a 24-bit color image of the Grand Canyon, along with its grayscale displays for the 8-bit RGB component images. The full color picture at the upper left is included in the color insert.

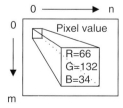

FIGURE 13.4   The 24-bit color image format.

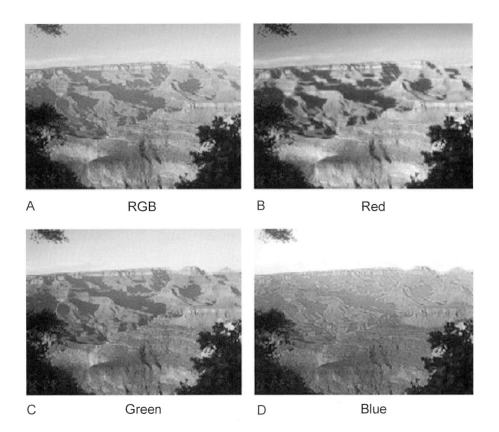

FIGURE 13.5 The 24-bit color image and its respective RGB components. (See color insert.)

13.1.3 8-Bit Color Images

The 8-bit color image is also a popular image format. Its pixel value is a color index that points to a color look-up table containing RGB components. We call this a *color indexed image*, and its format is shown in Figure 13.6. As an example in the figure, the color indexed image has a pixel index value of 5, which is the index for the entry of the color table, called the *color map*. At the index of location 5, there are three color components with RGB values of 66, 132, and 34, respectively. Each color component is encoded in 8 bits. There are only 256 different colors in the image. A 640×480 8-bit color image requires memory of 307.2 kilobytes for data storage and $3 \times 256 = 768$ bytes for color map storage. The 8-bit color image for the cruise ship shown in Figure 13.3 is displayed in Figure 13.7.

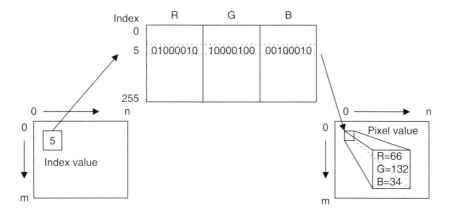

FIGURE 13.6 The 8-bit color indexed image format.

FIGURE 13.7 The 8-bit color indexed image. (See color insert.)

13.1.4 Intensity Images

As we noted in the first section, the grayscale image uses a pixel value ranging from 0 to 255 to present luminance, or the light intensity. A pixel value of 0 designates black, and a value 255 encodes for white.

In some processing environments such as MATLAB (*mat*rix *lab*oratory), floating-point operations are used. The grayscale image has an intensity value that is normalized to be in the range from 0 to 1.0, where 0 represents black and 1 represents white. We often change the pixel value to the normalized range to

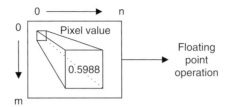

FIGURE 13.8   The grayscale intensity image format.

get the grayscale intensity image before processing it, then scale it back to the standard 8-bit range after processing for display. With the intensity image in the floating-point format, the digital filter implementation can be easily applied. Figure 13.8 shows the format of the grayscale intensity image, where the indicated pixel shows the intensity value of 0.5988.

13.1.5 Red, Green, Blue Components and Grayscale Conversion

In some applications, we need to convert a color image to a grayscale image so that storage space can be saved. As an example, the fingerprint image is stored in the grayscale format in the database system. In color image compression, as another example, the transformation converts the RGB color space to YIQ color space (Li and Drew, 2004; Rabbani and Jones, 1991), where Y is the luminance (Y) channel representing light intensity and the I (in-phase) and Q (quadrature) chrominance channels represent color details.

The luminance $Y(m, n)$ carries grayscale information with most of the signal energy (as much as 93%), and the chrominance channels $I(m, n)$ and $Q(m, n)$ carry color information with much less energy (as little as 7%). The transformation in terms of the standard matrix notion is given by

$$\begin{bmatrix} Y(m, n) \\ I(m, n) \\ Q(m, n) \end{bmatrix} = \begin{bmatrix} 0.299 & 0.587 & 0.114 \\ 0.596 & -0.274 & -0.322 \\ 0.212 & -0.523 & 0.311 \end{bmatrix} \begin{bmatrix} R(m, n) \\ G(m, n) \\ B(m, n) \end{bmatrix}. \quad (13.2)$$

As an example of data compression, after transformation, we can encode $Y(m,n)$ with a higher resolution using a larger number of bits, since it contains most of the signal energy, while we encode chrominance channels $I(m,n)$ and $Q(m,n)$ with less resolution using a smaller number of bits. Inverse transformation can be solved as

$$\begin{bmatrix} R(m, n) \\ G(m, n) \\ B(m, n) \end{bmatrix} = \begin{bmatrix} 1.000 & 0.956 & 0.621 \\ 1.000 & -0.272 & -0.647 \\ 1.000 & -1.106 & 1.703 \end{bmatrix} \begin{bmatrix} Y(m, n) \\ I(m, n) \\ Q(m, n) \end{bmatrix}. \quad (13.3)$$

To obtain the grayscale image, we simply convert each RGB pixel to a YIQ pixel and then keep its luminance channel and discard its IQ chrominance channels. Hence, the conversion formula is given by

$$Y(m, n) = 0.299 \cdot R(m, n) + 0.587 \cdot G(m, n) + 0.114 \cdot B(m, n). \quad (13.4)$$

Note that $Y(m, n)$, $I(m, n)$, and $Q(m, n)$ can be matrices that represent the luminance image and two color component images, respectively. Similarly, $R(m, n)$, $G(m, n)$, and $B(m, n)$ can be matrices for the RGB component images.

Example 13.1.

Given a pixel in an RGB image as follows:

$$R = 200,\ G = 10,\ B = 100,$$

a. Convert the pixel values to the YIQ values.

Solution:

a. Applying Equation (13.2), it follows that

$$\begin{bmatrix} Y \\ I \\ Q \end{bmatrix} = \begin{bmatrix} 0.299 & 0.587 & 0.114 \\ 0.596 & -0.274 & -0.322 \\ 0.212 & -0.523 & 0.311 \end{bmatrix} \begin{bmatrix} 200 \\ 10 \\ 100 \end{bmatrix}.$$

Carrying out the matrix operations leads to

$$\begin{bmatrix} Y \\ I \\ Q \end{bmatrix} = \begin{bmatrix} 0.299 \times 200 & 0.587 \times 10 & 0.114 \times 100 \\ 0.596 \times 200 & -0.274 \times 10 & -0.322 \times 100 \\ 0.212 \times 200 & -0.523 \times 10 & 0.311 \times 100 \end{bmatrix} = \begin{bmatrix} 77.07 \\ 84.26 \\ 68.27 \end{bmatrix}.$$

Rounding the values to integers, we have

$$\begin{bmatrix} Y \\ I \\ Q \end{bmatrix} = round \begin{bmatrix} 77.07 \\ 84.26 \\ 68.27 \end{bmatrix} = \begin{bmatrix} 77 \\ 84 \\ 68 \end{bmatrix}.$$

Now let us study the following example to convert the YIQ values back to the RGB values.

Example 13.2.

Given a pixel of an image in the YIQ color format as follows:

$$Y = 77,\ I = 84,\ Q = 68,$$

a. Convert the pixel values back to the RGB values.

Solution:

a. Applying Equation (13.3) yields

$$\begin{bmatrix} R \\ G \\ B \end{bmatrix} = \begin{bmatrix} 1.000 & 0.956 & 0.621 \\ 1.000 & -0.272 & -0.647 \\ 1.000 & -1.106 & 1.703 \end{bmatrix} \begin{bmatrix} 77 \\ 84 \\ 68 \end{bmatrix} = \begin{bmatrix} 199.53 \\ 10.16 \\ 99.90 \end{bmatrix}.$$

After rounding, it follows that

$$\begin{bmatrix} R \\ G \\ B \end{bmatrix} = round \begin{bmatrix} 199.53 \\ 10.16 \\ 99.9 \end{bmatrix} = \begin{bmatrix} 200 \\ 10 \\ 100 \end{bmatrix}.$$

Example 13.3.

Given the following 2×2 RGB image,

$$R = \begin{bmatrix} 100 & 50 \\ 200 & 150 \end{bmatrix} \quad G = \begin{bmatrix} 10 & 25 \\ 20 & 50 \end{bmatrix} \quad B = \begin{bmatrix} 10 & 5 \\ 20 & 15 \end{bmatrix},$$

a. Convert the RGB color image into a grayscale image.

Solution:

a. Since only Y components are kept in the grayscale image, we apply Equation (13.4) to each pixel in the 2×2 image and round the results to integers as follows:

$$Y = 0.299 \times \begin{bmatrix} 100 & 50 \\ 200 & 150 \end{bmatrix} + 0.587 \times \begin{bmatrix} 10 & 25 \\ 20 & 50 \end{bmatrix} + 0.114 \times \begin{bmatrix} 10 & 5 \\ 20 & 15 \end{bmatrix} = \begin{bmatrix} 37 & 30 \\ 74 & 76 \end{bmatrix}.$$

Figure 13.9 shows the grayscale image converted from the 24-bit full color image in Figure 13.5 using the RGB-to-YIQ transformation, where only the luminance information is retained.

13.1.6 MATLAB Functions for Format Conversion

The following list summarizes MATLAB functions for image format conversion:

imread = read image data file with the specified format
X = 8-bit grayscale image, 8-bit indexed image, or 24-bit RGB color image
map = color map table for the indexed image (256 entries)
imshow(X, map) = 8-bit image display

FIGURE 13.9 Grayscale image converted from the 24-bit color image in Figure 13.5 using RGB-to-YIQ transformation.

imshow(X) = 24-bit RGB color image display if image X is in a 24-bit RGB color format; grayscale image display if image X is in an 8-bit grayscale format
ind2gray = 8-bit indexed color image to 8-bit grayscale image conversion
ind2rgb = 8-bit indexed color image to 24-bit RGB color image conversion
rgb2ind = 24-bit RGB color image to 8-bit indexed color image conversion
rgb2gray = 24-bit RGB color image to 8-bit grayscale image conversion
im2double = 8-bit image to intensity image conversion
mat2gray = image data to intensity image conversion
im2uint8 = intensity image to 8-bit grayscale image conversion

Figure 13.10 on the next page outlines the applications of image format conversions.

13.2 Image Histogram and Equalization

An image histogram is a graph to show how many pixels are at each scale level, or at each index for the indexed color image. The histogram contains information needed for image equalization, where the image pixels are stretched to give a reasonable contrast.

13.2.1 Grayscale Histogram and Equalization

We can obtain a grayscale histogram by plotting pixel value distribution over the full grayscale range.

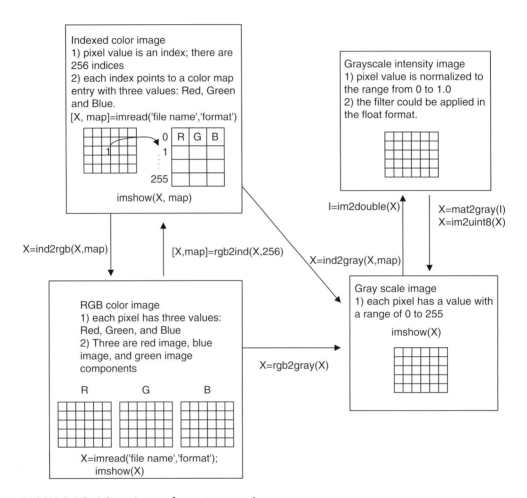

FIGURE 13.10 Image format conversions.

Example 13.4.

a. Produce a histogram given the following image (a matrix filled with integers) with the grayscale value ranging from 0 to 7, that is, with each pixel encoded into 3 bits.

$$\begin{bmatrix} 0 & 1 & 2 & 2 & 6 \\ 2 & 1 & 1 & 2 & 1 \\ 1 & 3 & 4 & 3 & 3 \\ 0 & 2 & 5 & 1 & 1 \end{bmatrix}$$

Solution:

a. Since the image is encoded using 3 bits for each pixel, we have the pixel value ranging from 0 to 7. The count for each grayscale is listed in Table 13.1.

Based on the grayscale distribution counts, the histogram is created as shown in Figure 13.11.

As we can see, the image has pixels whose levels are more concentrated in the dark scale in this example.

TABLE 13.1 Pixel counts distribution.

Pixel $p(m, n)$ Level	Number of Pixels
0	2
1	7
2	5
3	3
4	1
5	1
6	1
7	0

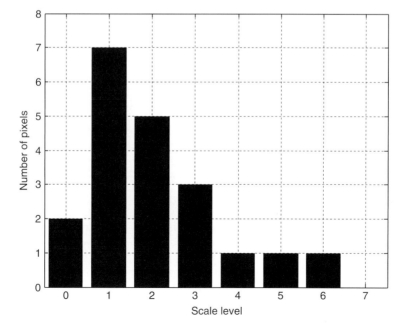

FIGURE 13.11 Histogram in Example 13.4.

With the histogram, the equalization technique can be developed. Equalization stretches the scale range of the pixel levels to the full range to give an improved contrast for the given image. To do so, the equalized new pixel value is redefined as

$$p_{eq}(m,n) = \frac{\text{Number of pixels with scale level} \leq p(m,n)}{\text{Total number of pixels}} \times (\text{maximum scale level})$$

(13.5)

The new pixel value is reassigned using the value obtained by multiplying the maximum scale level by the scaled ratio of the accumulative counts up to the current image pixel value over the total number of the pixels. Clearly, since the accumulate counts can range from 0 up to the total number of pixels, then the equalized pixel value can vary from 0 to the maximum scale level. It is due to the accumulation, the pixel values are spread over the whole range from 0 to the maximum scale level (255). Let us look at a simplified equalization example.

Example 13.5.

Given the following image (matrix filled with integers) with a grayscale value ranging from 0 to 7, that is, with each pixel encoded in 3 bits,

$$\begin{bmatrix} 0 & 1 & 2 & 2 & 6 \\ 2 & 1 & 1 & 2 & 1 \\ 1 & 3 & 4 & 3 & 3 \\ 0 & 2 & 5 & 1 & 1 \end{bmatrix},$$

a. Perform equalization using the histogram in Example 13.4, and plot the histogram for the equalized image.

Solution:

a. Using the histogram result in Table 13.1, we can compute an accumulative count for each grayscale level as shown in Table 13.2. The equalized pixel level using Equation (13.5) is given in the last column.
   To see how the old pixel level $p(m,n) = 4$ is equalized to the new pixel level $p_{eq}(m,n) = 6$, we apply Equation (13.5):

$$p_{eq}(m,n) = round\left(\frac{18}{20} \times 7\right) = 6.$$

The equalized image using Table 13.2 is finally obtained by replacing each old pixel value in the old image with its corresponding equalized new pixel value and given by

$$\begin{bmatrix} 1 & 3 & 5 & 5 & 7 \\ 5 & 3 & 3 & 5 & 3 \\ 3 & 6 & 6 & 6 & 6 \\ 1 & 5 & 7 & 3 & 3 \end{bmatrix}.$$

TABLE 13.2 Image equalization in Example 13.5.

Pixel $p(m, n)$ Level	Number of Pixels	Number of Pixels $\leq p(m, n)$	Equalized Pixel Level
0	2	2	1
1	7	9	3
2	5	14	5
3	3	17	6
4	1	18	6
5	1	19	7
6	1	20	7
7	0	20	7

TABLE 13.3 Pixel level distribution counts of the equalized image in Example 13.5.

Pixel $p(m, n)$ Level	Number of Pixels
0	0
1	2
2	0
3	6
4	0
5	5
6	4
7	2

To see how the histogram is changed, we compute the pixel level counts according to the equalized image. The result is given in Table 13.3, and Figure 13.12 shows the new histogram for the equalized image.

As we can see, the pixel levels in the equalized image are stretched to the larger scale levels. This technique woks for underexposed images.

Next, we apply the image histogram equalization to enhance a biomedical image of a human neck in Figure 13.13a, while Figure 13.13b shows the original image histogram. (The purpose of the arrow in Figure 13.13a will be explained later.) We see that there are many pixel counts residing at the lower scales in the histogram. Hence, the image looks rather dark and may be underexposed.

Figures 13.14a and 13.14b show the equalized grayscale image using the histogram method and its histogram, respectively. As shown in the histogram, the equalized pixels reside more on the larger scale, and hence the equalized image has improved contrast.

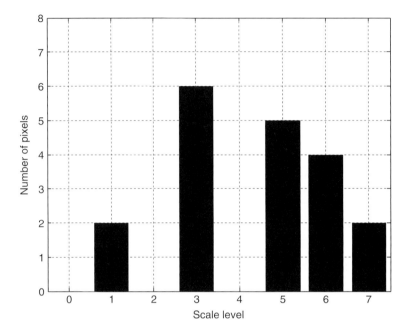

FIGURE 13.12 Histogram for the equalized image in Example 13.5.

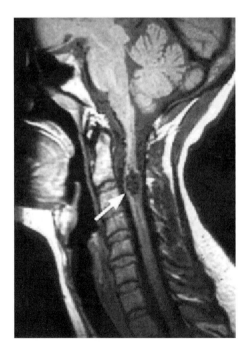

FIGURE 13.13A Original grayscale image.

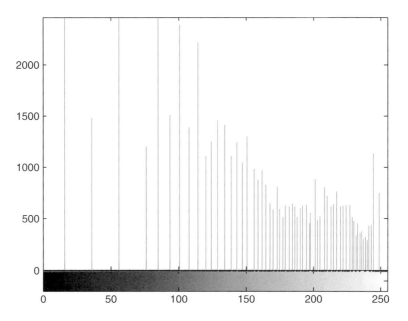

FIGURE 13.13B Histogram for the original grayscale image.

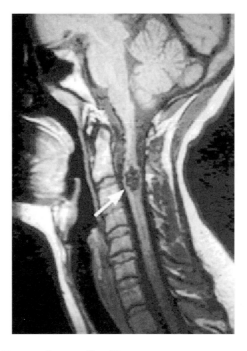

FIGURE 13.14A Grayscale equalized image.

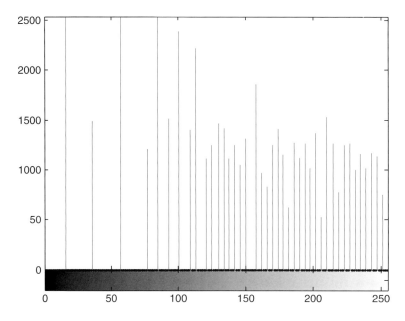

FIGURE 13.14B Histogram for the grayscale equalized image.

13.2.2 24-Bit Color Image Equalization

For equalizing the RGB image, we first transform RGB values to YIQ values, since the Y channel contains most of the signal energy, about 93%. Then the Y channel is equalized just like the grayscale image equalization to enhance the luminance. We leave the I and Q channels as they are, since these contain color information only and we do not equalize them. Next, we can repack the equalized Y channel back into the YIQ format. Finally, the YIQ values are transformed back to the RGB values for display. Figure 13.15 shows the procedure.

Figure 13.16a shows an original RGB color outdoors scene that is underexposed. Figure 13.16b shows the equalized RGB image using the method for equalizing the Y channel only. We can verify significant improvement with the equalized image showing much detailed information. The color print of the image is included in the color insert.

We can also use the histogram equalization method to equalize each of the R, G, and B channels, or their possible combinations. Figure 13.17 illustrates such a procedure.

Some color effects can be observed. Equalization of the R channel only would make the image look more red, since the red pixel values are stretched out to the full range. Similar observations can be made for equalizing the G channel or the B channel only. The equalized images for the R, G, and B

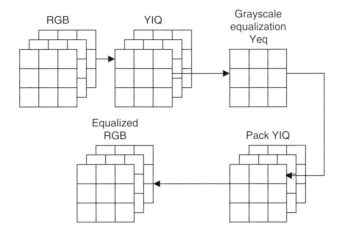

FIGURE 13.15 Color image equalization.

FIGURE 13.16A Original RGB color image. (See color insert.)

channels, respectively, are shown in Figure 13.18. The image from equalizing the R, G, and B channels simultaneously is shown in the upper left corner, which offers improved image contrast.

13.2.3 8-Bit Indexed Color Image Equalization

Equalization of the 8-bit color indexed image is more complicated. This is due to the fact that the pixel value is the index for color map entries, and there are three

FIGURE 13.16B Equalized RGB color image. (See color insert.)

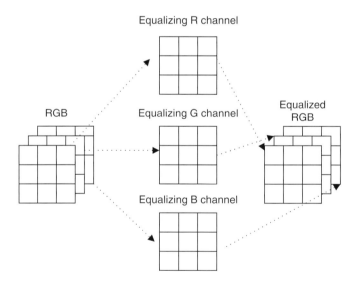

FIGURE 13.17 Equalizing RGB channels.

RGB color components for each entry. We expect that after equalization, the index for each pixel will not change from its location on the color map table. Instead, the RGB components in the color map are equalized and changed. The procedure is described and is shown in Figure 13.19.

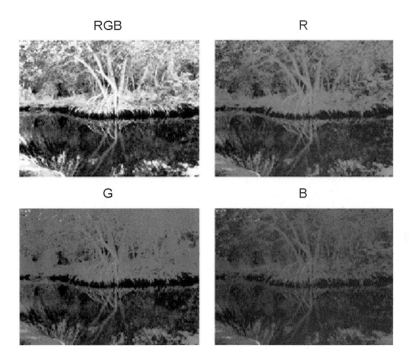

FIGURE 13.18   Equalization effects for RGB channels. (See color insert.)

Step 1. The RGB color map is converted to the YIQ color map. Note that there are only 256 color table entries. Since the image contains the index values, which point to locations on the color table containing RGB components, it is natural to convert the RGB color table to the YIQ color table.

Step 2. The grayscale image is generated using the Y channel value, so that grayscale equalization can be performed.

Step 3. Grayscale equalization is executed.

Step 4. The equalized 256 Y values are divided by their corresponding old Y values to obtain the relative luminance scale factors.

Step 5. Finally, the R, G, and B values are each scaled in the old RGB color table with the corresponding relative luminance scale factor and are normalized as new RGB channels in the color table in the correct range. Then the new RGB color map is the output.

Note that original index values are not changed; only the color map content is.

Using the previous outdoors picture for the condition of underexposure, Figure 13.20 shows the equalized indexed color image. We see that the equalized image displays much more detail. Its color version is reprinted in the color insert.

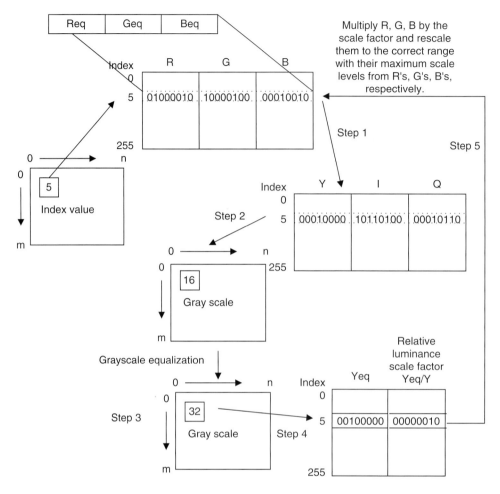

FIGURE 13.19 Equalization of 8-bit indexed color image.

13.2.4 MATLAB Functions for Equalization

Figure 13.21 lists MATLAB functions for performing equalization for the different image formats. The MATLAB functions are explained as follows:

histeq = grayscale histogram equalization, or 8-bit indexed color histogram equalization
imhist = histogram display
rgb2ntsc = 24-bit RGB color image to 24-bit YIQ color image conversion
ntsc2rgb = 24-bit YIQ color image to 24-bit RGB color image conversion

Examples using the MATLAB functions for image format conversion and equalization are given in Program 13.1.

Program 13.1. Examples of image format conversion and equalization.

```
disp('Read the rgb image');
XX=imread('trees','JPEG');           % Provided by the instructor
figure, imshow(XX); title('24-bit color');
disp('the grayscale image and histogram');
Y=rgb2gray(XX);          %RGB to grayscale conversion
figure, subplot(1, 2, 1);imshow(Y);
title('original');subplot(1, 2, 2);imhist(Y, 256);
disp('Equalization in grayscale domain');
Y=histeq(Y);
figure, subplot(1, 2, 1);imshow(Y);
title('EQ in grayscale-domain');subplot(1, 2, 2);imhist(Y, 256);
disp('Equalization of Y channel for RGB color image');
figure
subplot(1, 2, 1);imshow(XX);
title('EQ in RGB color');
subplot(1, 2, 2);imhist(rgb2gray(XX), 256);
Z1=rgb2ntsc(XX);         % Conversion from RGB to YIQ
Z1(:,:,1)=histeq(Z1(:,:,1));          %Equalizing Y channel
ZZ=ntsc2rgb(Z1);         %Conversion from YIQ to RGB
figure
subplot(1, 2, 1);imshow(ZZ);
title('EQ for Y channel for RGB color image');
subplot(1, 2, 2);imhist(im2uint8(rgb2gray(ZZ)), 256);
ZZZ=XX;
ZZZ(:,:,1)=histeq(ZZZ(:,:,1));        %Equalizing R channel
ZZZ(:,:,2)=histeq(ZZZ(:,:,2));        %Equalizing G channel
ZZZ(:,:,3)=histeq(ZZZ(:,:,3));        %Equalizing B channel
figure
subplot(1, 2, 1);imshow(ZZZ);
title('EQ for RGB channels');
subplot(1, 2, 2);imhist(im2uint8(rgb2gray(ZZZ)), 256);
disp('Eqlization in 8-bit indexd color');
[Xind, map]=rgb2ind(XX, 256);         % RGB to 8-bit index image conversion
newmap=histeq(Xind, map);
figure
subplot(1, 2, 1);imshow(Xind,newmap);
title('EQ in 8-bit indexed color');
subplot(1, 2, 2);imhist(ind2gray(Xind,newmap), 256);
```

13.3 Image Level Adjustment and Contrast

Image level adjustment can be used to linearly stretch the pixel level in an image to increase contrast and shift the pixel level to change viewing effects. Image

FIGURE 13.20   Equalized indexed 8-bit color image. (See color insert.)

level adjustment is also a requirement for modifying results from image filtering or other operations to an appropriate range for display. We will study this technique in the following subsections.

13.3.1   Linear Level Adjustment

Sometimes, if the pixel range in an image is small, we can adjust the image pixel level to make use of a full pixel range. Hence, contrast of the image is enhanced. Figure 13.22 illustrates linear level adjustment.

The linear level adjustment is given by the following formula:

$$p_{adjust}(m, n) = Bottom + \frac{p(m, n) - L}{H - L} \times (Top - Bottom), \qquad (13.6)$$

where $p(m, n)$ = original image pixel

$p_{adjust}(m, n)$ = desired image pixel
H = maximum pixel level in the original image
L = minimum pixel level in the original image
Top = maximum pixel level in the desired image
Bottom = minimum pixel level in the desired image

Besides adjusting the image level to a full range, we can also apply the method to shift the image pixel levels up or down.

13.3 Image Level Adjustment and Contrast 639

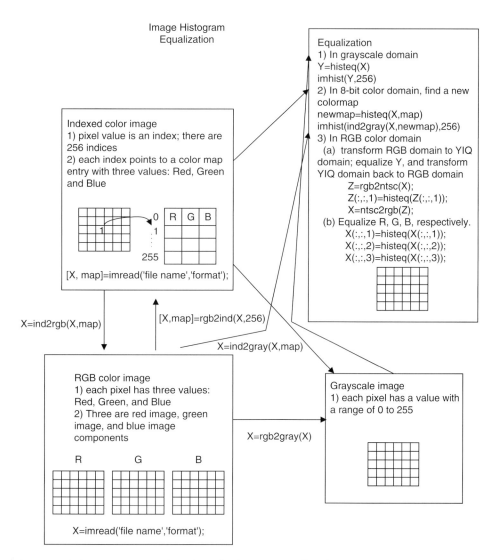

FIGURE 13.21 MATLAB functions for image equalization.

Example 13.6.

Given the following image (matrix filled with integers) with a grayscale value ranging from 0 to 7, that is, with each pixel encoded in 3 bits,

$$\begin{bmatrix} 3 & 4 & 4 & 5 \\ 5 & 3 & 3 & 3 \\ 4 & 4 & 4 & 5 \\ 3 & 5 & 3 & 4 \end{bmatrix},$$

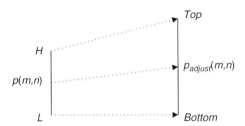

FIGURE 13.22 Linear level adjustment.

a. Perform level adjustment to a full range.
b. Shift the level to the range from 3 to 7.
c. Shift the level to the range from 0 to 3.

Solution:

a. From the given image, we set the following for level adjustment to the full range:

$$H = 5, L = 3, \text{Top} = 2^3 - 7, \text{Bottom} = 0.$$

Applying Equation (13.6) yields the second column in Table 13.4.

b. For the shift-up operation, it follows that

$$H = 5, L = 3, \text{Top} = 7, \text{Bottom} = 3.$$

c. For the shift-down operation, we set

$$H = 5, L = 3, \text{Top} = 3, \text{Bottom} = 0.$$

The results for (b) and (c) are listed in the third and fourth columns, respectively, of Table 13.4.

According to Table 13.4, we have three images:

$$\begin{bmatrix} 0 & 4 & 4 & 7 \\ 7 & 0 & 0 & 0 \\ 4 & 4 & 4 & 7 \\ 0 & 7 & 0 & 4 \end{bmatrix} \quad \begin{bmatrix} 3 & 5 & 5 & 7 \\ 7 & 3 & 3 & 3 \\ 5 & 5 & 5 & 7 \\ 3 & 7 & 3 & 5 \end{bmatrix} \quad \begin{bmatrix} 0 & 2 & 2 & 3 \\ 3 & 0 & 0 & 0 \\ 2 & 2 & 2 & 3 \\ 0 & 3 & 0 & 2 \end{bmatrix}$$

TABLE 13.4 Image adjustment results in Example 13.6.

Pixel $p(m, n)$ Level	Full Range	Range [3–7]	Range [0–3]
3	0	3	0
4	4	5	2
5	7	7	3

13.3 Image Level Adjustment and Contrast

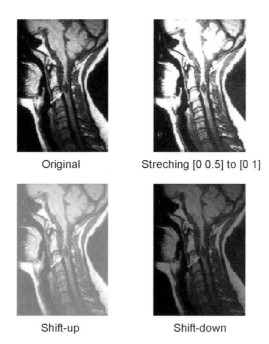

FIGURE 13.23 Image level adjustment.

Next, applying the level adjustment for the neck image of Figure 13.13(a), we get results as shown in Figure 13.23: the original image, the full-range stretched image, the level shift-up image, and the level shift-down image. As we can see, the stretching operation increases image contrast while the shift-up operation lightens the image and the shift-down operation darkens the image.

13.3.2 Adjusting the Level for Display

When two 8-bit images are added together or undergo other mathematical operations, the sum of two pixel values can be as low as 0 and as high as 510. We can apply the linear adjustment to scale the range back to 0–255 for display. The following addition of two 8-bit images:

$$\begin{bmatrix} 30 & 25 & 5 & 170 \\ 70 & 210 & 250 & 30 \\ 225 & 125 & 50 & 70 \\ 28 & 100 & 30 & 50 \end{bmatrix} + \begin{bmatrix} 30 & 255 & 50 & 70 \\ 70 & 3 & 30 & 30 \\ 50 & 200 & 50 & 70 \\ 30 & 70 & 30 & 50 \end{bmatrix} = \begin{bmatrix} 60 & 280 & 55 & 240 \\ 140 & 213 & 280 & 60 \\ 275 & 325 & 100 & 140 \\ 58 & 179 & 60 & 100 \end{bmatrix}$$

yields a sum that is out of the 8-bit range. To scale the combined image, modify Equation (13.6) as

$$p_{scaled}(m,n) = \frac{p(m,n) - \text{Minimum}}{\text{Maximum} - \text{Minimum}} \times (\text{Maximum scale level}). \qquad (13.7)$$

Note that in the image to be scaled,

$$\text{Maximum} = 325$$
$$\text{Minimum} = 55$$
$$\text{Maximum scale level} = 255,$$

we have after scaling:

$$\begin{bmatrix} 5 & 213 & 0 & 175 \\ 80 & 149 & 213 & 5 \\ 208 & 255 & 43 & 80 \\ 3 & 109 & 5 & 43 \end{bmatrix}.$$

13.3.3 Matlab Functions for Image Level Adjustment

Figure 13.24 lists applications of the MATLAB level adjustment function, which is defined as:

J = imajust(I, [bottom level, top level], [adjusted bottom, adjusted top], gamma)
I = input intensity image
J = output intensity image
gamma = 1 (linear interpolation function as we discussed in the text)
0 < gamma < 1 lightens image; and gamma > 1 darkens image.

13.4 Image Filtering Enhancement

As with one-dimensional digital signal processing, we can design a digital image filter such as lowpass, highpass, bandpass, and notch to process the image to obtain the desired effect. In this section, we discuss the most common ones: lowpass filters to remove noise, median filters to remove impulse noise, and edge detection filters to gain the boundaries of objects in the images. More advanced treatment of this subject can be explored in the well-known text by Gonzalez and Wintz (1987).

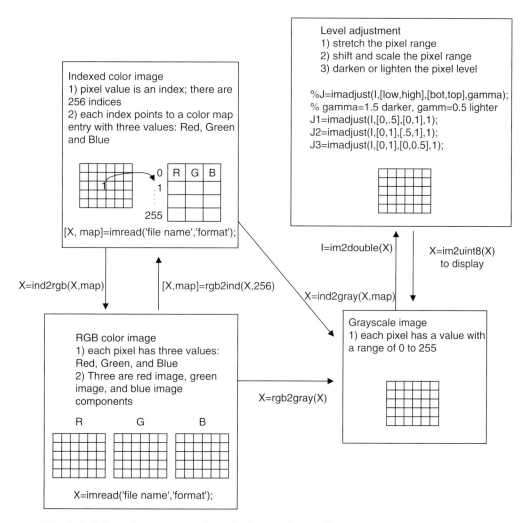

FIGURE 13.24 **MATLAB functions for image level adjustment.**

13.4.1 Lowpass Noise Filtering

One of the simplest lowpass filters is the average filter. The noisy image is filtered using the average convolution kernel with a size 3×3 block, 4×4 block, 8×8 block, and so on, in which the elements in the block have the same filter coefficients. The 3×3, 4×4, and 8×8 average kernels are as follows:

3×3 average kernel:

$$\frac{1}{9}\begin{bmatrix} 1 & 1 & 1 \\ 1 & 1 & 1 \\ 1 & 1 & 1 \end{bmatrix} \quad (13.8)$$

4×4 average kernel:

$$\frac{1}{16}\begin{bmatrix} 1 & 1 & 1 & 1 \\ 1 & 1 & 1 & 1 \\ 1 & 1 & 1 & 1 \\ 1 & 1 & 1 & 1 \end{bmatrix} \quad (13.9)$$

8×8 average kernel:

$$\frac{1}{64}\begin{bmatrix} 1 & 1 & 1 & 1 & 1 & 1 & 1 & 1 \\ 1 & 1 & 1 & 1 & 1 & 1 & 1 & 1 \\ 1 & 1 & 1 & 1 & 1 & 1 & 1 & 1 \\ 1 & 1 & 1 & 1 & 1 & 1 & 1 & 1 \\ 1 & 1 & 1 & 1 & 1 & 1 & 1 & 1 \\ 1 & 1 & 1 & 1 & 1 & 1 & 1 & 1 \\ 1 & 1 & 1 & 1 & 1 & 1 & 1 & 1 \\ 1 & 1 & 1 & 1 & 1 & 1 & 1 & 1 \end{bmatrix} \quad (13.10)$$

Each of the elements in the average kernel is 1, and the scale factor is the reciprocal of the total number of elements in the kernel. The convolution operates to modify each pixel in the image as follows. By passing the center of a convolution kernel through each pixel in the noisy image, we can sum each product of the kernel element and the corresponding image pixel value and multiply the sum by the scale factor to get the processed pixel. To understand the filter operation with the convolution kernel, let us study the following example.

Example 13.7.

a. Perform digital filtering on the noisy image using 2×2 convolutional average kernel, and compare the enhanced image with the original one given the following 8-bit grayscale original and corrupted (noisy) images.

13.4 Image Filtering Enhancement

4×4 original image: $\begin{bmatrix} 100 & 100 & 100 & 100 \\ 100 & 100 & 100 & 100 \\ 100 & 100 & 100 & 100 \\ 100 & 100 & 100 & 100 \end{bmatrix}$

4×4 corrupted image: $\begin{bmatrix} 99 & 107 & 113 & 96 \\ 92 & 116 & 84 & 107 \\ 103 & 93 & 86 & 108 \\ 87 & 109 & 106 & 107 \end{bmatrix}$

2×2 average kernel: $\dfrac{1}{4}\begin{bmatrix} 1 & 1 \\ 1 & 1 \end{bmatrix}$.

Solution:

a. In the following diagram, we pad edges with zeros in the last row and column before processing at the point where the first kernel and the last kernel are shown in dotted-line boxes, respectively.

99	107	113	96	0
92	116	84	107	0
103	93	86	108	0
87	109	106	107	0
0	0	0	0	0

To process the first element, we know that the first kernel covers the image elements as $\begin{bmatrix} 99 & 107 \\ 92 & 116 \end{bmatrix}$. Summing each product of the kernel element and the corresponding image pixel value, multiplying a scale factor of $1/4$, and rounding the result, it follows that

$$\frac{1}{4}(99 \times 1 + 107 \times 1 + 92 \times 1 + 116 \times 1) = 103.5$$

$round(103.5) = 104.$

In the processing of the second element, the kernel covers $\begin{bmatrix} 107 & 113 \\ 116 & 84 \end{bmatrix}$. Similarly, we have

$$\frac{1}{4}(107 \times 1 + 113 \times 1 + 116 \times 1 + 84 \times 1) = 105$$

$round(105) = 105.$

The process continues for the rest of the image pixels. To process the last element of the first row, 96, since the kernel covers only $\begin{bmatrix} 96 & 0 \\ 107 & 0 \end{bmatrix}$, we assume that the last two elements are zeros. Then:

$$\frac{1}{4}(96 \times 1 + 107 \times 1 + 0 \times 1 + 0 \times 1) = 50.75$$

$$round(50.75) = 51.$$

Finally, we yield the following filtered image:

$$\begin{bmatrix} 104 & 105 & 100 & 51 \\ 101 & 95 & 96 & 54 \\ 98 & 98 & 102 & 54 \\ 49 & 54 & 53 & 27 \end{bmatrix}.$$

As we know, due to zero padding for boundaries, the last-row and last-column values are in error. However, for a large image, these errors at boundaries can be neglected without affecting image quality. The first 3×3 elements in the processed image have values that are close to those of the original image. Hence, the image is enhanced.

Figure 13.25 shows the noisy image and enhanced images using the $3 \times 3, 4 \times 4$, and 8×8 average lowpass filter kernels, respectively. The average kernel removes noise. However, it also blurs the image. When using a large-sized kernel, the quality of the processed image becomes unacceptable.

The sophisticated large-size kernels are used for noise filtering. Although it is beyond the scope of this text, the Gaussian filter kernel with the standard deviation $\sigma = 0.9$, for instance, is given by the following:

$$\frac{1}{25}\begin{bmatrix} 0 & 2 & 4 & 2 & 0 \\ 2 & 15 & 27 & 15 & 2 \\ 4 & 27 & 50 & 27 & 4 \\ 2 & 15 & 27 & 15 & 2 \\ 0 & 2 & 4 & 2 & 0 \end{bmatrix}. \quad (13.11)$$

This kernel weights the center pixel to be processed most and weights less and less to the pixels away from the center pixel. In this way, the blurring effect can be reduced when filtering the noise. The plot of kernel values in the spacial domain looks like the bell-shape. The steepness of shape is controlled by the standard deviation of the Gaussian distribution function, in which the larger the standard deviation, the flatter the kernel; hence, the more blurring effect will occur.

Figure 13.26a shows the noisy image, while Figure 13.26b shows the enhanced image using the 5×5 Gaussian filter kernel. Clearly, the majority of the noise has been filtered, while the blurring effect is significantly reduced.

13.4.2 Median Filtering

The median filter is one type of nonlinear filters. It is very effective at removing impulse noise, the "pepper and salt" noise, in the image. The principle of the median filter is to replace the gray level of each pixel by the median of the gray

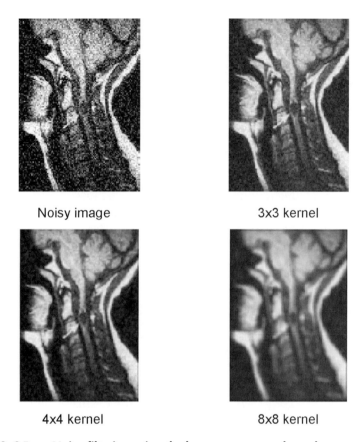

FIGURE 13.25 Noise filtering using the lowpass average kernels.

levels in a neighborhood of the pixel, instead of using the average operation. For median filtering, we specify the kernel size, list the pixel values covered by the kernel, and determine the median level. If the kernel covers an even number of pixels, the average of two median values is used. Before beginning median filtering, zeros must be padded around the row edge and the column edge. Hence, edge distortion is introduced at the image boundary. Let us look at Example 13.8.

Example 13.8.

Given a 3×3 median filter kernel and the following 8-bit grayscale original and corrupted (noisy) images,

$$4 \times 4 \text{ original image}: \begin{bmatrix} 100 & 100 & 100 & 100 \\ 100 & 100 & 100 & 100 \\ 100 & 100 & 100 & 100 \\ 100 & 100 & 100 & 100 \end{bmatrix}$$

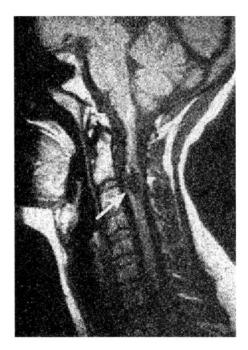

FIGURE 13.26A Noisy image of a human neck.

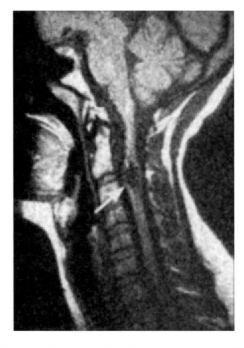

FIGURE 13.26B Enhanced image using Gaussian lowpass filter.

4×4 corrupted image by impulse noise:
$$\begin{bmatrix} 100 & 255 & 100 & 100 \\ 100 & 255 & 100 & 100 \\ 255 & 100 & 100 & 0 \\ 100 & 100 & 100 & 100 \end{bmatrix}$$

3×3 median filter kernel: $\begin{bmatrix} & & \\ & & \\ & & \end{bmatrix}_{3 \times 3}$

a. Perform digital filtering, and compare the filtered image with the original one.

Solution:

a. Step 1: The 3×3 kernel requires zero padding $3/2 = 1$ column of zeros at the left and right edges and $3/2 = 1$ row of zeros at the upper and bottom edges:

```
0   0    0    0    0   0
0  |100  255| 100  100 0
0  |100  255| 100  100 0
0  |255  100| 100  0   0
0  |100  100  100  100 0
0   0    0    0    0   0
```

Step 2: To process the first element, we cover the 3×3 kernel with the center pointing to the first element to be processed. The sorted data within the kernel are listed in terms of their values as

0, 0, 0, 0, 0, 100, 100, 255, 255.

The median value = median(0, 0, 0, 0, 0, 100, 100, 255, 255) = 0. Zero will replace 100.

Step 3: Continue for each element until the last is replaced. Let us see the element at the location (1, 1):

```
0   0    0    0    0    0
0  |100  255  100| 100  0
0  |100 [255] 100| 100  0
0  |255  100  100| 0    0
0  |100  100  100  100| 0
0   0    0    0    0    0
```

The values covered by the kernel are:

100, 100, 100, 100, 100, 100, 255, 255, 255.

The median value = median(100, 100, 100, 100, 100, 100, 255, 255, 255) = 100. The final processed image is

$$\begin{bmatrix} 0 & 100 & 100 & 0 \\ 100 & 100 & 100 & 100 \\ 0 & 100 & 100 & 0 \\ 100 & 100 & 100 & 100 \end{bmatrix}.$$

Some boundary pixels are distorted due to zero padding effect. However, for a large image, the distortion can be omitted versus the overall quality of the image. The 2×2 middle portion matches the original image exactly. The effectiveness of the median filter is verified via this example.

The image in Figure 13.27a is corrupted by "pepper and salt" noise. The median filter with a 3×3 kernel is used to filter this impulse noise. The enhanced image in Figure 13.27b has a significant quality improvement. Note that a larger size kernel is not appropriate for median filtering, because for a larger set of pixels the median value deviates from the desired pixel value.

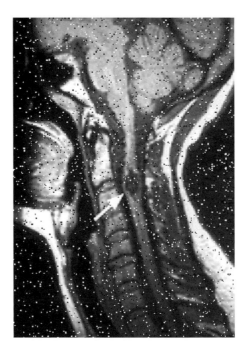

FIGURE 13.27A Noisy image (corrupted by "pepper and salt" noise).

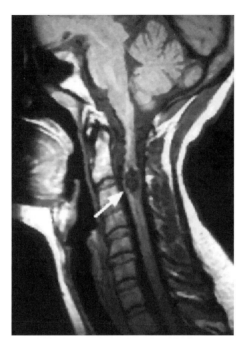

FIGURE 13.27B The enhanced image using the 3 X 3 median filter.

13.4.3 Edge Detection

In many applications, such as pattern recognition and fingerprint and iris biometric identification, image edge information is required. To obtain the edge information, a differential convolution kernel is used. Of these kernels, Sobel convolution kernels are used for horizontal and vertical edge detection. They are listed below:

Horizontal Sobel edge detector:

$$\begin{bmatrix} -1 & -2 & -1 \\ 0 & 0 & 0 \\ 1 & 2 & 1 \end{bmatrix}. \qquad (13.12)$$

The kernel subtracts the first row in the kernel from the third row to detect the horizontal difference.

Vertical Sobel edge detector:

$$\begin{bmatrix} -1 & 0 & 1 \\ -2 & 0 & 2 \\ -1 & 0 & 1 \end{bmatrix}. \qquad (13.13)$$

The kernel subtracts the first column in the kernel from the third column to detect the vertical difference.

A Laplacian edge detector is devised to tackle both vertical and horizontal edges. It is listed as follows:

Laplacian edge detector:

$$\begin{bmatrix} 0 & 1 & 0 \\ 1 & -4 & 1 \\ 0 & 1 & 0 \end{bmatrix}. \tag{13.14}$$

Example 13.9.

Given the following 8-bit grayscale image,

$$5 \times 4 \text{ original image:} \begin{bmatrix} 100 & 100 & 100 & 100 \\ 110 & 110 & 110 & 110 \\ 100 & 100 & 100 & 100 \\ 100 & 100 & 100 & 100 \\ 100 & 100 & 100 & 100 \end{bmatrix},$$

a. Use the Sobel horizontal edge detector to detect horizontal edges.

Solution:

a. We pad the image with zeros before processing as follows:

$$\begin{matrix}
0 & 0 & 0 & 0 & 0 & 0 \\
0 & 100 & 100 & 100 & 100 & 0 \\
0 & 110 & 110 & 110 & 110 & 0 \\
0 & 100 & 100 & 100 & 100 & 0 \\
0 & 100 & 100 & 100 & 100 & 0 \\
0 & 100 & 100 & 100 & 100 & 0 \\
0 & 0 & 0 & 0 & 0 & 0
\end{matrix}$$

After processing using the Sobel horizontal edge detector, we have:

$$\begin{bmatrix} 330 & 440 & 440 & 330 \\ 0 & 0 & 0 & 0 \\ -30 & -40 & -40 & -30 \\ 0 & 0 & 0 & 0 \\ -300 & -400 & -400 & -300 \end{bmatrix}.$$

Adjusting the scale level leads to

$$\begin{bmatrix} 222 & 255 & 255 & 222 \\ 121 & 121 & 121 & 121 \\ 112 & 109 & 109 & 112 \\ 121 & 121 & 121 & 121 \\ 30 & 0 & 0 & 30 \end{bmatrix}.$$

Disregarding the first row and first column and the last row and last column, since they are at image boundaries, we identify a horizontal line of 109 in the third row.

Figure 13.28 shows the results from edge detection.

Figure 13.29 shows the edge detection for the grayscale image of the cruise ship in Figure 13.3. Sobel edge detection can tackle only the horizontal edge or the vertical edge, as shown in Figure 13.29, where the edges of the image have both horizontal and vertical features. We can simply combine two horizontal and vertical edge-detected images and then rescale the resultant image in the full range. Figure 13.29(c) shows that the edge detection result is equivalent to that of the Laplacian edge detector.

Next, we apply a more sophisticated Laplacian of Gaussian filter for edge detection, which is a combined Gaussian lowpass filter and Laplacian derivative operator (highpass filter). The filter *smoothes* the image to suppress noise using the Gaussian lowpass filter, then performs Laplacian derivative operation for

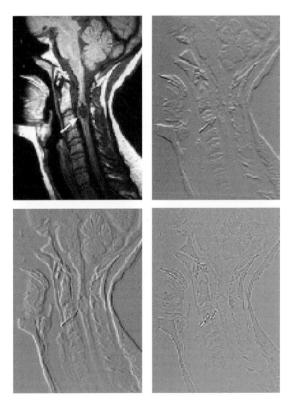

FIGURE 13.28 Image edge detection. (*Upper left*) Original image; (*upper right*) result from Sobel horizontal edge detector; (*lower left*) result from Sobel vertical edge detector; (*lower right*) result from Laplacian edge detector.

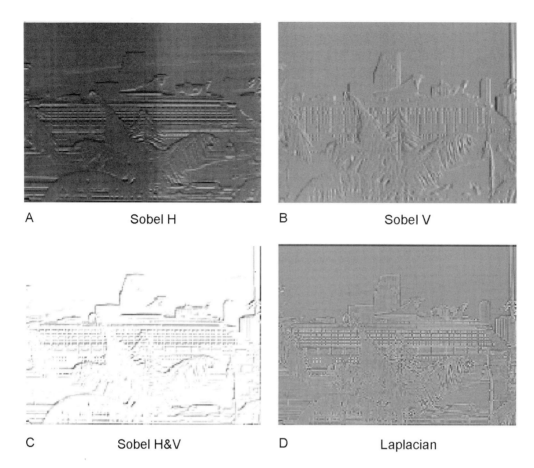

FIGURE 13.29 Edge detection. (H, horizontal, V, vertical, H&V, horizontal and vertical.)

edge detection, since the noisy image is very sensitive to the Laplacian derivative operation. As we discussed for the Gaussian lowpass filter, the standard deviation in the Gaussian distribution function controls degree of noise filtering before Laplacian derivative operation. A larger value of the standard deviation may blur the image; hence, some edge boundaries could be lost. Its selection should be based on the particular noisy image. The filter kernel with the standard deviation of $\sigma = 0.8$ is given by

$$\begin{bmatrix} 4 & 13 & 16 & 13 & 4 \\ 13 & 9 & -25 & 9 & 13 \\ 16 & -25 & -124 & -25 & 16 \\ 13 & 9 & -25 & 9 & 13 \\ 4 & 13 & 16 & 13 & 4 \end{bmatrix}. \qquad (13.15)$$

FIGURE 13.5   The 24-bit color image.

FIGURE 13.7   The 8-bit color indexed image.

FIGURE 13.16A Original RGB color image.

FIGURE 13.16B Equalized RGB color image.

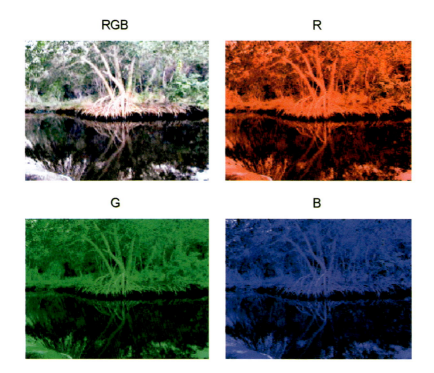

FIGURE 13.18 Equalization effects for RGB channels.

FIGURE 13.20 Equalized indexed 8-bit color image.

FIGURE 13.33B The pseudo-color image.

FIGURE 13.40   JPEG compressed color image.

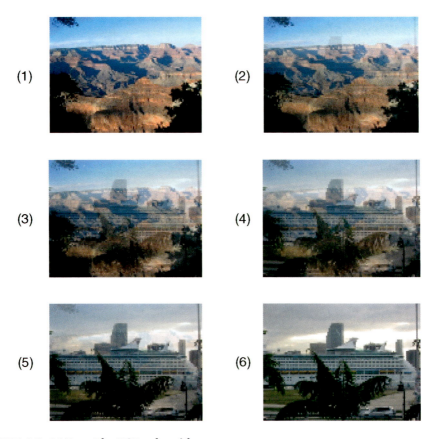

FIGURE 13.41B   The RGB color video sequence.

Laplacian of Gaussian filter 5x5 kernel

FIGURE 13.30   Image edge detection using Laplacian of Gaussian filter.

The processed edge detection using the Laplacian of Gaussian filter in Equation (13.15) is shown in Figure 13.30. We can further use a threshold value to convert the processed image to a black and white image, where the contours of objects can be clearly displayed.

13.4.4 MATLAB Functions for Image Filtering

MATLAB image filter design and implementation functions are summarized in Figure 13.31. MATLAB functions are explained as:

`X` = image to be processed
`fspecial('filter type', kernel size, parameter)` = convolution kernel generation
`H = FSPECIAL('gaussian',HSIZE,SIGMA)` returns a rotationally
symmetric Gaussian lowpass filter of size HSIZE with standard deviation SIGMA (positive).
`H = FSPECIAL('log',HSIZE,SIGMA)` returns a rotationally symmetric
Laplacian of Gaussian filter of size HSIZE with standard deviation SIGMA (positive).
`filter([convolution kernel], X)` = two-dimensional filter using the convolution kernel
`medfilt2(X, [row size, column size])` = two-dimensional median filter

Program 13.2 lists the sample MATLAB codes for filtering applications. Figure 13.31 outlines the applications of the MATLAB functions.

Program 13.2. Examples of Gaussian filtering, median filtering, and Laplacian of Gaussian filtering.

```
close all;clear all; clc;
X=imread('cruise','jpeg');          % Provided by the instructor
Y=rgb2gray(X);         % Convert the rgb image to the grayscale image
I=im2double(Y);          % Get the intensity image
image1_g=imnoise(I,'gaussian'); % Add random noise to the intensity image
ng=mat2gray(image1_g);          %Adjust the range
ng=im2uint8(ng);         % 8-bit corrupted image
%Linear Filtering
K_size= 5;         % Kernel size = 5 × 5
sigma =0.8; % sigma: the bigger, the smoother the processed image
h=fspecial('gaussian', K_size, sigma); % Determine Gaussian filter coefficients
%This command will construct a Gaussian filter
%of size 5x5 with a mainlobe width of 0.8.
image1_out=filter2(h,image1_g);         % Perform filtering
image1_out=mat2gray(image1_out);          % Adjust the range
image1_out=im2uint8(image1_out);            % Get the 8-bit image
subplot(1, 2, 1);imshow(ng),title('Noisy image');
subplot(1, 2, 2);imshow(image1_out);
title('5 × 5 Gaussian kernel');
%Median Filtering
image1_s=imnoise(I,'salt & pepper'); % Add ''pepper and salt'' noise to the image
mn=mat2gray(image1_s);          % Adjust the range
mn=im2uint8(mn);          % Get the 8-bit image
K_size=3;         % kernel size
image1_m=medfilt2(image1_s, [K_size, K_size]); % Perform median filtering
image1_m=mat2gray(image1_m);         % Adjust range
image1_m=im2uint8(image1_m);            % Get the 8-bit image
figure, subplot(1, 2, 1);imshow(mn)
title('Median noisy');
subplot(1, 2, 2);imshow(image1_m);
title('3 × 3 median kernel');
%Laplacian of Gaussian filtering
K_size =5;         % Kernel size
sigma =0.9;            % Sigma parameter
h=fspecial('log', K_size, sigma); % Determine the Laplacian of Gaussian filter kernel
image1_out=filter2(h,I);          % Perform filtering
image1_out=mat2gray(image1_out);          % Adjust the range
image1_out=im2uint8(image1_out);            % Get the 8-bit image
figure,subplot(1, 2, 1);imshow(Y)
title('Original');
subplot(1,2,2);imshow(image1_out);
title('Laplacian filter 5 × 5 kernel');
```

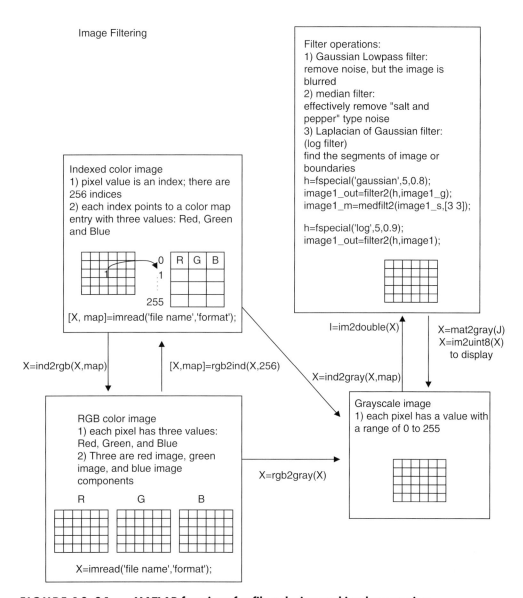

FIGURE 13.31 MATLAB functions for filter design and implementation.

13.5 Image Pseudo-Color Generation and Detection

We can apply certain transformations to the grayscale image so that it becomes a color image, and a wider range of pseudo-color enhancement can be obtained. In object detection, pseudo-color generation can produce the specific color for

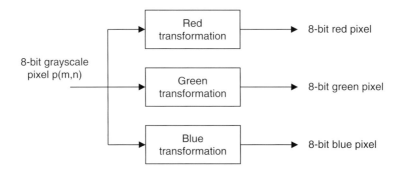

FIGURE 13.32 Block diagram for transforming a grayscale pixel to a pseudo-color pixel.

the object that is to be detected, say, red. This would significantly increase accuracy of the identification. To do so, we choose three transformations of the grayscale level to the RGB components, as shown in Figure 13.32.

As a simple choice, we choose three sine functions for RGB transformations, as shown in Figure 13.33(a). Changing the phase and period of one sine function can be easily done so that the grayscale pixel level of the object to be detected is aligned to the desired color with its component value as large as possible, while the other two functions transform the same grayscale level to have their color component values as small as possible. Hence, the single specified color object can be displayed in the image for identification. By carefully choosing the phase and period of each sine function, certain object(s) can be transformed to red, green, or blue with a favorable choice.

Example 13.10.

In the grayscale image in Figure 13.13a, the area pointed to by the arrow has a grayscale value approximately equal to 60. The background has a pixel value approximately equal to 10.

a. Generate the background to be as close to blue as possible, and make the area pointed to by the arrow as close to red as possible.

Solution:

a. The transformation functions are chosen as shown in Figure 13.33a, where the red value is largest at 60 and the blue and green values approach zero. At the grayscale of 10, the blue value is dominant. Figure 13.33b shows the processed pseudo-color image; it is included in the color insert.

Figure 13.34 illustrates the pseudo-color generation procedure.

13.5 Image Pseudo-Color Generation and Detection 659

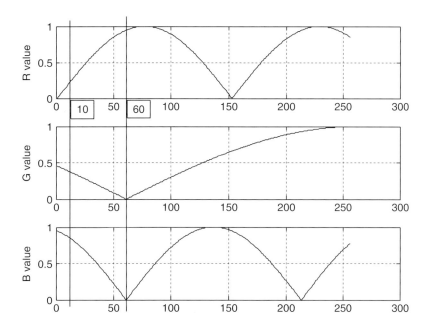

FIGURE 13.33A   Three sine functions for grayscale transformation.

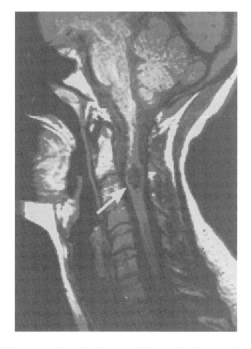

FIGURE 13.33B   The pseudo-color image. (See color insert.)

660 13 IMAGE PROCESSING BASICS

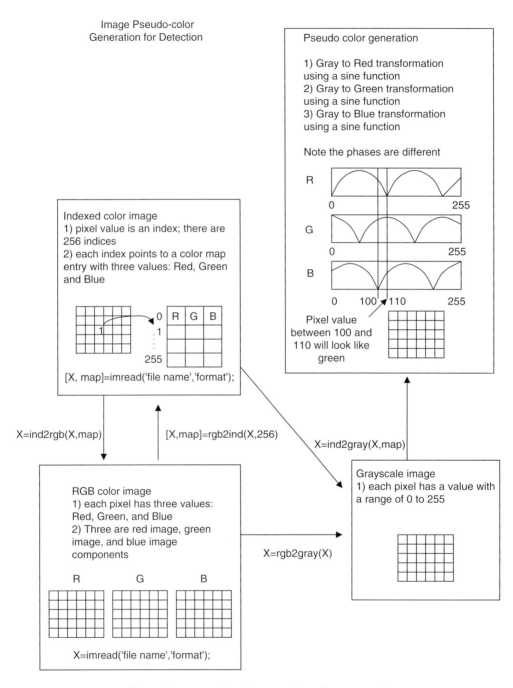

FIGURE 13.34 Illustrative procedure for pseudo-color generation.

Program 13.3 lists the sample MATLAB codes for pseudo-color generation for a grayscale image.

Program 13.3. Program examples for pseudo-color generation.

```
close all; clear all;clc
disp('Convert the grayscale image to the pseudo-color image1');
[X, map]=imread('clipim2','gif'); % Read 8-bit index image, provided by the instructor
Y=ind2gray(X,map); % 8-bit color image to the grayscale conversion
%Apply pseudo color functions using sinusoids
C_r =304;          % Cycle change for the red channel
P_r=0;             % Phase change for the red channel
C_b=804;           % Cycle change for the blue channel
P_b=60;            % Phase change for the blue channel
C_g=304;           % Cycle change for the green channel
P_g=60;            % Phase change for the green channel
r=abs(sin(2*pi*[-P_r:255-P_r]/C_r));
g=abs(sin(2*pi*[-P_b:255-P_b]/C_b));
b=abs(sin(2*pi*[-P_g:255-P_g]/C_g));
figure, subplot(3, 1, 1);plot(r,'r');grid;ylabel('R value')
subplot(3, 1, 2);plot(g,'g');grid;ylabel('G value');
subplot(3, 1, 3);plot(b,'b');grid;ylabel('B value');
figure, imshow(Y);
map=[r;g;b;]';             % Construct the color map
figure, imshow(Y, map);    % Display the pseudo-color image
```

13.6 Image Spectra

In one-dimensional digital signal processing such as for speech and other audio, we need to examine the frequency contents, check filtering effects, and perform feature extraction. Image processing is similar. However, we need to apply a two-dimensional discrete Fourier transform (2D-DFT) instead of a one-dimensional (1D) DFT. The spectrum including the magnitude and phase is also in two dimensions. The equations of the 2D-DFT are given by:

$$X(u,v) = \sum_{m=0}^{M-1} \sum_{n=0}^{N-1} p(m,n) W_M^{um} W_N^{vn}, \qquad (13.16)$$

where $W_M = e^{-j\frac{2\pi}{M}}$ and $W_N = e^{-j\frac{2\pi}{N}}$,
m and n = pixel locations
u and v = frequency indices.

Taking the absolute value of the 2D-DFT coefficients $X(u,v)$ and dividing the absolute value by $(M \times N)$, we get the magnitude spectrum as

$$A(u,v) = \frac{1}{(N \times M)} |X(u,v)|. \qquad (13.17)$$

Instead of going through the details of the 2D-DFT, we focus on application results via examples.

Example 13.11.

a. Determine the 2D-DFT coefficients and magnitude spectrum for the following 2×2 image:

$$\begin{bmatrix} 100 & 50 \\ 100 & -10 \end{bmatrix}.$$

Solution:

a. Since $M = N = 2$, applying Equation (13.16) leads to

$$X(u, v) = p(0, 0) e^{-j\frac{2\pi u \times 0}{2}} \times e^{-j\frac{2\pi v \times 0}{2}} + p(0, 1) e^{-j\frac{2\pi u \times 0}{2}} \times e^{-j\frac{2\pi v \times 1}{2}}$$
$$+ p(1, 0) e^{-j\frac{2\pi u \times 1}{2}} \times e^{-j\frac{2\pi v \times 0}{2}} + p(1, 1) e^{-j\frac{2\pi u \times 1}{2}} \times e^{-j\frac{2\pi v \times 1}{2}}.$$

For $u = 0$ and $v = 0$, we have

$$X(0, 0) = 100 e^{-j0} \times e^{-j0} + 50 e^{-j0} \times e^{-j0} + 100 e^{-j0} \times e^{-j0} - 10 e^{-j0} \times e^{-j0}.$$
$$= 100 + 50 + 100 - 10 = 240$$

For $u = 0$ and $v = 1$, we have

$$X(0, 1) = 100 e^{-j0} \times e^{-j0} + 50 e^{-j0} \times e^{-j\pi} + 100 e^{-j0} \times e^{-j0} - 10 e^{-j0} \times e^{-j\pi}.$$
$$= 100 + 50 \times (-1) + 100 - 10 \times (-1) = 160$$

Following similar operations,

$$X(1, 0) = 60 \quad \text{and} \quad X(1, 1) = -60.$$

Thus, we have DFT coefficients as

$$X(u, v) = \begin{bmatrix} 240 & 160 \\ 60 & -60 \end{bmatrix}.$$

Using Equation (13.17), we can calculate the magnitude spectrum as

$$A(u,v) = \begin{bmatrix} 60 & 40 \\ 15 & 15 \end{bmatrix}.$$

We can use the MATLAB function **fft2()** to verify the calculated DFT coefficients:

≫ X = fft2([100 50;100 − 10])

X =

240 160

60 − 60.

Example 13.12.

Given the following 200 × 200 grayscale image with a white rectangle (11 × 3 pixels) at its center and a black background, shown in Figure 13.35(a), we can compute the image's amplitude spectrum whose magnitudes are scaled in the range from 0 to 255. We can display the spectrum in terms of the grayscale. Figure 13.35(b) shows the spectrum image.

The displayed spectrum has four quarters. The left upper quarter corresponds to the frequency components, and the other three quarters are the image counterparts. From the spectrum image, the area of the upper left corner is white and hence has the higher scale value. So, the image signal has low-frequency dominant components. The first horizontal null line can be estimated as $200/11 = 18$, while the first vertical null line happens at $200/3 = 67$. Next, let us apply the 2D spectrum to understand image filtering effects in image enhancement.

FIGURE 13.35A   **A square image.**

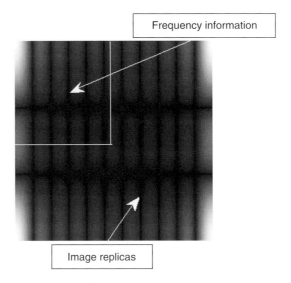

FIGURE 13.35B Magnitude spectrum for the square image.

Example 13.13.

Figure 13.36(a) is a biomedical image corrupted by random noise. Before we apply lowpass filtering, its 2D-DFT coefficients are calculated. We then compute its magnitude spectrum and scale it to the range of 0 to 255. To see noise spectral components, the spectral magnitude is further multiplied by a factor of 100. Once the spectral value is larger than 255, it is clipped to 255. The resultant spectrum is displayed in Figure 13.36(b), where we can see that noise occupies the entirety of the image.

To enhance the image, we apply a Gaussian lowpass filter. The enhanced image is shown in Figure 13.36(c), in which the enhancement is easily observed. Figure 13.36(d) displays the spectra for the enhanced image with the same scaling process as described. As we can see, the noise is significantly reduced compared with Figure 13.36(b).

13.7 Image Compression by Discrete Cosine Transform

Image compression is a must in our modern media systems, such as digital still and video cameras and computer systems. The purpose of compression is to reduce information storage or transmission bandwidth without losing image

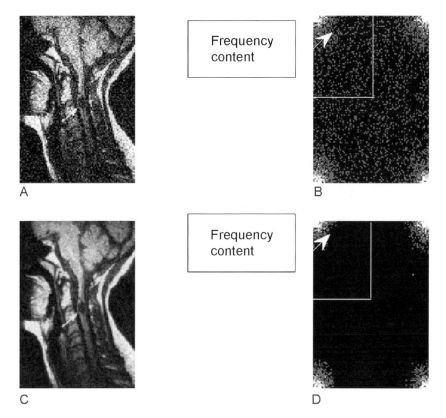

FIGURE 13.36 Magnitude spectrum plots for the noisy image and the noise-filtered image. (a) The noisy image, (b) magnitude spectrum of the noisy image, (c) noise-filtered image, (d) magnitude spectrum of the noise-filtered image.

quality or at least without losing it significantly. Image compression can be classified as lossless compression or lossy compression. Here we focus on lossy compression using discrete cosine transform (DCT).

The DCT is a core compression technology used in the industry standards JPEG (Joint Photographic Experts Group) for still-image compression and MPEG (Motion Picture Experts Group) for video compression, achieving compression ratios of 20:1 without noticeable quality degradation. JPEG standard image compression is used every day in real life.

The principle of the DCT is to transform the original image pixels to their DCT coefficients with the same number of the original image pixels, where the DCT coefficients have non-uniform distribution of direct-current (DC) terms representing the average values, and alternate-current (AC) terms representing fluctuations. The compression is achieved by applying the advantages of encoding DC terms (of a large dynamic range) with a larger number of bits and low-frequency AC terms (a few, with a reduced dynamic range) with a reduced number of bits, and neglecting some high-frequency AC terms having small dynamic ranges (most of which do not affect the visual quality of the picture).

13.7.1 Two-Dimensional Discrete Cosine Transform

Image compression uses 2D-DCT, whose transform pairs are defined as:

Forward DCT:

$$F(u, v) = \frac{2C(u)C(v)}{\sqrt{MN}} \sum_{i=0}^{M-1} \sum_{j=0}^{N-1} p(i, j) \cos\left(\frac{(2i+1)u\pi}{2M}\right) \cos\left(\frac{(2j+1)v\pi}{2N}\right) \quad (13.18)$$

Inverse DCT:

$$p(i, j) = \sum_{u=0}^{M-1} \sum_{v=0}^{N-1} \frac{2C(u)C(v)}{\sqrt{MN}} F(u, v) \cos\left(\frac{(2i+1)u\pi}{2M}\right) \cos\left(\frac{(2j+1)v\pi}{2N}\right) \quad (13.19)$$

where

$$C(m) = \begin{cases} \frac{\sqrt{2}}{2} & \text{if } m = 0 \\ 1 & \text{otherwise} \end{cases} \quad (13.20)$$

$p(i, j)$ = pixel level at the location (i, j)
$F(u, v)$ = DCT coefficient at the frequency indices (u, v).

JPEG divides an image into 8×8 image subblocks and applies DCT for each subblock individually. Hence, we simplify the general 2D-DCT in terms of 8×8 size. The equation for 2D 8×8 DCT is modified as:

$$F(u, v) = \frac{C(u)C(v)}{4} \sum_{i=0}^{7} \sum_{j=0}^{7} p(i, j) \cos\left(\frac{(2i+1)u\pi}{16}\right) \cos\left(\frac{(2j+1)v\pi}{16}\right). \quad (13.21)$$

The inverse of 2D 8×8 DCT is expressed as:

$$p(i, j) = \sum_{u=0}^{7} \sum_{v=0}^{7} \frac{C(u)C(v)}{4} F(u, v) \cos\left(\frac{(2i+1)u\pi}{16}\right) \cos\left(\frac{(2j+1)v\pi}{16}\right). \quad (13.22)$$

To become familiar with the 2D-DCT formulas, we study Example 13.14.

Example 13.14.

a. Determine the 2D-DCT coefficients for the following image:

$$\begin{bmatrix} 100 & 50 \\ 100 & -10 \end{bmatrix}.$$

Solution:

a. Applying $N = 2$ and $M = 2$ to Equation (13.18) yields

$$F(u, v) = \frac{2C(u)C(v)}{\sqrt{2 \times 2}} \sum_{i=0}^{1} \sum_{j=0}^{1} p(i, j) \cos\left(\frac{(2i+1)u\pi}{4}\right) \cos\left(\frac{(2j+1)v\pi}{4}\right).$$

For $u = 0$ and $v = 0$, we achieve:

$$F(0, 0) = c(0)c(0) \sum_{i=0}^{1} \sum_{j=0}^{1} p(i, j) \cos(0) \cos(0)$$

$$= \left(\frac{\sqrt{2}}{2}\right)^2 [p(0, 0) + p(0, 1) + p(1, 0) + p(1, 1)]$$

$$= \frac{1}{2}(100 + 50 + 100 - 10) = 120$$

For $u = 0$ and $v = 1$, we achieve:

$$F(0,1) = c(0)c(1) \sum_{i=0}^{1} \sum_{j=0}^{1} p(i,j) \cos(0) \cos\left(\frac{(2j+1)\pi}{4}\right)$$

$$= \left(\frac{\sqrt{2}}{2}\right) \times 1 \times \left(p(0,0)\cos\frac{\pi}{4} + p(0,1)\cos\frac{3\pi}{4} + p(1,0)\cos\frac{\pi}{4} + p(1,1)\cos\frac{3\pi}{4}\right)$$

$$= \frac{\sqrt{2}}{2}\left(100 \times \frac{\sqrt{2}}{2} + 50\left(-\frac{\sqrt{2}}{2}\right) + 100 \times \frac{\sqrt{2}}{2} - 10\left(-\frac{\sqrt{2}}{2}\right)\right) = 80$$

Similarly,

$$F(1, 0) = 30 \quad \text{and} \quad F(1, 1) = -30.$$

Finally, we get

$$F(u, v) = \begin{bmatrix} 120 & 80 \\ 30 & -30 \end{bmatrix}.$$

Applying the MATLAB function **dct2()** verifies the DCT coefficients as follows:

$$\gg F = dct2([100\ 50; 100\ -10])$$
$$F =$$
$$120.0000\ \ 80.0000$$
$$30.0000\ \ -30.0000.$$

Example 13.15.

Given the following DCT coefficients from a 2×2 image:

$$F(u, v) = \begin{bmatrix} 120 & 80 \\ 30 & -30 \end{bmatrix},$$

a. Determine the pixel $p(0, 0)$.

Solution:

a. Applying Equation (13.19) of the inverse 2D-DCT with $N = M = 2$, $i = 0$, and $j = 0$, it follows that

$$p(0,0) = \sum_{u=0}^{1} \sum_{v=0}^{1} c(u)c(v)F(u,v) \cos\left(\frac{u\pi}{4}\right) \cos\left(\frac{v\pi}{4}\right)$$

$$= \left(\frac{\sqrt{2}}{2}\right) \times \left(\frac{\sqrt{2}}{2}\right) \times F(0,0) + \left(\frac{\sqrt{2}}{2}\right) \times F(0,1) \times \left(\frac{\sqrt{2}}{2}\right)$$

$$+ \left(\frac{\sqrt{2}}{2}\right) \times F(1,0) \times \left(\frac{\sqrt{2}}{2}\right) + F(0,1)\left(\frac{\sqrt{2}}{2}\right) \times \left(\frac{\sqrt{2}}{2}\right)$$

$$= \frac{1}{2} \times 120 + \frac{1}{2} \times 80 + \frac{1}{2} \times 30 + \frac{1}{2}(-30) = 100$$

We apply the MATLAB function **idct2()** to verify the inverse DCT to get the pixel values as follows:

$$\gg p = idct2([120\ 80; 30\ -30])$$
$$p =$$
$$100.0000\ \ 50.0000$$
$$100.0000\ \ -10.0000.$$

13.7.2 Two-Dimensional JPEG Grayscale Image Compression Example

To understand JPEG image compression, we examine the 8×8 grayscale subblock. Table 13.5 shows a subblock of the grayscale image shown in Figure 13.37 to be compressed.

Applying 2D-DCT leads to Table 13.6.

These DCT coefficients have a big DC component of 1198 but small AC component values. These coefficients are further normalized (quantized) with a quality factor Q, defined in Table 13.7.

TABLE 13.5 8×8 subblock.

150	148	140	132	150	155	155	151
155	152	143	136	152	155	155	153
154	149	141	135	150	150	150	150
156	150	143	139	154	152	152	155
156	151	145	140	154	152	152	155
154	152	146	139	151	149	150	151
156	156	151	142	154	154	154	154
151	154	149	139	151	153	154	153

TABLE 13.6 DCT coefficients for the subblock image in Table 13.5.

1198	−10	26	24	−5	−16	0	12
−8	−6	3	8	0	0	0	0
0	−3	0	0	−8	0	0	0
0	0	0	0	0	0	0	0
0	−4	0	0	0	0	0	0
0	0	−1	0	0	0	0	0
−10	1	0	0	0	0	0	0
0	0	0	0	0	0	0	0

TABLE 13.7 The quality factor.

16	11	10	16	24	40	51	61
12	12	14	19	26	58	60	55
14	13	16	24	40	57	69	56
14	17	22	29	51	87	80	62
18	22	37	56	68	109	103	77
24	35	55	64	81	104	113	92
49	64	78	87	103	121	120	101
72	92	95	98	112	100	103	99

After normalization, as shown in Table 13.8, the DC coefficient is reduced to 75, and a few small AC coefficients exist, while most are zero. We can encode and transmit only nonzero DCT coefficients and omit transmitting zeros, since they do not carry any information. They can be easily recovered by resetting coefficients to zero during decoding. By this principle we achieve data compression.

As shown in Table 13.8, most nonzero coefficients reside at the upper left corner. Hence, the order of encoding for each value is based on the zigzag path in which the order is numbered, as in Table 13.9.

According to the order, we record the nonzero DCT coefficients as a JPEG vector, shown as:

$$\text{JPEG vector:} [75 \quad -1 \quad -1 \quad 0 \quad -1 \quad 3 \quad 2 \quad \text{EOB}].$$

where "EOB" = end of block coding. The JPEG vector can further be compressed by encoding the difference of DC values between subblocks, in *differential pulse code modulation* (DPCM), as discussed in Chapter 11, as well as by run-length coding of AC values and Huffman coding, which both belong to lossless compression techniques. We will pursue this in the next section.

TABLE 13.8 Normalized DCT coefficients.

75	−1	3	2	0	0	0	0
−1	−1	0	0	0	0	0	0
0	0	0	0	0	0	0	0
0	0	0	0	0	0	0	0
0	0	0	0	0	0	0	0
0	0	0	0	0	0	0	0
0	0	0	0	0	0	0	0
0	0	0	0	0	0	0	0

TABLE 13.9 The order to scan DCT coefficients.

0	1	5	6	14	15	27	28
2	4	7	13	16	26	29	42
3	8	12	17	25	30	41	43
9	11	18	24	31	40	44	53
10	19	23	32	39	45	52	54
20	22	33	38	46	51	55	60
21	34	37	47	50	56	59	61
35	36	48	49	57	58	62	63

TABLE 13.10 The recovered image subblock.

153	145	138	139	147	154	155	153
153	145	138	139	147	154	155	153
155	147	139	140	148	154	155	153
157	148	141	141	148	155	155	152
159	150	142	142	149	155	155	152
161	152	143	143	149	155	155	152
162	153	144	144	150	155	154	151
163	154	145	144	150	155	154	151

TABLE 13.11 The coding error of the image subblock.

3	−3	−2	7	−3	−1	0	2
−2	−7	−5	3	−5	−1	0	0
1	−2	−2	5	−2	4	5	3
1	−2	−2	2	−6	3	3	−3
3	−1	−3	2	−5	3	3	−3
7	0	−3	4	−2	6	5	1
6	−3	−7	2	−4	1	0	−3
12	0	−4	5	−1	2	0	−2

During the decoding stage, the JPEG vector is recovered first. Then the quantized DCT coefficients are recovered according to the zigzag path. Next, the recovered DCT coefficients are multiplied by a quality factor to obtain the estimate of the original DCT coefficients. Finally, we apply the inverse DCT to achieve the recovered image subblock, the recovered subblock of which is shown in Table 13.10.

For comparison, the errors between the recovered image and the original image are calculated and listed in Table 13.11.

The original and compressed images are displayed in Figures 13.37 and 13.38, respectively. We do not see any noticeable difference between these two grayscale images.

13.7.3 JPEG Color Image Compression

This section is devoted to reviewing JPEG standard compression and examines the steps briefly. We focus on the encoder, since decoding is just the reverse process of encoding. The block diagram for the JPEG encoder is in Figure 13.39.

The JPEG encoder has the following main steps:

1. Transform RGB to YIQ or YUV (U and V = chrominance components).
2. Perform DCT on blocks.

FIGURE 13.37 Original image.

FIGURE 13.38 JPEG compressed image.

3. Perform quantization.
4. Perform zigzag ordering, DPCM, and run-length encoding.
5. Perform entropy encoding (Huffman coding).

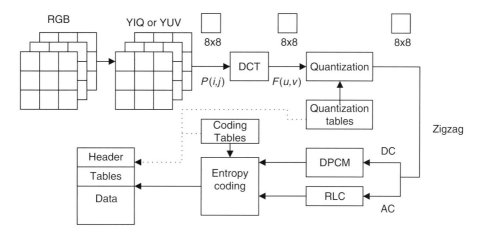

FIGURE 13.39 Block diagram for JPEG encoder.

RGB to YIQ Transformation

The first transformation is of the RGB image to a YIQ or YUV image. Transformation from RGB to YIQ has previously been discussed. The principle is that in YIQ format, the luminance channel carries more signal energy, up to 93%, while the chrominance channels carry up to only 7% of signal energy. After transformation, more effort can be spent on coding the luminance channel.

DCT on Image Blocks

Each image is divided into 8×8 blocks. 2D-DCT is applied to each block to obtain the 8×8 DCT coefficient block. Note that there are three blocks: Y, I, and Q.

Quantization

Quantization is operated using the 8×8 quantization matrix. Each DCT coefficient is quantized, divided by the corresponding value given in the quantization matrix. In this way, a smaller number of bits can be used for encoding the DCT coefficients. There are two different quantization tables, one for luminance (which is the same as the one in the last section and is listed here again for comparison) and the other for chrominance.

We can see that the chrominance table has numbers with larger values, so that small values of DCT coefficients will result and hence a smaller number of bits are required for encoding each DCT coefficient. Zigzag ordering to produce the JPEG vector is similar to the grayscale case, except that there are three JPEG vectors.

TABLE 13.12 The quality factor for luminance.

16	11	10	16	24	40	51	61
12	12	14	19	26	58	60	55
14	13	16	24	40	57	69	56
14	17	22	29	51	87	80	62
18	22	37	56	68	109	103	77
24	35	55	64	81	104	113	92
49	64	78	87	103	121	120	101
72	92	95	98	112	100	103	99

TABLE 13.13 The quality factor for chrominance.

17	18	24	47	99	99	99	99
18	21	26	66	99	99	99	99
24	26	56	99	99	99	99	99
47	66	99	99	99	99	99	99
99	99	99	99	99	99	99	99
99	99	99	99	99	99	99	99
99	99	99	99	99	99	99	99
99	99	99	99	99	99	99	99

Differential Pulse Code Modulation on Direct-Current Coefficients

Since each 8×8 image block has only one DC coefficient, which can be a very large number and varies slowly, we make use of DPCM for coding DC coefficients. As an example for the first five image blocks, DC coefficients are 200, 190, 180, 160, and 170. DPCM with a coding rule of $d(n) = DC(n) - DC(n-1)$ with initial condition $d(0) = DC(0)$ produces a DPCM sequence as

$$200, -10, -10, -20, 10.$$

Hence, the reduced signal range of these values is feasible for entropy coding.

Run-Length Coding on Alternating-Current Coefficients

The run-length method encodes the pair of

- the number of zeros to skip and
- the next nonzero value.

The zigzag scan of the 8×8 matrix makes up a vector of 64 values with a long run of zeros. For example, the quantized DCT coefficients are scanned as

13.7 Image Compression by Discrete Cosine Transform

$$[75, -1, 0, -1, 0, 0, -1, 3, 0, 0, 0, 2, 0, 0, \ldots, 0],$$

with one run, two runs, and three runs of zeros in the middle and 52 extra zeros toward the end. The run-length method encodes AC coefficients by producing a pair (run length, value), where the run length is the number of zeros in the run, while the value is the next nonzero coefficient. A special pair (0, 0) indicates EOB. Here is the result from a run-length encoding of AC coefficients:

$$(0, -1), (1, -1), (2, -1), (0,3), (3,2), (0,0).$$

Lossless Entropy Coding

The DC and AC coefficients are further compressed using entropy coding. JPEG allows Huffman coding and arithmetic coding. We focus on Huffman coding here.

Coding DC Coefficients

Each DPCM-coded DC coefficient is encoded by a pair of symbols (size, amplitude) with the size (4-bit code) designating the number of bits for the coefficient as shown in Table 13.14, while the amplitude is encoded by the actual bits. For the negative number of the amplitude, 1's complement is used.

For example, we can code the DPCM-coded DC coefficients 200, –10, –10, –20, 10 as

(8, 11001000), (4, 0101), (4, 0101), (5, 01011), (4, 1010).

Since there needs to be 4 bits for encoding each size, we can use 45 bits in total for encoding the DC coefficients for these five subblocks.

TABLE 13.14 Huffman coding table.

Size	Amplitude
1	–1, 1
2	–3, –2, 2, 3
3	–7, ..., –4, 4, ..., 7
4	–15, ..., –8, 8, ..., 15
5	–31, ..., –16, 16, ..., 31
.	.
.	.
.	.
10	–1023, ..., –512, 512, ..., 1023

Coding AC Coefficients

The run-length AC coefficients have the format (run length, value). The value can be further compressed using the Huffman coding method, similar to coding the DPCM-coded DC coefficients. The run length and the size are each encoded by 4 bits and packed into a byte.

$$\text{Symbol 1: (run length, size)}$$
$$\text{Symbol 2: (amplitude).}$$

The 4-bit run length can tackle only the number of runs of zeros from one to fifteen. If the run length of zeros is longer than fifteen, then a special code (15, 0) is used for Symbol 1. Symbol 2 is the amplitude in Huffman coding as shown in Table 13.14, while the encoded Symbol 1 is kept in its format:

$$\text{(run length, size, amplitude).}$$

Let us code the following run-length code of AC coefficients:

$$(0, -1), (1, -1), (2, -1), (0, 3), (3, 2), (0, 0).$$

We can produce a bit stream for AC coefficients as:

(0000, 0001, 0), (0001, 0001, 0), (0010, 0001, 0),
(0000, 0010, 11), (0011, 0010, 10), (0000, 0000).

There are 55 bits in total. Figure 13.40 shows a JPEG compressed color image (included in the color insert). The decompressed image is indistinguishable from the original image after comparison.

FIGURE 13.40 JPEG compressed color image. (See color insert.)

13.8 Creating a Video Sequence by Mixing Two Images

In this section, we introduce a method to mix two images to generate an image (video) sequence. Applications of mixing two images may include fading in and fading out images, blending two images, or overlaying text on an image.

In mixing two images in a video sequence, a smooth transition is required from fading out one image of interest to fading in another image of interest. We want to fade out the first image and gradually fade in the second. This cross-fade scheme is implemented using the following operation:

$$\text{Mixed image} = (1 - \alpha) \times \text{image}_1 + \alpha \times \text{image}_2, \quad (13.23)$$

where α = fading in proportionally to the weight of the second image (value between 0 and 1), and $(1 - \alpha)$ = fading out proportionally to the weight of the second image.

The video sequence in Figure 13.41a, b consisting of six frames is generated using $\alpha = 0.0$, $\alpha = 0.2$, $\alpha = 0.4$, $\alpha = 0.6$, $\alpha = 0.8$, and $\alpha = 1.0$, respectively, for two images. The equations for generating these frames are listed as follows:

$$\text{Mixed image}_1 = 1.0 \times \text{image}_1 + 0.0 \times \text{image}_2$$
$$\text{Mixed image}_2 = 0.8 \times \text{image}_1 + 0.2 \times \text{image}_2$$
$$\text{Mixed image}_3 = 0.6 \times \text{image}_1 + 0.4 \times \text{image}_2$$
$$\text{Mixed image}_4 = 0.4 \times \text{image}_1 + 0.6 \times \text{image}_2$$
$$\text{Mixed image}_5 = 0.2 \times \text{image}_1 + 0.8 \times \text{image}_2$$
$$\text{Mixed image}_6 = 0.0 \times \text{image}_1 + 1.0 \times \text{image}_2.$$

The sequence begins with the Grand Canyon image and fades in with the cruise ship image. At frame 4, 60% of the cruise ship is faded in, and the image begins to be discernible as such. The sequence ends with the cruise ship in 100% fade-in. Figure 13.41a displays the generated grayscale sequence. Figure 13.41b shows the color RGB video sequence (also given in the color insert).

13.9 Video Signal Basics

Video signals generally can be classified as component video, composite video, and S-video. In *component video,* three video signals—such as the red, green, and blue channels or the Y, I, and Q channels—are used. Three wires are required for connection to a camera or TV. Most computer systems use component video signals. *Composite video* has intensity (luminance) and two-color (chrominance) components that modulate the carrier wave. This signal is used in broadcast

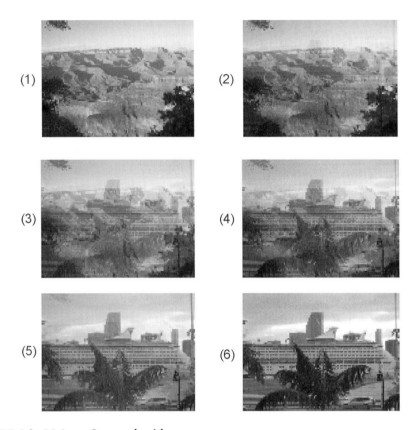

FIGURE 13.41A Grayscale video sequence.

color TV. The standard by the U.S.-based National Television System Committee (NTSC) combines channel signals into a chroma signal, which is modulated to a higher frequency for transmission. Connecting TVs or VCRs requires only one wire, since both video and audio are mixed into the same signal. *S-video* sends luminance and chrominance separately, since the luminance presenting black-and-white intensity contains most of the signal information for visual perception.

13.9.1 Analog Video

In computer systems, progressive scanning traces a whole picture, called *frame via row-wise*. A higher-resolution computer uses 72 frames per second (fps). The video is usually played at a frame rate varying from 15 to 30 fps.

FIGURE 13.41B Color RGB video sequence. (See color insert.)

In TV reception and some monitors, *interlaced scanning* is used in a cathode-ray tube display, or raster. The odd-numbered lines are traced first, and the even-numbered lines are traced next. We then get the odd-field and even-field scans per frame. The interlaced scheme is illustrated in Figure 13.42, where the odd lines are traced, such as A to B, then C to D, and so on, ending in the middle at E. The even field begins at F in the middle of the first line of the even field and ends at G. The purpose of using interlaced scanning is to transmit a full frame quickly to reduce flicker. Trace jumping from B to C is called horizontal retrace, while trace jumping from E to F or G to A is called vertical retrace.

The video signal is amplitude modulated. The modulation levels for NTSC video are shown in Figure 13.43. In the United States, negative modulation is used, considering that less amplitude comes from a brighter scene, while more amplitude comes from a darker one. This is due to the fact that most pictures contain more white than black levels. With negative modulation, possible power

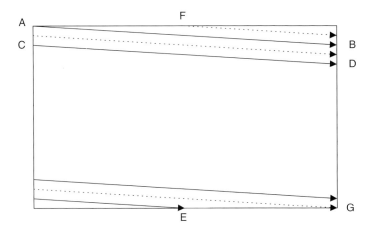

FIGURE 13.42 Interlaced raster scanning.

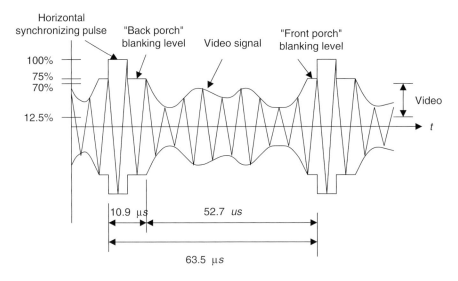

FIGURE 13.43 Video-modulated waveform.

efficiency can be achieved for transmission. The reverse process will apply for display at the receiver.

The horizontal synchronizing pulse controls the timing for the horizontal retrace. The blanking levels are used for synchronizing as well. The "back porch" (Figure 13.43) of the blanking also contains the color subcarrier burst for color demodulation.

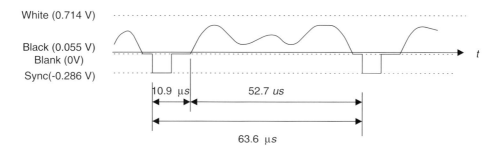

FIGURE 13.44 The demodulated signal level for one NTSC scan line.

The demodulated electrical signal can be seen in Figure 13.44, where a typical electronic signal for one scan line is depicted. The white intensity has a peak value of 0.714 volt, and the black has a voltage level of 0.055 volt, which is close to zero. The blank corresponds to zero voltage, and the synchronizing pulse is at the level of -0.286 volt. The time duration for synchronizing is 10.9 microseconds; that of the video occupies 52.7 microseconds; and that of one entire scan line occupies 63.6 microseconds. Hence, the line scan rate can be determined as 15.75 kHz.

Figure 13.45 describes vertical synchronization. A pulse train is generated at the end of each field. The pulse train contains 6 equalizing pulses, 6 vertical synchronizing pulses, and another 6 equalizing pulses at the rate of twice the size of the line scan rate (31.5 kHz), so that the timing for sweeping half the width of the field is feasible. In NTSC, vertical retrace takes the time interval of 20 horizontal lines designated for control information at the beginning of each field. The 18 pulses of the vertical blanking occupy the time interval that is equivalent to 9 lines. This leaves lines 10 to 20 for other uses.

A color subcarrier resides at the back porch, as shown in Figure 13.45. The 8 cycles of the color subcarrier are recovered via a delayed gating circuit triggered by the horizontal sync pulse. Synchronization includes the color burst frequency and phase information. The color subcarrier is then applied to demodulate the color (chrominance).

Let us summarize NTSC video signals. The NTSC TV standard uses an aspect ratio of 4:3 (ratio of picture width to height), and 525 scan lines per frame at 30 fps. Each frame has an odd field and an even field. So there are $525/2 = 262.5$ lines per field. NTSC actually uses 29.97 fps. The horizontal sweep frequency is $525 \times 29.97 = 15{,}734$ lines per second, and each line takes $1/15{,}734 = 63.6\,\mu\text{sec}$. Horizontal retrace takes $10.9\,\mu\text{sec}$, while the line signal takes $52.7\,\mu\text{sec}$ for one line of image display. Vertical retrace and sync are also needed so that the first 20 lines for each field are reserved to be used. The active video lines per frame are 485. The layout of the video data, retrace, and sync data is shown in Figure 13.46.

682 13 IMAGE PROCESSING BASICS

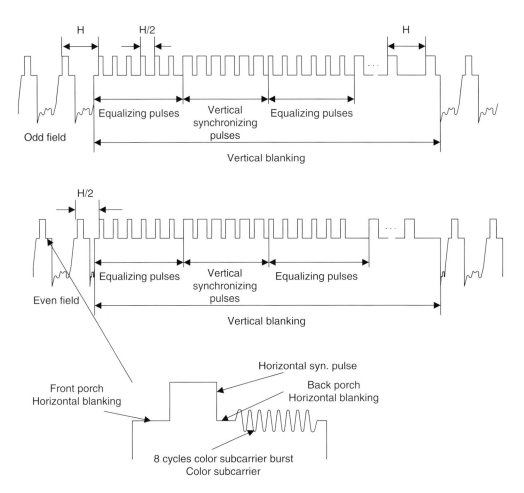

FIGURE 13.45 Vertical synchronization for each field and the color subcarrier burst.

Blanking regions can be used for V-chip information, stereo audio channel data, and subtitles in various languages. The active line is then sampled for display. A pixel clock divides each horizontal line of video into samples. For example, vertical helical scan (VHS) uses 240 samples per line; Super VHS, 400–425 samples per line.

Figure 13.47 shows the NTSC video signal spectra. The NTSC standard assigns a bandwidth of 4.2 MHz for luminance Y, 1.6 MHz for I, and 0.6 for Q, due to human perception of color information. Since the human eye has higher resolution to the I color component than to the Q color component, the wider bandwidth for I is allowed.

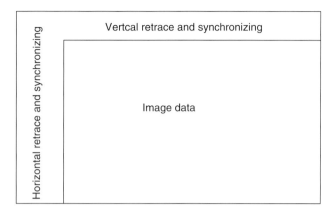

FIGURE 13.46 Video data, retrace, and sync layout.

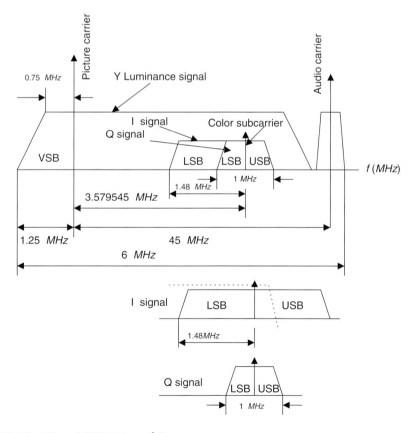

FIGURE 13.47 NTSC Y, I, and Q spectra.

As shown in Figure 13.47, *vestigial sideband modulation* (VSB) is employed for the luminance, with a picture carrier of 1.25 MHz relative to the VSB left edge. The space between the picture carrier and the audio carrier is 4.5 MHz.

The audio signal containing the frequency range from 50 Hz to 15 kHz is stereo frequency modulated (FM), using a peak frequency deviation of 25 kHz. Therefore, the stereo FM audio requires a transmission bandwidth of 80 kHz, with an audio carrier located at 4.5 MHz relative to the picture carrier.

The color burst carrier is centered at 3.58 MHz above the picture carrier. The two color components I and Q undergo *quadrature amplitude modulation* (QAM) with modulated component I output, which is VSB filtered to remove two-thirds of the upper sideband, so that all chroma signals fall within a 4.2 MHz video bandwidth. The color burst carrier of 3.58 MHz is chosen such that the chroma signal and luminance are interleaved in the frequency domain to reduce interference between them.

Generating a chroma signal with QAM gives

$$C = I\cos(2\pi f_{sc}t) + Q\sin(2\pi f_{sc}t), \tag{13.24}$$

where C = chroma component and f_{sc} = color subcarrier = 3.58 MHz. The NTSC signal is further combined into a composite signal:

$$Composite = Y + C = Y + I\cos(2\pi f_{sc}t) + Q\sin(2\pi f_{sc}t). \tag{13.25}$$

At decoding, the chroma signal is obtained by separating Y and C first. Generally, the lowpass filters located at the lower end of the channel can be used to extract Y. The comb filters may be employed to cancel interferences between the modulated luminance signal and the chroma signal (Li and Drew, 2004). Then we perform demodulation for I and Q as follows:

$$\begin{aligned} C \times 2\cos(2\pi f_{sc}t) &= I2\cos^2(2\pi f_{sc}t) + Q \times 2\sin(2\pi f_{sc}t)\cos(2\pi f_{sc}t) \\ &= I + I \times \cos(2 \times 2\pi f_{sc}t) + Q\sin(2 \times 2\pi f_{sc}t). \end{aligned} \tag{13.26}$$

Applying a lowpass filter yields the I component. Similar operation by applying a carrier signal of $2\sin(2\pi f_{sc}t)$ for demodulation recovers the Q component.

PAL Video

The phase alternative line (PAL) system uses 625 scan lines per frame at 25 fps, with an aspect ratio of 4:3. It is widely used in western Europe, China, and India. PAL uses the YUV color model, with an 8-MHz channel in which Y has 5.5 MHz and U and V each have 1.8 MHz with the color subcarrier frequency of 4.43 MHz relative to the picture carrier. The U and V are the color difference signals (chroma signals) of the B-Y signal and R-Y signal, respectively. The chroma signals have alternate signs (e.g., +V and −V) in successive scan lines.

TABLE 13.15 Analog broadband TV systems.

TV System	Frame Rate (fps)	Number of Scan Lines	Total Bandwidth (MHz)	Y Bandwidth (MHz)	U or I Bandwidth (MHz)	V or Q Bandwidth (MHz)
NTSC	29.97	525	6.0	4.2	1.6	0.6
PAL	25	625	8.0	5.5	1.8	1.8
SECAM	25	625	8.0	6.0	2.0	2.0

Source: Li and Drew, 2004.

Hence, the signal and its sign reversed one in the consecutive lines are averaged to cancel out the phase errors that could be displayed as the color errors.

SECAM Video

The SECAM (Séquentiel Couleur à Mémoire) system uses 625 scan lines per frame at 25 fps, with an aspect ratio of 4:3 and interlaced fields. The YUV color model is employed, and U and V signals are modulated using separate color subcarriers of 4.25 and 4.41 MHz, respectively. The U and V signals are sent on each line alternatively. In this way, the quadrature multiplexing and the possible cross-coupling of color signals could be avoided by halving the color resolution in the vertical dimension.

13.9.2 Digital Video

Digital video has become dominant over the long-standing analog method in modern systems and devices because it offers high image quality; flexibility of storage, retrieval, and editing capabilities; digital enhancement of video images; encryption; channel noise tolerance; and multimedia system applications.

Digital video formats are developed by the Consultative Committee for International Radio (CCIR). A most important standard is CCIR-601, which became ITU-R-601, an international standard for professional video applications.

In CCIR-601, chroma subsampling is carried for digital video. Each pixel is in YCbCr color space, where Y is the luminance, and Cb and Cr are the chrominance. Subsampling schemes include 4:4:4 (no chroma subsampling), 4:2:2, 4:1:1, and 4:2:0, as illustrated in Figure 13.48.

In a 4:4:4 video format in each frame, the number of values for each chrominance component, Cb or Cr, is the same as that for luminance, Y, both horizontally and vertically. This format finds applications in computer graphics, in which the chrominance resolution is required for both horizontal and vertical dimensions. The format is not widely used in video applications due to a huge storage requirement.

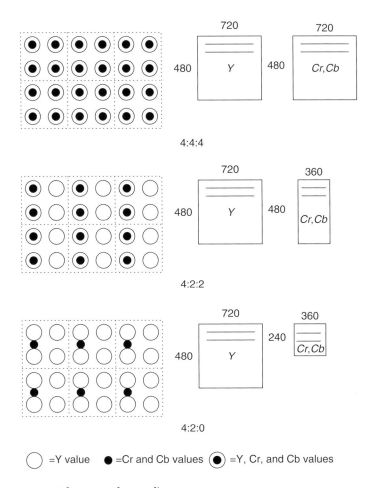

FIGURE 13.48   Chroma subsampling.

As shown in Figure 13.48, for each frame in the 4:2:2 video format, the number of chrominance components for Cr or Cb is half the number of luminance components for Y. The resolution is full vertically, and the horizontal resolution is downsampled by a factor of 2. Considering the first line of six pixels, transmission occurs in the following form: (Y0, Cb0), (Y1, Cr0), (Y2, Cb2), (Y3, Cr2), (Y4, Cb4), (Y5, Cr4), and so on. Six Y values are sent for every two Cb and Cr values that are sent.

In the 4:2:0 video format, the number of values for each chrominance Cb and Cr is half the number of luminance Y for both horizontal and vertical directions. That is, the chroma is downsampled horizontally and vertically by a factor of 2. The location for both Cb and Cr is shown in Figure 13.48. Digital video specifications are given in Table 13.16.

TABLE 13.16 Digital video specifications.

	CCIR-601 525/60 NTSC	CCR-601 625/50 PAL/SECAM	CIF	QCIF
Luminance resolution	720×480	720×576	352×288	176×144
Chrominance resolution	360×480	360×576	176×144	88×72
Color subsampling	4:2:2	4:2:2	4:2:0	4:2:0
Aspect ratio	4:3	4:3	4:3	4:3
Fields/sec	60	50	30	30
Interlaced	Yes	Yes	No	No

Source: Li and Drew, 2004. CCR, comparison category rating; CIF, common intermediate format; QCIF, quarter-CIF.

TABLE 13.17 High-definition TV (HDTV) formats.

Number of Active Pixels per Line	Number of Active Lines	Aspect Ratio	Picture Rate
1920	1080	16:9	60I 30P 24P
1280	720	16:9	60P 30P 24 P
704	480	16:9 and 4:3	60I 60P 30P 24P
640	480	4:3	60I 60P 30P 24P

Source: Li and Drew, 2004.

CIF was specified by the Comité Consultatif International Téléphonique et Télégraphique (CCITT), which is now the International Telecommunications Union (ITU). CIF produces a low bit rate video and supports a progressive scan. QCIF achieves an even lower bit rate video. Neither format supports the interlaced scan mode.

Table 13.17 outlines the high-definition TV (HDTV) formats supported by the Advanced Television System Committee (ATSC), where "I" means interlaced scan and "P" indicates progressive scan. MPEG compressions of video and audio are employed.

13.10 Motion Estimation in Video

In this section, we study motion estimation, since this technique is widely used in the MPEG video compression. A video contains a time-ordered sequence of frames. Each frame consists of image data. When the objects in an image are still, the pixel values do not change under constant lighting conditions. Hence, there is no motion between the frames. However, if the objects are moving, then the pixels are moved. If we can find the motions, which are the pixel displacements, with *motion vectors,* the frame data can be recovered from the reference frame by copying and pasting at locations specified by the motion vectors. To explore such an idea, let us look at Figure 13.49.

As shown in Figure 13.49, the reference frame is displayed first, and the next frame is the target frame containing a moving object. The image in the target frame is divided into $N \times N$ macroblocks (20 macroblocks). A macroblock match is searched within the search window in the reference frame to find the closest match between a macroblock under consideration in the target frame and the macroblock in the reference frame. The differences between two locations (motion vectors) for the matched macroblocks are encoded.

The criteria for searching the best matching can be chosen using the mean absolute difference (MAD) between the reference frame and the target frame:

$$MAD(i, j) = \frac{1}{N^2} \sum_{k=0}^{N-1} \sum_{l=0}^{N-1} |T(m+k, n+l) - R(m+k+i, n+l+j)| \quad (13.27)$$

$$u = i, v = j \text{ for } MAD(i, j) = \text{minimum, and} - p \leq i, j \leq p. \quad (13.28)$$

There are many search methods for finding motion vectors, including optimal sequential, or brute force, searches; and suboptimal searches such as 2D-logarithmic and hierarchical searches. Here we examine the sequential search to understand the basic idea.

The sequential search for finding motion vectors employs methodical "brute force" to search the entire $(2p+1) \times (2p+1)$ search window in the reference frame. The macrobock in the target frame compares each macroblock centered at each pixel in the search window of the reference frame. Comparison proceeds pixel by pixel to find the best match in which the vector (i, j) produces the

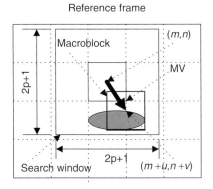
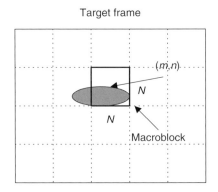

FIGURE 13.49 Macroblocks and motion vectors in the reference frame and target frame.

smallest MAD. Then the motion vector (MV(u,v)) is found to be u = i, and v = j. The algorithm is described as:

```
min_MAD=large value
for i = -p, ..., p
  for j = -p, ..., p
    cur_MAD=MDA(i, j);
    if cur_MAD < min_MAD
      min_MAD=cur_MAD;
      u = i;
      v = j;
    end
  end
end
```

The sequential search provides the best matching with the least MAD. However, it requires a huge amount of computations. Other suboptimal search methods can be employed to reduce the computational requirement, but with sacrifices of image quality. These topics are beyond our scope.

Example 13.16

The 80×80 reference frame, target frame, and their difference are displayed in Figure 13.50. The macroblock with a size of 16×16 is used, and the search window has a size of 32×32. The target frame is obtained by moving the reference frame to the right by 6 pixels and to the bottom by 4 pixels. The sequential search method is applied to find all the motion vectors. The reconstructed target frame using the motion vectors and reference image is given in Figure 13.50.

Since $80 \times 80/(16 \times 16) = 25$, there are 25 macroblocks in the target frame and 25 motion vectors in total. The motion vectors are found to be:

Horizontal direction =

$$-6 \quad -6 \quad -6 \quad -6 \quad -6 \quad -6 \quad -6 \quad -6 \quad -6 \quad -6 \quad -6 \quad -6 \quad -6 \quad -6$$
$$-6 \quad -6 \quad -6 \quad -6 \quad -6 \quad -6 \quad -6 \quad -6 \quad -6 \quad -6 \quad -6 \quad \quad \quad \quad \quad \quad (13.29)$$

Vertical direction =

$$-4 \quad -4 \quad -4 \quad -4 \quad -4 \quad -4 \quad -4 \quad -4 \quad -4 \quad -4 \quad -4 \quad -4 \quad -4 \quad -4 \quad -4$$
$$-4 \quad -4 \quad -4 \quad -4 \quad -4 \quad -4 \quad -4 \quad -4 \quad -4 \quad -4 \quad \quad \quad \quad \quad \quad (13.30)$$

The motion vector comprises the pixel displacements from the target frame to the reference frame. Hence, given the reference frame, directions specified in the motion vector should be switched to indicate the motion towards the target

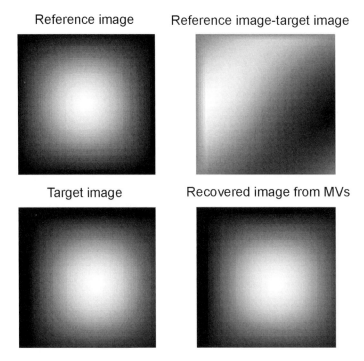

FIGURE 13.50 Reference frame, target frame, their difference, and the reconstructed frame by the motion vectors.

frame. As indicated by the obtained motion vectors, the target image is the version of the reference image moving to the right by 6 pixels and down by 4 pixels.

Summary

1. A digital image consists of pixels. For a grayscale image, each pixel is assigned a grayscale level that presents the luminance of the pixel. For an RGB color image, each pixel is assigned a red component, a green component, and a blue component. For an indexed color image, each pixel is assigned an address that is the location of the color table (map) made up of the red, green, and blue components.

2. Common image data formats are 8-bit grayscale, 24-bit color, and 8-bit indexed color.

3. The larger the number of pixels in an image, or the larger the numbers of the RGB components, the finer is the spatial resolution in the image. Similarly, the more scale levels used for each pixel, the better the scale-level image resolution. The more pixels and more bits used for the scale levels in the image, the more storage is required.

4. The RGB color pixels can be converted to YIQ color pixels. Y component is the luminance occupying 93% of the signal energy, while the I and Q components represent the color information of the image, occupying the remainder of the energy.

5. The histogram for a grayscale image shows the number of pixels at each grayscale level. The histogram can be modified to enhance the image. Image equalization using the histogram can improve the image contrast and effectively enhances contrast for image underexposure. Color image equalization can be done only in the luminance channel or RGB channels.

6. Image enhancement techniques such as average lowpass filtering can filter out random noise in the image; however, it also blurs the image. The degree of blurring depends on the kernel size. The bigger the kernel size, the more blurring occurs.

7. Median filtering effectively removes the "pepper and salt" noise in an image.

8. The edge detection filter with Sobel convolution, Laplacian, and Laplacian of Gaussian kernels can detect the image boundaries.

9. The grayscale image can be made into a facsimile of the color image by pseudo-color image generation, using the red, green, and blue transformation functions.

10. RGB-to-YIQ transformation is used to obtain the color image in YIQ space, or vice versa. It can also be used for color-to-grayscale conversion, that is, keeping only the luminance channel after the transformation.

11. 2D spectra can be calculated and are used to examine filtering effects.

12. JPEG compression uses the 2D-DCT transform for both grayscale and color images in YIQ color space. JPEG uses different quality factors to normalize DCT coefficients for the luminance (Y) channel and the chrominance (IQ) channels.

13. The mixing of two images, in which two pixels are linearly interpolated using the weights $1 - \alpha$ and α, can produce video sequences that have effects such as fading in and fading out of images, blending of two images, and overlaying of text on an image.

14. Analog video uses interlaced scanning. A video frame contains odd and even fields. Analog video standards include NTSC, PAL, and SECAM.

15. Digital video carries the modulated information for each pixel in YCbCr color space, where Y is the luminance and Cb and Cr are the chrominance. Chroma subsampling creates various digital video formats. The industrial standards include CCIR-601, CCR-601, CIF, and QCIF.

16. The motion compensation of the video sequence produces motion vectors for all the image blocks in the target video frame, which contain displacements of these image blocks relative to the reference video frame. Recovering the target frame involves simply copying each image block of the reference frame to the target frame at the location specified in the motion vector. Motion compensation is a key element in MPEG video.

13.12 Problems

13.1. Determine the memory storage requirement for each of the following images:

    a. 320×240 8-bit grayscale

    b. 640×480 24-bit color image

    c. 1600×1200 8-bit indexed image

13.2. Determine the number of colors for each of the following images:

    a. 320×240 16-bit indexed image

    b. 200×100 24-bit color image

13.3. Given a pixel in an RGB image as follows:

$$R = 200, G = 120, B = 100,$$

convert the RGB values to the YIQ values.

13.4. Given a pixel of an image in YIQ color format as follows:

$$Y = 141, I = 46, Q = 5,$$

convert the YIQ values back to RGB values.

13.5. Given the following 2×2 RGB image,

$$R = \begin{bmatrix} 100 & 50 \\ 100 & 50 \end{bmatrix} \quad G = \begin{bmatrix} 20 & 40 \\ 10 & 30 \end{bmatrix} \quad B = \begin{bmatrix} 100 & 50 \\ 200 & 150 \end{bmatrix},$$

convert the image into grayscale.

13.6. Produce a histogram of the following image, which has a grayscale value ranging from 0 to 7, that is, each pixel is encoded in 3 bits.

$$\begin{bmatrix} 0 & 1 & 2 & 2 & 0 \\ 2 & 1 & 1 & 2 & 1 \\ 1 & 1 & 4 & 2 & 3 \\ 0 & 2 & 5 & 6 & 1 \end{bmatrix}$$

13.7. Given the following image with a grayscale value ranging from 0 to 7, that is, each pixel being encoded in 3 bits,

$$\begin{bmatrix} 0 & 1 & 2 & 2 & 0 \\ 2 & 1 & 1 & 2 & 1 \\ 1 & 1 & 4 & 2 & 3 \\ 0 & 2 & 5 & 6 & 1 \end{bmatrix},$$

perform equalization using the histogram in Problem 13.6, and plot the histogram for the equalized image.

13.8. Given the following image with a grayscale value ranging from 0 to 7, that is, each pixel being encoded in 3 bits,

$$\begin{bmatrix} 2 & 4 & 4 & 2 \\ 2 & 3 & 3 & 3 \\ 4 & 4 & 4 & 2 \\ 3 & 2 & 3 & 4 \end{bmatrix},$$

perform level adjustment to the full range, shift the level to the range from 3 to 7, and shift the level to the range from 0 to 3.

13.9. Given the following 8-bit grayscale original and noisy images, and 2×2 convolution average kernel,

4×4 original image: $\begin{bmatrix} 100 & 100 & 100 & 100 \\ 100 & 100 & 100 & 100 \\ 100 & 100 & 100 & 100 \\ 100 & 100 & 100 & 100 \end{bmatrix}$

4×4 corrupted image: $\begin{bmatrix} 93 & 116 & 109 & 96 \\ 92 & 107 & 103 & 108 \\ 84 & 107 & 86 & 107 \\ 87 & 113 & 106 & 99 \end{bmatrix}$

2×2 average kernel: $\frac{1}{4}\begin{bmatrix} 1 & 1 \\ 1 & 1 \end{bmatrix}$,

perform digital filtering on the noisy image, and compare the enhanced image with the original image.

13.10. Given the following 8-bit grayscale original and noisy images, and 3×3 median filter kernel,

$$4 \times 4 \text{ original image: } \begin{bmatrix} 100 & 100 & 100 & 100 \\ 100 & 100 & 100 & 100 \\ 100 & 100 & 100 & 100 \\ 100 & 100 & 100 & 100 \end{bmatrix}$$

$$4 \times 4 \text{ corrupted image by impulse noise: } \begin{bmatrix} 100 & 255 & 100 & 100 \\ 0 & 255 & 255 & 100 \\ 100 & 0 & 100 & 0 \\ 100 & 255 & 100 & 100 \end{bmatrix}$$

$$3 \times 3 \text{ average kernel: } \begin{bmatrix} \quad \\ \quad \end{bmatrix},$$

perform digital filtering, and compare the filtered image with the original image.

13.11. Given the following 8-bit 5×4 original grayscale image,

$$\begin{bmatrix} 110 & 110 & 110 & 110 \\ 110 & 100 & 100 & 110 \\ 110 & 100 & 100 & 110 \\ 110 & 110 & 110 & 110 \\ 110 & 110 & 110 & 110 \end{bmatrix},$$

apply the following edge detectors to the image:

a. Sobel vertical edge detector

b. Laplacian edge detector,

and scale the resultant image pixel value to the range of 0 to 255.

13.12. In Example 13.10, if we switch the transformation functions between the red function and the green function, what is the expected color for the area pointed to by the arrow, and what is the expected background color?

13.13. In Example 13.10, if we switch the transformation functions between the red function and the blue function, what is the expected color for the area pointed to by the arrow, and what is the expected background color?

13.14. Given the following grayscale image $p(i, j)$:

$$\begin{bmatrix} 100 & -50 & 10 \\ 100 & 80 & 100 \\ 50 & 50 & 40 \end{bmatrix},$$

determine the 2D-DFT coefficient $X(1,2)$ and the magnitude spectrum $A(1, 2)$.

13.15. Given the following grayscale image $p(i, j)$:

$$\begin{bmatrix} 10 & 100 \\ 200 & 150 \end{bmatrix},$$

determine the 2D-DFT coefficients $X(u,v)$ and magnitude $A(u, v)$.

13.16. Given the following grayscale image $p(i, j)$:

$$\begin{bmatrix} 10 & 100 \\ 200 & 150 \end{bmatrix},$$

apply the 2D-DCT to determine the DCT coefficients.

13.17. Given the following DCT coefficients $F(u, v)$:

$$\begin{bmatrix} 200 & 10 \\ 10 & 0 \end{bmatrix},$$

apply the inverse 2D-DCT to determine 2D data.

13.18. In JPEG compression, DCT DC coefficients from several blocks are 400, 390, 350, 360, and 370. Use DPCM to produce the DPCM sequence, and use the Huffman table to encode the DPCM sequence.

13.19. In JPEG compression, DCT coefficients from an image subblock are

[175, −2, 0, 0, 0, 4, 0, 0, −3, 7, 0, 0, 0, 0, −2, 0, 0, . . . , 0].

a. Generate the run-length codes for AC coefficients.

b. Perform entropy coding for the run-length codes using the Huffman table.

13.20. Explain the difference between horizontal retrace and vertical retrace. Which one would take more time?

13.21. What is the purpose of using interlaced scanning in the traditional NTSC TV system?

13.22. What is the bandwidth in the traditional NTSC TV broadcast system? What is the bandwidth to transmit luminance Y, and what are the required bandwidths to transmit Q and I channels, respectively?

13.23. What type of modulation is used for transmitting audio signals in the NTSC TV system?

13.24. Given the composite NTSC signal

$$Composite = Y + C = Y + I\cos(2\pi f_{sc}t) + Q\sin(2\pi f_{sc}t),$$

show demodulation for the Q channel.

13.25. Where does the color subcarrier burst reside? What is the frequency of the color subcarrier, and how many cycles does the color burst have?

13.26. Compare differences among the NTSC, PAL, and SECAM video systems in terms of the number of scan lines, frame rates, and total bandwidths required for transmission.

13.27. In the NTSC TV system, what is the horizontal line scan rate? What is the vertical synchronizing pulse rate?

13.28. Explain which of the following digital video formats achieves the most data transmission efficiency.

   a. 4:4:4

   b. 4:2:2

   c. 4:2:0

13.29. What is the difference between an interlaced scan and a progressive scan? Which of the following video systems uses the progressive scan?

   a. CCIR-601

   b. CIF

13.30. In motion compensation, which of the following would require more computation? Explain.

   a. Finding the motion vector using a sequential search

   b. Recovering the target frame with the motion vectors and reference frame

13.31. Given a reference frame and target frame of size 80×80, a macroblock size of 16×16, and a search window size of 32×32, estimate the number of subtraction, absolute value, and addition operations for searching all the motion vectors using the sequential search method.

MATLAB Problems

Use MATLAB to solve Problems 13.32 to 13.37.

13.32. Given the image data "trees.jpg," use MATLAB functions to perform each of the following processing:

   a. Use MATLAB to read and display the image.

   b. Convert the image to the grayscale image.

   c. Perform histogram equalization for the grayscale image in (b) and display the histogram plots for both original grayscale image and equalized grayscale image.

   d. Perform histogram equalization for the color image in (a) and display the histogram plots of Y channel for both original color image and equalized color image.

13.33. Given the image data "cruise.jpg," perform the following linear filtering:

   a. Convert the image to grayscale image and then create the 8-bit noisy image by adding Gaussian noise using the following code:

   noise_image=imnoise(I,'gaussian');

   where I is the intensity image obtained from normalizing the grayscale image.

   b. Process the noisy image using the Gaussian filter with the following parameters: the convolution kernel size = 4, SIGMA = 0.8, and compare the filtered image with the noisy image.

13.34. Given the image data "cruise.jpg," perform the following filtering process:

   a. Convert the image to grayscale image and then create the 8-bit noisy image by adding "pepper and salt" noise using the following code:

   noise_image=imnoise(I,'salt & pepper');

   where I is the intensity image obtained from normalizing the grayscale image.

   b. Process the noisy image using the median filtering with a convolution kernel size of 4×4.

13.35. Given the image data "cruise.jpg," convert the image to the grayscale image and detect the image boundaries using Laplacian of Gaussian filtering with the following parameters:

　　a. Kernel size = 4 and SIGMA = 0.9

　　b. Kernel size = 10 and SIGMA = 10

Compare the results.

13.36. Given the image data "clipim2.gif," perform the following process:

　　a. Convert the indexed image to the grayscale image.

　　b. Adjust the color transformation functions (sine functions) to make the object indicated by the arrow in the image to be red and the background color to be green.

13.37. Given the image data "cruiseorg.tiff," perform JPEG compression by completing the following steps:

　　a. Convert the image to grayscale image.

　　b. Write a MATLAB program for encoding: (1) Dividing the image into 8x8 blocks. (2) Transforming each block using the discrete-cosine transform. (3) Scaling and rounding DCT coefficients with the standard quality factor. Note that lossless-compressing the quantized DCT coefficients is omitted here for a simple simulation.

　　c. Continue to write the MALAB program for decoding: (1) inversing the scaling process for quantized DCT coefficients. (2) Performing the inverse DCT for each 8 X 8 image block. (3) Recovering the image.

　　d. Run the developed MATLAB program to examine the image quality using

　　　　I. The quality factor
　　　　II. The quality factor X 5
　　　　III. The quality factor X 10

References

Gonzalez, R. C., and Wintz, P. (1987). *Digital Image Processing*, 2nd ed. Reading, MA: Addison-Wesley Publishing Company.

Li, Z.-N., and Drew, M. S. (2004). *Fundamentals of Multimedia*. Upper Saddle River, NJ: Pearson Prentice Hall.

Rabbani, M., and Jones, P. W. (1991). *Digital Image Compression Techniques*. Presentation to the Society of Photo-Optical Instrumentation Engineers (SPIE), Bellingham, Washington.

Introduction to the MATLAB Environment

Matrix Laboratory (MATLAB) is used extensively in this book for simulations. The goal here is to help students acquire familiarity with MATLAB and build basic skills in the MATLAB environment. Hence, Appendix A serves the following objectives:

1. Learn to use the help system to study basic MATLAB command and syntax
2. Learn array indexing
3. Learn basic plotting utilities
4. Learn to write script m-files
5. Learn to write functions in MATLAB.

A.1 Basic Commands and Syntax

MATLAB has an accessible help system by the help command. By issuing the MATLAB help command following the topic or function (routine), MATLAB will return the text information of the topic and show how to use the function. For example, by entering **help** on the MATLAB prompt to question a function

sum( ), we see the text information (listed partially here) to show how to use the MATLAB function **sum( )**.

```
» help sum
SUM Sum of the elements.
   For vectors, SUM(X) is the sum of the elements of X.
   For matrices, SUM(X) is a row vector with the sum over
   each column.
»
```

The following examples are given to demonstrate the usage:

```
» x = [1 2 3 1.5 −1.5 −2];           %Initialize an array
» sum(x) %Call MATLAB function sum
ans =
4            %Display the result
»
» x = [1 2 3; −1.5 1.5 2.5;4 5 6] %Initialize the 3 × 3 matrix
x =          %Display the contents of the 3 × 3 matrix
   1.0000 2.0000 3.0000
  −1.5000 1.5000 2.5000
   4.0000 5.0000 6.0000
» sum(x) %Call MATLAB function sum
ans =
   3.5000 8.5000 11.5000           %Display the results
»
```

MATLAB can be used like a calculator to work with numbers, variables, and expressions in the command window. The following are the basic syntax examples:

```
   » sin(pi/4)
ans =
  0.7071
» pi*4
ans =
  12.5664
```

A.1 Basic Commands and Syntax

In MATLAB, variable names can store values, vectors, and matrices. See the following examples.

```
» x = cos(pi/8)
x =
  0.9239
» y = sqrt(x) − 2∧2
y =
  −3.0388
» z = [1 −2 1 2]
z =
  1 −2 1 2
» zz=[1 2 3; 4 5 6]
zz =
  1 2 3
  4 5 6
```

Complex numbers are natural in MATLAB. See the following examples.

```
» z = 3 + 4i  % Complex number
z =
  3.0000 + 4.0000i
» conj(z)  % Complex conjugate of z
ans =
  3.0000 − 4.0000i
» abs(z)  % Magnitude of z
ans =
  5
» angle(z)  % Angle of z (radians)
ans =
  0.9273
» real(z)  % Real part of a complex number z
ans =
  3
» imag(z)  % Imaginary part of a complex number z
ans =
  4
» exp(j*pi/4)  % Polar form of a complex number
ans =
  0.7071 + 0.7071i
```

The following shows examples of array operations. Krauss, Shure, and Little (1994) and Stearns (2003) give the detailed explanation for each operation.

```
» x = [1 2;3 4]  %Initialize a 2 × 2 matrix
x =
  1 2
  3 4
» y = [−4 3; −2 1]
  y =
  −4 3
  −2 1
» x+y % Add two matrices
ans =
  −3 5
  1 5
» x*y % Matrix product
  ans =
  −8 5
  −20 13
» x.*y % Array element product
  ans =
  −4 6
  −6 4
  » x′ % Matrix transpose
ans =
  1 3
  2 4
» 2.∧x % Exponentiation: matrix x contains each exponent.
  ans =
  2 4
  8 16
» x.∧3 % Exponentiation: power of 3 for each element in matrix x
  ans =
  1 8
  27 64
  » y.∧x % Exponentiation: each element in y has a power
% specified by a corresponding element in matrix x
  ans =
−4 9
−8 1
» x = [0 1 2 3 4 5 6] % Initialize a row vector
x =
  0 1 2 3 4 5 6
» y = x.*x − 2*x % Use array element product to compute a quadratic function
y =
  0 −1 0 3 8 15 24
»
» z = [1 3]′ % Initialize a column vector
  z =
  1
  3
  »w = x\z % Inverse matrix x, then multiply it by the column vector z
  w =
  1
  0
```

A.2 MATLAB Array and Indexing

Let us look at the syntax to create an array as follows:

Basic syntax: **x** = **begin : step : end**

An array *x* is created with the first element value of **begin**. The value increases by a value of **step** for each next element in the array and stops when the next stepped value is beyond the specified end value of **end**. In simulation, we may use this operation to create the sample indices or array of time instants for digital signals. The values **begin**, **step**, and **end** can be assigned to be integers or floating-point numbers.

The following examples are given for illustrations:

```
» n=1:3:8 % Create a vector containing n=[1 4 7]
n =
   1 4 7
» m = 9:-1:2 % Create a vector m=[9 8 7 6 5 4 3 2]
m =
   9 8 7 6 5 4 3 2
» x=2:(1/4):4 % Create x=[2 2.25 2.5 2.75 3 3.25 3.5 3.75 4]
x =
   Columns 1 through 7
   2.0000 2.2500 2.5000 2.7500 3.0000 3.2500 3.5000
   Columns 8 through 9
   3.7500 4.0000
» y = 2:-0.5:-1 % Create y = [2 1.5 1 0.5 0 -0.5 -1]
y =
   2.0000 1.5000 1.0000 0.5000 0 -0.5000 -1.0000
```

Next, we examine creating a vector and extracting numbers in a vector:

```
» xx = [1 2 3 4 5 [5:-1:1]] % create xx=[ 1 2 3 4 5 5 4 3 2 1]
xx =
   1 2 3 4 5 5 4 3 2 1
» xx(3:7) % Show [3 4 5 5 4]
ans =
   3 4 5 5 4
» length(xx) % Return of the number of elements in vector xx
ans =
   10
» yy=xx(2:2:length(xx)) % Display even indexed numbers in array xx
yy =
   2 4 5 3 1
```

A.3 Plot Utilities: Subplot, Plot, Stem, and Stair

The following are common MATLAB plot functions for digital signal processing (DSP) simulation:

subplot opens subplot windows for plotting.
plot produces an *x-y* plot. It can also create multiple plots.
stem produces discrete-time signals.
stair produces staircase (sample-and-hold) signals.

The following program contains different MATLAB plot functions:

```
t=0:0.01:0.4; %Create time vector for time instants from 0 to 0.4 second
xx = 4.5* sin (2*pi*10*t + pi/4); % Calculate a sine function with the frequency 10 Hz
yy = 4.5* cos (2*pi*5*t − pi/3); %Calculate a cos function with the frequency 5 Hz
subplot(4,1,1), plot(t,xx);grid % Plot a sine function in window 1
subplot(4,1,2), plot(t,xx, t,yy,'-.');grid; % Plot sine and cos functions in window 2
subplot(4,1,3), stem(t,xx);grid % Plot a sine function in the discrete-time form
subplot(4,1,4), stairs(t,yy);grid % Plot a cos function in the sample-and-hold form
xlabel('Time (sec.)');
```

Each plot is shown in Figure A.1. Notice that dropping the semicolon at the end of the MATLAB syntax will display values on the MATLAB prompt.

A.4 MATLAB Script Files

We can create a MATLAB script file using the built-in MATLAB editor (or Windows Notepad) to write MATLAB source code. The script file is named "filename.m" and can be run by typing the file name at the MATLAB prompt and hitting the return key. The following script file called **test.m** containing the MATLAB code is listed here for illustration.

At the MATLAB prompt, run the program

≫which test % show the folder where test.m resides

Go to the folder that contains test.m, and run your script from MATLAB.

≫test % run the test.m
≫type test % display the contents of test.m

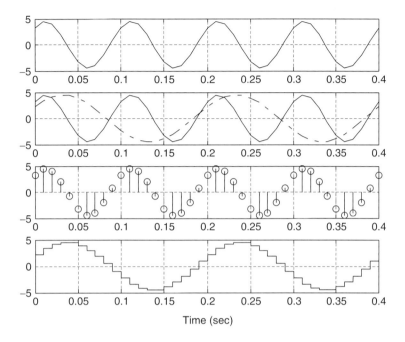

FIGURE A.1 Illustration of MATLAB plot functions.

test.m

```
t=0:0.01:1;
x = sin (2*pi*2*t);
y = 0.5* cos (2*pi*5*t – pi/4);
plot(t,x), grid on
title('Test plots of sinusoids')
ylabel('Signal amplitudes');
xlabel('Time (sec.)');hold on
plot(t,y,'-.');
```

A.5 MATLAB Functions

A MATLAB function is often used for replacing the repetitive portions of the MATLAB codes. It is created using the MATLAB script file. However, the code begins with the key word **function**, following the function declaration, comments for the help system, and program codes. A function **sumsub.m** that computes the addition and subtraction of two numbers is listed here for illustration.

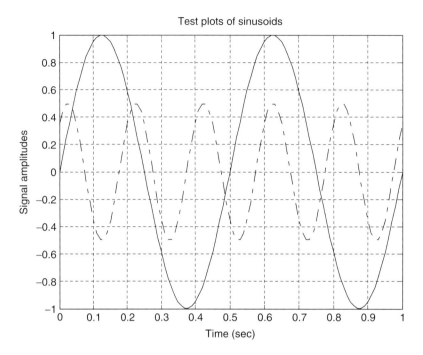

FIGURE A.2   Illustration for the MATLAB script file test.m.

sumsub.m

```
function [sum, sub]=sumsub(x1,x2)
%sumsub: Function to add and subtract two numbers
% Usage:
% [sum, sub] = sumsub(x1, x2)
% x1 = the first number
% x2= the second number
% sum = x1 + x2;
% sub = x1 − x2
sum = x1 + x2;          %Add two numbers
sub= x1 − x2;           %Subtract x2 from x1
```

To use the MATLAB function, go to the folder that contains **sumsub.m**. At the MATLAB prompt, try the following:

```
»help sumsub % Display usage information on MATLAB prompt
sumsub: Function to add and subtract two numbers
Usage:
[sum, sub] = sumsub(x1, x2)
x1 = the first number
x2= the second number
sum = x1 + x2;
sub = x1 − x2
```

Run the function as follows:

```
» [x1, x2]=sumsub(3, 4-3i);          % Call function sumsub
» x1 %Display the result of sum
x1 =
7.0000 −3.0000i
» x2 %Display the result of subtraction
x2 =
−1.0000 + 3.0000i
```

The MATLAB function can also be used inside the m-file. More MATLAB exercises for introduction to DSP can be explored in McClellan, Schafer, and Yoder (1998) and Stearns (2003).

References

Krauss, T. P., Shure, L., and Little, J. N. (1994). *Signal Processing TOOLBOX for Use with MATLAB.* Natick, MA: The MathWorks Inc.

McClellan, J. H., Schafer, R. W., and Yoder, M. A. (1998). *DSP First—A Multimedia Approach.* Upper Saddle River, NJ: Prentice Hall.

Stearns, S. D. (2003). *Digital Signal Processing with Examples in MATLAB.* Boca Raton, FL: CRC Press LLC.

B

Review of Analog Signal Processing Basics

B.1 Fourier Series and Fourier Transform

In electronics applications, we have been familiar with some periodic signals such as the square wave, rectangular wave, triangular wave, sinusoid, sawtooth wave, and so on. These periodic signals can be analyzed in frequency domain with the help of the Fourier series expansion. According to Fourier theory, a periodic signal can be represented by a Fourier series that contains the sum of a series of sine and/or cosine functions (harmonics) plus a direct-current (DC) term. There are three forms of Fourier series: (1) sine-cosine, (2) amplitude-phase, and (3) complex exponential. We will review each of them individually in the following text. Comprehensive treatments can be found in Ambardar (1999), Soliman and Srinath (1998), and Stanley (2003).

B.1.1 Sine-Cosine Form

The Fourier series expansion of a periodic signal $x(t)$ with a period of T via the sine-cosine form is given by

$$x(t) = a_0 + \sum_{n=1}^{\infty} a_n \cos(n\omega_0 t) + \sum_{n=1}^{\infty} b_n \sin(n\omega_0 t), \tag{B.1}$$

whereas $\omega_0 = 2\pi/T_0$ is the fundamental angular frequency in radians per second, while the fundamental frequency in terms of Hz is $f_0 = 1/T_0$. The

Fourier coefficients of $a_0, a_n,$ and b_n may be found according to the following integral equations:

$$a_0 = \frac{1}{T_0} \int_{T_0} x(t) dt \qquad (B.2)$$

$$a_n = \frac{2}{T_0} \int_{T_0} x(t) \cos(n\omega_0 t) dt \qquad (B.3)$$

$$b_n = \frac{2}{T_0} \int_{T_0} x(t) \sin(n\omega_0 t) dt. \qquad (B.4)$$

Notice that the integral is performed over one period of the signal to be expanded. From Equation (B.1), the signal $x(t)$ consists of a DC term and sums of sine and cosine functions with their corresponding harmonic frequencies. Again, note that $n\omega_0$ is the nth harmonic frequency.

B.1.2 Amplitude-Phase Form

From the sine-cosine form, we notice that there is a sum of two terms with the same frequency and that one from the first sum is $a_n \cos(n\omega_0 t)$ and the other is $b_n \sin(n\omega_0 t)$. We can combine these two terms and modify the sine-cosine form into the amplitude-phase form:

$$x(t) = A_0 + \sum_{n=1}^{\infty} A_n \cos(n\omega_0 t + \phi_n). \qquad (B.5)$$

The DC term is the same as before; that is,

$$A_0 = a_0, \qquad (B.6)$$

and the amplitude and phase are given by

$$A_n = \sqrt{a_n^2 + b_n^2} \qquad (B.7)$$

$$\phi_n = \tan^{-1}\left(\frac{-b_n}{a_n}\right), \qquad (B.8)$$

respectively. The amplitude-phase form provides very useful information for spectral analysis. With the calculated amplitude and phase for each harmonic frequency, we can create the spectral plots. One depicts a plot of the amplitude versus its corresponding harmonic frequency, called the amplitude spectrum, while the other plot shows each phase versus its harmonic frequency, called the phase spectrum. Note that the spectral plots are one-sided, since amplitudes and phases are plotted versus the positive harmonic frequencies. We will illustrate these via Example B.1.

B.1.3 Complex Exponential Form

The complex exponential form is developed based on expanding sine and cosine functions in the sine-cosine form into their exponential expressions using Euler's formula and regrouping these exponential terms. Euler's formula is given by

$$e^{\pm jx} = \cos(x) \pm j\sin(x),$$

which can be written as two separate forms:

$$\cos(x) = \frac{e^{jx} + e^{-jx}}{2}$$

$$\sin(x) = \frac{e^{jx} - e^{-jx}}{2j}.$$

We will focus on the interpretation and application rather than its derivation. Thus the complex exponential form is expressed as

$$x(t) = \sum_{n=-\infty}^{\infty} c_n e^{jn\omega_0 t}, \tag{B.9}$$

where c_n represents the complex Fourier coefficients, which may be found from

$$c_n = \frac{1}{T_0} \int_{T_0} x(t) e^{-jn\omega_0 t} dt. \tag{B.10}$$

The relationship between the complex Fourier coefficients c_n and the coefficients of the sine-cosine form are

$$c_0 = a_0 \tag{B.11}$$

$$c_n = \frac{a_n - jb_n}{2}, \text{ for } n > 0. \tag{B.12}$$

for $n < 0$, and c_n is equal to the complex conjugate of $\bar{c}_n$ for $n > 0$. It follows that

$$c_{-n} = \bar{c}_n = \frac{a_n + jb_n}{2}, \text{ for } n > 0. \tag{B.13}$$

Since c_n is a complex value, which can be further written in the magnitude-phase form, we obtain

$$c_n = |c_n| \angle \phi_n, \tag{B.14}$$

where $|c_n|$ is the magnitude and φ_n is the phase of the complex Fourier coefficient. Similar to the amplitude-phase form, we can create the spectral plots for $|c_n|$ and φ_n. Since the frequency index n goes from $-\infty$ to ∞, the plots of resultant spectra are two-sided.

Example B.1.

Considering the following square waveform $x(t)$, shown in Figure B.1, where T_0 represents a period, find the Fourier series expansions in terms of (a) the sine-cosine form, (b) the amplitude-phase form, and (c) the complex exponential form.

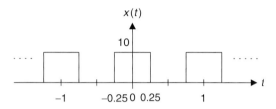

FIGURE B.1 Square waveform in Example B.1.

Solution:

From Figure B.1, we notice that $T_0 = 1$ second and $A = 10$. The fundamental frequency is

$$f_0 = 1/T_0 = 1 \text{ Hz or } \omega_0 = 2\pi \times f_0 = 2\pi \text{ rad/sec}.$$

a. Using Equations (B.1) to (B.3) yields

$$a_0 = \frac{1}{T_0} \int_{-T_0/2}^{T_0/2} x(t)dt = \frac{1}{1} \int_{-0.25}^{0.25} 10 dt = 5$$

$$a_n = \frac{2}{T_0} \int_{-T_0/2}^{T_0/2} x(t) \cos(n\omega_0 t) dt$$

$$= \frac{2}{1} \int_{-0.25}^{0.25} 10 \cos(n 2\pi t) dt$$

$$= \frac{2}{1} \frac{10 \times \sin(n 2\pi t)}{n 2\pi}\bigg|_{-0.25}^{0.25} = 10 \frac{\sin(0.5\pi n)}{0.5\pi n}$$

$$b_n = \frac{2}{T_0} \int_{-T_0/2}^{T_0/2} x(t) \sin(n\omega_0 t) dt$$

$$= \frac{2}{1} \int_{-0.25}^{0.25} 10 \times \sin(n 2\pi t) dt$$

$$= \frac{2}{1} \frac{-10 \cos(n 2\pi t)}{n 2\pi}\bigg|_{-0.25}^{0.25} = 0$$

Thus, the Fourier series expansion in terms of the sine-cosine form is written as

$$x(t) = 5 + \sum_{n=1}^{\infty} 10 \frac{\sin(0.5\pi n)}{0.5\pi n} \cos(n2\pi t)$$

$$= 5 + \frac{20}{\pi}\cos(2\pi t) - \frac{20}{3\pi}\cos(6\pi t) + \frac{4}{\pi}\cos(10\pi t) - \frac{20}{7\pi}\cos(14\pi t) + \ldots$$

b. Making use of the relations between the sine-cosine form and the amplitude-phase form, we obtain

$$A_0 = a_0 = 5$$

$$A_n = \sqrt{a_n^2 + b_n^2} = |a_n| = 10 \times \left|\frac{\sin(0.5\pi n)}{0.5\pi n}\right|.$$

Again, noting that $-\cos(x) = \cos(x + 180°)$, the Fourier series expansion in terms of the amplitude-phase form is

$$x(t) = 5 + \frac{20}{\pi}\cos(2\pi t) + \frac{20}{3\pi}\cos(6\pi t + 180°) + \frac{4}{\pi}\cos(10\pi t)$$

$$+ \frac{20}{7\pi}\cos(14\pi t + 180°) + \ldots$$

c. First let us find the complex Fourier coefficients using the formula, that is,

$$c_n = \frac{1}{T_0} \int_{-T_0/2}^{T_0/2} x(t) e^{-jn\omega_0 t} dt$$

$$= \frac{1}{1} \int_{-0.25}^{0.25} A e^{-jn2\pi t} dt$$

$$= 10 \times \left.\frac{e^{-jn2\pi t}}{-jn2\pi}\right|_{-0.25}^{0.25} = 10 \times \frac{(e^{-j0.5\pi n} - e^{j0.5\pi n})}{-jn2\pi}$$

Applying Euler's formula yields

$$c_n = 10 \times \frac{\cos 0.5\pi n - j\sin(0.5\pi n) - [\cos(0.5\pi n) + j\sin(0.5\pi n)]}{-jn2\pi}$$

$$= 5 \frac{\sin(0.5\pi n)}{0.5\pi n}.$$

Second, using the relationship between the sine-cosine form and the complex exponential form, it follows that

$$c_n = \frac{a_n - jb_n}{2} = \frac{a_n}{2} = 5\frac{\sin(0.5n\pi)}{(0.5n\pi)}.$$

Certainly, the result is identical to that obtained directly from the formula. Note that c_0 cannot be evaluated directly by substituting $n = 0$, since we have the indeterminate term $\frac{0}{0}$. Using L'Hospital's rule, described in Appendix F, leads to

$$c_0 = \lim_{n \to 0} 5\frac{\sin(0.5n\pi)}{(0.5n\pi)} = \lim_{n \to 0} 5\frac{\frac{d(\sin(0.5n\pi))}{dn}}{\frac{d(0.5n\pi)}{dn}}$$

$$= \lim_{n \to 0} 5\frac{0.5\pi \cos(0.5n\pi)}{0.5\pi} = 5$$

Finally, the Fourier expansion in terms of the complex exponential form is shown as follows:

$$x(t) = \ldots + \frac{10}{\pi}e^{-j2\pi t} + 5 + \frac{10}{\pi}e^{j2\pi t} - \frac{10}{3\pi}e^{j6\pi t} + \frac{2}{\pi}e^{j10\pi t} - \frac{10}{7\pi}e^{j14\pi t} + \ldots$$

B.1.4 Spectral Plots

As previously discussed, the amplitude-phase form can provide information to create a one-sided spectral plot. The amplitude spectrum is obtained by plotting A_n versus the harmonic frequency $n\omega_0$, and the phase spectrum is obtained by plotting φ_n versus $n\omega_0$, both for $n \geq 0$. Similarly, if the complex exponential form is used, the two-sided amplitude and phase spectral plots of $|c_n|$ and φ_n versus $n\omega_0$ for $-\infty < n < \infty$ can each be achieved, respectively. We illustrate this by the following example.

Example B.2.

a. Plot the amplitude spectrum of Example B.1 for a one-sided amplitude spectrum and a two-sided amplitude spectrum, respectively.

Solution:

a. Based on the solution of A_n, the one-sided amplitude spectrum is shown as follows:

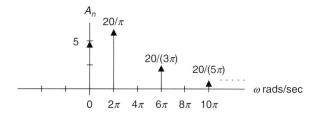

FIGURE B.2 One-sided spectrum of the square waveform in Example B.2.

According to the solution of the complex exponential form, the two-sided amplitude spectrum is demonstrated as follows:

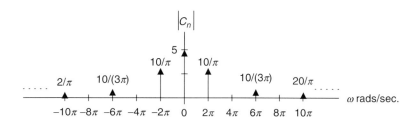

FIGURE B.3 Two-sided spectrum of the square waveform in Example B.2.

A general pulse train $x(t)$ with a period T_0 seconds and a pulse width τ seconds is shown in Figure B.4.

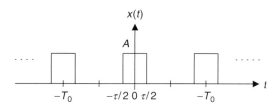

FIGURE B.4 Rectangular waveform (pulse train).

The Fourier series expansions for sine-cosine and complex exponential forms can be derived similarly and are given on the next page.

Sine-cosine form:

$$x(t) = \frac{\tau A}{T_0} + \frac{2\tau A}{T_0}\left(\frac{\sin(\omega_0\tau/2)}{(\omega_0\tau/2)}\cos(\omega_0 t) + \frac{\sin(2\omega_0\tau/2)}{(2\omega_0\tau/2)}\cos(2\omega_0 t)\right.$$
$$\left. + \frac{\sin(3\omega_0\tau/2)}{(3\omega_0\tau/2)}\cos(3\omega_0 t) + \ldots\right) \quad (B.15)$$

Complex exponential form:

$$x(t) = \ldots + \frac{\tau A}{T_0}\frac{\sin(\omega_0\tau/2)}{(\omega_0\tau/2)}e^{-j\omega_0 t} + \frac{\tau A}{T_0} + \frac{\tau A}{T_0}\frac{\sin(\omega_0\tau/2)}{(\omega_0\tau/2)}e^{j\omega_0 t}$$
$$+ \frac{\tau A}{T_0}\frac{\sin(2\omega_0\tau/2)}{(2\omega_0\tau/2)}e^{j2\omega_0 t} + \ldots, \quad (B.16)$$

where $\omega_0 = 2\pi f_0 = 2\pi/T_0$ is the fundamental angle frequency of the periodic waveform. The reader can derive the one-sided amplitude spectrum A_n and the two-sided amplitude spectrum $|c_n|$. The expressions for one-sided amplitude and two-sided amplitude spectra are given by the following:

$$A_0 = \frac{\tau}{T_0}A \quad (B.17)$$

$$A_n = \frac{2\tau}{T_0}A\left|\frac{\sin(n\omega_0\tau/2)}{(n\omega_0\tau/2)}\right|, \text{ for } n = 1, 2, 3\ldots \quad (B.18)$$

$$|c_n| = \frac{\tau}{T_0}A\left|\frac{\sin(n\omega_0\tau/2)}{(n\omega_0\tau/2)}\right|, -\infty < n < \infty. \quad (B.19)$$

Example B.3.

a. In Figure B.4, if $T_0 = 1\,\text{ms}$, $\tau = 0.2\,\text{ms}$, and $A = 10$, use Equations (B.17) to (B.19) to derive the amplitude one-sided spectrum and two-sided spectrum for each of the first four harmonic frequency components.

Solution:

a. The fundamental frequency is

$$\omega_0 = 2\pi f_0 = 2\pi \times (1/0.001) = 2000\pi\,\text{rad/sec}.$$

Using Equations (B.17) and (B.18) yields the one-sided spectrum as

$$A_0 = \frac{\tau}{T_0}A = \frac{0.0002}{0.001} \times 10 = 2, \text{ for } n = 0, n\omega_0 = 0.$$

For $n = 1$, $n\omega_0 = 2000\pi$ rad/sec:

$$A_1 = \frac{2 \times 0.0002}{0.001} \times 10 \times \left|\frac{\sin(1 \times 2000\pi \times 0.0002/2)}{(1 \times 2000\pi \times 0.0002/2)}\right| = 4\frac{\sin(0.2\pi)}{(0.2\pi)} = 3.7420,$$

for $n = 2$, $n\omega_0 = 4000\pi$ rad/sec:

$$A_2 = \frac{2 \times 0.0002}{0.001} \times 10 \times \left|\frac{\sin(2 \times 2000\pi \times 0.0002/2)}{(2 \times 2000\pi \times 0.0002/2)}\right| = 4\frac{\sin(0.4\pi)}{(0.4\pi)} = 3.0273,$$

for $n = 3$, $n\omega_0 = 6000\pi$ rad/sec:

$$A_3 = \frac{2 \times 0.0002}{0.001} \times 10 \times \left|\frac{\sin(3 \times 2000\pi \times 0.0002/2)}{(3 \times 2000\pi \times 0.0002/2)}\right| = 4\frac{\sin(0.6\pi)}{(0.6\pi)} = 2.0182,$$

for $n = 4$, $n\omega_0 = 8000\pi$ rad/sec:

$$A_4 = \frac{2 \times 0.0002}{0.001} \times 10 \times \left|\frac{\sin(4 \times 2000\pi \times 0.0002/2)}{(4 \times 2000\pi \times 0.0002/2)}\right| = 4\frac{\sin(0.8\pi)}{(0.8\pi)} = 0.9355.$$

The one-sided amplitude spectrum is plotted in Figure B.5.

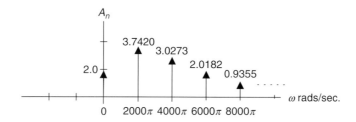

FIGURE B.5 One-sided spectrum in Example B.3.

Similarly, applying Equation (B.19) leads to

$$|c_0| = \frac{0.0002}{0.001} \times 10 \times \left|\lim_{n \to 0} \frac{\sin(n \times 2000\pi \times 0.0002/2)}{(n \times 2000\pi \times 0.0002/2)}\right| = 2 \times |1| = 2.$$

Note: we use the fact $\lim_{x \to 0} \frac{\sin(x)}{x} = 1.0$ (see L'Hospital's rule in Appendix F)

$$|c_1| = |c_{-1}| = \frac{0.0002}{0.001} \times 10 \times \left|\frac{\sin(1 \times 2000\pi \times 0.0002/2)}{(1 \times 2000\pi \times 0.0002/2)}\right| = 2 \times \left|\frac{\sin(0.2\pi)}{0.2\pi}\right| = 1.8710$$

$$|c_2| = |c_{-2}| = \frac{0.0002}{0.001} \times 10 \times \left|\frac{\sin(2 \times 2000\pi \times 0.0002/2)}{(2 \times 2000\pi \times 0.0002/2)}\right| = 2 \times \left|\frac{\sin(0.4\pi)}{0.4\pi}\right| = 1.5137$$

$$|c_3| = |c_{-3}| = \frac{0.0002}{0.001} \times 10 \times \left|\frac{\sin(3 \times 2000\pi \times 0.0002/2)}{(3 \times 2000\pi \times 0.0002/2)}\right| = 2 \times \left|\frac{\sin(0.6\pi)}{0.6\pi}\right| = 1.0091$$

$$|c_4| = |c_{-4}| = \frac{0.0002}{0.001} \times 10 \times \left|\frac{\sin(4 \times 2000\pi \times 0.0002/2)}{(4 \times 2000\pi \times 0.0002/2)}\right| = 2 \times \left|\frac{\sin(0.8\pi)}{0.8\pi}\right| = 0.4677$$

Figure B.6 shows the two-sided amplitude spectral plot.

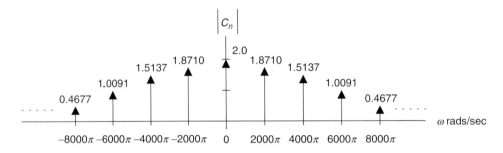

FIGURE B.6 Two-sided spectrum in Example B.3.

The following example illustrates the use of table information to determine the Fourier series expansion of the periodic waveform. Table B.1 consists of the Fourier series expansions for common periodic signals in the sine-cosine form while Table B.2 shows the expansions in the complex exponential form.

Example B.4.

In the sawtooth waveform shown in Table B.1, if $T_0 = 1$ ms and $A = 10$, use the formula in the table to determine the Fourier series expansion in amplitude-phase form, and determine the frequency f_3 and amplitude value of A_3 for the third harmonic.

a. Write the Fourier series expansion also in a complex exponential form, and determine $|c_3|$ and $|c_{-3}|$ for the third harmonic.

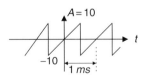

FIGURE B.7 Sawtooth waveform in Example B.4.

Solution:

a. Based on the information in Table B.1, we have

$$x(t) = \frac{2A}{\pi}\left(\sin \omega_0 t - \frac{1}{2}\sin 2\omega_0 t + \frac{1}{3}\sin 3\omega_0 t - \frac{1}{4}\sin 4\omega_0 t + \ldots\right).$$

TABLE B.1 Fourier series expansions for some common waveform signals in the sine-cosine form.

Time domain signal $x(t)$	Fourier series expansion
Positive square wave 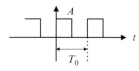	$x(t) = \dfrac{A}{2} + \dfrac{2A}{\pi}\left(\sin \omega_0 t + \dfrac{1}{3}\sin 3\omega_0 t \right.$ $\left. + \dfrac{1}{5}\sin 5\omega_0 t + \dfrac{1}{7}\sin 7\omega_0 t + \cdots \right)$
Square wave 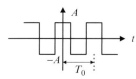	$x(t) = \dfrac{4A}{\pi}\left(\cos \omega_0 t - \dfrac{1}{3}\cos 3\omega_0 t \right.$ $\left. + \dfrac{1}{5}\cos 5\omega_0 t - \dfrac{1}{7}\cos 7\omega_0 t + \cdots \right)$
Triangular wave 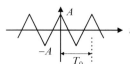	$x(t) = \dfrac{8A}{\pi^2}\left(\cos \omega_0 t + \dfrac{1}{9}\cos 3\omega_0 t \right.$ $\left. + \dfrac{1}{25}\cos 5\omega_0 t + \dfrac{1}{49}\cos 7\omega_0 t + \cdots \right)$
Sawtooth wave 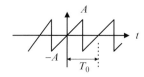	$x(t) = \dfrac{2A}{\pi}\left(\sin \omega_0 t - \dfrac{1}{2}\sin 2\omega_0 t \right.$ $\left. + \dfrac{1}{3}\sin 3\omega_0 t - \dfrac{1}{4}\sin 4\omega_0 t + \cdots \right)$
Rectangular wave (Pulse train) Duty cycle $= d = \dfrac{\tau}{T_0}$	$x(t) = Ad + 2Ad\left(\dfrac{\sin \pi d}{\pi d}\right)\cos \omega_0 t$ $+ 2Ad\left(\dfrac{\sin 2\pi d}{2\pi d}\right)\cos 2\omega_0 t$ $+ 2Ad\left(\dfrac{\sin 3\pi d}{3\pi d}\right)\cos 3\omega_0 t + \cdots$
Ideal impulse train 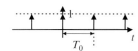	$x(t) = \dfrac{1}{T_0} + \dfrac{2}{T_0}(\cos \omega_0 t + \cos 2\omega_0 t$ $+ \cos 3\omega_0 t + \cos 4\omega_0 t + \cdots)$

TABLE B.2 Fourier series expansions for some common waveform signals in the complex exponential form.

Time domain signal $x(t)$	Fourier series expansion
Positive square wave 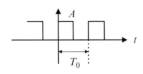	$x(t) = \cdots - \dfrac{A}{j3\pi}e^{-j3\omega_0 t} - \dfrac{A}{j\pi}e^{-j\omega_0 t} + \dfrac{A}{2}$ $+ \dfrac{A}{j\pi}e^{j\omega_0 t} + \dfrac{A}{j3\pi}e^{j3\omega_0 t} + \dfrac{A}{j5\pi}e^{j5\omega_0 t} + \cdots)$
Square wave 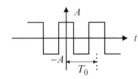	$x(t) = \dfrac{2A}{\pi}\left(\cdots + \dfrac{1}{5}e^{-j5\omega_0 t} - \dfrac{1}{3}e^{-j3\omega_0 t} + e^{-j\omega_0 t} \right.$ $\left. + e^{j\omega_0 t} - \dfrac{1}{3}e^{j3\omega_0 t} + \dfrac{1}{5}e^{j5\omega_0 t} - \cdots \right)$
Triangular wave 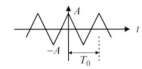	$x(t) = \dfrac{4A}{\pi^2}\left(\cdots + \dfrac{1}{25}e^{j5\omega_0 t} + \dfrac{1}{9}e^{-j3\omega_0 t} \right.$ $\left. + e^{-j\omega_0 t} + e^{j\omega_0 t} + \dfrac{1}{9}e^{j3\omega_0 t} + \dfrac{1}{25}e^{j5\omega_0 t} + \cdots \right)$
Sawtooth wave 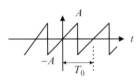	$x(t) = \dfrac{A}{j\pi}\left(\cdots - \dfrac{1}{3}e^{-j3\omega_0 t} + \dfrac{1}{2}e^{-j2\omega_0 t} - e^{-j\omega_0 t} \right.$ $\left. + e^{j\omega_0 t} - \dfrac{1}{2}e^{j2\omega_0 t} + \dfrac{1}{3}e^{j3\omega_0 t} + \cdots \right)$
Rectangular wave (Pulse train) Duty cycle $= d = \dfrac{\tau}{T_0}$	$x(t) = \cdots + Ad\left(\dfrac{\sin \pi d}{\pi d}\right)e^{-j\omega_0 t} + Ad$ $+ Ad\left(\dfrac{\sin \pi d}{\pi d}\right)e^{j\omega_0 t} + Ad\left(\dfrac{\sin 2\pi d}{2\pi d}\right)e^{j2\omega_0 t}$ $+ Ad\left(\dfrac{\sin 3\pi d}{3\pi d}\right)e^{j3\omega_0 t} + \cdots$
Ideal impulse train	$x(t) = \dfrac{1}{T_0}\left(\cdots + e^{-j3\omega_0 t} + e^{-j2\omega_0 t} + e^{-j\omega_0 t} + 1 \right.$ $\left. + e^{j\omega_0 t} + e^{j2\omega_0 t} + e^{j3\omega_0 t} + \cdots \right)$

Since $T_0 = 1$ ms, the fundamental frequency is
$$f_0 = 1/T_0 = 1000 \text{ Hz, and } \omega_0 = 2\pi f_0 = 2000\pi \text{ rad/sec.}$$
Then, the expansion is determined as
$$x(t) = \frac{2 \times 10}{\pi}\left(\sin 2000\pi t - \frac{1}{2}\sin 4000\pi t + \frac{1}{3}\sin 6000\pi t\right.$$
$$\left. -\frac{1}{4}\sin 8000\pi t + \ldots\right).$$

Using trigonometric identities:
$$\sin x = \cos(x - 90°) \text{ and } -\sin x = \cos(x + 90°),$$
and simple algebra, we finally obtain
$$x(t) = \frac{20}{\pi}\cos(2000\pi t - 90°) + \frac{10}{\pi}\cos(4000\pi t + 90°)$$
$$+ \frac{20}{3\pi}\cos(6000\pi t - 90°) + \frac{5}{\pi}\cos(8000\pi t + 90°) + \ldots.$$

From the amplitude-phase form, we then determine f_3 and A_3 as follows:
$$f_3 = 3 \times f_0 = 3000 \text{ Hz, and } A_3 = \frac{20}{3\pi} = 2.1221.$$

b. From Table B.2, the complex exponential form is
$$x(t) = \frac{10}{j\pi}\left(\ldots - \frac{1}{3}e^{-j6000\pi t} + \frac{1}{2}e^{-j4000\pi t} - e^{-j2000\pi t} + e^{j2000\pi t}\right.$$
$$\left. - \frac{1}{2}e^{j4000\pi t} + \frac{1}{3}e^{j6000\pi t} + \ldots\right).$$

From the expression, we have
$$|c_3| = \left|\frac{10}{j\pi} \times \frac{1}{3}\right| = \left|\frac{1.061}{j}\right| = 1.061 \text{ and}$$
$$|c_{-3}| = \left|-\frac{10}{j\pi} \times \frac{1}{3}\right| = \left|-\frac{1.061}{j}\right| = 1.061.$$

B.1.5 Fourier Transform

Fourier transform is a mathematical function that provides the frequency spectral analysis for a nonperiodic signal. The Fourier transform pair is defined as

$$\text{Fourier transform: } X(\omega) = \int_{-\infty}^{\infty} x(t)e^{-j\omega t} dt \qquad (B.20)$$

Inverse Fourier transform: $\quad x(t) = \dfrac{1}{2\pi}\displaystyle\int_{-\infty}^{\infty} X(\omega)e^{j\omega t}d\omega, \qquad$ (B.21)

where $x(t)$ is a nonperiodic signal and $X(\omega)$ is a two-sided continuous spectrum versus the continuous frequency variable ω, where $-\infty < \omega < \infty$. Again, the spectrum is a complex function that can be further written as

$$X(\omega) = |X(\omega)|\angle\phi(\omega), \qquad (B.22)$$

where $|X(\omega)|$ is the continuous amplitude spectrum, while $\angle\phi(\omega)$ designates the continuous phase spectrum.

Example B.5.

Let $x(t)$ be a single rectangular pulse, shown in Figure B.8, where the pulse width is $\tau = 0.5$ second.

a. Find its Fourier transform and sketch the amplitude spectrum.

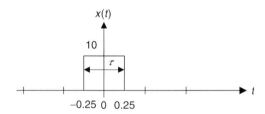

FIGURE B.8 Rectangular pulse in Example B.5.

Solution:

a. Applying Equation (B.21) and using Euler's formula, we have

$$X(\omega) = \int_{-\infty}^{\infty} x(t)e^{-j\omega t}dt = \int_{-0.25}^{0.25} 10e^{-j\omega}dt$$

$$= 10\dfrac{e^{-j\omega t}}{-j\omega}\bigg|_{-0.25}^{0.25} = 10 \times \dfrac{(e^{-j0.25\omega} - e^{j0.25\omega})}{-j\omega}$$

$$= 10 \times \dfrac{\cos(0.25\omega) - j\sin(0.25\omega) - [\cos(0.25\omega) + j\sin(0.25\omega)]}{-j\omega}$$

$$= 5\dfrac{\sin(0.25\omega)}{0.25\omega}$$

where the amplitude spectrum is expressed as

$$|X(\omega)| = 5 \times \left|\dfrac{\sin(0.25\omega)}{0.25\omega}\right|.$$

Or using $\omega = 2\pi f$, we can express the spectrum in terms of Hz as

$$|X(f)| = 5 \times \left|\frac{\sin(0.5\pi f)}{0.5\pi f}\right|.$$

The amplitude spectrum is shown in Figure B.9. Note that the first null point is at $\omega = 2\pi/0.5 = 4\pi$ rad/sec, and the spectrum is symmetric.

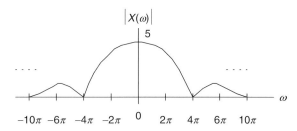

FIGURE B.9 Amplitude spectrum in Example B.5.

Example B.6.

Let $x(t)$ be an exponential function given by

$$x(t) = 10e^{-2t}u(t) = \begin{cases} 10e^{-2t} & t \geq 0 \\ 0 & t < 0 \end{cases}.$$

a. Find its Fourier transform.

Solution:

a. According to the definition of the Fourier transform,

$$X(\omega) = \int_0^\infty 10e^{-2t}u(t)e^{-j\omega t}dt = \int_0^\infty 10e^{-(2+j\omega)t}dt$$

$$= \left.\frac{10e^{-(2+j\omega)t}}{-(2+j\omega)}\right|_0^\infty = \frac{10}{2+j\omega}$$

$$X(\omega) = \frac{10}{\sqrt{2^2+\omega^2}} \angle -\tan^{-1}\left(\frac{\omega}{2}\right)$$

Using $\omega = 2\pi f$, we get

$$X(f) = \frac{10}{2+j2\pi f} = \frac{10}{\sqrt{2^2+(2\pi f)^2}} \angle -\tan^{-1}(\pi f).$$

The Fourier transforms for some common signals are listed in Table B.3. Some useful properties of the Fourier transform are summarized in Table B.4.

TABLE B.3 Fourier transforms for some common signals.

Time domain signal $x(t)$	Fourier spectrum $X(f)$
Rectangular pulse 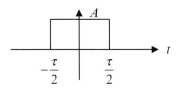	$X(f) = A\tau \frac{\sin \pi f \tau}{\pi f \tau}$
Triangular pulse 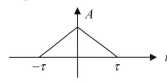	$X(f) = A\tau \left(\frac{\sin \pi f \tau}{\pi f \tau}\right)^2$
Cosine pulse 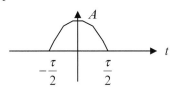	$X(f) = \frac{2A\tau}{\pi} \left(\frac{\cos \pi f \tau}{1 - 4f^2 \tau^2}\right)$
Sawtooth pulse 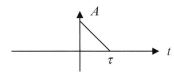	$X(f) = \frac{jA}{2\pi f} \left(\frac{\sin \pi f \tau}{\pi f \tau} e^{-j\pi f \tau} - 1\right)$
Exponential function $\alpha = \frac{1}{\tau}$ 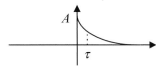	$X(f) = \frac{A}{\alpha + j2\pi f}$
Impulse function 	$X(f) = A$

TABLE B.4 Properties of Fourier transform.

Line	Time Function	Fourier Transform
1	$\alpha x_1(t) + \beta x_2(t)$	$\alpha X_1(f) + \beta X_2(f)$
2	$\frac{dx(t)}{dt}$	$j2\pi f X(f)$
3	$\int_{-\infty}^{t} x(t)dt$	$\frac{X(f)}{j2\pi f}$
4	$x(t - \tau)$	$e^{-j2\pi f \tau} X(f)$
5	$e^{j2\pi f_0 t} x(t)$	$X(f - f_0)$
6	$x(at)$	$\frac{1}{a} X\left(\frac{f}{a}\right)$

Example B.7.

Find the Fourier transforms of the following functions:

a. $x(t) = \delta(t)$, where $\delta(t)$ is an impulse function defined by

$$\delta(t) = \begin{cases} \neq 0 & t = 0 \\ 0 & elsewhere \end{cases}$$

with a property given as

$$\int_{-\infty}^{\infty} f(t)\delta(t - \tau)dt = f(\tau)$$

b. $x(t) = \delta(t - \tau)$

Solution:

a. We first use the Fourier transform definition and then apply the delta function property,

$$X(\omega) = \int_{-\infty}^{\infty} \delta(t)e^{-j\omega t} dt = e^{-j\omega t}\Big|_{t=0} = 1.$$

b. Similar to (a), we obtain

$$X(\omega) = \int_{-\infty}^{\infty} \delta(t - \tau)e^{-j\omega t} dt = e^{-j\omega t}\Big|_{t=\tau} = e^{-j\omega \tau}.$$

Example B.8 shows how to use the table information to determine the Fourier transform of the nonperiodic signal.

Example B.8.

a. Use Table B.3 to determine the Fourier transform of the following cosine pulse.

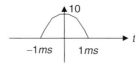

FIGURE B.10 Cosine pulse in Example B.8.

Solution:

a. According to the graph, we can identify

$$\frac{\tau}{2} = 1\,ms, \text{ and } A = 1.$$

τ is given by

$$\tau = 2 \times 1\,ms = 0.002 \text{ second}.$$

Applying the formula from Table B.3 gives

$$X(f) = \frac{2 \times 10 \times 0.002}{\pi} \left(\frac{\cos \pi f 0.002}{1 - 4f^2 0.002^2} \right) = \frac{0.04}{\pi} \left(\frac{\cos 0.002 \pi f}{1 - 4 \times 0.002^2 f^2} \right).$$

B.2 Laplace Transform

In this section, we will review Laplace transform and its applications.

B.2.1 Laplace Transform and Its Table

Laplace transform plays an important role in analysis of continuous signals and systems. We define Laplace transform pairs as

$$X(s) = L\{x(t)\} = \int_0^\infty x(t)e^{-st}dt \qquad (B.23)$$

$$x(t) = L^{-1}\{X(s)\} = \frac{1}{2\pi j} \int_{\gamma-j\infty}^{\gamma+j\infty} X(s)e^{st}ds. \qquad (B.24)$$

Notice that the symbol $L\{\}$ denotes the forward Laplace operation, while the symbol $L^{-1}\{\}$ indicates the inverse Laplace operation. Some common Laplace transform pairs are listed in Table B.5.

In Examples B.9, we examine Laplace transform in light of its definition.

Example B.9.

a. Derive the Laplace transform of the unit step function.

Solution:

a. By the definition in Equation (B.23),

$$X(s) = \int_0^\infty u(t)e^{-st}dt$$

$$= \int_0^\infty e^{-st}dt = \frac{e^{-st}}{-s}\bigg|_0^\infty = \frac{e^{-\infty}}{-s} - \frac{e^0}{-s} = \frac{1}{s}.$$

The answer is consistent with the result listed in Table B.5. Now we use the results in Table B.5 to find the Laplace transform of a function.

Example B.10.

Perform the Laplace transform for each of the following functions.

a. $x(t) = 5\sin(2t)u(t)$

b. $x(t) = 5e^{-3t}\cos(2t)u(t)$.

Solution:

a. Using line 5 in Table B.5 and noting that $\omega = 2$, the Laplace transform immediately follows that

$$X(s) = 5L\{2\sin(2t)u(t)\}$$

$$= \frac{5 \times 2}{s^2 + 2^2} = \frac{10}{s^2 + 4}.$$

b. Applying line 9 in Table B.5 with $\omega = 2$ and $a = 3$ yields

$$X(s) = 5L\{e^{-3t}\cos(2t)u(t)\}$$

$$= \frac{5(s+3)}{(s+3)^2 + 2^2} = \frac{5(s+3)}{(s+3)^2 + 4}.$$

B.2.2 Solving Differential Equations Using Laplace Transform

One of the important applications of Laplace transform is to solve the differential equation. Using the differential property in Table B.5, we can transform the differential equation from the time domain to the Laplace domain. This will change the differential equation into an algebraic equation, and we then solve

TABLE B.5 Laplace transform table.

Line	Time Function $x(t)$	Laplace Transform $X(s) = L(x(t))$
1	$\delta(t)$	1
2	1 or $u(t)$	$\frac{1}{s}$
3	$tu(t)$	$\frac{1}{s^2}$
4	$e^{-at}u(t)$	$\frac{1}{s+a}$
5	$\sin(\omega t)u(t)$	$\frac{\omega}{s^2+\omega^2}$
6	$\cos(\omega t)u(t)$	$\frac{s}{s^2+\omega^2}$
7	$\sin(\omega t + \theta)u(t)$	$\frac{s\sin(\theta) + \omega\cos(\theta)}{s^2+\omega^2}$
8	$e^{-at}\sin(\omega t)u(t)$	$\frac{\omega}{(s+a)^2+\omega^2}$
9	$e^{-at}\cos(\omega t)u(t)$	$\frac{s+a}{(s+a)^2+\omega^2}$
10	$\left(A\cos(\omega t) + \frac{B-aA}{\omega}\sin(\omega t)\right)e^{-at}u(t)$	$\frac{As+B}{(s+a)^2+\omega^2}$
11a	$t^n u(t)$	$\frac{n!}{s^{n+1}}$
11b	$\frac{1}{(n-1)!}t^{n-1}u(t)$	$\frac{1}{s^n}$
12a	$e^{-at}t^n u(t)$	$\frac{n!}{(s+a)^{n+1}}$
12b	$\frac{1}{(n-1)!}e^{-at}t^{n-1}u(t)$	$\frac{1}{(s+a)^n}$
13	$(2\text{Real}(A)\cos(\omega t) - 2\text{Imag}(A)\sin(\omega t))$	$\frac{A}{s+\alpha-j\omega} + \frac{A^*}{s+\alpha+j\omega}$
14	$\frac{dx(t)}{dt}$	$sX(s) - x(0^-)$
15	$\int_0^t x(t)dt$	$\frac{X(s)}{s}$
16	$x(t-a)u(t-a)$	$e^{-as}X(s)$
17	$e^{-at}x(t)u(t)$	$X(s+a)$

the algebraic equation. Finally, the inverse Laplace transform is processed to yield the time domain solution.

Example B.11.

a. Solve the following differential equation using Laplace transform:

$$\frac{dy(t)}{dt} + 10y(t) = x(t) \text{ with an initial condition: } y(0) = 0,$$

where the input is $x(t) = 5u(t)$.

Solution:

a. Applying the Laplace transform on both sides of the differential equation and using the differential property (line 14 in Table B.5), we get

$$sY(s) - y(0) + 10Y(s) = X(s).$$

Note that

$$X(s) = L\{5u(t)\} = \frac{5}{s}.$$

Substituting the initial condition yields

$$Y(s) = \frac{5}{s(s+10)}.$$

Then we use a partial fraction expansion by writing

$$Y(s) = \frac{A}{s} + \frac{B}{s+10}$$

where $A = sY(s)|_{s=0} = \left.\frac{5}{s+10}\right|_{s=0} = 0.5$

and $B = (s+10)Y(s)|_{s=-10} = \left.\frac{5}{s}\right|_{s=-10} = -0.5.$

Hence,

$$Y(s) = \frac{0.5}{s} - \frac{0.5}{s+10}.$$

$$y(t) = L^{-1}\left\{\frac{0.5}{s}\right\} - L^{-1}\left\{\frac{0.5}{s+10}\right\}.$$

Finally, applying the inverse Laplace transform leads to using the results listed in Table B.5, we obtain the time domain solution as

$$y(t) = 0.5u(t) - 0.5e^{-10t}u(t).$$

B.2.3 Transfer Function

The linear analog system can be described using the Laplace transfer function. The transfer function relating the input and output of the linear system is depicted as

$$Y(s) = H(s)X(s), \quad \text{(B.25)}$$

where $X(s)$ and $Y(s)$ are the system input and response (output), respectively, in the Laplace domain, and the transfer function is defined as a ratio of the Laplace response of the system to the Laplace input given by

$$H(s) = \frac{Y(s)}{X(s)}. \quad \text{(B.26)}$$

The transfer function will allow us to study the system behavior. Considering an impulse function as the input to a linear system, that is, $x(t) = \delta(t)$, whose Laplace transform is $X(s) = 1$, we then find the system output due to the impulse function to be

$$Y(s) = H(s)X(s) = H(s). \quad \text{(B.27)}$$

Therefore, the response in time domain $y(t)$ is called the impulse response of the system and can be expressed as

$$h(t) = L^{-1}\{H(s)\}. \quad \text{(B.28)}$$

The analog impulse response can be sampled and transformed to obtain a digital filter transfer function as one of the applications. This topic is covered in Chapter 8.

Example B.12.

Consider a linear system described by the differential equation shown in Example B.11. $x(t)$ and $y(t)$ designate the system input and system output, respectively.

a. Derive the transfer function and the impulse response of the system.

Solution:

a. Tacking the Laplace transform on both sides of the differential equation yields

$$L\left\{\frac{dy(t)}{dt}\right\} + L\{10y(t)\} = L\{x(t)\}.$$

Applying the differential property and substituting the initial condition, we have

$$Y(s)(s + 10) = X(s).$$

Thus, the transfer function is given by

$$H(s) = \frac{Y(s)}{X(s)} = \frac{1}{s+10}.$$

The impulse response can be found by taking the inverse Laplace transform as

$$h(t) = L^{-1}\left\{\frac{1}{s+10}\right\} = e^{-10t}u(t).$$

B.3 Poles, Zeros, Stability, Convolution, and Sinusoidal Steady-State Response

This section is a review of analog system analysis.

B.3.1 Poles, Zeros, and Stability

To study the system behavior, the transfer function is written in a general form, given by

$$H(s) = \frac{N(s)}{D(s)} = \frac{b_m s^m + b_{m-1} s^{m-1} + \ldots + b_0}{a_n s^n + a_{n-1} s^{n-1} + \ldots + a_0}. \quad \text{(B.29)}$$

It is a ratio of the numerator polynomial of degree of m to the denominator polynomial of degree n. The numerator polynomial is expressed as

$$N(s) = b_m s^m + b_{m-1} s^{m-1} + \ldots + b_0, \quad \text{(B.30)}$$

while the denominator polynomial is given by

$$D(s) = a_n s^n + a_{n-1} s^{n-1} + \ldots + a_0. \quad \text{(B.31)}$$

Again, the roots of $N(s)$ are called zeros, while the roots of $D(s)$ are called poles of the transfer function $H(s)$. Notice that zeros and poles can be real numbers or complex numbers.

Given a system transfer function, the poles and zeros can be found. Further, a pole-zero plot can be created on the s-plane. Having the pole-zero plot, the stability of the system is determined by the following rules:

1. The linear system is stable if the rightmost pole(s) is(are) on the left-hand half plane (LHHP) on the s-plane.

2. The linear system is marginally stable if the rightmost pole(s) is(are) simple order (first order) on the $j\omega$ axis, including the origin on the s-plane.

3. The linear system is unstable if the rightmost pole(s) is(are) on the right-hand half plane (RHHP) of the s-plane or if the rightmost pole(s) is(are) multiple order on the $j\omega$ axis on the s-plane.
4. Zeros do not affect the system stability.

Example B.13.

Determine whether each of the following transfer functions is stable, marginally stable, or unstable.

a. $H(s) = \dfrac{s+1}{(s+1.5)(s^2+2s+5)}$

b. $H(s) = \dfrac{(s+1)}{(s+2)(s^2+4)}$

c. $H(s) = \dfrac{s+1}{(s-1)(s^2+2s+5)}$

Solution:

a. A zero is found to be $s = -1$.
   The poles are calculated as $s = -1.5$, $s = -1+j2$, $s = -1-j2$.
   The pole-zero plot is shown in Figure B.11a. Since all the poles are located on the LHHP, the system is stable.

b. A zero is found to be $s = -1$.
   The poles are calculated as $s = -2$, $s = j2$, $s = -j2$.
   The pole-zero plot is shown in Figure B.11b. Since the first-order poles $s = \pm j2$ are located on the $j\omega$ axis, the system is marginally stable.

c. A zero is found to be $s = -1$.
   The poles are calculated as $s = 1$, $s = -1+j2$, $s = -1-j2$.
   The pole-zero plot is shown in Figure B.11c. Since there is a pole $s = 1$ located on the RHHP, the system is unstable.

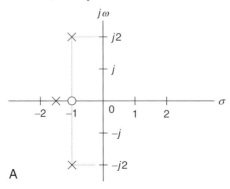

A

FIGURE B.11A Pole-zero plot for (a).

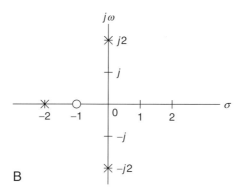

FIGURE B.11B   Pole-zero plot for (b).

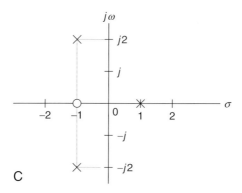

FIGURE B.11C   Pole-zero plot for (c).

B.3.2 Convolution

As we discussed before, the input and output relationship of a linear system in Laplace domain is shown as

$$Y(s) = H(s)X(s). \tag{B.32}$$

It is apparent that in Laplace domain, the system output is the product of the Laplace input and transfer function. But in time domain, the system output is given as

$$y(t) = h(t)*x(t), \tag{B.33}$$

where $*$ denotes linear convolution of the system impulse response $h(t)$ and the system input $x(t)$. The linear convolution is further expressed as

$$y(t) = \int_0^\infty h(\tau)x(t-\tau)d\tau. \tag{B.34}$$

Example B.14.

As you have seen in Examples B.11 and B.12, for a linear system, the impulse response and the input are given, respectively, by

$$h(t) = e^{-10t}u(t) \text{ and } x(t) = 5u(t).$$

a. Determine the system response $y(t)$ using the convolution method.

Solution:

a. Two signals $h(\tau)$ and $x(\tau)$ that are involved in the convolution integration are displayed in Figure B.12. To evaluate the convolution, the time-reversed signal $x(-\tau)$ and the shifted signal $x(t-\tau)$ are also plotted for reference. Figure B.12 shows an overlap of $h(\tau)$ and $x(t-\tau)$. According to the overlapped (shaded) area, the lower limit and the upper limit of the convolution integral are determined to be 0 and t, respectively. Hence,

$$y(t) = \int_0^t e^{-10\tau} \cdot 5 d\tau = \frac{5}{-10} e^{-10\tau} \Big|_0^t$$
$$= -0.5 e^{-10t} - (-0.5 e^{-10 \times 0}).$$

Finally, the system response is found to be

$$y(t) = 0.5u(t) - 0.5e^{-10t}u(t).$$

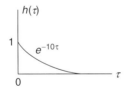
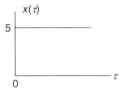
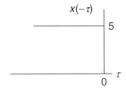
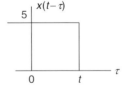
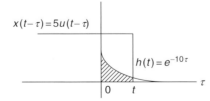

FIGURE B.12 Convolution illustration for Example B.14.

The solution is the same as that obtained using the Laplace transform method described in Example B.11.

B.3.3 Sinusoidal Steady-State Response

For linear analog systems, if the input to system is a sinusoid of radian frequency ω, the steady-state response of the system will also be a sinusoid of the same frequency. Therefore, the transfer function, which provides the relationship between a sinusoidal input and a sinusoidal output, is called the steady-state transfer function. The steady-state transfer function is obtained from the Laplace transfer function by substituting $s = j\omega$, as shown in the following:

$$H(j\omega) = H(s)\big|_{s=j\omega}. \tag{B.35}$$

Thus we have a system relationship in a sinusoidal steady state as

$$Y(j\omega) = H(j\omega)X(j\omega). \tag{B.36}$$

Since $H(j\omega)$ is a complex function, we may write it in a phasor form:

$$H(j\omega) = A(\omega)\angle\beta(\omega), \tag{B.37}$$

where the quantity $A(\omega)$ is the amplitude response of the system defined as

$$A(\omega) = |H(j\omega)|, \tag{B.38}$$

and the phase angle $\beta(\omega)$ is the phase response of the system. The following example is presented to illustrate the application.

Example B.15.

Consider a linear system described by the differential equation shown in Example B.12, where $x(t)$ and $y(t)$ designate the system input and system output, respectively. The transfer function has been derived as

$$H(s) = \frac{10}{s+10}.$$

a. Derive the steady-state transfer function.

b. Derive the amplitude response and phase response.

c. If the input is given as a sinusoid, that is, $x(t) = 5\sin(10t + 30°)u(t)$, find the steady-state response $y_{ss}(t)$.

Solution:

a. By substituting $s = j\omega$ into the transfer function in terms of a suitable form, we get the steady-state transfer function as

$$H(j\omega) = \frac{1}{\frac{s}{10} + 1} = \frac{1}{\frac{j\omega}{10} + 1}.$$

b. The amplitude response and phase response are found to be

$$A(\omega) = \frac{1}{\sqrt{\left(\frac{\omega}{10}\right)^2 + 1}}$$

$$\beta(\omega) = \angle - \tan^{-1}\left(\frac{\omega}{10}\right).$$

c. When $\omega = 10 \text{ rad/sec}$, the input sinusoid can be written in terms of the phasor form as

$$X(j10) = 5\angle 30°.$$

For the amplitude and phase of the steady-state transfer function at $\omega = 10$, we have

$$A(10) = \frac{1}{\sqrt{\left(\frac{10}{10}\right)^2 + 1}} = 0.7071$$

$$\beta(10) = -\tan^{-1}\left(\frac{10}{10}\right) = -45°.$$

Hence, we yield

$$H(j10) = 0.7071\angle - 45°.$$

Using Equation (B.36), the system output in the phasor form is obtained as:

$$Y(j10) = H(j10)X(j10) = (1.4141\angle - 45°)(5\angle 30°)$$
$$Y(j10) = 3.5355\angle - 15°.$$

Converting the output in the phasor form back to the time domain results in the steady-state system output as

$$y_{ss}(t) = 3.5355 \sin(10t - 15°)u(t).$$

B.4 Problems

B.1. Develop equations for the amplitude spectra, that is, A_n (one sided) and $|c_n|$ (two sided), of the following pulse train $x(t)$.

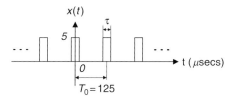

FIGURE B.13 Pulse train in problem B.1.

where $\tau = 10\,\mu$ sec.

  a. Plot and label the one-sided amplitude spectrum up to 4 harmonic frequencies including DC.

  b. Plot and label the two-sided amplitude spectrum up to 4 harmonic frequencies including DC.

B.2. In the waveform shown in Figure B.14, $T_0 = 1$ ms and $A = 10$. Use the formula in Table B.1 to write a Fourier series expansion in magnitude-phase form. Determine the frequency f_3 and amplitude value of A_3 for the third harmonic.

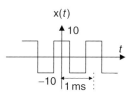

FIGURE B.14 Square wave in problem B.2.

B.3. In the waveform shown in Figure B.15, $T_0 = 1$ ms, $\tau = 0.2$ ms, and $A = 10$.

  a. Use the formula in Table B.1 to write a Fourier series expansion in magnitude-phase form.

  b. Determine the frequency f_2 and amplitude value of A_2 for the second harmonic.

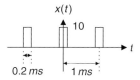

FIGURE B.15 Rectangular wave in problem B.3.

B.4. Find the Fourier transform $X(\omega)$ and sketch the amplitude spectrum for the following rectangular pulse $x(t)$.

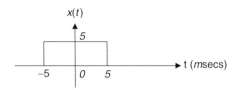

FIGURE B.16   Rectangular pulse in problem B.4.

B.5. Use Table B.3 to determine the Fourier transform of the following pulse.

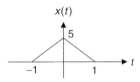

FIGURE B.17   Triangular pulse in problem B.5.

B.6. Use Table B.3 to determine the Fourier transform of the following pulse.

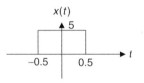

FIGURE B.18   Rectangular pulse in problem B.6.

B.7. Determine the Laplace transform $X(s)$ for each of the following time domain functions using the Laplace transform in Table B.5.

   a. $x(t) = 10\delta(t)$
   b. $x(t) = -100tu(t)$
   c. $x(t) = 10e^{-2t}u(t)$
   d. $x(t) = 2u(t-5)$
   e. $x(t) = 10\cos(3t)u(t)$
   f. $x(t) = 10\sin(2t + 45°)u(t)$
   g. $x(t) = 3e^{-2t}\cos(3t)u(t)$
   h. $x(t) = 10t^5 u(t)$

B.8. Determine the inverse Laplace transform of analog signal $x(t)$ for each of the following given functions using Table B.5 and partial fraction expansion.

a. $X(s) = \dfrac{10}{s+2}$

b. $X(s) = \dfrac{100}{(s+2)(s+3)}$

c. $X(s) = \dfrac{100s}{s^2 + 7s + 10}$

d. $X(s) = \dfrac{25}{s^2 + 4s + 29}$

B.9. Solve the following differential equation using the Laplace transform method.

$$2\dfrac{dx(t)}{dt} + 3x(t) = 15u(t) \quad \text{with} \quad x(0) = 0$$

a. Determine $X(s)$.

b. Determine the continuous signal $x(t)$ by taking the inverse Laplace transform of $X(s)$.

B.10. Solve the following differential equation using the Laplace transform method.

$$\dfrac{d^2 x(t)}{dt^2} + 3\dfrac{dx(t)}{dt} + 2x(t) = 10u(t) \text{ with } x'(0) = 0 \text{ and } x(0) = 0$$

a. Determine $X(s)$.

b. Determine $x(t)$ by taking the inverse Laplace transform of $X(s)$.

B.11. Determine the locations of all finite zeros and poles. In each case, make an s-plane plot of the poles and zeros, and determine whether the given transfer function is stable, unstable, or marginally stable.

a. $H(s) = \dfrac{(s-3)}{(s^2 + 4s + 4)}$

b. $H(s) = \dfrac{s(s^2 + 5)}{(s^2 + 9)(s^2 + 2s + 4)}$

c. $H(s) = \dfrac{(s^2 + 1)(s+1)}{s(s^2 + 7s - 8)(s+3)(s+4)}$

B.12. Given the transfer function of a system as

$$H(s) = \frac{5}{s+5},$$

and the input $x(t) = u(t)$,

a. determine the system impulse response $h(t)$;

b. determine the system Laplace output based on $Y(s) = H(s)X(s)$;

c. determine the system response $y(t)$ in time domain by taking the inverse Laplace transform of $Y(s)$.

B.13. Given the transfer function of a system as

$$H(s) = \frac{5}{s+5},$$

a. determine the steady-state transfer function;

b. determine the amplitude response and phase response in terms of the frequency ω;

c. determine the steady-state response of the system output $y_{ss}(t)$ in time domain using the results that you obtained in (b), given an input to the system of $x(t) = 5\sin(2t)u(t)$.

B.14. Given the transfer function of a system as

$$H(s) = \frac{5}{s+5},$$

and the input $x(t) = u(t)$, determine the system output $y(t)$ using the convolution method; that is, $y(t) = h(t)^*x(t)$.

References

Ambardar, A. (1999). *Analog and Digital Signal Processing*, 2nd ed. Pacific Grove, CA: Brooks/Cole Publishing Company.

Soliman, S. S., and Srinath, M. D. (1998). *Continuous and Discrete Signals and Systems*. Upper Saddle River, NJ: Prentice Hall.

Stanley, W. D. (2003). *Network Analysis with Applications*, 4th ed. Upper Saddle River, NJ: Prentice Hall.

C

Normalized Butterworth and Chebyshev Functions

C.1 Normalized Butterworth Function

The normalized Butterworth squared magnitude function is given by

$$|P_n(\omega)|^2 = \frac{1}{1 + \varepsilon^2(\omega)^{2n}}, \qquad (C.1)$$

where n is the order and ε is the specified ripple on filter passband. The specified ripple in dB is expressed as $\varepsilon_{dB} = 20 \cdot \log_{10}\left(\sqrt{1 + \varepsilon^2}\right)$ dB.

To develop the transfer function $P_n(s)$, we first let $s = j\omega$ and then substitute $\omega^2 = -s^2$ into Equation (C.1) to obtain

$$P_n(s)P_n(-s) = \frac{1}{1 + \varepsilon^2(-s^2)^n}. \qquad (C.2)$$

Equation (C.2) has $2n$ poles, and $P_n(s)$ has n poles on the left-hand half plane (LHHP) on the s-plane, while $P_n(-s)$ has n poles on the right-hand half plane (RHHP) on the s-plane. Solving for poles leads to

$$(-1)^n s^{2n} = -1/\varepsilon^2. \qquad (C.3)$$

If n is an odd number, Equation (C.3) becomes

$$s^{2n} = 1/\varepsilon^2$$

and the corresponding poles are solved as

$$p_k = \varepsilon^{-1/n} e^{j\frac{2\pi k}{2n}} = \varepsilon^{-1/n}[\cos(2\pi k/2n) + j\sin(2\pi k/2n)], \quad \text{(C.4)}$$

where $k = 0, 1, \ldots, 2n - 1$. Thus in the phasor form, we have

$$r = \varepsilon^{-1/n}, \text{ and } \theta_k = 2\pi k/(2n) \text{ for } k = 0, 1, \ldots, 2n - 1. \quad \text{(C.5)}$$

When n is an even number, it follows that

$$s^{2n} = -1/\varepsilon^2$$

$$p_k = \varepsilon^{-1/n} e^{j\frac{2\pi k+\pi}{2n}} = \varepsilon^{-1/n}[\cos((2\pi k + \pi)/2n) + j\sin((2\pi k + \pi)/2n)], \quad \text{(C.6)}$$

where $k = 0, 1, \ldots, 2n - 1$. Similarly, the phasor form is given by

$$r = \varepsilon^{-1/n}, \text{ and } \theta_k = (2\pi k + \pi)/(2n) \text{ for } k = 0, 1, \ldots, 2n - 1. \quad \text{(C.7)}$$

When n is an odd number, we can identify the poles on the LHHP as

$$p_k = -r, \ k = 0 \text{ and}$$
$$p_k = -r\cos(\theta_k) + jr\sin(\theta_k), \ k = 1, \ldots, (n-1)/2. \quad \text{(C.8)}$$

Using complex conjugate pairs, we have

$$p_k^* = -r\cos(\theta_k) - jr\sin(\theta_k).$$

Notice that

$$(s - p_k)(s - p_k^*) = s^2 + (2r\cos(\theta_k))s + r^2,$$

and from a factor from the real pole $(s + r)$, it follows that

$$P_n(s) = \frac{K}{(s+r)\prod_{k=1}^{(n-1)/2}(s^2 + (2r\cos(\theta_k))s + r^2)} \quad \text{(C.9)}$$

and

$$\theta_k = 2\pi k/(2n) \text{ for } k = 1, \ldots, (n-1)/2.$$

Setting $P_n(0) = 1$ for the unit passband gain leads to

$$K = r^n = 1/\varepsilon.$$

When n is an even number, we can identify the poles on the LHHP as

$$p_k = -r\cos(\theta_k) + jr\sin(\theta_k), \ k = 0, 1, \ldots, n/2 - 1. \quad \text{(C.10)}$$

Using complex conjugate pairs, we have

$$p_k^* = -r\cos(\theta_k) - jr\sin(\theta_k).$$

The transfer function is given by

$$P_n(s) = \frac{K}{\prod_{k=1}^{n/2}(s^2 + (2r\cos(\theta_k))s + r^2)} \qquad \text{(C.11)}$$

$$\theta_k = (2\pi k + \pi)/(2n) \text{ for } k = 0, 1, \ldots, n/2 - 1.$$

Setting $P_n(0) = 1$ for the unit passband gain, we have

$$K = r^n = 1/\varepsilon.$$

Let us examine the following examples.

Example C.1.

a. Compute the normalized Butterworth transfer function for the following specifications:

Ripple = 3 dB
$n = 2$

Solution:

a. $n/2 = 1$
$\theta_k = (2\pi \times 0 + \pi)/(2 \times 2) = 0.25\pi$
$\varepsilon^2 = 10^{0.1 \times 3} - 1$,
$r = 1$, and $K = 1$.

Applying Equation (C.11) leads to

$$P_2(s) = \frac{1}{s^2 + 2 \times 1 \times \cos(0.25\pi)s + 1^2} = \frac{1}{s^2 + 1.4141s + 1}.$$

Example C.2.

a. Compute the normalized Butterworth transfer function for the following specifications:

Ripple = 3 dB
$n = 3$

Solution:

a. $(n-1)/2 = 1$
$\varepsilon^2 = 10^{0.1 \times 3} - 1$,
$r = 1$, and $K = 1$
$\theta_k = (2\pi \times 1)/(2 \times 3) = \pi/3.$

From Equation (C.9), we have

$$P_3(s) = \frac{1}{(s+1)(s^2 + 2 \times 1 \times \cos(\pi/3)s + 1^2)}$$
$$= \frac{1}{(s+1)(s^2 + s + 1)}.$$

For the unfactored form, we can carry out

$$P_3(s) = \frac{1}{s^3 + 2s^2 + 2s + 1}.$$

Example C.3.

a. Compute the normalized Butterworth transfer function for the following specifications:

Ripple $= 1.5\,\text{dB}$
$n = 3$

Solution:

a. $(n-1)/2 = 1$
$\varepsilon^2 = 10^{0.1 \times 1.5} - 1$,
$r = 1.1590$, and $K = 1.5569$
$\theta_k = (2\pi \times 1)/(2 \times 3) = \pi/3$.

Applying Equation (C.9), we achieve the normalized Butterworth transfer function as

$$P_3(s) = \frac{1}{(s + 1.1590)(s^2 + 2 \times 1.1590 \times \cos(\pi/3)s + 1.1590^2)}$$
$$= \frac{1}{(s+1)(s^2 + 1.1590s + 1.3433)}.$$

For the unfactored form, we can carry out

$$P_3(s) = \frac{1.5569}{s^3 + 2.3180s^2 + 2.6866s + 1.5569}.$$

C.2 Normalized Chebyshev Function

Similar to analog Butterworth filter design, the transfer function is derived from the normalized Chebyshev function, and the result is usually listed in the table for design reference. The Chebyshev magnitude response function with an order of n and the normalized cutoff frequency $\omega = 1$ radian per second is given by

$$|B_n(\omega)| = \frac{1}{\sqrt{1 + \varepsilon^2 C_n^2(\omega)}}, \quad n \geq 1, \tag{C.12}$$

where the function $C_n(\omega)$ is defined as

$$C_n(\omega) = \begin{cases} \cos(n \cos^{-1}(\omega)) & \omega \leq 1 \\ \cosh(n \cosh^{-1}(\omega)) & \omega > 1, \end{cases} \tag{C.13}$$

and ε is the ripple specification on the filter passband. Notice that

$$\cosh^{-1}(x) = \ln(x + \sqrt{x^2 - 1}). \tag{C.14}$$

To develop the transfer function $B_n(s)$, we let $s = j\omega$ and substitute $\omega^2 = -s^2$ into Equation (C.12) to obtain

$$B_n(s) B_n(-s) = \frac{1}{1 + \varepsilon^2 C_n^2(s/j)}. \tag{C.15}$$

The poles can be found from

$$1 + \varepsilon^2 C_n^2(s/j) = 0$$

or

$$C_n(s/j) = \cos(n \cos^{-1}(s/j)) = \pm j1/\varepsilon. \tag{C.16}$$

Introducing a complex variable $v = \alpha + j\beta$ such that

$$v = \alpha + j\beta = \cos^{-1}(s/j), \tag{C.17}$$

we can then write

$$s = j \cos(v). \tag{C.18}$$

Substituting Equation (C.17) into Equation (C.16) and using trigonometric identities, it follows that

$$\begin{aligned} C_n(s/j) &= \cos(n \cos^{-1}(s/j)) \\ &= \cos(nv) = \cos(n\alpha + jn\beta) \\ &= \cos(n\alpha) \cosh(n\beta) - j \sin(n\alpha) \sinh(n\beta) = \pm j1/\varepsilon. \end{aligned} \tag{C.19}$$

To solve Equation (C.19), the following conditions must be satisfied:

$$\cos(n\alpha) \cosh(n\beta) = 0 \tag{C.20}$$

$$-\sin(n\alpha) \sinh(n\beta) = \pm 1/\varepsilon. \tag{C.21}$$

Since $\cosh(n\beta) \geq 1$ in Equation (C.20), we must let

$$\cos(n\alpha) = 0, \tag{C.22}$$

which therefore leads to

$$\alpha_k = (2k+1)\pi/(2n), \; k = 0, 1, 2, \ldots, 2n-1. \quad (C.23)$$

With Equation (C.23), we have $\sin(n\alpha_k) = \pm 1$. Then Equation (C.21) becomes

$$\sinh(n\beta) = 1/\varepsilon. \quad (C.24)$$

Solving Equation (C.24) gives

$$\beta = \sinh^{-1}(1/\varepsilon)/n. \quad (C.25)$$

Again from Equation (C.18),

$$s = j\cos(v) = j[\cos(\alpha_k)\cosh(\beta) - j\sin(\alpha_k)\sinh(\beta)] \quad (C.26)$$
$$\text{for } k = 0, 1, \ldots, 2n-1.$$

The poles can be found from Equation (C.26):

$$p_k = \sin(\alpha_k)\sinh(\beta) + j\cos(\alpha_k)\cosh(\beta) \quad (C.27)$$
$$\text{for } k = 0, 1, \ldots, 2n-1.$$

Using Equation (C.27), if n is an odd number, the poles on the left side are solved to be

$$p_k = -\sin(\alpha_k)\sinh(\beta) + j\cos(\alpha_k)\cosh(\beta), \; k = 0, 1, \ldots, (n-1)/2 - 1. \quad (C.28)$$

Using complex conjugate pairs, we have

$$p_k^* = -\sin(\alpha_k)\sinh(\beta) - j\cos(\alpha_k)\cosh(\beta) \quad (C.29)$$

and a real pole

$$p_k = -\sinh(\beta), \; k = (n-1)/2. \quad (C.30)$$

Notice that

$$(s - p_k)(s - p_k^*) = s^2 + b_k s + c_k \quad (C.31)$$

and a factor from the real pole $[s + \sinh(\beta)]$, it follows that

$$B_n(s) = \frac{K}{[s + \sinh(\beta)] \prod_{k=0}^{(n-1)/2 - 1} (s^2 + b_k s + c_k)}, \quad (C.32)$$

where $\alpha_k = (2k+1)\pi/(2n)$ for $k = 0, 1, \ldots, (n-1)/2 - 1$ \quad (C.33)
$$b_k = 2\sin(\alpha_k)\sinh(\beta) \quad (C.34)$$
$$c_k = [\sin(\alpha_k)\sinh(\beta)]^2 + [\cos(\alpha_k)\cosh(\beta)]^2. \quad (C.35)$$

For the unit passband gain and the filter order as an odd number, we set $B_n(0) = 1$. Then

$$K = \sinh(\beta) \prod_{k=0}^{(n-1)/2-1} c_k \tag{C.36}$$

$$\beta = \sinh^{-1}(1/\varepsilon)/n \tag{C.37}$$

$$\sinh^{-1}(x) = \ln(x + \sqrt{x^2 + 1}). \tag{C.38}$$

Following a similar procedure for the even number of n, we have

$$B_n(s) = \frac{K}{\prod_{k=0}^{n/2-1} (s^2 + b_k s + c_k)} \tag{C.39}$$

where $\alpha_k = (2k+1)\pi/(2n)$ for $k = 0, 1, \ldots, n/2 - 1$ \hfill (C.40)

$$b_k = 2\sin(\alpha_k)\sinh(\beta) \tag{C.41}$$

$$c_k = [\sin(\alpha_k)\sinh(\beta)]^2 + [\cos(\alpha_k)\cosh(\beta)]^2. \tag{C.42}$$

For the unit passband gain and the filter order as an even number, we require that $B_n(0) = 1/\sqrt{1+\varepsilon^2}$, so that the maximum magnitude of the ripple on passband equals 1. Then we have

$$K = \prod_{k=0}^{n/2-1} c_k / \sqrt{1+\varepsilon^2} \tag{C.43}$$

$$\beta = \sinh^{-1}(1/\varepsilon)/n \tag{C.44}$$

$$\sinh^{-1}(x) = \ln(x + \sqrt{x^2 + 1}). \tag{C.45}$$

Equations (C.32) to (C.45) are applied to compute the normalized Chebyshev transfer function. Now let us look at the following illustrative examples.

Example C.4.

    a. Compute the normalized Chebyshev transfer function for the following specifications:

    Ripple = 0.5 dB
    $n = 2$

Solution:

    a. $n/2 = 1$.

Applying Equations (C.39) to (C.45), we obtain

$\alpha_0 = (2 \times 0 + 1)\pi/(2 \times 2) = 0.25\pi$
$\varepsilon^2 = 10^{0.1 \times 0.5} - 1 = 0.1220, \ 1/\varepsilon = 2.8630$
$\beta = \sinh^{-1}(2.8630)/n = \ln(2.8630 + \sqrt{2.8630^2 + 1})/2 = 0.8871$
$b_0 = 2\sin(0.25\pi)\sinh(0.8871) = 1.4256$
$c_0 = [\sin(0.25\pi)\sinh(0.8871)]^2 + [\cos(0.25\pi)\cosh(0.8871)]^2 = 1.5162$
$K = 1.5162/\sqrt{1 + 0.1220} = 1.4314.$

Finally, the transfer function is derived as

$$B_2(s) = \frac{1.4314}{s^2 + 1.4256s + 1.5162}.$$

Example C.5.

a. Compute the normalized Chebyshev transfer function for the following specifications:

Ripple = 1 dB
$n = 3$

Solution:

a. $(n-1)/2 = 1.$

Applying Equations (C.32) to (C.38) leads to

$\alpha_0 = (2 \times 0 + 1)\pi/(2 \times 3) = \pi/6$
$\varepsilon^2 = 10^{0.1 \times 1} - 1 = 0.2589, \ 1/\varepsilon = 1.9653$
$\beta = \sinh^{-1}(1.9653)/n = \ln(1.9653 + \sqrt{1.9653^2 + 1})/3 = 0.4760$
$b_0 = 2\sin(\pi/6)\sinh(0.4760) = 0.4942$
$c_0 = [\sin(\pi/6)\sinh(0.4760)]^2 + [\cos(\pi/6)\cosh(0.4760)]^2 = 0.9942$
$\sinh(\beta) = \sinh(0.4760) = 0.4942$
$K = 0.4942 \times 0.9942 = 0.4913.$

We can conclude the transfer function as

$$B_3(s) = \frac{0.4913}{(s + 0.4942)(s^2 + 0.4942s + 0.9942)}.$$

Finally, the unfactored form is found to be

$$B_3(s) = \frac{0.4913}{s^3 + 0.9883s^2 + 1.2384s + 0.4913}.$$

D

Sinusoidal Steady-State Response of Digital Filters

D.1 Sinusoidal Steady-State Response

Analysis of the sinusoidal steady-state response of digital filters will lead to the development of the magnitude and phase responses of digital filters. Let us look at the following digital filter with a digital transfer function $H(z)$ and a complex sinusoidal input

$$x(n) = Ve^{j(\Omega n + \varphi_x)}, \tag{D.1}$$

where $\Omega = \omega T$ is the normalized digital frequency, while T is the sampling period and $y(n)$ denotes the digital output, as shown in Figure D.1.

The z-transform output from the digital filter is then given by

$$Y(z) = H(z)X(z). \tag{D.2}$$

Since $X(z) = \frac{Ve^{j\varphi_x}z}{z - e^{j\Omega}}$, we have

$$Y(z) = \frac{Ve^{j\varphi_x}z}{z - e^{j\Omega}} H(z). \tag{D.3}$$

Based on the partial fraction expansion, $Y(z)/z$ can be expanded as the following form:

$$\frac{Y(z)}{z} = \frac{Ve^{j\varphi_x}}{z - e^{j\Omega}} H(z) = \frac{R}{z - e^{j\Omega}} + \text{sum of the rest of partial fractions.} \tag{D.4}$$

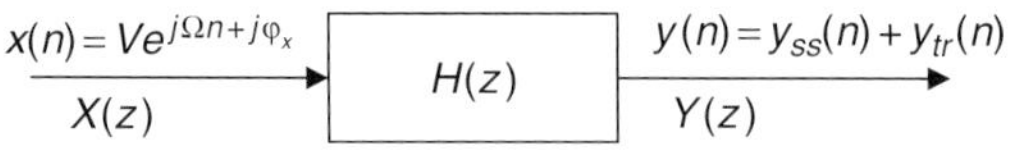

D RESPONSE OF DIGITAL FILTERS

$$x(n) = Ve^{j\Omega n + j\varphi_x} \quad X(z) \longrightarrow \boxed{H(z)} \longrightarrow y(n) = y_{ss}(n) + y_{tr}(n) \quad Y(z)$$

FIGURE D.1 Steady-state response of the digital filter.

Multiplying the factor $(z - e^{j\Omega})$ on both sides of Equation (D.4) yields

$$Ve^{j\phi_x} H(z) = R + (z - e^{j\Omega})(\text{sum of the rest of partial fractions}). \quad (D.5)$$

Substituting $z = e^{j\Omega}$, we get the residue as

$$R = Ve^{j\phi_x} H(e^{j\Omega}).$$

Then substituting $R = Ve^{j\varphi} \times H(e^{j\Omega})$ back into Equation (D.4) results in

$$\frac{Y(z)}{z} = \frac{Ve^{j\varphi_x} H(e^{j\Omega})}{z - e^{j\Omega}} + \text{sum of the rest of partial fractions}, \quad (D.6)$$

and multiplying z on both sides of Equation (D.6) leads to

$$Y(z) = \frac{Ve^{j\varphi_x} H(e^{j\Omega}) z}{(z - e^{j\Omega})} + z \times \text{sum of the rest of partial fractions}. \quad (D.7)$$

Taking the inverse z-transform leads to two parts of the solution:

$$y(n) = Ve^{j\varphi_x} H(e^{j\Omega}) e^{j\Omega n} + Z^{-1}(z \times \text{sum of the rest of partial fractions}). \quad (D.8)$$

From Equation (D.8), we have the steady-state response

$$y_{ss}(n) = Ve^{j\varphi_x} H(e^{j\Omega}) e^{j\Omega n} \quad (D.9)$$

and the transient response

$$y_{tr}(n) = Z^{-1}(z \times \text{sum of the rest of partial fractions}). \quad (D.10)$$

Note that since the digital filter is a stable system, and the locations of its poles must be inside the unit circle on the z-plane, the transient response will be settled to zero eventually. To develop filter magnitude and phase responses, we write the digital steady-state response as

$$y_{ss}(n) = V|H(e^{j\Omega})|e^{j\Omega + j\varphi_x + \angle H(e^{j\Omega})}. \quad (D.11)$$

Comparing Equation (D.11) and Equation (D.1), it follows that

$$\text{Magnitude response} = \frac{\text{Amplitude of the steady-state output}}{\text{Amplitude of the sinusoidal input}}$$
$$= \frac{V|H(e^{j\Omega})|}{V} = |H(e^{j\Omega})| \quad (D.12)$$

$$\text{Phase response} = \frac{e^{j\varphi_x + \angle H(e^{j\Omega})}}{e^{j\varphi_x}} = e^{j\angle H(e^{j\Omega})} = \angle H(e^{j\Omega}). \qquad (D.13)$$

Thus we conclude that

$$\text{Frequency response} = H(e^{j\Omega}) = H(z)\big|_{z=e^{j\Omega}}. \qquad (D.14)$$

Since $H(e^{j\Omega}) = |H(e^{j\Omega})| \angle H(e^{j\Omega})$

$$\text{Magnitude response} = |H(e^{j\Omega})| \qquad (D.15)$$

$$\text{Phase response} = \angle H(e^{j\Omega}). \qquad (D.16)$$

D.2 Properties of the Sinusoidal Steady-State Response

From Euler's identity and trigonometric identity, we know that

$$\begin{aligned}
e^{j(\Omega + k2\pi)} &= \cos(\Omega + k2\pi) + j\sin(\Omega + k2\pi) \\
&= \cos\Omega + j\sin\Omega = e^{j\Omega},
\end{aligned} \qquad (D.17)$$

where k is an integer taking values of $k = 0, \pm 1, \pm 2, \ldots$. Then:

$$\text{Frequency response: } H(e^{j\Omega}) = H(e^{j(\Omega + k2\pi)}) \qquad (D.18)$$

$$\text{Magnitude frequency response: } |H(e^{j\Omega})| = |H(e^{j(\Omega + k2\pi)})| \qquad (D.19)$$

$$\text{Phase response: } \angle H(e^{j\Omega}) = \angle H(e^{j\Omega + 2k\pi}). \qquad (D.20)$$

Clearly, the frequency response is periodical, with a period of 2π. Next, let us develop the symmetric properties. Since the transfer function is written as

$$H(z) = \frac{Y(z)}{X(z)} = \frac{b_0 + b_1 z^{-1} + \ldots + b_M z^{-M}}{1 + a_1 z^{-1} + \ldots + a_N z^{-N}}, \qquad (D.21)$$

substituting $z = e^{j\Omega}$ into Equation (D.21) yields

$$H(e^{j\Omega}) = \frac{b_0 + b_1 e^{-j\Omega} + \ldots + b_M e^{-jM\Omega}}{1 + a_1 e^{-j\Omega} + \ldots + a_N e^{-jN\Omega}}. \qquad (D.22)$$

Using Euler's identity, $e^{-j\Omega} = \cos\Omega - j\sin\Omega$, we have

$$H(e^{j\Omega}) = \frac{(b_0 + b_1 \cos\Omega + \ldots + b_M \cos M\Omega) - j(b_1 \sin\Omega + \ldots + b_M \sin M\Omega)}{(1 + a_1 \cos\Omega + \ldots + a_N \cos N\Omega) - j(a_1 \sin\Omega + \ldots + a_N \sin N\Omega)}. \qquad (D.23)$$

Similarly,

$$H(e^{-j\Omega}) = \frac{(b_0 + b_1\cos\Omega + \ldots + b_M\cos M\Omega) + j(b_1\sin\Omega + \ldots + b_M\sin M\Omega)}{(1 + a_1\cos\Omega + \ldots + a_N\cos N\Omega) + j(a_1\sin\Omega + \ldots + a_N\sin N\Omega)}.$$
(D.24)

Then the magnitude response and phase response can be expressed as

$$|H(e^{j\Omega})| = \frac{\sqrt{(b_0 + b_1\cos\Omega + \ldots + b_M\cos M\Omega)^2 + (b_1\sin\Omega + \ldots + b_M\sin M\Omega)^2}}{\sqrt{(1 + a_1\cos\Omega + \ldots + a_N\cos N\Omega)^2 + (a_1\sin\Omega + \ldots + a_N\sin N\Omega)^2}}$$
(D.25)

$$\angle H(e^{j\Omega}) = \tan^{-1}\left(\frac{-(b_1\sin\Omega + \ldots + b_M\sin M\Omega)}{b_0 + b_1\cos\Omega + \ldots + b_M\cos M\Omega}\right)$$
$$- \tan^{-1}\left(\frac{-(a_1\sin\Omega + \ldots + a_N\sin N\Omega)}{1 + a_1\cos\Omega + \ldots + a_N\cos N\Omega}\right).$$
(D.26)

Based on Equation (D.24), we also have the magnitude and phase response for $H(e^{-j\Omega})$ as

$$|H(e^{-j\Omega})| = \frac{\sqrt{(b_0 + b_1\cos\Omega + \ldots + b_M\cos M\Omega)^2 + (b_1\sin\Omega + \ldots + b_M\sin M\Omega)^2}}{\sqrt{(1 + a_1\cos\Omega + \ldots + a_N\cos N\Omega)^2 + (a_1\sin\Omega + \ldots + a_N\sin N\Omega)^2}}$$
(D.27)

$$\angle H(e^{-j\Omega}) = \tan^{-1}\left(\frac{b_1\sin\Omega + \ldots + b_M\sin M\Omega}{b_0 + b_1\cos\Omega + \ldots + b_M\cos M\Omega}\right)$$
$$- \tan^{-1}\left(\frac{a_1\sin\Omega + \ldots + a_N\sin N\Omega}{1 + a_1\cos\Omega + \ldots + a_N\cos N\Omega}\right).$$
(D.28)

Comparing (D.25) with (D.27) and (D.26) with (D.28), respectively, we conclude the symmetric properties as

$$|H(e^{-j\Omega})| = |H(e^{j\Omega})|$$
(D.29)

$$\angle H(e^{-j\Omega}) = -\angle H(e^{j\Omega}).$$
(D.30)

E

Finite Impulse Response Filter Design Equations by the Frequency Sampling Design Method

Recall from Section 7.5 in Chapter 7 on the "Frequency Sampling Design Method":

$$h(n) = \frac{1}{N} \sum_{k=0}^{N-1} H(k) W_N^{-kn}, \quad \text{(E.1)}$$

where $h(n)$, $0 \leq n \leq N-1$, is the causal impulse response that approximates the finite impulse response (FIR) filter, and $H(k)$, $0 \leq k \leq N-1$, represents the corresponding coefficients of the discrete Fourier transform (DFT), and $W_N = e^{-j\frac{2\pi}{N}}$. We further write DFT coefficients $H(k)$, $0 \leq k \leq N-1$, into the polar form:

$$H(k) = H_k e^{j\varphi_k}, \quad 0 \leq k \leq N-1, \quad \text{(E.2)}$$

where H_k and φ_k are the kth magnitude and the phase angle, respectively. The frequency response of the FIR filter is expressed as

$$H(e^{j\Omega}) = \sum_{n=0}^{N-1} h(n) e^{-jn\Omega}. \quad \text{(E.3)}$$

Substituting (E.1) into (E.3) yields

$$H(e^{j\Omega}) = \sum_{n=0}^{N-1} \frac{1}{N} \sum_{k=0}^{N-1} H(k) W_N^{-kn} e^{-j\Omega n}. \quad \text{(E.4)}$$

Interchanging the order of the summation in Equation (E.4) leads to

$$H(e^{j\Omega}) = \frac{1}{N} \sum_{k=0}^{N-1} H(k) \sum_{n=0}^{N-1} \left(W_N^{-k} e^{-j\Omega}\right)^n. \tag{E.5}$$

Since $W_N^{-k} e^{-j\Omega} = \left(e^{-j2\pi/N}\right)^{-k} e^{-j\Omega} = e^{-(j\Omega - 2\pi k/N)}$,

and using the identity $\sum_{n=0}^{N-1} r^n = 1 + r + r^2 + \cdots + r^{N-1} = \frac{1 - r^N}{1 - r}$,

we can write the second summation in Equation (E.5) as

$$\sum_{n=0}^{N-1} \left(W_N^{-k} e^{-j\Omega}\right)^n = \frac{1 - e^{-j(\Omega - 2\pi k/N)N}}{1 - e^{-j(\Omega - 2\pi k/N)}}. \tag{E.6}$$

Using the Euler formula leads Equation (E.6) to

$$\sum_{n=0}^{N-1} \left(W_N^{-k} e^{-j\Omega}\right)^n = \frac{e^{-jN(\Omega - 2\pi k/N)/2} (e^{jN(\Omega - 2\pi k/N)/2} - e^{-jN(\Omega - 2\pi k/N)/2})/2j}{e^{-j(\Omega - 2\pi k/N)/2} (e^{j(\Omega - 2\pi k/N)/2} - e^{-j(\Omega - 2\pi k/N)/2})/2j}$$

$$= \frac{e^{-jN(\Omega - 2\pi k/N)/2} \sin\left[N(\Omega - 2\pi k/N)/2\right]}{e^{-j(\Omega - 2\pi k/N)/2} \sin\left[(\Omega - 2\pi k/N)/2\right]}. \tag{E.7}$$

Substituting Equation (E.7) into Equation (E.5) leads to

$$H(e^{j\Omega}) = \frac{1}{N} e^{-j(N-1)\Omega/2} \sum_{k=0}^{N-1} H(k) e^{j(N-1)k\pi/N} \frac{\sin\left[N(\Omega - 2\pi k/N)/2\right]}{\sin\left[(\Omega - 2\pi k/N)/2)\right]}. \tag{E.8}$$

Let $\Omega = \Omega_m = \frac{2\pi m}{N}$, and substituting it into Equation (E.8) we get

$$H(e^{j\Omega_m}) = \frac{1}{N} e^{-j(N-1)2\pi m/(2N)} \sum_{k=0}^{N-1} H(k) e^{j(N-1)k\pi/N} \frac{\sin\left[N(2\pi m/N - 2\pi k/N)/2\right]}{\sin\left[(2\pi m/N - 2\pi k/N)/2)\right]}. \tag{E.9}$$

Clearly, when $m \neq k$, the last term of the summation in Equation (E.9) becomes

$$\frac{\sin\left[N(2\pi m/N - 2\pi k/N)/2\right]}{\sin\left[(2\pi m/N - 2\pi k/N)/2)\right]} = \frac{\sin(\pi(m-k))}{\sin(\pi(m-k)/N)} = \frac{0}{\sin(\pi(m-k)/N)} = 0.$$

When $m = k$, and using L'Hospital's rule, we have

$$\frac{\sin\left[N(2\pi m/N - 2\pi k/N)/2\right]}{\sin\left[(2\pi m/N - 2\pi k/N)/2)\right]} = \frac{\sin(N\pi(m-k)/N)}{\sin(\pi(m-k)/N)} = \lim_{x \to 0} \frac{\sin(Nx)}{\sin(x)} = N.$$

Then Equation (E.9) is simplified to

$$H(e^{j\Omega_k}) = \frac{1}{N} e^{-j(N-1)\pi k/N} H(k) e^{j(N-1)k\pi/N} N = H(k).$$

That is, $H(e^{j\Omega_k}) = H(k)$, $0 \leq k \leq N-1$, (E.10)

where $\Omega_k = \frac{2\pi k}{N}$, corresponding to the kth DFT frequency component. The fact is that if we specify the desired frequency response, $H(e^{j\Omega_k})$, $0 \leq k \leq N-1$, at the equally spaced sampling frequency determined by $\Omega_k = \frac{2\pi k}{N}$, they are actually the DFT coefficients; that is, $H(k)$, $0 \leq k \leq N-1$, via Equation (E.10), and furthermore, the inverse DFT of (E.10) will give the desired impulse response, $h(n)$, $0 \leq n \leq N-1$.

To devise the design procedure, we substitute Equation (E.2) into Equation (E.8) to obtain

$$H(e^{j\Omega}) = \frac{1}{N} e^{-j(N-1)\Omega/2} \sum_{k=0}^{N-1} H_k e^{j\varphi_k + j(N-1)k\pi/N} \frac{\sin[N(\Omega - 2\pi k/N)/2]}{\sin[(\Omega - 2\pi k/N)/2]}. \quad (E.11)$$

It is required that the frequency response of the designed FIR filter expressed in Equation (E.11) be a linear phase. This can easily be accomplished by setting

$$\varphi_k + (N-1)k\pi/N = 0, \; 0 \leq k \leq N-1 \quad (E.12)$$

in Equation (E.11) so that the summation part becomes a real value, thus resulting in the linear phase of $H(e^{j\Omega})$, since only one complex term, $e^{-j(N-1)\Omega/2}$, is left, which presents the constant time delay of the transfer function. Second, the sequence $h(n)$ must be real. To proceed, let $N = 2M + 1$, and due to the properties of DFT coefficients, we have

$$\bar{H}(k) = H(N-k), \; 1 \leq k \leq M, \quad (E.13)$$

where the bar indicates complex conjugate. Note the fact that

$$\bar{W}_N^{-k} = W_N^{-(N-k)}, \; 1 \leq k \leq M. \quad (E.14)$$

From Equation (E.1), we write

$$h(n) = \frac{1}{N} \left(H(0) + \sum_{k=1}^{M} H(k) W_N^{-kn} + \sum_{k=M+1}^{2M} H(k) W_N^{-kn} \right). \quad (E.15)$$

Equation (E.15) is equivalent to

$$h(n) = \frac{1}{N} \left(H(0) + \sum_{k=1}^{M} H(k) W_N^{-kn} + \sum_{k=1}^{M} H(N-k) W_N^{-(N-k)n} \right).$$

Using Equations (E.13) and (E.14) in the last summation term leads to

$$h(n) = \frac{1}{N}\left(H(0) + \sum_{k=1}^{M} H(k)W_N^{-kn} + \sum_{k=1}^{M} \bar{H}(k)\bar{W}_N^{-kn}\right)$$

$$= \frac{1}{2M+1}\left(H(0) + \sum_{k=1}^{M}\left(H(k)W_N^{-kn} + \bar{H}(k)\bar{W}_N^{-kn}\right)\right).$$

Combining the last two summation terms, we achieve

$$h(n) = \frac{1}{2M+1}\left\{H(0) + 2\operatorname{Re}\left(\sum_{k=1}^{M} H(k)W_N^{-kn}\right)\right\}, 0 \le n \le N-1. \quad \text{(E.16)}$$

Solving Equation (E.12) gives

$$\varphi_k = -(N-1)k\pi/N, 0 \le k \le N-1. \quad \text{(E.17)}$$

Again, note that Equation (E.13) is equivalent to

$$H_k e^{-j\varphi_k} = H_{N-k} e^{j\varphi_{N-k}}, 1 \le k \le M. \quad \text{(E.18)}$$

Substituting (E.17) into (E.18) yields

$$H_k e^{j(N-1)k\pi/N} = H_{N-k} e^{-j(N-1)(N-k)\pi/N}, 1 \le k \le M. \quad \text{(E.19)}$$

Simplification of Equation (E.19) leads to the following result:

$$H_k = H_{N-k} e^{-j(N-1)\pi} = (-1)^{N-1} H_{N-k}, 1 \le k \le M. \quad \text{(E.20)}$$

Since we constrain the filter length to be $N = 2M + 1$, Equation (E.20) can be further reduced to

$$H_k = (-1)^{2M} H_{2M+1-k} = H_{2M+1-k}, 1 \le k \le M. \quad \text{(E.21)}$$

Finally, by substituting (E.21) and (E.17) into (E.16), we obtain a very simple design equation:

$$h(n) = \frac{1}{2M+1}\left\{H_0 + 2\sum_{k=1}^{M} H_k \cos\left(\frac{2\pi k(n-M)}{2M+1}\right)\right\}, 0 \le n \le 2M. \quad \text{(E.22)}$$

Thus the design procedure is simply summarized as follows: Given the filter length, $2M + 1$, and the specified frequency response, H_k at $\Omega_k = \frac{2\pi k}{(2M+1)}$ for $k = 0, 1, \ldots, M$, FIR filter coefficients can be calculated via Equation (E.22).

F

Some Useful Mathematical Formulas

Form of a complex number:

$$\text{Rectangular form: } a + jb, \text{ where } j = \sqrt{-1} \tag{F.1}$$

$$\text{Polar form: } Ae^{j\theta} \tag{F.2}$$

$$\text{Euler formula: } e^{\pm jx} = \cos x \pm j \sin x \tag{F.3}$$

Conversion from the polar form to the rectangular form:

$$\begin{aligned} Ae^{j\theta} &= A\cos\theta + jA\sin\theta = a + jb, \\ \text{where } a &= A\cos\theta, \text{ and } b = A\sin\theta. \end{aligned} \tag{F.4}$$

Conversion from the rectangular form to the polar form:

$$\begin{aligned} a + jb &= Ae^{j\theta}, \\ \text{where } A &= \sqrt{a^2 + b^2}. \end{aligned} \tag{F.5}$$

We usually specify the principal value of the angle such that $-180° < \theta \leq 180°$. The angle value can be determined as:

$$\theta = \tan^{-1}\left(\frac{b}{a}\right) \text{ if } a \geq 0$$

(that is, the complex number is in the first or fourth quadrant in the rectangular coordinate system);

$$\theta = 180° + \tan^{-1}\left(\frac{b}{a}\right) \text{ if } a < 0 \text{ and } b \geq 0$$

(that is, the complex number is in the second quadrant in the rectangular coordinate system); and

$$\theta = -180° + \tan^{-1}\left(\frac{b}{a}\right) \text{ if } a < 0 \text{ and } b \leq 0$$

(that is, the complex number is in the third quadrant in the rectangular coordinate system). Note that

$$\theta \text{ radian} = \frac{\theta \text{ degree}}{180°} \times \pi$$

$$\theta \text{ degree} = \frac{\theta \text{ radian}}{\pi} \times 180°.$$

Complex numbers:

$$e^{\pm j\pi/2} = \pm j \tag{F.6}$$

$$e^{\pm j2n\pi} = 1 \tag{F.7}$$

$$e^{\pm j(2n+1)\pi} = -1 \tag{F.8}$$

Complex conjugate of $a + jb$:

$$(a + jb)^* = conj(a + jb) = a - jb \tag{F.9}$$

Complex conjugate of $Ae^{j\theta}$:

$$(Ae^{j\theta})^* = conj(Ae^{j\theta}) = Ae^{-j\theta} \tag{F.10}$$

Complex number addition and subtraction:

$$(a_1 + jb_1) \pm (a_2 + jb_2) = (a_1 \pm a_2) + j(b_1 \pm b_2) \tag{F.11}$$

Complex number multiplication:

Rectangular form:

$$(a_1 + jb_1) \times (a_2 + jb_2) = a_1 a_2 - b_1 b_2 + j(a_1 b_2 + a_2 b_1) \tag{F.12}$$

$$(a + jb) \cdot conj(a + jb) = (a + jb)(a - jb) = a^2 + b^2 \tag{F.13}$$

Polar form:

$$A_1 e^{j\theta_1} A_2 e^{j\theta_2} = A_1 A_2 e^{j(\theta_1 + \theta_2)} \tag{F.14}$$

Complex number division:

Rectangular form:

$$\frac{a_1 + jb_1}{a_2 + jb_2} = \frac{(a_1 + jb_1)(a_2 - jb_2)}{(a_2 + jb_2)(a_2 - jb_2)}$$

$$= \frac{(a_1 a_2 + b_1 b_2) + j(a_2 b_1 - a_1 b_2)}{(a_2)^2 + (b_2)^2} \tag{F.15}$$

Polar form:
$$\frac{A_1 e^{j\theta_1}}{A_2 e^{j\theta_2}} = \left(\frac{A_1}{A_2}\right) e^{j(\theta_1 - \theta_2)} \tag{F.16}$$

Trigonometric identities:

$$\sin x = \frac{e^{jx} - e^{-jx}}{2j} \tag{F.17}$$

$$\cos x = \frac{e^{jx} + e^{-jx}}{2} \tag{F.18}$$

$$\sin(x \pm 90°) = \pm \cos x \tag{F.19}$$

$$\cos(x \pm 90°) = \mp \sin x \tag{F.20}$$

$$\sin x \cos x = \frac{1}{2} \sin 2x \tag{F.21}$$

$$\sin^2 x + \cos^2 x = 1 \tag{F.22}$$

$$\sin^2 x = \frac{1}{2}(1 - \cos 2x) \tag{E.23}$$

$$\cos^2 x = \frac{1}{2}(1 + \cos 2x) \tag{F.24}$$

$$\sin(x \pm y) = \sin x \cos y \pm \cos x \sin y \tag{F.25}$$

$$\cos(x \pm y) = \cos x \cos y \mp \sin x \sin y \tag{F.26}$$

$$\sin x \cos y = \frac{1}{2}(\sin(x + y) + \sin(x - y)) \tag{F.27}$$

$$\sin x \sin y = \frac{1}{2}(\cos(x - y) - \cos(x + y)) \tag{F.28}$$

$$\cos x \cos y = \frac{1}{2}(\cos(x - y) + \cos(x + y)) \tag{F.29}$$

Series of exponentials:

$$\sum_{k=0}^{N-1} a^k = \frac{1 - a^N}{1 - a}, \quad a \neq 1 \tag{F.30}$$

$$\sum_{k=0}^{\infty} a^k = \frac{1}{1 - a}, \quad |a| < 1 \tag{F.31}$$

$$\sum_{k=0}^{\infty} k a^k = \frac{1}{(1 - a)^2}, \quad |a| < 1 \tag{F.32}$$

$$\sum_{k=0}^{N-1} e^{\left(j \frac{2\pi n k}{N}\right)} = \begin{cases} 0 & 1 \leq n \leq N - 1 \\ N & n = 0, N \end{cases} \tag{F.33}$$

L'Hospital's rule:

If $\lim_{x \to a} \dfrac{f(x)}{g(x)}$ results in the undetermined form $\dfrac{0}{0}$ or $\dfrac{\infty}{\infty}$, then

$$\lim_{x \to a} \frac{f(x)}{g(x)} = \lim_{x \to a} \frac{f'(x)}{g'(x)} \tag{F.34}$$

where $f'(x) = \dfrac{df(x)}{dx}$ and $g'(x) = \dfrac{dg(x)}{dx}$.

Solution of the quadratic equation:

For a quadratic equation expressed as

$$ax^2 + bx + c = 0, \tag{F.35}$$

the solution is given by

$$x = \frac{-b \pm \sqrt{b^2 - 4ac}}{2a}. \tag{F.36}$$

Bibliography

Ahmed, N., and Natarajan, T. (1983). *Discrete-Time Signals and Systems*. Reston, VA: Reston Publishing Co.

Akansu, A. N., and Haddad, R. A. (1992). *Multiresolution Signal Decomposition: Transforms, Subbands, and Wavelets*. Boston: Academic Press.

Alkin, O. (1993). *Digital Signal Processing: A Laboratory Approach Using PC-DSP*. Englewood Cliffs, NJ: Prentice Hall.

Ambardar, A. (1999). *Analog and Digital Signal Processing*, 2nd ed. Pacific Grove, CA: Brooks/Cole Publishing Company.

Brandenburg, K. (1997). Overview of MPEG audio: Current and future standards for low-bit-rate audio coding. *Journal of Audio Engineering Society*, 45 (1/2).

Carr, J. J., and Brown, J. M. (2001). *Introduction to Biomedical Equipment Technology*, 4th ed. Upper Saddle River, NJ: Prentice Hall.

Chen, W. (1986). *Passive and Active Filters: Theory and Implementations*. New York: John Wiley & Sons.

Dahnoun, N. (2000). *Digital Signal Processing Implementation Using the TMS320C6000TM DSP Platform*. Englewood Cliffs, NJ: Prentice Hall.

Deller, J. R., Proakis, J. G., and Hansen, J. H. L., *Discrete-Time Processing of Speech Signals*, Macmillian Publishing Company, 1993.

El-Sharkawy, M. (1996). *Digital Signal Processing Applications with Motorola's DSP56002 Processor*. Upper Saddle River, NJ: Prentice Hall.

Embree, P. M. (1995). *C Algorithms for Real-Time DSP*. Upper Saddle River, NJ: Prentice Hall.

Gonzalez, R. C., and Wintz, P. (1987). *Digital Image Processing*, 2nd ed. Reading, MA: Addison-Wesley Publishing Company.

Grover, D., and Deller, J. R. (1998). *Digital Signal Processing and the Microcontroller*. Upper Saddle River, NJ: Prentice-Hall.

Haykin, S. (1991). *Adaptive Filter Theory*, 2nd ed. Englewood Cliffs, NJ: Prentice Hall.

Ifeachor, E. C., and Jervis, B. W. (2002). *Digital Signal Processing: A Practical Approach*, 2nd ed. Upper Saddle River, NJ: Prentice Hall.

Kehtarnavaz, N., and Simsek, B. (2000). *C6X-Based Digital Signal Processing*. Upper Saddle River, NJ: Prentice Hall.

Krauss, T. P., Shure, L., and Little, J. N. (1994). *Signal Processing TOOLBOX for Use with MATLAB*. Natick, MA: The MathWorks, Inc.

Li, Z.-N., and Drew, M. S. (2004). *Fundamentals of Multimedia*. Upper Saddle River, NJ: Pearson Prentice Hall.

Lipshitz, S. P., Wannamaker, R. A., and Vanderkooy, J. (1992). Quantization and dither: A theoretical survey. *Journal of the Audio Engineering Society*, 40 (5): 355–375.

Lynn, P. A., and Fuerst, W. (1999). *Introductory Digital Signal Processing with Computer Applications*, 2nd ed. Chichester and New York: John Wiley & Sons.

Maher, Robert C. (1992). On the nature of granulation noise in uniform quantization systems. *Journal of the Audio Engineering Society*, 40 (1/2): 12–20.

McClellan, J. H., Oppenheim, A. V., Schafer, R. W., Burrus, C. S., Parks, T. W., and Schuessler, H. (1998). *Computer Based Exercises for Signal Processing Using MATLAB 5*. Upper Saddle River, NJ: Prentice Hall.

McClellan, J. H., Schafer, R. W., and Yoder, M. A. (1998). *DSP First—A Multimedia Approach*. Upper Saddle River, NJ: Prentice Hall.

Nelson, M. (1992). *The Data Compression Book*. Redwood City, CA: M&T Publishing.

Oppenheim, A. V., and Schafer, R. W. (1975). *Discrete-Time Signal Processing*. Englewood Cliffs, NJ: Prentice Hall.

Oppenheim, A. V., Schafer, R. W., and Buck, J. R. (1999). *Discrete-Time Signal Processing*, 2nd ed. Upper Saddle River, NJ: Prentice Hall.

Pan, D. (1995). A tutorial on MPEG/audio compression. *IEEE Multimedia*, 2: 60–74.

Phillips, C. L., and Harbor, R. D. (1991). *Feedback Control Systems*, 2nd ed. Englewood Cliffs, NJ: Prentice Hall.

Phillips, C. L., and Nagle, H. T. (1995). *Digital Control System Analysis and Design*. Englewood Cliffs, NJ: Prentice Hall.

Porat, B. (1997). *A Course in Digital Signal Processing*. New York: John Wiley & Sons.

Princen, J., and Bradley, A. B. (1986). Analysis/synthesis filter bank design based on time domain aliasing cancellation. *IEEE Transactions on Acoustics, Speech, and Signal Processing*, ASSP 34 (5).

Proakis, J. G., and Manolakis, D. G. (1996). *Digital Signal Processing: Principles, Algorithms, and Applications*, 3rd ed. Upper Saddle River, NJ: Prentice Hall.

Rabbani, M., and Jones, P. W. (1991). *Digital Image Compression Techniques*. Presentation to the Society of Photo-Optical Instrumentation Engineers (SPIE), Bellingham, Washington.

Rabiner, L. R., and Schafer, R. W. (1978). *Digital Processing of Speech Signals*. Englewood Cliffs, NJ: Prentice Hall.

Roddy, D., and Coolen, J. (1997). *Electronic Communications*, 4th ed. Englewood Cliffs, NJ: Prentice Hall.

Sayood, K. (2000). *Introduction to Data Compression*, 2nd ed. San Francisco: Morgan Kaufmann Publishers.

Soliman, S. S., and Srinath, M. D. (1998). *Continuous and Discrete Signals and Systems*, 2nd ed. Upper Saddle River, NJ: Prentice Hall.

Sorensen, H. V., and Chen, J. P. (1997). *A Digital Signal Processing Laboratory Using TMS320C30*. Upper Saddle River, NJ: Prentice Hall.

Stanley, W. D. (2003). *Network Analysis with Applications*, 4th ed. Upper Saddle River, NJ: Prentice Hall.

Stearns, S. D. (2003). *Digital Signal Processing with Examples in MATLAB*. Boca Raton, FL: CRC Press LLC.

Stearns, S. D., and David, R. A. (1996). *Signal Processing Algorithms in MATLAB*. Upper Saddle River, NJ: Prentice Hall.

Stearns, S. D., and Hush, D. R. (1990). *Digital Signal Analysis*, 2nd ed. Englewood Cliffs, NJ: Prentice Hall.

Texas Instruments. (1991). *TMS320C3x User's Guide*, Dallas, TX: Author.

Texas Instruments. (1998). *TMS320C6x CPU and Instruction Set Reference Guide*, Literature ID# SPRU 189C, Dallas, TX: Author.

Texas Instruments. (2001). *Code Composer Studio: Getting Started Guide*. Dallas, TX: Author.

Tomasi, W. (2004). *Electronic Communications Systems: Fundamentals Through Advanced*, 5th ed. Upper Saddle River, NJ: Pearson Prentice Hall.

Van der Vegte, J. (2002). *Fundamentals of Digital Signal Processing*. Upper Saddle River, NJ: Prentice Hall.

Vetterli, M., and Kovacevic, J. (1995). *Wavelets and Subband Coding*. Englewood Cliffs, NJ: Prentice Hall.

Webster, J. G. (1998). *Medical Instrumentation: Application and Design*, 3rd ed. New York: John Wiley & Sons, Inc.

Yost, W. A. (1994). *Fundamentals of Hearing: An Introduction*, 3rd ed. San Diego: Academic Press.

Answers to Selected Problems

Chapter 2

2.1.

$$5\cos(2\pi \times 1000t) = 5 \cdot \left(\frac{e^{j2\pi \times 1500t} + e^{-j2\pi \times 1000t}}{2}\right)$$
$$= 2.5e^{j2\pi \times 1500t} + 2.5e^{-j2\pi \times 1500t}$$
$$c_1 = 2.5 \text{ and } c_{-1} = 2.5$$

a.

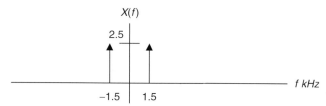

b.

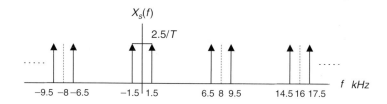

766 ANSWERS TO SELECTED PROBLEMS

2.3.

$$x(t) = e^{-j2\pi \cdot 4500t} + 2.5e^{-j2\pi \cdot 2500t} + 2.5e^{j2\pi \cdot 2500t} + e^{j2\pi \cdot 4500t}$$

a.

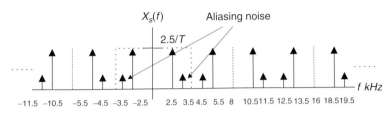

b.

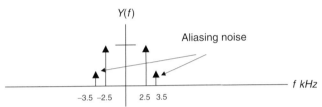

c. the aliasing frequency = 3.5 kHz

2.5.

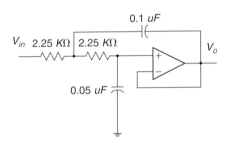

2.7.

    a. % aliasing noise level = 57.44%

    b. % aliasing noise level = 20.55%

2.9.

    a. % distortion = 24.32%

    b. % distortion = 5.68%

2.11.

    b1b0 = 01

2.13.

    a. $L = 2^4 = 16$ levels

    b. $\Delta = \dfrac{x_{max} - x_{min}}{L} = \dfrac{5}{16} = 0.3125$

    c. $x_q = 3.125$

    d. binary code $= 1010$

    e. $e_q = -0.075$

2.15.

    a. $L = 2^6 = 64$ levels

    b. $\Delta = \dfrac{x_{max}-x_{min}}{L} = \dfrac{20}{64} = 0.3125$

    c. $SNR_{dB} = 1.76 + 6.02 \times 6 = 37.88$ dB

Chapter 3

3.1.

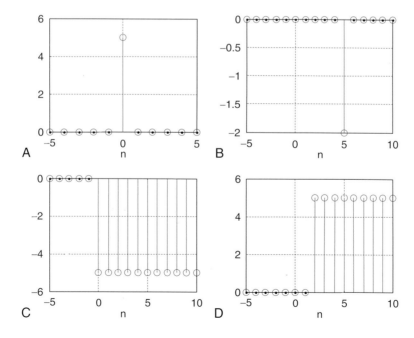

3.3.

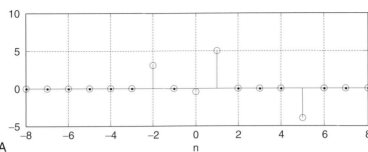

A

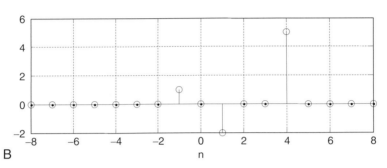

B

3.5.

a. $x(n) = e^{-0.5n}u(n) = (0.6065)^n u(n)$

b. $x(n) = 5 \sin(0.2\pi n)u(n)$

c. $x(n) = 10 \cos(0.4\pi n + \pi/6)u(n)$

d. $x(n) = 10e^{-n} \sin(0.15\pi n)u(n) = 10(0.3679)^n \sin(0.15\pi n)u(n)$

3.7.

a. time invariant.

b. time variant.

3.9.

a. causal system

b. noncausal system

c. causal system

3.10.

a. $h(n) = 0.5\delta(n) - 0.5\delta(n-2)$

b. $h(n) = (0.75)^n; n \geq 0$

c. $h(n) = 1.25\delta(n) - 1.25(-0.8)^n; n \geq 0$

3.11.

a. $h(n) = 5\delta(n - 10)$

b. $h(n) = \delta(n) + 0.5\delta(n - 1)$

3.13.

a. stable

b. unstable

3.15.

$y(0) = 4$, $y(1) = 6$, $y(2) = 8$, $y(3) = 6$, $y(4) = 5$, $y(5) = 2$, $y(6) = 1$,
$y(n) = 0$ for $n \geq 7$

3.17.

$y(0) = 0$, $y(1) = 1$, $y(2) = 2$, $y(3) = 1$, $y(4) = 0$
$y(n) = 0$ for $n \geq 4$

Chapter 4

4.1.

$X(0) = 1$, $X(1) = 2 - j$, $X(2) = -1$, $X(3) = 2 + j$

4.3.

From 4.2: $X(0) = 10$, $X(1) = 2 - 2j$, $X(2) = 2$, $X(3) = 2 + 2j$

Ans: $x(0) = 4$, $x(1) = 3$, $x(2) = 2$, $x(3) = 1$

4.5.

From 4.4: $X(0) = 10$, $X(1) = 3.5 - 4.3301j$, $X(2) = 2.5 - 0.8660j$,
$X(3) = 2$, $X(3) = 2.5 + 0.8660j$, $X(3) = 3.5 + 4.3301j$

Ans: $\bar{x}(0) = 4$, $\bar{x}(4) = 0$

4.7.

$N = 4096$, $\Delta f = 0.488$ Hz

4.9.

a. w = [0.0800 0.2532 0.6424 0.9544 0.9544 0.6424 0.2532 0.0800]

b. w = [0.1170 0.4132 0.7500 0.9698 0.9698 0.7500 0.4132 0.1170]

4.11.

    a. $A_0 = 0.1667$, $A_1 = 0.3727$, $A_2 = 0.5$, $A_3 = 0.3727$

       $\varphi_0 = 0^0$, $\varphi_1 = 154.43^0$, $\varphi_2 = 0^0$, $\varphi_3 = -154.43^0$

       $P_0 = 0.0278$, $P_1 = 0.1389$, $P_2 = 0.25$, $P_3 = 0.1389$

    b. $A_0 = 0.2925$, $A_1 = 0.3717$, $A_2 = 0.6375$, $A_3 = 0.3717$

       $\varphi_0 = 0^0$, $\varphi_1 = 145.13^0$, $\varphi_2 = 0^0$, $\varphi_3 = -145.13^0$

       $P_0 = 0.0586$, $P_1 = 0.1382$, $P_2 = 0.4064$, $P_3 = 0.1382$

    c. $A_0 = 0.6580$, $A_1 = 0.3302$, $A_2 = 0.9375$, $A_3 = 0.3302$

       $\varphi_0 = 0^0$, $\varphi_1 = 108.86^0$, $\varphi_2 = 0^0$, $\varphi_3 = -108.86^0$

       $P_0 = 0.4330$, $P_1 = 0.1091$, $P_2 = 0.8789$, $P_3 = 0.1091$

4.13. Hint:

$X(0) = 10$, $X(1) = 2 - 2j$, $X(2) = 2$, $X(3) = 2 + 2j$, 4 complex multiplications

4.15.

$X(0) = 10$, Hint: $X(1) = 2 - 2j$, $X(2) = 2$, $X(3) = 2 + 2j$, 4 complex multiplications

Chapter 5

5.1.

    a. $X(z) = \dfrac{4z}{z - 1}$

    b. $X(z) = \dfrac{z}{z + 0.7}$

    c. $X(z) = \dfrac{4z}{z - e^{-2}} = \dfrac{4z}{z - 0.1353}$

    d. $X(z) = \dfrac{4z[z - 0.8 \times \cos(0.1\pi)]}{z^2 - [2 \times 0.8z \cos(0.1\pi)] + 0.8^2} = \dfrac{4z(z - 0.7608)}{z^2 - 1.5217z + 0.64}$

    e. $X(z) = \dfrac{4e^{-3}\sin(0.1\pi)z}{z^2 - 2e^{-3}z\cos(0.1\pi) + e^{-6}} = \dfrac{0.06154z}{z^2 - 0.0947z + 0.00248}$

5.3.
   a. $X(z) = 15z^{-3} - 6z^{-5}$
   b. $x(n) = 15\delta(n-3) - 6\delta(n-5)$

5.5.
   a. $X(z) = -25 + \dfrac{5z}{z-0.4} + \dfrac{20z}{z+0.1}$,
   $x(n) = -25\delta(n) + 5(0.4)^n u(n) + 20(-0.1)^n u(n)$

   b. $X(z) = \dfrac{1.6667z}{z-0.2} - \dfrac{1.6667z}{z+0.4}$, $x(n) = 1.6667(0.2)^n u(n) - 1.6667(-0.4)^n u(n)$

   c. $X(z) = \dfrac{1.3514z}{z+0.2} + \dfrac{Az}{z-P} + \dfrac{A^*z}{z-P^*}$,

   where $P = 0.5 + 0.5j = 0.707\angle 45°$, and $A = 1.1625\angle -125.54°$
   $x(n) = 1.3514(-0.2)^n u(n) + 2.325(0.707)^n \cos(45° \times n - 125.54°)$

   d. $X(z) = \dfrac{4.4z}{z-0.6} + \dfrac{-0.4z}{z-0.1} + \dfrac{-1.2z}{(z-0.1)^2}$,
   $x(n) = 4.4(0.6)^n u(n) - 0.4(0.1)^n u(n) - 12n(0.1)^n u(n)$

5.7.
   $Y(z) = \dfrac{9.84z}{z-0.2} + \dfrac{-29.46z}{z-0.3} + \dfrac{20z}{z-0.4}$
   $y(n) = 9.84(0.2)^n u(n) - 29.46(0.3)^n u(n) + 20(0.4)^n u(n)$

5.9.
   a. $Y(z) = \dfrac{Az}{z-P} + \dfrac{A^*z}{z-P^*}$,
   $P = 0.2 + 0.5j = 0.5385\angle 68.20°$, $A = 0.8602\angle -54.46°$
   $y(n) = 1.7204(0.5382)^n \cos(n \times 68.20° - 54.46°)$

   b. $Y(z) = \dfrac{1.6854z}{z-1} + \dfrac{Az}{z-P} + \dfrac{A^*z}{z-P^*}$,
   where $P = 0.2 + 0.5j = 0.5385\angle 68.20°$, $A = 0.4910\angle -136.25°$
   $y(n) = 1.6845u(n) + 0.982(0.5382)^n \cos(n \times 68.20° - 136.25°)$

Chapter 6

6.1.
  a. $y(0) = -2$, $y(1) = 2.3750$, $y(2) = -1.0312$,
     $y(3) = 0.7266$, $y(4) = -0.2910$
  b. $y(0) = 0$, $y(1) = 1$, $y(2) = -0.2500$, $y(3) = 0.3152$, $y(4) = -0.0781$

6.2.
  a. $H(z) = (0.5 + 0.5z^{-1})$,
  b. $y(n) = 2\delta(n) + 2\delta(n-1)$, $y(n)$
  c. $y(n) = -5\delta(n) + 10u(n)$

6.3.
  a. $H(z) = \dfrac{1}{1 + 0.5z^{-1}}$,
  b. $y(n) = (-0.5)^n u(n)$,
  c. $y(n) = 0.6667 u(n) + 0.3333(-0.5)^n u(n)$

6.5.
  $H(z) = 1 - 0.3z^{-1} + 0.28z^{-2}$, $A(z) = 1$, $B(z) = 1 - 0.3z^{-1} + 0.28z^{-2}$

6.7.
  $H(z) = \dfrac{(z + 0.4)(z - 0.4)}{(z + 0.2)(z + 0.5)}$

6.9. Hint:
  a. zero: $z = 0.5$,
     poles: $z = -0.25$ ($|z| = 0.25$), $z = -0.5 \pm 0.7416j$ ($|z| = 0.8944$), stable
  b. zeros: $z = \pm 0.5j$,
     poles: $z = 0.5$ ($|z| = 0.5$), $z = -2 \pm 1.7321j$ ($|z| = 2.6458$), unstable
  c. zero: $z = -0.95$,
     poles: $z = 0.2$ ($|z| = 0.2$), $z = -0.7071 \pm 0.7071j$ ($|z| = 1$),
     marginally stable
  d. zeros: $z = -0.5$, $z = -0.5$, poles: $z = 1$ ($|z| = 1$),
     $z = -1$, $z = -1$ ($|z| = 1$), $z = 0.36$ ($|z| = 0.36$), unstable

6.11.
  a. $H(z) = \dfrac{1}{1 + 0.5z^{-2}}$

b. Hint: $H(e^{j\Omega}) = \dfrac{1}{1 + 0.5e^{-j2\Omega}}$

c. bandpass filter $|H(e^{j\Omega})| = \dfrac{1}{\sqrt{(1 + 0.5\cos 2\Omega)^2 + (0.5\sin 2\Omega)^2}}$,

$\angle H(e^{j\Omega}) = -\tan^{-1}\left(\dfrac{-0.5\sin 2\Omega}{1 + 0.5\cos 2\Omega}\right)$

6.13.

a. (1) $H(z) = 0.5 + 0.5z^{-1}$

(2) $H(z) = 0.5 - 0.5z^{-1}$

(3) $H(z) = 0.5 + 0.5z^{-2}$

(4) $H(z) = 0.5 - 0.5z^{-2}$

b. Hint:

(1) $H(e^{j\Omega}) = 0.5 + 0.5e^{-j\Omega}$

(2) $H(e^{j\Omega}) = 0.5 - 0.5e^{-j\Omega}$

(3) $H(e^{j\Omega}) = 0.5 + 0.5e^{-j2\Omega}$

(4) $H(e^{j\Omega}) = 0.5 - 0.5e^{-j2\Omega}$

(1) $|H(e^{j\Omega})| = \sqrt{(0.5 + 0.5\cos \Omega)^2 + (0.5\sin \Omega)^2}$,

$\angle H(e^{j\Omega}) = \tan^{-1}\left(\dfrac{-0.5\sin \Omega}{0.5 + 0.5\cos \Omega}\right)$

(2) $|H(e^{j\Omega})| = \sqrt{(0.5 - 0.5\cos \Omega)^2 + (0.5\sin \Omega)^2}$,

$\angle H(e^{j\Omega}) = \tan^{-1}\left(\dfrac{0.5\sin \Omega}{0.5 - 0.5\cos \Omega}\right)$

(3) $|H(e^{j\Omega})| = \sqrt{(0.5 + 0.5\cos 2\Omega)^2 + (0.5\sin 2\Omega)^2}$,

$\angle H(e^{j\Omega}) = \tan^{-1}\left(\dfrac{-0.5\sin 2\Omega}{0.5 + 0.5\cos 2\Omega}\right)$

(4) $|H(e^{j\Omega})| = \sqrt{(0.5 - 0.5\cos 2\Omega)^2 + (0.5\sin 2\Omega)^2}$,

$\angle H(e^{j\Omega}) = \tan^{-1}\left(\dfrac{0.5\sin 2\Omega}{0.5 - 0.5\cos 2\Omega}\right)$

c. (1) lowpass
   (2) highpass
   (3) bandstop
   (4) bandpass

6.15.

a. $H(z) = \dfrac{0.5}{1 + 0.7z^{-1} + 0.1z^{-2}}$

b. $y(n) = 0.5556u(n) - 0.111(-0.2)^n u(n) + 0.5556(-0.5)^n u(n)$

6.17.

a. Hint: $y(n) = x(n) - 0.9x(n-1) - 0.1x(n-2) - 0.3y(n-1) + 0.04y(n-2)$

b. Hint: $w(n) = x(n) - 0.3w(n-1) + 0.04w(n-2)$
   $y(n) = w(n) - 0.9w(n-1) - 0.1w(n-2)$

c. Hint: $H(z) = \dfrac{(z-1)(z+0.1)}{(z+0.4)(z-0.1)}$
   $w_1(n) = x(n) - 0.4w_1(n-1)$
   $y_1(n) = w_1(n) - w_1(n-1)$
   $w_2(n) = y_1(n) + 0.1w_2(n-1)$
   $y(n) = w_2(n) + 0.1w_2(n-1)$

d. Hint: $H(z) = 2.5 + \dfrac{2.1z}{z+0.4} - \dfrac{3.6z}{z-0.1}$
   $y_1(n) = 2.5x(n)$
   $w_2(n) = x(n) - 0.4w_2(n-1)$
   $y_2(n) = 2.1w_2(n)$
   $w_3(n) = x(n) + 0.1w_3(n-1)$
   $y_3(n) = -3.6w_3(n)$
   $y(n) = y_1(n) + y_2(n) + y_3(n)$

6.18.

a. $y(n) = x(n) - 0.5x(n-1)$
   $y(n) = x(n) - 0.7x(n-1)$
   $y(n) = x(n) - 0.9x(n-1)$

b. Filter $H(z) = 1 - 0.9z^{-1}$ emphasizes high frequency components most.

Chapter 7

7.1. Hint:

   a. $H(z) = 0.2941 + 0.3750z^{-1} + 0.2941z^{-2}$

   b. $H(z) = 0.0235 + 0.3750z^{-1} + 0.0235z^{-2}$

7.3. Hint:

   a. $H(z) = -0.0444 + 0.0117z^{-1} + 0.0500z^{-2} + 0.0117z^{-3} - 0.0444z^{-4}$

   b. $H(z) = -0.0035 + 0.0063z^{-1} + 0.0500z^{-2} + 0.0063z^{-3} - 0.0035z^{-4}$

7.5.

   a. Hanning window

   b. filter length = 63

   c. cutoff frequency = 1000 Hz

7.7.

   a. Hamming window

   b. filter length = 45

   c. lower cutoff frequency = 1,500 Hz, upper cutoff frequency = 2,300 Hz

7.9.

   a. $y(n) = 0.25x(n) - 0.5x(n-1) + 0.25x(n-2)$

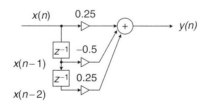

   b. $y(n) = 0.25[x(n) + x(n-2)] - 0.5x(n-1)$

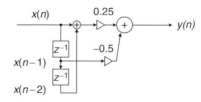

776 ANSWERS TO SELECTED PROBLEMS

7.11.
$$H(z) = -0.1236 + 0.3236z^{-1} + 0.6z^{-2} + 0.3236z^{-3} - 0.1236z^{-4}$$

7.13.
$$H(z) = 0.1718 - 0.2574z^{-1} - 0.0636z^{-2} + 0.2857z^{-3}$$
$$-0.0636z^{-4} - 0.2574z^{-5} + 0.1781z^{-6}$$

7.15. Hint:

Hamming window, filter length = 33, lower cutoff frequency = 3,500 Hz.

7.16.

Hamming window, filter length = 53. Lower cutoff frequency =1,250 Hz; upper cutoff frequency = 2,250 Hz.

7.17.

a. Lowpass filter: Hamming window

Highpass filter: Hamming window

b.

lowpass: Filter length = 91

highpass: Filter length = 91

c.

lowpass: Cutoff frequency = 2,000 Hz

highpass: Cutoff frequency = 2,000 Hz

Chapter 8

8.1. $H(z) = \dfrac{0.3333 + 0.3333z^{-1}}{1 - 0.3333z^{-1}}$

$y(n) = 0.3333x(n) + 0.3333x(n-1) + 0.3333y(n-1)$

8.3.

a. $H(z) = \dfrac{0.6625 - 0.6625z^{-1}}{1 - 0.3249z^{-1}}$

$y(n) = 0.6225x(n) - 0.6225x(n-1) + 0.3249y(n-1)$

Answers to Selected Problems

8.5.
   a. $H(z) = \dfrac{0.2113 - 0.2113z^{-2}}{1 - 0.8165z^{-1} + 0.5774z^{-2}}$

   $y(n) = 0.2113x(n) - 0.2113x(n-2) + 08165y(n-1)$
   $\qquad -0.5774y(n-2)$

8.7.
   a. $H(z) = \dfrac{0.1867 + 0.3734z^{-1} + 0.1867z^{-2}}{1 - 0.4629z^{-1} + 0.2097z^{-2}}$

   $y(n) = 0.1867x(n) + 0.3734x(n-1)$
   $\qquad + 0.1867x(n-2) + 0.4629y(n-1) - 0.2097y(n-2)$

8.9.
   a. $H(z) = \dfrac{0.0730 - 0.0730z^{-2}}{1 + 0.8541z^{-2}}$

   $y(n) = 0.0730x(n) - 0.0730x(n-2) - 0.8541y(n-2)$

8.11.
   a. $H(z) = \dfrac{0.5677 + 0.5677z^{-1}}{1 + 0.1354z^{-1}}$

   $y(n) = 0.5677x(n) + 0.5677x(n-1) - 0.1354y(n-2)$

8.13.
   a. $H(z) = \dfrac{0.1321 - 0.3964z^{-1} + 0.3964z^{-2} - 0.1321z^{-3}}{1 + 0.3432z^{-1} + 0.6044z^{-2} + 0.2041z^{-3}}$

   $y(n) = 0.1321x(n) - 0.3964x(n-1) + 0.3964x(n-2) - 0.1321x(n-3)$
   $\qquad -0.3432y(n-1) - 0.6044y(n-2) - 0.2041y(n-3)$

8.15.
   a. $H(z) = \dfrac{0.9609 + 0.7354z^{-1} + 0.9609z^{-2}}{1 + 0.7354z^{-1} + 0.9217z^{-2}}$

   $y(n) = 0.9609x(n) + 0.7354x(n-1) + 0.9609x(n-2)$
   $\qquad - 0.7354y(n-1) - 0.9217y(n-2)$

8.17.
   a. $H(z) = \dfrac{0.0242 + 0.0968z^{-1} + 0.1452z^{-2} + 0.0968z^{-3} + 0.0242z^{-4}}{1 - 1.5895z^{-1} + 1.6690z^{-2} - 0.9190z^{-3} + 0.2497z^{-4}}$

   $y(n) = 0.0242x(n) + 0.0968x(n-1) + 0.1452x(n-2)$
   $\qquad + 0.0968x(n-3) + 0.0242x(n-4)$
   $\qquad + 1.5895y(n-1) - 1.6690y(n-2) + 0.9190y(n-3)$
   $\qquad - 0.2497y(n-4)$

778   ANSWERS TO SELECTED PROBLEMS

8.19.

a. $H(z) = \dfrac{1}{1 - 0.3679z^{-1}}$

$y(n) = x(n) + 0.3679y(n-1)$

8.21.

a. $H(z) = \dfrac{0.1 - 0.09781z^{-1}}{1 - 1.6293z^{-1} + 0.6703z^{-2}}$

$y(n) = 0.1x(n) - 0.0978x(n-1) + 1.6293y(n-1) - 0.6703y(n-2)$

8.23.

$H(z) = \dfrac{0.9320 - 1.3180z^{-1} + 0.9320z^{-2}}{1 - 1.3032z^{-1} + 0.8492}$

$y(n) = 0.9320x(n) - 1.3180x(n-1) + 0.9329x(n-2)$
$\quad\quad + 1.3032y(n-1) - 0.8492y(n-2)$

8.25.

$H(z) = \dfrac{0.9215 + 0.9215z^{-1}}{1 + 0.8429z^{-1}}$

$y(n) = 0.9215x(n) + 0.9215x(n-1) - 0.8429y(n-1)$

8.27.

$H(z) = \dfrac{0.9607 - 0.9607z^{-1}}{1 - 0.9215z^{-1}}$

$y(n) = 0.9607x(n) - 0.9607x(n-1) + 0.9215y(n-1)$

8.29.

b. for section 1: $w_1(n) = x(n) - 0.7075w_1(n-1) - 0.7313w_1(n-2)$

$y_1(n) = 0.3430w_1(n) + 0.6859w_1(n-1)$
$\quad\quad + 0.3430w_1(n-2)$

for section 2: $w_2(n) = y_1(n) + 0.1316w_2(n-1) - 0.1733w_2(n-2)$

$y_2(n) = 0.4371w_2(n) + 0.8742w_2(n-1) + 0.4371w_2(n-2)$

8.31.

Chebyshev notch filter 1: order = 2

$H(z) = \dfrac{0.9915 - 1.9042z^{-1} + 0.9915z^{-2}}{1.0000 - 1.9042z^{-1} + 0.9830z^{-2}}$

Chebyshev notch filter 2: order = 2

$H(z) = \dfrac{0.9917 - 1.3117z^{-1} + 0.9917z^{-2}}{1.0000 - 1.3117z^{-1} + 0.9835z^{-2}}$

8.33.

Filter order = 4;

$$H(z) = \frac{0.1103 + 0.4412z^{-1} + 0.6618z^{-2} + 0.4412z^{-3} + 0.1103z^{-4}}{1.0000 + 0.1509z^{-1} + 0.8041z^{-2} - 0.1619z^{-3} + 0.1872z^{-4}}$$

8.35.

Filter order = 4;

$$H(z) = \frac{0.0300 - 0.0599z^{-2} + 0.0300z^{-4}}{1.0000 - 0.6871z^{-1} + 1.5741z^{-2} - 0.5176z^{-3} + 0.5741z^{-4}}$$

8.37.

a. $H(z) = \dfrac{0.5878z^{-1}}{1 - 1.6180z^{-1} + z^2}$

b. $y(n) = 0.5878x(n-1) + 1.6180y(n-1) - y(n-2)$

8.39.

a. $X(0) = 1$

b. $|X(0)|^2 = 1$

c. $A_0 = 0.25$

d. $X(1) = 1 - j2$

e. $|X(1)|^2 = 5$

f. $A_1 = 1.12$

Chapter 9

9.1.

0.2560123 (decimal) = 0.0 1 0 0 0 0 0 1 1 0 0 0 1 0 1 (Q-15)

9.3.

1.0 1 0 1 0 1 1 1 0 1 0 0 0 1 0 (Q-15) = -0.6591186

9.5.

1.1 0 1 0 1 0 1 1 1 0 0 0 0 0 1 + 0.0 1 0 0 0 1 1 1 1 0 1 1 0 1 0
= 1.1 1 1 1 0 0 1 1 0 0 1 1 0 1 1

9.7.

1101 011100011011(floating) = 0.8881835×2^{-3} (decimal)

0100 101111100101(floating) = -0.5131835×2^{4} (decimal)

0.8881835×2^{-3} (decimal)

$= 0.0069389 \times 2^{4}$ (decimal) = 0100 000000001110 (floating)

0100 101111100101 (floating) + 0100 000000001110 (floating)

= 0100 101111110010 (floating) = -8.1094 (decimal)

9.9.

$(-1)^0 \times 1.625 \times 2^{1536-1023} = 4.3575 \times 10^{154}$

9.11.

$C = 2, S = 1$

$x_s(n) = x(n), y_s(n) = 0.675x_s(n) + 0.15y_f(n-1), y_f(n) = 2y_s(n), y(n) = y_f(n)$

9.12.

$S = 8, A = 2, B = 4$

$x_s(n) = x(n)/8, w_s(n) = 0.5x_s(n) - 0.675w(n-1) - 0.25w(n-2),$

$w(n) = 2w_s(n)$

$y_s(n) = 0.18w(n) + 0.355w(n-1) + 0.18w(n-2), y(n) = 32y_s(n)$

Chapter 10

10.1.

$w^* = 2$, and $J_{\min} = 10$

10.3.

$w^* \approx w_2 = 1.984$, and $J_{\min} = 10.0026$

10.5.

a. $y(n) = w(0)x(n) + w(1)x(n-1)$

$e(n) = d(n) - y(n)$

$w(0) = w(0) + 0.2 \times e(n)x(n)$

$w(1) = w(1) + 0.2 \times e(n)x(n-1)$

b. for $n = 0$

$$y(0) = 0$$
$$e(0) = 3$$
$$w(0) = 1.8$$
$$w(1) = 1$$

for $n = 1$

$$y(1) = 1.2$$
$$e(1) = -3.2$$
$$w(0) = 2.44$$
$$w(1) = -0.92$$

for $n = 2$

$$y(2) = 5.8$$
$$e(2) = -4.8$$
$$w(0) = 0.52$$
$$w(1) = 0.04$$

10.7.

a. $n(n) = 0.5 \cdot x(n - 5)$

b. $xx(n) = 5000 \cdot \delta(n)$,
$yy(n) = 0.7071xx(n-1) + 1.4141yy(n-1) - yy(n-2)$

c. $d(n) = yy(n) - n(n)$

d. for $i = 0, \ldots, 24$, $w(i) = 0$

$$y(n) = \sum_{i=0}^{24} w(i)x(n-i)$$

$$e(n) = d(n) - y(n)$$

for $i = 0, \ldots, 24$

$$w(i) = w(i) + 2\mu e(n)x(n-i)$$

10.8.

30 coefficients

10.9.

For $i = 0, \ldots, 19$, $w(i) = 0$

$$y(n) = \sum_{i=0}^{19} w(i)x(n-i)$$

$$e(n) = d(n) - y(n)$$

For $i = 0, \ldots, 19$

$$w(i) = w(i) + 2\mu e(n)x(n-i)$$

Chapter 11

11.1.

    a. $\Delta = 0.714$

    b. for $x = 1.6$ volts, binary code = 110, $x_q = 1.428$ volts, and $e_q = -0.172$ volts

    for $x = -0.2$ volt, binary code = 000, $x_q = 0$ volts, and $e_q = 0.2$ volt

11.3.

For $x = 1.6$ volts, binary code = 111, $x_q = 1.132$ volts, and $e_q = -0.468$ volt

For $x = -0.2$ volt, binary code = 010, $x_q = -0.224$ volt, and $e_q = -0.024$ volt

11.5.

    a. 0 0 0 1 0 1 0 1

    b. 1 1 1 0 0 1 1 1

11.7.

010, 001, 010

11.9.

    a. 1:1

    b. 2:1

    c. 4:1

11.10.

    a. 128 kilo bits per sample (kbps)

    b. 64 kbps

    c. 32 kbps

11.11.
- a. 12 channels
- b. 24 channels
- c. 48 channels

11.13.
- a. inverse DCT: 10.0845 6.3973 13.6027 −2.0845
- b. Quantized DCT coefficients: 16, 8, −8, 8
  recovered inverse DCT: 11.3910 8.9385 15.0615 −3.3910
- c. quantization error: 1.3066 2.5412 1.4588 −1.3066

11.15.
- a. −9.0711 −0.5858

  −13.3137 −0.0000

  −7.8995 0.5858
- b. 3, 4, 5, 4

  Same as the original data

Chapter 12

12.1.
- a.

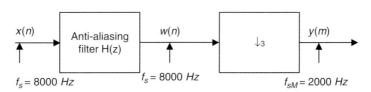

- b. Hamming window, $N = 133$, $f_c = 900$ Hz

12.3.
- a.

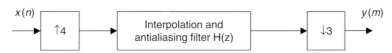

- b. Combined filter $H(z)$: Hamming window, $N = 133$, $f_c = 2700$ Hz

12.5.

    a.

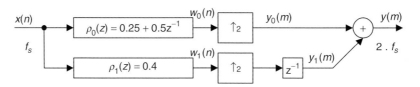

    b.

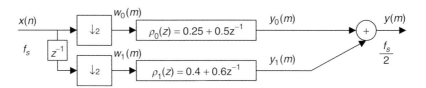

12.7.

    a. $f_s = 2f_{max}2^{2(n-m)} = 2 \times 15 \times 2^{2 \times (16-12)} = 7680\,\text{RHz}$

    b.

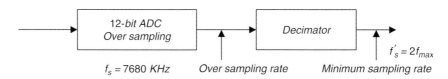

12.9.

    a.

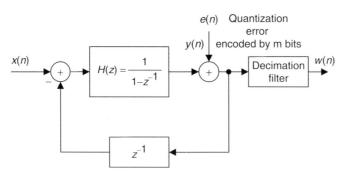

    b. $n = 1 + 1.5 \times \log_2\left(\frac{128}{2 \times 4}\right) - 0.86 \approx 6\,\text{bits}$

12.11.

a. $f_c/B = 6 =$ even number, which is the case 1, $f_s = 10\,\text{kHz}$

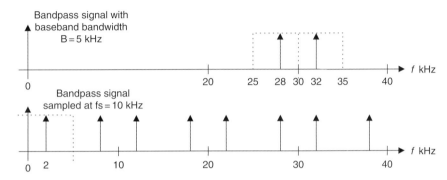

b. $f_c/B = 5 =$ odd number, which is case 2, $f_s = 10\,\text{kHz}$

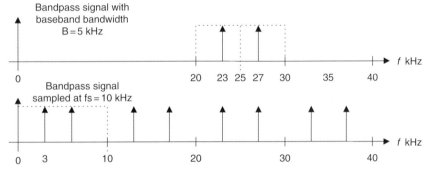

c. $f_c/B = 6.6 =$ non integer. Extended band width $\bar{B} = 5.5\,\text{kHz}$, $f_c/\bar{B} = 6$ and $f_s = 2\bar{B} = 11\,\text{kHz}$.

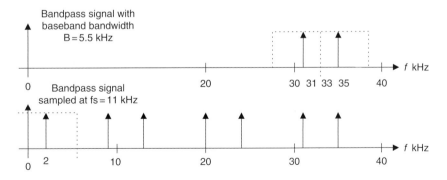

Chapter 13

13.1.
- a. 76.8 Kbytes
- b. 921.6 Kbytes
- c. 1920.768 Kbytes

13.3.
$$Y = 142, I = 54, Q = 11$$

13.5.
$$\begin{bmatrix} 53 & 44 \\ 59 & 50 \end{bmatrix}$$

13.7.
$$\begin{bmatrix} 1 & 4 & 6 & 6 & 1 \\ 6 & 4 & 4 & 6 & 4 \\ 4 & 4 & 6 & 6 & 6 \\ 1 & 6 & 7 & 7 & 4 \end{bmatrix}$$

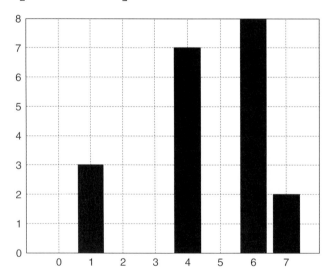

13.9. Hint:
$$\begin{bmatrix} 102 & 109 & 104 & 51 \\ 98 & 101 & 101 & 54 \\ 98 & 103 & 100 & 51 \\ 50 & 55 & 51 & 25 \end{bmatrix}$$

13.10. Hint:
$$\begin{bmatrix} 0 & 100 & 100 & 0 \\ 0 & 100 & 100 & 100 \\ 0 & 100 & 100 & 100 \\ 0 & 100 & 0 & 0 \end{bmatrix}$$

13.11.

a. $\begin{bmatrix} 225 & 125 & 130 & 33 \\ 249 & 119 & 136 & 6 \\ 249 & 119 & 136 & 6 \\ 255 & 125 & 130 & 0 \\ 255 & 128 & 128 & 30 \end{bmatrix}$

b. $\begin{bmatrix} 0 & 106 & 106 & 0 \\ 106 & 255 & 255 & 106 \\ 106 & 255 & 255 & 106 \\ 117 & 223 & 223 & 117 \\ 0 & 117 & 117 & 0 \end{bmatrix}$

13.13.

Blue color is dominant in the area pointed to by the arrow; red color is dominant in the background.

13.15.
$$X(u, v) = \begin{bmatrix} 460 & -40 \\ -240 & -140 \end{bmatrix} \text{ and } A(u, v) = \begin{bmatrix} 115 & 10 \\ 60 & 35 \end{bmatrix}$$

13.16.
$$\text{Forward DCT: } F(u, v) = \begin{bmatrix} 230 & -20 \\ -120 & -70 \end{bmatrix}$$

13.17.
$$\text{Inverse DCT: } p(i, j) = \begin{bmatrix} 110 & 100 \\ 100 & 90 \end{bmatrix}$$

13.19.

a. (0, −2), (3, 4), (2, −3), (0, 7), (4, −2), (0, 0)

b. (0000, 0010, 01), (0011, 0011, 100), (0010, 0010, 00), (0000, 0011, 111), (0100, 0010, 01), (0000, 0000)

13.21.

The interlaced scanning contains the odd field and even field per frame. The purpose of using the interlaced scanning is to transmit a full frame quickly to reduce flicker. See Section 13.9.1.

13.23.

Frequency modulated (FM) using a peak frequency deviation of 25 kHz. Assuming the audio baseband bandwidth is 15 kHz, the stereo FM audio requires a transmission bandwidth of 80 kHz with an audio carrier located at 4.5 MHz relative to the picture carrier. FM (frequency modulation). See Section 13.9.1.

13.24.

Hint:

Composite $\times 2\sin(2\pi f_{sc}t) = Y \times 2\sin(2\pi f_{sc}t)$
$+ I\cos(2\pi f_{sc}t) \times 2\sin(2\pi f_{sc}t) + Q \times 2\sin^2(2\pi f_{sc}t)$
$= Y \times 2\sin(2\pi f_{sc}t) + I\sin(2 \times 2\pi f_{sc}t) + Q - Q\cos(2 \times 2\pi f_{sc}t)$

Then apply lowpass filtering

13.25.

The back porch of the blanking contains the color sub-carrier burst for the color demodulation. The color burst carrier is centered at 3.58 MHz above the picture carrier and has the 8 cycles. See Section 13.9.1.

13.27.

The scan line rate 525 lines per frame $\times$ 30 frames per second = 15.75 kHz

The vertical synchronizing pulse rate (used with equalization pulses to provide timing) = 31.5 kHz. See Section 13.9.1.

13.29.

The progressive scanning traces a whole picture which is called the frame via row-wise, the interlaced scanning retraces the odd field and even field in each frame alternatively. CIF uses the progressive scan. See Section 13.9.2.

13.31.

$$\frac{80 \times 80}{16 \times 16}(16^2 \times 32^2 \times 3) = 19.661 \times 10^6 \text{ operations}$$

Appendix B

B.1. Hint:

$A_0 = 0.4$, $A_1 = 0.7916$, $A_2 = 0.7667$, $A_3 = 0.7263$, $A_4 = 0.6719$

$|c_0| = 0.4$, $|c_1| = |c_{-1}| = 0.3958$,

$|c_2| = |c_{-2}| = 0.3834$, $|c_3| = |c_{-3}| = 0.3632$, $|c_4| = |c_{-4}| = 0.3359$

B.3.
  a. $x(t) = 2 + 3.7420 \times \cos(2000\pi t) + 3.0273 \times \cos(4000\pi t)$
  $+ 2.0182 \times \cos(6000\pi t) + 0.9355 \times \cos(8000\pi t) + \cdots$

  b. $f_2 = 2000$ Hz, $A_2 = 3.0273$

B.5.
$$X(f) = 5\left(\frac{\sin \pi f}{\pi f}\right)^2$$

B.7.
  a. $X(s) = 10$
  b. $X(s) = -100/s^2$
  c. $X(s) = \dfrac{10}{s+2}$
  d. $X(s) = \dfrac{2e^{-5s}}{s}$
  e. $X(s) = \dfrac{10s}{s^2+9}$
  f. $X(s) = \dfrac{14.14 + 7.07s}{s^2+9}$
  g. $X(s) = \dfrac{3(s+2)}{(s+2)^2+9}$
  h. $X(s) = \dfrac{12000}{s^6}$

B.9.
  a. $X(s) = \dfrac{7.5}{s(s+1.5)}$
  b. $x(t) = 5u(t) - 5e^{-1.5t}u(t)$

B.11.
- a. Zero: $s = 3$, poles: $s = -2$, $s = -2$, stable
- b. Zeros: $s = 0$, $s = \pm 2.236j$,
  poles: $s = \pm 3j$, $s = -1 \pm 1.732j$, marginally stable
- c. Zeros: $s = \pm j$, $s = -1$,
  poles: $s = 0$, $s = -3$, $s = -4$, $s = -8$, $s = 1$, unstable

B.13.
- a. $H(j\omega) = \dfrac{1}{\dfrac{j\omega}{5} + 1}$
- b. $A(\omega) = \dfrac{1}{\sqrt{1 + \left(\dfrac{\omega}{5}\right)^2}}$ $\quad \beta(\omega) = \angle - \tan\left(\dfrac{\omega}{5}\right)$
- c. $Y(j2) = 4.6424 \angle -21.80^\circ$ that is, $y_{ss}(t) = 4.6424 \sin(2t - 21.80^\circ)u(t)$

Index

AC. *See* Alternating current
AC-2 coding, 533
AC-3 coding, 533
Acoustics, audio equalizer and, 346
Active noise control, 463
Active suspension systems, 11t
Adaptive DM, 515
Adaptive DPCM (ADPCM), 515, 537
   decoding, 526f
      MATLAB function for, 536–544
   encoding, 510f
      MATLAB function for, 536–544
   original speech and, 520f
   quantization error and, 520f
   quantized speech and, 520f
   quantizer and, 532t
   scale factor and, 515
Adaptive filter, 463–496
   for ECG interference cancellation, 490–491
   line enhancement with, 473–491, 484f
   LMS for, 484
   for noise cancellation, 473–494
   one-tap, 464f
   output from
      spectrum for, 479f
      waveform for, 478f
   periodic interference cancellation with, 486–494, 488f
   for system modeling, 479–493, 479f
   for telephone echo cancellation, 463–491
   Wiener filter, 467–492
ADC. *See* Analog-to-digital conversion
Address generators, 418, 438–440, 460
Adjustable filter coefficients, 463
ADPCM. *See* Adaptive DPCM
Advanced Television System Committee (ATSC), 687
Aliasing, 15–16, 19, 354. *See also* Anti-aliasing
   distortion from, 3, 9–16, 579

Alternating current (AC), 131, 522
   coding coefficients for, 675
   DCT and, 522–554
   run-length coding and, 674–676
Alternation theorem, 278–279
ALUs. *See* Arithmetic logic units
AM. *See* Amplitude-modulated
Amplifier, 2, 37
   chopper buffer, 598
   gain, 598
Amplitude, of impulse response, 219
Amplitude spectrum, 6f, 87–133
   computation of, 123
   with DFT, 87–131, 108f
   Hamming window function and, 119f, 121f
   one-sided, 99–100, 103–121, 120f–121f
   of periodic digital signal, 90f
   window functions and, 117, 119f, 121f
Amplitude-modulated (AM), 602
   spectrum for, 608f, 609f
   video signal as, 679
Analog broadband TV systems, 685t
Analog Devices, 420, 437, 439
Analog domain, mapping from, 315f
Analog filter, 2
   to lowpass prototype, 323t
   with lowpass prototype transformation, 306–310
Analog signal
   conversion of, 57
   sampling of, 98
Analog video, 678–685, 691
   PAL and, 685–687
   SECAM and, 685
Analog-to-digital conversion (ADC), 2, 35
   adaptive filters and, 463
   flash, 36–37
   oversampling of, 557, 589–614, 595f

Analog-to-digital conversion (ADC) (Cont'd)
  power gain of, 448
  process of, 41f
  quantization and, 497
  real-time processing and, 451
  resolution, 593–597
  sampling and, 25–47
  SDM in, 593
  sigma-delta, 36
  successive approximation, 36
  traditional, 591f
  z-transform and, 174
Answering machines, 11t
Anti-aliasing, 25–29
  in CD recording, 9
  decimation and, 588f
  downsampling and, 558, 560f–562f
  Nyquist sampling theorem and, 558
  spectrum after, 573f
  stopband frequency edge for, 580f
Anti-image filter, 22. See also Reconstruction filter
  design of, 32–33
  interpolation and, 614
  signal reconstruction and, 20, 53f
  upsampling and, 609
Application-specific integrated circuit (ASIC), 420
ARAUs. See Auxiliary register arithmetic units
Arithmetic logic units (ALUs), 438, 439
ASIC. See Application-specific integrated circuit
ATRAC, 533
ATSC. See Advanced Television System Committee
Audio crossover system
  frequency response for, 259f
  impulse response for, 260f
Audio equalizer
  acoustics and, 346
  bandpass filter and, 346, 397
  Butterworth bandpass filter and, 346
  center frequency and, 346
  filter gain and, 346
  frequency boosting and, 346
  IIR and, 303
  magnitude frequency response for, 348f
  MATLAB function for, 347–348
  noise and, 346

  specifications for, 347t
Audio recorders, compression in, 8
Audio signals, spectrums of, 4
Autocorrelation, 469
Auxiliary register arithmetic units (ARAUs), 440
Average lowpass filter, image enhancement and, 691

Back porch, 680, 681
Band edges, 278–280, 279t, 288
Band reject filter, 290, 291
  original speech and, 253f
Bandpass digital Butterworth filter, 346
Bandpass digital Chebyshev filter, 340–342
Bandpass filter, 2, 189, 294. See also Butterworth bandpass filter; Second-order bandpass filter
  amplitude spectral plots for, 204f
  audio equalizer and, 346, 401
  BLT and, 374
  crossover audio systems and, 205
  cutoff frequency for, 242, 305
  ECG and, 370
  equalizing and, 205
  filter length for, 242
  folding frequency for, 274
  frequency boosting and, 205
  frequency response of, 207, 248f
  frequency warping and, 316
  Hamming window function and, 248f
  linear phase, 266, 399t
  lowpass prototype transformation and, 306f
  magnitude frequency response for, 290
  magnitude response of, 189f
  MATLAB for, 207
  multirate digital signal processing and, 557, 589
  noise reduction and, 205
  original speech and, 250f
  passband ripple for, 274
  speech enhancement and, 202
  stopband attenuation for, 274
Bandpass signals
  plots of, 604, 605f
  undersampling of, 557, 601–610
Bandstop filter, 189
  Butterworth, 336–338
  cutoff frequency for, 242

filter length for, 242
frequency warping and, 316
lowpass prototype transformation and, 307, 308f
magnitude response of, 189f
Bartlett window function. *See* Triangular window function
begin function, in MATLAB array indexing, 699
Bessel filter, 599
BIBO. *See* Bounded-in-and-bounded out stability
Bilinear transformation (BLT)
bandpass filter and, 372
frequency warping and, 310–317, 401
for IIR, 303, 305f
vs. impulse invariant design, 400f
Laplace transfer function and, 353
mapping and, 313f
MATLAB and, 319
vs. pole-zero placement, 400f
procedure for, 317–322
s-plane and, 312
z-plane and, 312
Biomedical signal enhancement, 463
Bipolar quantizer, 497
characteristics of, 40f
Bit reversal process, 125, 125f, 441
Blackman window function, 229, 291
filter length and, 250
FIR filter coefficients for, 251t
frequency response for, 240f, 252f
Blanking levels, 680
Block edge artifacts, 530
BLT. *See* Bilinear transformation
Blue component
equalization and, 632–635
in image processing, 622–627
Blurring, 691
Bounded-in-and-bounded out stability (BIBO), 72–82, 176
Buffering. *See also* Linear buffering
circular, 418, 418f, 419f, 460
FIR and, 419
Buses, 450
Butterworth bandpass filter, 346
Butterworth bandstop filter, 338–340
for frequency response, 340f
Butterworth filter, 25, 33, 54, 205

bandpass filter design, 326, 336
bandstop filter design, 326, 336
filter order for, 397
lowpass filter and, 326, 398
lowpass prototype function and, 322–326
magnitude frequency response of, 27f
second-order bandpass, 336–338
second-order lowpass, 331
Butterworth fourth-order lowpass filter, 343–345
frequency response for, 345f
Butterworth lowpass prototype, 307, 322, 323t
Butterworth prototype functions, in cascade form, 344t
Butterworth transfer function, normalized, 743–744

Cameras, 1, 11t
Capacitors, 2
Cardiac monitoring, 11t
Carrier wave, 677
Cascade realization, 195, 197–198, 198f, 200f
in higher-order IIR design, 365–368
of IIR direct form I, 367f
of IIR direct form II, 367f
Cathode-ray tube display, 679
Causal sequence, 135
z-transform and, 135
Causality, linear time-invariant causal systems and, 64
CCIR. *See* Consultative Committee for International Radio
CCITT. *See* Comité Consultatif International Téléphonique et Télégraphique
CCS. *See* Code Composer Studio
CD. *See* Compact-disc
Cellular telephones, 1, 11t
Center frequency, 338, 398
audio equalizer and, 346
for Chebyshev bandpass filter, 342
Central processing unit (CPU), 413–414, 448
of TMS320C67x, 448
Chebyshev bandpass filter
passband width for, 338
stopband attenuation for, 338
Chebyshev filter, 25, 26, 268
bandpass filter design, 340–342
bandstop filter design, 340
filter order for, 397

Chebyshev filter (Cont'd)
  highpass filter and, 326–329
  lowpass filter and, 326–329, 398
  lowpass prototype function and, 322–325, 397
Chebyshev function, normalized, 744–747
Chebyshev lowpass filter, 379–397
  MATLAB for, 380
  second-order, 379
Chebyshev lowpass prototype, 307, 323t, 324–326
Chebyshev prototype functions, 322
  in cascade form, 344t
Chebyshev transfer function, 747–748
Chopper buffer amplifier, 598
Chrominance
  in composite video, 677
  demodulation of, 680
  S-video and, 677
  in YCbCr color space, 685–692
CIF, 687, 692
Circular addressing, 441
Circular buffering, 418, 418f, 460
C-language, 452
Clean speech
  spectrum for, 479f
  waveform for, 478f
Code Composer Studio (CCS), 451
Coding. See also Decoding; Encoding; Waveform
  of AC coefficients, 675
  AC-2, 533
  AC-3, 533
  of DC coefficients, 522–525
  with DCT, 529f
  EOB, 670
  lossless entropy, 675
  in MPEG, 530–534
  run-length, 674–676
  of speech, 7, 11t
  with W-MDCT, 529f
Coefficient accuracy effects, on FIR, 283
Coefficient quantization effects, 377–380
Color burst frequency, 681
Color demodulation, 680
Color image, equalization of, 632–633, 633f
Color indexed image, 620
Color subcarrier burst, 680
  vertical synchronization and, 682f

Comité Consultatif International Téléphonique et Télégraphique (CCITT), 515, 687
Commutative model, for polyphase interpolation filter, 588, 589f
Compact-disc (CD), 1, 11t, 346
  decoder for, 601f
  Hamming window function and, 576
  interpolation and, 578–586
  MATLAB function for, 578
  multi-rate signal processing and, 557
  Nyquist frequency and, 586
  oversampling in, 591f
  recording system for, 9
  SDM and, 599–612
  upsampling and, 577–583
Complement integer format, 420
Complex numbers, 99, 757–758
Component video, 677
Composite video, 677
Compression, 497–552. See also Decompression; Waveform
  in audio recorders, 8
  of data, 7–8, 8f, 11t, 534
  DCT and, 522–530
  DPCM and, 510–516
  lossless, 665
  MDCT and, 527–533
  μ-law, 505f
    analog, 533
    digital, 533
  μ-law companding and, 501–506
  in MP3, 8
  of MPEG, 687
  of speech, 501–554
  two-dimensional JPEG grayscale, 669–671
  in voice recorders, 8
Compression ratio (CR), 520–521
Cone speakers, 256
Constant system coefficients, 139
Consultative Committee for International Radio (CCIR), 685, 692.
Continuous-time. See Analog
Contrast
  image level adjustment and, 637–643
  improvement in, 632
  pixels and, 637
Convergence factor, 491

Conversion. *See also* Analog-to-digital conversion; Digital-to-analog conversion; Grayscale conversion
  of analog signal, 62
  of image format, 624–626, 626f
  from RGB to grayscale, 622–625
Convolution, 74–82, 140–142, 155, 442, 731–740
  formula method for, 80
  graphical method for, 78, 78t, 83
  in image processing, 644
  of sequences, 79f
  steps of, 81t
  trapezoidal shape and, 81
  z-transform, 139–143, 142t
Convolution sum, 70
  plot of, 80f
Corrupted image, 647
Corrupted signal, 370, 473
  in noise cancellation, 467f
Corrupted speech
  spectrum for, 479f
  waveform for, 478f
Cosine pulse, 726f, 724t
CPU. *See* Central processing unit
CR. *See* Compression ratio
CRC. *See* Cyclic redundancy check
Crossover audio systems, 6, 11t
  bandpass filtering and, 205
Cutoff frequency, 28–29, 35, 188, 242, 287
  for bandpass filter, 242, 307, 483
  for bandstop filter, 242
  calculation of, 219, 234
  downsampling and, 565
  FIR and, 215
  for first-order lowpass filter, 359
  for Hanning window function, 244
  for highpass digital processing, 242
  for highpass filter, 307
  for lowpass filter, 242, 264, 306, 320
  for multistage decimation, 578
  for noise reduction, 253
  for second-order bandpass digital Butterworth filter, 336–337
  for speech noise reduction, 255
Cyclic redundancy check (CRC), 531

DAC. *See* Digital-to-analog conversion
Data
  compression of, 7–8, 8f 11t, 534
  decompression of, 8, 8f
  encoding of, 3
  encryption of, 11t
  frames, 531
Daughter cards, 448
DC. *See* Direct current
DCT. *See* Discrete cosine transform
dct2() function, 668
dct() function, 526–527
Debugging, 451
Decimation, 614. *See also* Downsampling
  anti-aliasing and, 591f
  MATLAB function for, 566–571
  multistage, 579–583
  polyphase filter for, 588f
  z-transform and, 559
Decimation-in-frequency, 122–126
Decimation-in-time, 122, 127–130
  eight-point FFT and, 129f
  eight-point IFFT and, 130f
  first iteration with, 129f
  four-point FFT and, 130f
  four-point IFFT and, 130f
  input sequences and, 127
  second iteration with, 129f
Decoding
  ADPCM, 516f
  in CD recording, 9f, 600f, 601f
  MDCT, 526–534
  midtread quantizer, MATLAB function for, 537
Decompression
  of data, 8, 8f
  μ-law companding and, 505–506
Delta modulation (DM), 513–515, 533
Demodulation, 680
  of chrominance, 681
  of color, 680
  for NTSC scan line, 681f
Desired impulse response, 218
Determined sequence, 146t
DFT. *See* Discrete Fourier transform
Difference equations, 294–296, 444
  coefficients for, 68
  DSP and, 152–153
  filtering and, 159–165
  format for, 68
  transfer function and, 165–169

Difference equations (Cont'd)
  with z-transform, 151–155
Differential equations, Laplace transform and, 727–731
Differential pulse code modulation (DPCM), 510–515, 670
  DC and, 674
  decoder of, 510f
  encoder of, 510f
Digital domain, mapping to, 315f
Digital filter. See Filter(s)
Digital integration method, 311f
Digital linear system, 64f
  stability of, 72–74, 73f
Digital sequences, 122
  plot of, 61f, 63f
Digital signal(s)
  BIBO with, 72–73
  common digital sequences in, 58–62
  difference equations with, 68–69
  digital convolution and, 74–82
  generation of, 62–64
  impulse responses with, 68–69
  linear time-invariant causal systems and, 67–69
  notation for, 57, 58f
  periodic, 90f
  systems and, 57–82
Digital signal processing (DSP), 1
  applications of, 11t
  difference equations and, 152–153
  filtering and, 159–209
  hardware for, 416–460
  multirate, 557–610
  software for, 416–460
Digital signal (DS) processor, 13
  fixed-point, 437–439
  floating-point, 58
  manufacturers of, 419–420
Digital video, 685–687
Digital-to-analog conversion (DAC), 2
  power gain of, 448
  process of, 42f
  real-time processing and, 451
  sampling and, 16, 31–45
Direct current (DC), 92, 99, 103, 131, 307, 325, 534
  coding coefficients for, 674
  DCT and, 526, 665

  DPCM and, 674
  Fourier series coefficients and, 709
Direct form I
  cascade realization of, 369f
  of FIR, 442f
  of IIR, 443f
  realization in, 195–196, 196f, 199f, 365–366
Direct form II
  cascade realization of, 369f
  Goertzel algorithm and, 387
  of IIR, 443f
  realization in, 195–197, 197f, 199f, 365–366
Discontinuity, 110, 111
Discrete cosine transform (DCT), 497, 522–530, 665
  AC and, 666
  DC and, 666
  Image compression and, 664–675
  JPEG color image compression and, 670–675
  quantization with, 525–531, 673
  two-dimensional, 666–669
  two-dimensional JPEG grayscale compression and, 669–671
Discrete Fourier transform (DFT), 87–131, 260, 522
  amplitude spectrum and, 98–113, 108f–109f
  applications of, 99f
  FFT algorithms for, 103–104
  formulas for, 92–96, 93f
  Fourier series coefficients and, 88–92
  Goertzel algorithm and, 386–391
  one-dimensional, 661
  power spectrum and, 98–109, 108f–109f
  spectral estimation and, 110–121
  speech spectral estimation and, 121
  two-dimensional, 661
  window functions and, 110–117
Discrete-time analog integrator, 593
Displacement currents, 488
Display
  adjusting level for, 641
  cathode-ray tube, 679
Distortion, 30
  from aliasing, 579
  sample-and-hold effect and, 31f

DM. *See* Delta modulation
DMA, 450
Domain sequence, 88, 95
Double precision format, 436–437
Downsampling, 558–563, 613
   anti-aliasing and, 559, 561f–562f
   cutoff frequency and, 561
   FIR and, 561
   folding frequency and, 558
   lowpass filter and, 565
   multistage decimation and, 578–582
   Nyquist sampling theorem and, 579
   spectrum after, 560f–562f, 568f
   spectrum before, 561f–562f, 568f
DPCM. *See* Differential pulse code modulation
DS. *See* Digital signal processor
DSP. *See* Digital signal processing
DTMF. *See* Dual-tone multifrequency
Dual-tone multifrequency (DTMF), 11t
   detector, 392f
   frequency bins and, 392t
   Goertzel algorithm and, 381–391, 392f
   IIR and, 303
   MATLAB and, 385
   modified Goertzel algorithm and, 386–391
   oscillation and, 454
   single-tone generator and, 382–402
   tone generator, 384–385
   tone specifications for, 382f
DVD's, 1

ECG. *See* Electrocardiography
Echo cancellation, 11t, 463
   in telephones, 489–491
Edge detection, 651–655, 653f, 654f
8-bit color images, 620–621
8-bit color indexed image, 623f
   equalization of, 633–637
   format for, 622f
8-bit compressed code, 509
   format for, 507t
8-bit gray level images, 618, 618f, 690
Eight-point FFT, 126f
   decimation-in-time and, 129f
   first iteration of, 124f
   inverse of, 126f
   second iteration of, 129f

Eight-point IFFT, decimation-in-time and, 129f
Electrocardiography (ECG), 1, 11t
   60 Hz interference in, 479
   bandpass filter and, 372
   heart rate detection with, 370–377
   IIR and, 303, 365–377
   interference cancellation in, 7, 8f, 488–491
   LMS and, 491
   pulse characteristics in, 371f
   signal enhancement system, 372f
   60 Hz hum eliminator with, 370–376
   60 Hz hum with, 370–376, 488f
EMIF. *See* External memory interface
Encoding
   ADPCM, 515f
   in CD recording, 9f
end function, in MATLAB array indexing, 703
End of block coding (EOB), 670
Enhancement. *See also* Line enhancement
   adaptive filter and, 484–486, 484f
   of biomedical signal, 463
   of ECG, 372f
   of image, 648f
      average lowpass filtering and, 691
      by Gaussian filter kernel, 648f
      by median filter, 651
      photographic, 10
   of signal, 372f, 466, 485f
   of speech, 202–207
EOB. *See* End of block coding
Equalization. *See also* Histogram
   of eight8-bit indexed color image, 633
   of color image, 633–636, 634f
   of grayscale histogram, 625–632
   MATLAB function for, 636–637, 639f
   of RGB, 632–636, 635f
Equalized image
   color, 634f
   grayscale, 631f
   histogram for, 632f
   indexed 8-bit color, 638f
Equalizer, 11t. *See also* Audio equalizer
   bandpass filtering and, 205
   in communication channels, 463
   graphical, 10
   IIR and, 346
   implementation of, 33f
   signal reconstruction and, 29–35
   for speech signal, 290

Equalizer magnitude frequency, sample-and-hold effect and, 33f
Equiripples, 280
Error output, waveform for, 482f
Euler's identity, 23, 181, 221, 234, 388
  in Fourier complex exponential form, 711
Execution cycle, 415
  on Harvard architecture, 416f
  on Von Neumann architecture, 416f
Expectation operator, 43
Exponential function, 60, 723
  plot of, 60f
  sample values from, 60f
Exponential sequence, 136
External memory interface (EMIF), 448–449, 450
Extremal frequencies, 278
Extremal points, 278, 279, 279f, 280

Fast Fourier transform (FFT), 4, 87, 121–129
  applications of, 99f
  bit reversal process in, 125, 125f
  decimation-in-frequency and, 122–127
  decimation-in-time and, 127–130, 129f, 130f
  for DFT coefficients, 103–104
  eight-point, 124f, 126f
  four-point, 127
  functions of, 94t
  index mapping for, 125f
  MPEG and, 532
  pre-emphasis filter and, 202
  pre-emphasized speech and, 202
  radix-2 algorithms, 122
  radix-4 algorithms, 122
  split radix algorithms, 122
Fetch cycle, 415
FFT. *See* Fast Fourier transform
FIFO. *See* First-in/first-out
Filter(s), 3–4. *See also* Adaptive filter; Bandpass filter; Bandstop filter; Butterworth filter; Chebyshev filter; Finite impulse response filter; Highpass filter; Infinite impulse response filter; Lowpass filter; Polyphase filter
  average lowpass, 691
  band reject, 290, 291
    original speech and, 253f
  Bessel, 599
  cascade, 195

coefficients with
  in adaptive filters, 463
  adjustable, 465
  in Q-15 format, 458t
design of, MATLAB function for, 657f
difference equations and, 159–169
digital convolution and, 74
direct form I, 195
direct form II, 195
in DSP systems, 159–169
frequency response and, 179–188
impulse response and, 169–171
input and, 160f
lowpass, 188
output and, 160f
parallel, 195
realization of, 195–202
Sallen-Key lowpass, 26
sinusoidal steady-state response of, 749–751, 750f
speech enhancement and, 202–207
stability and, 171–177
steady-state frequency response of, 179f
step response and, 169–171
system response and, 169–171
transfer function of, 166f
types of, 188–194
z-plane pole-zero plot and, 171–179
Filter gain, 277–307
  audio equalizer and, 346
Filter length, 216
  for bandpass filter, 242
  for bandstop filter, 242
  Blackman window function and, 250
  for highpass digital processing, 242
  for lowpass filter, 242
  for noise reduction, 253
Filter order, 35, 287, 288
  for Butterworth filter, 397
  for Chebyshev filter, 397
filter() function, 162–165
filtic() function, 164
Finite impulse response (FIR) filter, 70, 188, 189, 209, 215–293
  adaptive filters, 464–466
  Blackman window function and, 251t
  buffering and, 419
  coefficients accuracy effects on, 283–285
  coefficients of, 231f, 239t, 245t, 248

direct-form I of, 442f
downsampling and, 563
filter format, 215–217
in fixed-point processor, 441–447
Fourier transform, 217–228
frequency sampling and, 260–268, 753–756
ideal impulse response for, 220t
linear buffering in, 452–456
linear phase property of, 225f
MATLAB and, 236t
noise reduction and, 253–260
optimal design method, 268–280
realization structures of, 280–283
17-tap, 224, 224t
speech noise reduction, 255–256
transfer function and, 216
two-band digital crossover, 256–259
window method, 229–253, 231f
FIR. *See* Finite impulse response filter firfs function, 263
illustrative usage for, 264t
First-in/first-out (FIFO), 418–419
First-order complex poles, 146–147, 175
First-order highpass filter, 364–365
pole-zero placement for, 364f
First-order lowpass filter
folding frequency for, 362
pole-zero placement for, 362, 362f
First-order lowpass prototype, 306–307, 339
firwd function, 236, 238
5-bit midtread uniform quantizer, 505f
Fixed-point format, 420–429
Fixed-point processor, 283, 420–429, 437–439
by Analog Devices, 420, 437
FIR in, 441–447
IIR in, 441–447
by Motorola, 420, 437
sample C programs with, 455–460
by TI, 420
Flash ADC, 36–37
Floating point format, 429–434, 430t
double precision format, 436–437
IEEE and, 434–436
multiplication rule for, 431–432
overflow with, 433, 447
underflow with, 434
Floating point numbers, 58, 283
Floating point processor, 429–434, 439–441
by Analog Devices, 420

IEEE, 434–439
sample C programs with, 455
by TI, 420
FM. *See* Frequency modulated
Folding frequency, 19, 89, 103, 106, 353
for bandpass filter, 274
downsampling and, 558
for first-order lowpass filter, 359
for lowpass filter, 271
For speech noise reduction, lowpass filter and, 255
Formula method, for digital convolution, 80
4-bit bipolar quantizer, quantization error and, 48
Fourier series coefficients, 19–20
amplitude-phase form, 712
for common waveforms, 720t
complex exponential form, 711–716
DC and, 709
harmonic frequency and, 710
for ideal impulse train, 720t
for positive square waveform, 720t
for rectangular waveform, 720t
for sawtooth waveform, 720t
sine-cosine form, 709–710
for square waveform, 720t
for triangular waveform, 720t
Fourier transform, *See also* 721–726
Discrete Fourier transform; Fast Fourier transform
for common signals, 724t
for cosine pulse, 726f, 724t
for exponential function, 724t
FIR and, 217–228
for impulse function, 724t
properties of, 725t
for rectangular pulse, 724t
for sawtooth pulse, 724t
for triangular pulse, 724t
window functions and, 287
Four-point FFT, 127f
decimation-in-time and, 130f
Four-point IFFT, decimation-in-time and, 130f
Fourth-order bandpass IIR filter, 481
Fourth-order lowpass Butterworth filter, 343–345
fps. *See* Frames per second

Frame via row-wise, 678
Frames per second (fps), 678
freqs() function, 308
Frequency. *See also* Cutoff frequency; Dual-tone multifrequency; Folding frequency; Frequency response; Nyquist frequency
  boosting of
    audio equalizer and, 346
    bandpass filtering and, 205
  domain representation, 87–88, 97–98, 243, 481
  index, 99, 131
  resolution, 90, 97, 103, 105, 131
    spectral leakage and, 121
  sampling of
    FIR and, 260–268, 753–756
    frequency responses for, 265f, 268f
    MATLAB functions and, 293
  spacing, 90, 100, 131
  warping
    bandpass filter and, 318
    bandstop filter and, 318
    bilinear transformation and, 305–319
    BLT and, 305–319, 399
    lowpass filter and, 317
Frequency bins, 131
  DTMF and, 392t
Frequency mapping. *See* Mapping
Frequency modulated (FM), 684
Frequency response, 228f, 273f, 276f. *See also* Magnitude frequency response; Steady-state frequency response
  for audio crossover system, 260f
  of bandpass filter, 205
  for Blackman window function, 240f, 252f
  for Butterworth fourth-order lowpass filter, 345
  calculations of, 184t, 187t, 222t, 234t
  of cascade notched filters, 374f
  for Chebyshev bandpass filter, 340
  for Chebyshev lowpass filter, 335
  for frequency sampling, 265f, 268f
  for Goertzel filter bank, 394f
  for Hamming window function, 239f, 240f, 248f
  for Hanning window function, 240f, 245f
  for highpass filter, 245f, 259, 260f
  for lowpass filter, 218f, 259, 260f, 270f
  magnitude, 223f
  MATLAB function for, 191
  with Parks-McClellan algorithm, 270f
  for passband ripple, 273f, 276f
  periodicity of, 182
  phase, 223f, 319
  plots of, 193f–194f
  pole-zero placement and, 358f
  of pre-emphasis filter, 203f
  for rectangular window function, 239f
  for sampling rate, 205
  for second-order bandpass digital Butterworth filter, 337
  for second-order lowpass Chebyshev filter, 335
  symmetry of, 182
  system transient, 179f, 180–181
  for unknown system, 482f
freqz() function, 190–191
Fundamental frequency, 91

Gain amplifier, 598
Gaussian filter kernel, 646
  enhanced image by, 648f
Gaussian noise, 477
Gibbs effect, 222, 226, 228, 268, 291
  Hamming window function and, 232
  windows functions and, 240
Goertzel algorithm, 381–391, 396 *See also* Modified Goertzel algorithm
  DFT and, 386–391
  direct form II and, 387
  DTMF and, 381–392, 392f
  second-order, 387f
Goertzel filter bank, frequency response for, 394f
Graphical equalizer, 10
Graphical method, for digital convolution, 78, 78t, 80
Graphical user interface (GUI), 451
Grayscale conversion
  image of, 626f
  in image processing, 626
Grayscale equalized image, 631f
  histogram for, 631f, 636
Grayscale histogram equalization, 626–639
Grayscale image, 626f, 636f
  format for, 619f

histogram for, 631f
pixels and, 660
Grayscale intensity image, format for, 626f
Grayscale transformation, three sine functions for, 659f
Green component
  equalization and, 633–636
  in image processing, 622–624
GUI. *See* Graphical user interface

Halfing process, 37
Hamming window function, 113, 114, 115, 229, 230–232, 235, 248, 284t, 290
  amplitude spectrum and, 118, 120f
  bandpass filter and, 248f
  CD and, 575–576
  computation of, 117
  FIR filter coefficients for, 241t, 245t
  frequency response for, 239f, 240f, 248f
  Gibbs effect and, 232
  for highpass filter, 258
  interpolation and, 573
  for lowpass filter, 258, 284
  for multistage decimation, 578–583
  noise reduction and, 253
  one-sided amplitude spectrum and, 121f
  for speech data, 120f
  for speech noise reduction, 255
  for two-band digital crossover, 256
Hanning window function, 113, 229, 291
  computation of, 117
  cutoff frequency for, 245
  FIR filter coefficients for, 245t
  frequency response for, 240f
  highpass digital processing and, 245f
  one-sided amplitude spectrum and, 120f
Hard disk drives, 11t
Hardware
  address generators, 418
  DS processors, 419–420
  for DSP, 413–462
  fixed-point processors, 420–429, 437–439
  floating-point processors, 429–434, 439–441
  MAC, 416–417
  shifters, 417
  Texas Instruments, 447–451
  TMS320C67X DSK, 447–451
Harmonic frequency, 89, 110
  of 60 Hz, 370, 370f

Fourier series coefficients and, 709
SDM and, 601
Harvard architecture, 414–416, 439, 460
  DSP with, 416
  execution cycle on, 416f
Heart rate detection, 370–377. *See also* Electrocardiography
High-definition TV (HDTV), 689
Higher-order design, 343–346
  cascade realization in, 369
Highpass digital processing, 2, 187
  in audio system, 6
  cutoff frequency for, 242
  filter length for, 242
  Hanning window function and, 245f
  magnitude response of, 189f
  original speech and, 247f
Highpass filter, 291
  Chebyshev filter and, 327, 330–333
  cutoff frequency with, 309
  first-order, 362–364
  frequency response for, 258, 260f
  Hamming window function for, 258
  impulse response for, 260f
  lowpass filter with, 258
  lowpass prototype transformation and, 305, 306f
  magnitude frequency response for, 309
  for noise shaping, 600
  two-band digital crossover and, 258
Histogram
  equalization of, 10–11, 633–639
    of color image, 639
    8-bit indexed color image equalization and, 639
    grayscale, 626–633
  for equalized image, 633f
  for grayscale equalized image, 631f
  for original grayscale image, 631f
  pixels and, 626
Horizontal retrace, 683
Horizontal Sobel edge detector, 653–654, 653f
Horizontal synchronizing pulse, 680
Horns (speakers), 256
Host port interface (HPI), 450
HPI. *See* Host port interface
Huffman coding, 532–534, 675t, 676

802 INDEX

I channel. *See* In-phase
IBM, 414
IC. *See* Integrated circuits
idct2() function, 668
idct() function, 524–525
Ideal impulse train, 720t
IDFT. *See* Inverse of DFT
IEEE. *See* Institute of Electrical and
   Electronics Engineers
IEEE floating-point format, 434, 437, 460
   double precision format, 436–437, 436f
   single precision format, 434–436
IEEE single precision floating-point standard,
   434–435, 435f, 461
IFFT function, 97
IIR. *See* Infinite impulse response
Image. *See also* Equalized image; Grayscale
   image
   color indexed, 620
   compression of, DCT and, 664–676
   8-bit color, 620–621
   8-bit color indexed, 623f
      equalization of, 633–637
      format for, 622f
   8-bit gray level, 618, 618f, 690
   enhancement of, 10–11, 642–657
      average lowpass filtering and, 691
      edge detection in, 651–655
      lowpass noise filtering in, 643–646
      median filtering in, 646–651
   filtering of, MATLAB functions for,
      655–657
   format conversion for, 626f
      MATLAB function for, 624–625
      from RGB to grayscale, 622–624
   level adjustment of, 641f
      for display, 641–642
      linear level adjustment in, 638–641
      MATLAB function for, 642, 643f
   mixing of, 677
   processing of, 11t, 617–698
      analog video and, 678–685
      blue component in, 622–624
      color equalization and, 632–636
      compression in, 664–676
      contrast and, 637–642
      convolution in, 644
      data formats for, 617–625
      DCT and, 665–674
      digital video and, 685–687
      edge detection in, 651–655
      8-bit color images, 620–621
      8-bit gray images, 618
      8-bit indexed color image equalization
         and, 633–637
      grayscale conversion in, 622–624
      green component in, 622–624
      histogram equalization for, 625–637
      image filtering enhancement and, 642–657
      image level adjustment and, 637–642
      image pseudo-color generation and
         detection in, 657–661
      image spectra in, 661–664
      intensity images, 621–622
      level adjustment for display and, 637–642
      linear level adjustment in, 638–641
      lowpass noise filtering in, 643–646
      median filtering in, 646–651
      by mixing two images, 677
      motion estimation in, 687–690
      notation for, 617–625
      PAL and, 684–685
      red component in, 622–624
      scale factor in, 644
      SECAM and, 685
      24-bit color images, 619–620
      zero padding effect in, 646
   spectra of, 661–644
Image pseudo-color generation and detection,
   657–661
imdctf() function, 528
Impulse input sequence, 455
   IIR with, 456f
Impulse invariant design, 350–357, 351f, 397
   vs. BLT, 400f
   vs. pole-zero placement, 400f
   sampling interval effect in, 355f
Impulse response, 154, 172f
   amplitude of, 219
   for audio crossover system, 259f
   desired, 218
   filtering and, 169–171
   for FIR, 219, 236t
   for highpass filter, 258f
   for lowpass filter, 219f, 260f
   system representation and, 69–73
impz() function, 444
imshow() function, 624

In-band frequency range, 590
Index mapping, for FFT, 125f
Index matching, 124
Indexed 8-bit color image, 638f
Infinite impulse response (IIR) filter, 72, 188, 209, 303–396
  adaptive filters, 466
  analog filters and, 306–309
  audio equalizer, 346–348
  bilinear transformation and, 303–322
  Butterworth filter design, 326–343
  cascade realization in, 369f,
  Chebyshev filter design and, 322–343
  coefficient quantization effects on, 377–380
  design methods, 399f
  design selection, 396–398
  direct form I of, 364–366, 367f, 443f
  direct form II of, 365–367, 369f, 445f
  DTMF and, 381–391
  ECG and, 370–374
  first-order highpass filter, 362–364
  in fixed-point processor, 337–441
  format for, 303
  Goertzel algorithm and, 381–396
  heart rate detection, 370–377
  higher-order design, 343–346, 368–370
  with impulse input sequence, 456f
  impulse invariant design, 350–357
  linear buffering in, 452–455, 453f, 454f
  modified Goertzel algorithm and, 386–391
  oscillation with, 454–455
  pole-zero form and, 358–365
  realization structures of, 365–370, 367f, 369f
  second-order bandpass filter, 359–360
  second-order bandstop filter, 360–362
  60 Hz hum eliminator, 370–377
  60 Hz interference cancellation in, 303
  two-band digital crossover and, 253
Information display, 3
In-phase (I channel), 624
Input, 160f, 304
  FIR and, 215–217
  plots of, 162f
Input and output (I/O), 414
Input sequences
  decimation-in-time and, 127
  index matching with, 124
  plot of, 78f

Input signal, spectrum for, 482f
Instant encoded values, 62
Institute of Electrical and Electronics Engineers (IEEE), 434–437, 460
Integer factor, sampling rate and, 564–575
Integrated circuits (IC), 2
Intensity image, 621–622
  format for, 622f
Interference cancellation, in ECG, 7, 8f, 488–489
Interlaced scanning, 679, 691
  raster, 680f
Internal buses, 450
  of TMS320C67x, 448
International Telecommunications Union (ITU), 687
Internet phones, 11t
Interpolated spectrum, 105
Interpolation, 577. *See also* Upsampling
  anti-image filters and, 609
  CD and, 575–583
  and Hamming window function, 574
  MATLAB function for, 568–571
  polyphase filter for, 586f
  spectrum after, 572f, 576f
  spectrum before, 572f
Interrupt service thread (IST), 457
Inverse of DFT (IDFT), 261
Inverse z-transform, 140–151, 233, 444
I/O. *See* Input and output
IST. *See* Interrupt service thread
ITU. *See* International Telecommunications Union
ITU-R-601, 685

Joint Photographic Experts Group. *See* JPEG
JPEG (Joint Photographic Experts Group), 665
  color image compression, 671–676, 676f
    quantization with, 673
  compressed image, 674f, 672f
  two-dimensional grayscale compression, 669–671

Kaiser window function, 229
Kernel, 643–647, 691
  Gaussian filter, 646, 655f
  lowpass average, 647f

Laplace domain, 733
Laplace shift property, 174
Laplace transfer function, 179, 311, 352, 353–355, 730
    BLT and, 397
    MATLAB for, 352
Laplace transform, 143, 174, 311, 726–731
    differential equations and, 727–729
    inverse of, 350, 351, 397
    table of, 728t
    unit step function of, 727
    z-transform and, 174f
Laplacian edge detector, 652–655, 655f, 656, 657f
Law of probability, 43
Least mean square (LMS) algorithm, 464–466, 489
    for adaptive filter, 484
    for ECG interference cancellation, 489
    line enhancement and, 484
    noise cancellation and, 473–479
Least-significant bit (LSB), 42, 425t
Left-hand half plane (LHHP), 175
    of s-plane, 731–732, 741
LHHP. *See* Left-hand half plane
L'Hospital's rule, 714, 760
Line enhancement, 484–486
    with adaptive filter, 484f
    enhanced signal and, 485f
    LMS and, 484
    MATLAB function for, 486
    noisy signal and, 487f
Linear buffering
    in FIR filtering, 452–453, 453f
    in IIR filtering, 452–453, 454f
Linear convolution. *See* Convolution
Linear level adjustment, 638–641, 641f
Linear midtread quantizer, MATLAB function for, 534
Linear phase bandpass filter, 266, 399
Linear phase delay, 254
Linear phase property, 224, 280
    of FIR, 225f
Linear phase realization structure, 280, 282–283, 283f
Linear phase requirement, 262, 287
Linear phase response, 226f
Linear prediction
    with line enhancement, 484–486

periodic interference cancellation with, 486–491
Linear time-invariant causal system, 64–67, 159, 160
    causality and, 64
    illustration of, 66f
    linearity and, 64
    time invariance and, 65–67
    unit-impulse response and, 69–73
    unit-impulse sequence of, 69f
Linearity
    linear time-invariant causal systems and, 64
    z-transform and, 142t
LMS. *See* Least mean square algorithm
Log-PCM coding, 501
Long-distance telephone, echo cancellation in, 489–491
Lossless compression, 665
Lossless entropy coding, 675
Lowpass average kernels, noise filtering by, 647f
Lowpass filter, 3, 25, 184, 241f, 292
    in audio system, 6
    Butterworth filter and, 322–335
    in CD recording, 9
    Chebyshev filter and, 322–335
    coefficient calculation for, 219–220
    cutoff frequency for, 242, 262, 305, 320
    downsampling and, 562, 565
    effect, 31f
    filter length for, 241
    folding frequency for, 271
    frequency response for, 217f, 259, 270f
    frequency warping and, 316
    with Hamming window function, 258, 284
    with highpass filter, 258
    impulse response for, 219f, 260f
    lowpass prototype transformation into, 306f
    magnitude response of, 188f
    of noise, 646–649
    noise reduction and, 253
    original speech and, 243f, 244f
    passband ripple for, 271
    periodicity of, 218f
    stopband attenuation for, 271
    two-band digital crossover and, 256
    upsampling and, 575
Lowpass prototype

analog filters to, 325t
Butterworth filter and, 326, 329, 397
Chebyshev filter and, 222–223, 397
transformation of
analog filters with, 305–306
to highpass filter, 307, 308f
into lowpass filter, 306f
Lowpass reconstruction filters, 21
LSB. *See* Least-significant bit
Luminance (Y Channel), 622, 632, 685
in composite video, 677
pixels and, 690
S-video and, 677
VSB for, 684

m bits, 43–44, 594–597
MAC. *See* Multiplier and accumulator
Maclaurin series, 444, 594
Macroblocks, 688–689
in reference frame, 688f
in target frame, 688f
MAD. *See* Mean absolute difference
Magnetic induction, 488
Magnitude bits, 424
Magnitude frequency response, 223f, 299
for audio equalizer, 347f
for bandpass filter, 292
for highpass filter, 290
for second-order lowpass digital Butterworth filter, 331
Magnitude response, 183–189
of bandpass filter, 189f
of bandstop filter, 190f
of highpass digital processing, 189f
of lowpass filter, 188f
periodicity of, 185f
pole-zero placement and, 400f
Magnitude spectrum
for noise-filtered image, 665f
for noisy image, 665f
for square image, 664f
Mantissa, 420, 429
Mapped frequencies, 97
Mapping
from analog domain, 315f
to digital domain, 315f
of s-plane, 312
of z-plane, 285f
Mark 1 relay-based computers, 414

MATLAB (Matrix Laboratory) function, 45, 94, 105–106, 117, 134, 148–151, 202, 238, 246, 251, 265, 267, 272, 319, 326, 328, 330, 703–707
for adaptive line enhancement, 486
for adaptive noise cancellation, 473, 477
for adaptive system identification, 483
for ADPCM decoding, 540, 544
for ADPCM encoding, 552–553
array indexing in, 699
for audio equalizer, 346–349
for bandpass filtering, 207
begin in, 705
BLT and, 319t
for Butterworth bandstop filter, 336
for Butterworth fourth-order lowpass filter, 343
for CD, 583
for Chebyshev bandpass filter, 340
for Chebyshev lowpass filter, 379
dct2() function, 668
dct() function, 526–527
for decimation, 562–563
for DTMF tone generation, 386
end function, 705
for equalization, 636–637, 639f
fft() function, 94t
for filter design, 657f
filter() function, 162, 164
filttic () function, 164–165
for FIR, 236t, 291–293
firfs function of, 263, 264t
firwd function, 236
freqs() function, 308
for frequency response, 191, 226–228
frequency sampling and, 293
freqz() function, 190–191
for generating sinusoid, 384
help in, 700
histeq function, 637
idct2() function, 668
idct() function, 524–525
ifft() function, 94t, 97
for IIR direct-form II representation, 445
illustrative usage for, 264t
for image filtering, 655–656
for image format conversion, 624–625
for image level adjustment, 642, 643f
imdctf() function, 528

MATLAB (Matrix Laboratory) function (Cont'd)
impz() function, 444
for interpolation, 572–573
for Laplace transfer function, 352, 355
mdcth() function, 528
for midtread quantizer, 499–500, 505f, 537
for midtread quantizer decoding, 537
for midtread quantizer encoding, 537
for μ-law companding, 536
for μ-law decoding, 539
for μ-law encoding, 535, 538
for noise filtering, 256
with non-integer factor, 570
partial fractions with, 148–149
phase response and, 191
plot functions in, 704–705
for pre-emphasis speech, 204
for pseudo-color generation, 661
for realization of IIR direct form I, 367f, 366
for realization of IIR direct form II, 365, 367f
remez() function, 272, 274, 290
residue function, 150
for sampling rate change, 575
script files in, 704
for second-order bandpass digital Butterworth filter, 336
for second-order lowpass Chebyshev filter, 379
for signal to quantization noise ratio, 51
SNR and, 46–47, 538, 540
stair function, 704
stem function, 704
step in, 703
subplot function, 704
sum function, 700
sumsub.m function, 705–706
for system response, 162
test.m function, 705, 706f
for uniform quantization coding, 50
for uniform quantization decoding, 50
zero-crossing and, 375
z-transform and, 148–149, 155
Matrix Laboratory. See MATLAB
MAX1402, 599
Maxim Integrated Products, 598
McBSP. See Multichannel buffered serial port

mdcth() function, 528
Mean absolute difference (MAD), 688
Mean square between (MSB), 499
Mean square error (MSE), 468f
minimization of, 470
noise cancellation and, 474, 474f
for Wiener filter, 470
Media Player, 10
Median filter, 646–647
enhanced imaging by, 651f
Memory, of TMS320C67x, 448
Midtread quantizer, 449–450, 537
MATLAB function for, 537
3-bit, 498f, 499t
Midtread quantizer decoding, MATLAB function for, 537
Midtread quantizer encoding, MATLAB function for, 537
Million instruction sets per second (MIPS), 439
Minimax filters, 269
Minimum detectable voltage, 42
MIPS. See Million instruction sets per second
Mixing two video images, 677, 691
μ= 255
12-bit decoded speech, 509f
compression characteristics, 507f
compressors, 509f
decoding table, 508t
8-bit compressed data, 509f
encoding table, 508t
expanders, 509f
quantization error, 509f
12-bit speech data, 509f
μ-law companding
analog, 501
characteristics of, 507f
digital, 506–508
MATLAB function for, 536–538
μ-law compression, 505f
analog, 533
digital, 538
original speech with, 505f
quantization error with, 505f
quantized speech, 505f
μ-law decoding, MATLAB function for, 539
μ-law encoding, MATLAB function for, 535, 538
μ-law expander, 505f

characteristics of, 507f
original speech with, 505f
quantization error with, 505f
quantized speech with, 505f
Modems, high speed, 11t
Modified discrete cosine transform (MDCT), 522
Modified Goertzel algorithm, 390–391
second-order, 390f
Modulation. *See also* Amplitude-modulated; Demodulation; Sigma-delta modulation
DM, 517–518, 533
FM, 684
negative, 679
QAM, 684
spectrum of, 606f
VSB, 684
Most-significant bit (MSB), 425, 425t
Motion compensation, 692
Motion estimation, 687–689
Motion Picture Experts Group. *See* MPEG
Motion vectors, 688, 688f, 690f, 690
in reference frame, 688f
in target frame, 688f
Motorola, 420, 437
MP3 (MPEG-1 layer 3), 1, 497, 526, 534
compression in, 8
MDCT and, 529, 534
MPEG (Motion Picture Experts Group), 526, 534
audio frame formats of, 531f
audio frame size of, 530f
compression of, 687
FFT and, 532
MDCT and, 529
Nyquist frequency and, 530
transform coding in, 530
MPEG-1 layer 3. *See* MP3
MPEG-2, 533
MSB. *See* Mean square between; Most-significant bit
MSE. *See* Mean square error
Multichannel buffered serial port (McBSP), 450
Multiple-order poles, 175
Multiplication rule, for floating point format, 431–432
Multiplier and accumulator (MAC), 415, 416–417, 417f, 438, 439

Multirate digital signal processing, 557
ADC, 595–601
bandpass filters and, 606–614
CD and, 610
decimation and, 580–588
downsampling and, 558–563
integer factor and, 570–571, 575
non-integer factor and, 570
oversampling and, 590–601
polyphase filters, 587–589
sampling rate and, 568–572
undersampling and, 604, 607, 609, 610
Multistage decimation, 578, 582f
passband ripple for, 580–581
plot of, 577f
stopband attenuation for, 581, 582

National Television System Committee (NTSC), 678, 695
Negative modulation, 679
Negative-indexed frequency components, 102
Noise. *See also* Signal-to-noise
adaptive filters and, 465
audio equalizer and, 346
filtering of, 647f
Gaussian, 477
in image, 643–644
lowpass filter of, 643–648
pepper and salt, 646, 650, 691
reference, 478f
signal with, 254t, 487f
vehicle, 487
Noise cancellation, 463
adaptive filter for, 466t, 475–477
corrupted signal in, 467f
LMS and, 473–474
MATLAB function for, 479–480, 477, 483
MSE and, 476, 476f
one-tap adaptive filter for, 473f
reference noise in, 467f
Wiener filter for, 469
Noise filtering, for MATLAB function, 256
Noise reduction, 11t
for speech, 255–256
bandpass filtering and, 205
cutoff frequency for, 254
filter length for, 254
FIR and, 253–256
Hamming window function and, 254

Noise reduction (Cont'd)
 lowpass filter and, 254
 passband ripple, 255
 stopband attenuation for, 255
 two-band digital crossover and, 253, 256
Noise shaping filter, 597
 for quantization noise, 601f
Noise shaping, highpass filter for, 597
Noise-filtered image, magnitude spectrum for, 665f
Noisy image, 648f, 650f
 magnitude spectrum for, 665f
Noncausal sequence, 139
Non-integer factor MATLAB function with, 577
 sampling rate and, 575–576, 614
Nonlinear phase response, 226f
Nonzero attenuation, 25
Nonzero coefficients, 68–69, 216
Normalized Butterworth function, 744
Normalized Chebyshev function, 748
Notch filter. See Second-order bandstop filter
NTSC. See National Television System Committee
Nyquist frequency, 19, 21, 24, 255, 288, 353
 CD and, 599
 for first-order lowpass filter, 359
 MPEG and, 533–534
 polyphase filter and, 587
 reconstruction filter and, 577
 upsampling and, 569
Nyquist sampling theorem anti-aliasing and, 561
 downsampling and, 561

Ohm's law, 37
1D-DFT. See One-dimensional discrete Fourier transform
1-Hz sine wave, sampling of, 110
One-dimensional discrete Fourier transform (1D-DFT), 663
One-sided amplitude spectrum, 99–100, 103, 103f, 117
 Hamming window function and, 121f
 Hanning window function and, 120f
 window functions and, 120f, 121f
One-sided z-transform, 135
One-tap adaptive filter, 464f
 for noise cancellation, 474f

Opcode, 414
Operand, 414
Original RGB color image, 633f
Original signal, quantized signal and, 46f
Original speech ADPCM and, 520f
 amplitude spectral plots for, 204f, 207
 band reject filter and, 252f
 bandpass filter and, 250f
 highpass digital processing and, 247f
 lowpass filter and, 243f, 244f
 with µ-law compressor, 506f
 with µ-law expander, 506f
 plots of, 206f, 504, 505f
 pre-emphasis speech and, 203f
 pre-emphasized speech and, 203f
 quantized speech and, 48f
 spectrum for, 479f
 waveform for, 482f
Oscillation (ripple), 222, 291
 DTMF and, 455
 with IIR, 455–456
 Remez exchange algorithm and, 280
 for two-band digital crossover, 256
Output, 160f, 300
 for adaptive filter, 484f
 FIR and, 215–216
 plots of, 162f
Output frequency, index matching with, 124
Output sequence, 155
Overflow, 422–423, 442
 with floating-point format, 433–434, 447
 Q-15 format and, 460
Oversampling, 32
 of ADC, 589–593, 595f
 ADC resolution and, 595–599
 in CD recording system, 600f, 601f

PAL. See Phase alternative line
Parallel realization, 198, 201, 198f, 201f
 partial fraction expansion and, 201
Parks-McClellan algorithm, 268, 288
 design procedure for, 270
 frequency response with, 270f
 popularity of, 280
 with Remez exchange algorithm, 291
Partial fraction expansion, 155, 171, 397
 with MATLAB, 148–151
 parallel realization and, 201
 z-transform and, 142–144, 142t

Passband frequency edge, 322
  in multistage decimation, 578
Passband gain, 338, 356, 365, 455
Passband ripple, 240, 242, 243, 270, 271, 322, 323, 398. *See also* Oscillation
  for bandpass filter, 274
  frequency response for, 273f, 276f, 277f
  for lowpass filter, 271
  for multistage decimation, 578
  for speech noise reduction, 255
Passband width
  for Butterworth bandstop filter, 336
  for Chebyshev bandpass filter, 338
Pepper and salt noise, 646, 650, 691
periodic digital signal, 88–92, 89f
  two-sided spectrum for, 92f
Periodic interference cancellation
  with adaptive filter, 488f
  with linear prediction, 487
Periodicity
  of frequency response, 179–180
  of lowpass filter, 217f
  of magnitude response, 185f
  of phase response, 185f
Phase alternative line (PAL), 684, 696
Phase distortion, 224
Phase frequency response, 223f, 319
Phase response, 185–186, 273f, 276f
  linear, 226f
  MATLAB and, 191
  periodicity of, 185f
  for second-order lowpass digital Butterworth filter, 326
Phase spectrum, 99, 100
  computation of, 105
Photo image enhancement, 10
Pipelining, 416
Pixels, 617–618
  contrast and, 637
  counts distributions of, 627t, 629t, 631t
  grayscale image and, 690
  histograms and, 625
  notation for, 618f
  resolution and, 690
  stretching of, 641
  zero padding effect and, 650
Plot functions, in MATLAB, 704–705
Poles, 731–732

pole-zero form, 169
  first-order highpass filter, 364–365
  IIR and, 358–359
  plots of, 178f, 191f
  second-order bandpass filter and, 359–360, 359f
Pole-zero placement, 396
  vs. BLT, 400f
  for first-order highpass filter, 364f
  for first-order lowpass filter, 359, 362f
  frequency response and, 354f
  impulse invariant design *vs.*, 400f
  magnitude response and, 358f
  second-order bandpass filter and, 359f, 398
  second-order bandstop filter and, 359f, 398
Polyphase filter, 583–588, 612
  commutative model for, 587
  implementation of, 588f
  for interpolation, 584f, 586f
  Nyquist frequency and, 586
Positive-indexed frequency components, 102
Power gain
  of ADC, 447
  of DAC, 448
Power lines, 371
  interference from, 488
Power spectrum, 87
  computation of, 105
  DFT and, 98–101, 108f
Power-down units, 450
Pre-emphasis filter
  FFT and, 202
  frequency response of, 203f
  of speech, 202–205
Pre-emphasis speech
  amplitude spectral plots for, 204f
  MATLAB for, 202, 204
  original speech and, 203f
Probability, 43
Pseudo-color generation, 657
  illustrative procedure for, 660f
  MATLAB function for, 661
Pseudo-color image, 659f
Psycho-acoustic mode, 534
Pulse train, 16

Q channel. *See* Quadrature
Q-15 format
  coding notations for, 460f

Q-15 format (Contd)
   filter coefficients in, 459t
   overflow and, 460
QAM. See Quadrature amplitude modulation
QCIF, 692
Q-format number representation, 424, 428, 439, 441–442
   Q-15 format, 424f, 425t
QRS complex, 371
Quadratic equations, 760
Quadrature (Q channel), 624. See also Chrominance
Quadrature amplitude modulation (QAM), 686
Quantization, 35, 401, 497–501
   ADC and, 501
   DCT and, 522–526, 674
   DPCM and, 510–511
   of infinite precision filter coefficients, 377
   with JPEG color image compression, 673
   MDCT and, 529–533
   sampling and, 32, 49
   of signals, 13–50
Quantization error, 38, 43, 501, 509
   ADPCM and, 520f
   4-bit bipolar quantizer and, 48f
   $\mu = 255$, 513f
   with μ-law compressor, 506f
   with μlaw expander, 506f
   plots of, 504
   sample-and-hold effect and, 504
Quantization noise, 594
   noise shaping filter for, 595f
Quantization step size, 504
Quantized signal, original signal and, 46
Quantized speech
   ADPCM and, 520f
   with μ-law compressor, 506f
   with μ-law expander, 506f
   original speech and, 48f
   plots of, 504, 505f
Quantizer. See also Midtread quantizer
   ADPCM and, 521t
   bipolar, 40f, 497
   5-bit midtread uniform, 505f
   4-bit bipolar, 48f
   3-bit bipolar, 39, 40t
   3-bit midtread, 498f, 499t
   unipolar, 38, 39f, 497

R-2R ladder, 37, 37f
Radix-2 FFT algorithms, 122, 131
Radix-4 FFT algorithms, 122
RAM. See Random access memory
Random access memory (RAM), 438
Random variables, 43
Read-only memory (ROM), 438
Realization. See also Cascade realization
   in direct form I, 195–196, 196f, 199f, 365–366
   in direct form II, 195–197, 197f, 199f, 365–366
   of filtering, 202
   FIR and, 280–282f
   IIR and, 367f
   parallel, 195, 198–199, 198f, 201f
   series, 197–198
Realization structures
   of IIR, 365
   linear phase, 282–283, 283f
RealPlayer, 10
Real-time processing, 451, 451f
Reconstruction filter, 2–3, 8
   CD and, 575
   Nyquist frequency and, 575
   sample-and-hold effect and, 577
Recovery system, 34f
Rectangular waveform, 715f, 720t
Rectangular window function, 117, 229, 291
   FIR filter coefficients for, 239t
   frequency response with, 239f
Red component
   equalization and, 632–635
   in image processing, 619–623
Red, green, and blue (RGB), 633f
   in component video, 677
   equalization of, 635f, 636f
   24-bit color image and, 620f
   YIQ and, 69
Reference frame, 688–690, 690f
   macroblocks in, 688f
   motion vectors in, 688f
   in target frame, 688f
Reference noise
   in noise cancellation, 467f
   waveform for, 478f
Region of convergence, 135, 136
Rejection band, 337
Relay latches, 414

Remez exchange algorithm, 269, 271, 275, 278, 288, 290
  with Parks-McClellan algorithm, 291
  ripples and, 275
  3-tap FIR filter coefficients with, 277f
Remez() function, 272, 274, 290
Replicas, 17
Residue function, 150, 180
Resistors, 2
Resolution, 617–618. *See also* Frequency
  pixels and, 691
Retracing, 681
RHHP. *See* Right-hand half plane
Right-hand half plane (RHHP), 175
  of s-plane, 731, 741
Ripple. *See* Oscillation
RMS. *See* Root mean squared
ROM. *See* Read-only memory
Root mean squared (RMS), 43
Round-off error, 427
Run-length coding, 674–676

Sallen-Key lowpass filter, 26, 26f, 35
Sample C programs
  fixed-point implementation, 455, 458
  floating-point implementation, 455, 460
Sample number, 87
Sample-and-hold effect, 16, 15f, 30f
  distortion and, 31f
  equalizer magnitude frequency and, 33f
  overcoming of, 32
  quantization error and, 504
  reconstruction filter and, 577
  spectral shaping and, 34f
Sampled signal spectrum, 17–18, 18f, 606f, 607f
  for AM, 608f, 609f
  plots of, 607, 605f
Sampling, 13–50. *See also*
    Downsampling; 13–50
    Frequency; Oversampling; Upsampling
  ADC and, 35–42, 44, 49
  of analog signal, 98
  DAC and, 16, 35–49
  multivariate signal processing and, 557
  of 1-Hz sine wave, 110
  quantization and, 35–48
  signal reconstruction and, 20–21
  spectral leakage and, 111f
Sampling frequency, 336

  for Chebyshev bandpass filter, 338
Sampling interval effect, 355f
Sampling period, 98
Sampling rate, 10, 398
  in CD, 576f
  frequency response for, 207
  integer factor and, 564, 570, 575
  MATLAB function for, 575
  non-integer factor and, 570, 607, 610, 614
  sampling theorem and minimum, 19
Sampling theorem, 15
  minimum sampling rate in, 19
  violations of, 24
Sampling time constant, 87
Satellite, 1, 11t
Sawtooth waveform, 186, 719, 719f, 720t
Scale factor, 430
  ADPCM and, 520
  in image processing, 644
Scale-factor selection information (SCFSI), 532
SCFSI. *See* Scale-factor selection information
Script files, in MATLAB, 704
SDM. *See* Sigma-delta modulation
SDRAM. *See* Synchronous dynamic random access memory
SECAM. *See* Séquentiel Couleur à Mémoire
Second-order bandpass Butterworth filter, 336
Second-order bandpass filter, 359
  butterworth, 331–332
  IIR filter, 358–359
  pole-zero form and, 358, 357f
  pole-zero placement and, 359f, 397
Second-order bandstop filter, 359–360
  IIR filter, 358–359
  pole-zero placement and, 359f, 397
Second-order filter module, 202
Second-order Goertzel IIR filter, 387f
Second-order lowpass Butterworth filter, 331
Second-order lowpass Chebyshev filter, MATLAB and, 379
Second-order unit gain, 35
Sensors, 2
Sequence, output, 155
Sequences. *See also* Digital sequences; Input sequences
  causal, 135, 155
  convolution of, 79f

Sequences. *See also* Digital sequences; Input sequences (*Cont'd*)
  determined, 146t
  domain, 94–95
  exponential, 136
  impulse input, 454, 556f
  noncausal, 139
  plot of, 77f, 78f
  reversal of, 75–76, 76f
  shifted, 75, 77
  unit impulse, 59, 59f, 69f
  unit step, 59, 59f
  video, 677
  window, 114f
  window impulse, 287
Séquentiel Couleur à Mémoire (SECAM), 685, 696
Series realization, 197–198
Shannon sampling theorem, 19
  violations of, 22
Shaped-in-band noise power, 596
Shift operation, 61
Shift theorem, 139–140
  z-transform and, 139–140, 142t
Shift unit, 438, 439, 460
Shift-down operation, 640
Shifters, 417
Shift-up operation, 640
Sigma-delta ADC, 36
  CD and, 599
  diagram for, 599
Sigma-delta modulation (SDM)
  in ADC conversion, 597, 41
  for CD recording, 599, 600f, 601f
  first-order, 595f
  harmonic frequency and, 598
  second-order, 598
  z-transform for, 597
Sign bit extended Q-30 format, 429f
Signal(s). *See also* Bandpass signals; Digital signals; Periodic digital signal
  analog, 62, 98
  audio, 4
  corrupted, 370, 370f, 469
  enhancement of, 372f, 463, 484f 487f
  Fourier transform for common, 725t
  with noise, 254f
  original, 46
  quantization of, 13, 35–50

reconstruction and, 20–23
quantized, 46
reconstruction of, 20–23
  anti-image filters and, 30, 33–35
  equalizer and, 29–35
sampling of, 13–50
  anti-aliasing filters and, 25–29
  reconstruction and, 20–23
spectrum for, 482f
speech, equalizer for, 290
undersampling of, 557, 610
video, as amplitude modulated, 679
Signal frequency analysis, 4–6
Signal (DS) processor, 2
Signal-to-noise (SNR), 43, 499
  MATLAB function for, 46–47, 541–544
Sine-cosine form, 715–716
Single precision format, 434–436
Single-tone generator, 382, 383f
Sinusoidal function, 179, 454
  sample values from, 60f
  transfer function with, 735–736
  z-transform of, 382–383
Sinusoidal steady-state response, 735
  of filters, 753, 755–756
  properties of, 755
Sinusoidal waveform, 43, 44, 60
  plot of, 60f
60 Hz hum, 400
  with ECG, 370–375, 479, 481, 488f, 490f
  harmonic frequency of, 370, 370f
60 Hz hum eliminator, 370–376, 370f
60 Hz interference cancellation, IIR and, 303
Smooth filter. *See* Reconstruction filter
SNR. *See* Signal-to-noise
Sobel edge detector, 651–653, 653f
Software
  CCS, 451
  for DSP, 413–462
Spatial resolution, 617–618
Speakers, drivers for, 256, 258
Spectral estimation, with window functions, 110–117
Spectral leakage, 87, 111
  frequency resolution and, 119
  sampling and, 110f
  window functions and, 113f, 119
Spectral overlap, 19, 491

INDEX 813

Spectral shaping, sample-and-hold effect and, 34f
Spectrum. *See also* Amplitude spectrum; Onesided amplitude spectrum; Power spectrum; Sampled signal spectrum
  for adaptive filter output, 482f
  for amplitude-modulated, 612f, 613f
  after anti-aliasing, 573f
  of audio signals, 4
  for clean/corrupted speech, 479f
  after downsampling, 560–562f, 573f
  before downsampling, 561f–562f
  image, 661–666
  for input signal, 482f
  after interpolation, 572f,
  before interpolation, 568f
  magnitude, 664f, 665f,
  of modulated signal, 602f
  one-sided, 715f, 717f
  for original speech, 479f
  phase, 99, 100, 101
  two-sided, 92f, 106, 121, 718f
  of undersampled signal, 607
  for unknown system, 482f
  after upsampling, 568f, 572f, 576f,
  before upsampling, 571f, 572f
  for YIQ, 682
Spectrum analysis, 4–6
  speech and, 257f, 479f
Speech. *See also* Original speech; Quantized speech
  clean
    spectrum for, 482f
    waveform for, 482f
  coding of, 7, 11t
  compression of, 501–554
  corrupted
    spectrum for, 482f
    waveform for, 482f
  enhancement of, 202–207
    filtering and, 202–207
    pre-emphasis and, 202–204
  formants, 5
  pre-emphasis of, 202–204
  spectrum analysis and, 257f
  synthesis of, 11t
Speech noise reduction
  bandpass filtering and, 205
  cutoff frequency for, 255

filter length for, 254
FIR and, 251–256
Hamming window function for, 254
lowpass filter and, 254
passband ripple for, 255
stopband attenuation for, 255
two-band digital crossover and, 253–259
Speech recognition, 3, 11t
Speech signal, equalizer for, 290
Speech spectral estimation, 121
S-plane
  BLT and, 310
  LHHP of, 732, 742, 745
  mapping of, 311f
  RHHP of, 732, 741
  z-plane and, 175f
Split radix FFT algorithms, 122
Square image, 663f
  magnitude spectrum for, 664f
Square waveform, 712, 715f, 720t
Stability, 731–735
  BIBO, 72–73, 176
  of digital linear system, 73f
  filtering and, 171–179
  illustrations of, 177f
stair function, 704
Steady-state frequency response, 179f, 180–181
  of filters, 754f
  properties of, 751–752
Steady-state transfer function, 735–736
Steepest descent algorithm, 470–471, 470f, 493, 495
stem function, 704
step function, 703
Step response, 169–171, 172f
  filtering and, 169–171
Step size, 504
Stereo, 11t
  in TV, 681
Stop frequency edge, 565
Stopband attenuation, 240, 242, 243, 270, 272, 275
  for bandpass filter, 274
  for Chebyshev bandpass filter, 340
  for lowpass filter, 271
  for multistage decimation, 578
  for speech noise reduction, 255
  for two-band digital crossover, 256

Stopband frequency edge, for anti-aliasing filter, 580f
Stopband width, 338
subplot function, 704
Successive approximation ADC, 36
sum function, 700
Summers, 37
sumsub.m function, 705–707
Super VHS (S-VHS), 682
Surround sound, 11t
S-VHS. *See* Super VHS
S-video, 677
Symmetry, 221–222, 233, 235, 262
  of frequency response, 181–182
Synchronizing pulse, 680, 681
Synchronous dynamic random access memory (SDRAM), 447
System input. *See* Input
System modeling, adaptive filter for, 479–478, 479f
System output. *See* Output
System representation
  impulse response and, 68–72
  with unit-impulse response, 82
System response, 154, 172f
  filtering and, 169–171
System transient frequency response, 179f, 180–181

Tape hum, 487
Taps, 481, 587
Target frame, 688–690, 690f
  motion vectors in, 690f
  reference frame in, 690f
Telephones
  cellular, 1, 11t
  echo cancellation in, 463, 489–491, 490f, 492f
  internet, 11t
  long-distance, 489–490
  touchpads for, 381–382
Television (TV), 1
  aspect ratio of, 681
  high-definition, 687
  stereo in, 684
test.m function, 704, 706f
Texas Instruments (TI), 420, 437, 439, 447–451
  Veloci architecture of, 450
  website for, 439

Text-to-voice technologies, 11t
Three sine functions, for grayscale transformation, 659f
3-bit 2's complement system
  with fractional representation, 423t
  with number representation, 421t
3-bit bipolar quantizer, 39
  quantization table for, 39f, 40f
3-bit midtread quantizer, 499f
  quantization table for, 499f
3-bit quantizer, quantization table for, 511t
TI. *See* Texas Instruments
TigerSHARC, 439
Time domain representation, 87–88, 243
  CD and, 578
Time invariance, linear time-invariant causal systems and, 65–67
TMS320C30 processor, 440f, 441
TMS320C54x processor, 438–439
  architecture of, 438f
TMS320C67x, 447–448
  diagram of, 449f
  registers of, 449f
Tone specifications, for DTMF, 382f
Touchpads, 382
Trace jumping, 679
Transducers, 2
Transfer function, 455. *See also* Laplace transfer function
  Butterworth, 743–744
  Chebyshev, 747–748
  difference equations and, 165–169
  of filter, 166f
  FIR and, 216
  with sinusoidal function, 735
  steady-state, 735–736
Transient response effects, 254
Transistors, 2
Transversal form, 280, 281f
Trapezoidal shape, 81
Triangular waveform, 720t
Triangular window function, 113, 117f, 229, 291
  computation of, 121
Trigonometric identity, 181, 745, 759
Turntable rumble, 487
TV. *See* Television
Tweeters, 258
12-bit linear code, 507

24-bit color image, 619–620, 692
  format for, 621f
  RGB components and, 620f
Twiddle factor, 93, 122
2-bit circular buffer, 418f
2-bit flash ADC, 36f
Two-band digital crossover, 7f, 258f
  FIR and, 253–256
  Hamming window function for, 258
  highpass filter and, 258
  HR and, 258
  lowpass filter and, 258
  noise reduction and, 253–255
  ripple for, 258
  stopband attenuation for, 258
D-DFT. *See* Two-dimensional discrete Fourier transform
Two-dimensional discrete cosine transform (2D-DCT), 666
Two-dimensional discrete Fourier transform (2D-DFT), 661
Two-dimensional JPEG grayscale compression, 669
Two's complement number system, 420–423
Two-sided spectrum, 121, 718f
  for periodic digital signal, 92f

Underflow, 429, 434
Undersampling, 610
  of bandpass signals, 557, 601–613
  spectrum of, 607
Unilateral transform, 135
Unipolar quantizer, 38, 497
  characteristics of, 39f
Unit impulse response, linear time-invariant causal systems and, 64–67
Unit impulse sequence, 59
  of linear time-invariant system, 69f
  shifted, 59f
Unit step function, 154
  of Laplace transform, 726
Unit step sequences, 59
  shifted, 59f
Unknown system
  frequency response for, 481f
  spectrum for, 482f
  waveform for, 482f
Unnormalized value, 436
Upsampling, 564–572, 572f, 609
  anti-image filters and, 610
  CD and, 575–578
  lowpass filter and, 572
  Nyquist frequency and, 575
  spectrum after, 572f, 573f, 576f
  spectrum before, 568f

V-chip, 682
Vehicle noise, 487
Veloci architecture, 450
Vertical helical scan (VHS), 682
Vertical retrace, 681
Vertical Sobel edge detector, 651–653, 653f
Vertical synchronization, 681
  color subcarrier burst and, 682f
Very long instruction word architecture (VLIW), 450
Vestigial sideband modulation (VSB), 684
VHS. *See* Vertical helical scan
Video. *See also* Analog video
  component, 677
  composite, 677
  digital, 685–687
  sequence, by mixing two images, 677
  signal, as amplitude modulated, 679
  S-video, 678
Video-modulated waveform, 680f
VLIW. *See* Very long instruction word architecture
Voice recorders, compression in, 8
Voice-to-text technologies, 11t
Von Neumann architecture, 413–414, 460
  execution cycle on, 416f
von Neumann, John, 413
VSB. *See* Vestigial sideband modulation

Waveform
  for adaptive filter output, 484f
  for clean speech, 479f
  coding of, 6–7
    with DCT, 533
    with W-MDCT, 533
  compression of, 501–554
    DCT and, 525–529
    DPCM and, 533
    MCDT and, 527–528
  for corrupted speech, 479f
  for error output, 482f
  Fourier series coefficients for, 720t

Waveform (Cont'd)
  for original speech, 479f
  quantization of, 501–554
    DCT and, 525–529, 674
    DPCM and, 533
    MCDT and, 527–528
  rectangular, 717, 709, 720t
  for reference noise, 478f
  sawtooth, 186, 719, 719f
  sinusoidal, 43, 44, 60, 60f
  triangular, 724t
  for unknown system, 482f
Weight function, 269, 272, 288
  in Wiener filter, 469–470
Wiener filter, 469–470, 492
  MSE for, 471
  for noise cancellation, 473
Window functions, 224. *See also* Blackman window function; Hamming window function; Hanning window function; Rectangular window function; Triangular window function
  amplitude spectrum and, 113, 120f, 121f
  Fourier transform and, 287
  illustration of, 112f
  Kaiser, 229
  one-sided amplitude spectrum and, 120f, 121f
  shapes of, 230f
  spectral estimation with, 110–117
  spectral leakage and, 113f, 117
Window impulse sequence, 287
Window sequences, 114f
Windowed MDCT (W-MDCT), 526–533
Windows Media Player, 10
Wireless local area networking, 11t
W-MDCT. *See* Windowed MDCT
Woofers, 256

X-rays, 1, 11t

Y channel. *See* Luminance
YCbCr color space, 685, 692
YIQ color space, 622–635
  RGB and, 692
YUV color model, 684–685

Zero padding effect, 104, 104f, 106, 107, 131
  in image processing, 647
  pixels and, 650
Zero-crossing, 374
  MATLAB and, 374
Zero-initial condition, 72, 153
Zeros, 731–735
Z-plane
  BLT and, 310
  mapping of, 313f
  s-plane and, 175f
  zeros in, 352
Z-plane pole-zero plot, 173f
  filtering and, 187–188
Z-transform, 135–155, 304, 310, 312, 395, 397
  ADC and, 174
  causal sequence and, 155
  convolution and, 140–142, 142t
  decimation and, 563
  definition of, 135–136
  difference equations with, 151–155
  FIR and, 215–216
  inverse z-transform and, 142–148, 233, 444
  Laplace transform and, 174f
  MATLAB and, 148–151, 155
  pairs of, 137t
  partial fractions and, 143–144, 144t, 148–149
  properties of, 139–142, 142t
  for SDM, 600
  shift theorem and, 139–140, 142t
  of sinusoidal function, 382